MICROSTRIP AND PRINTED ANTENNAS

MICROSTRIP AND PRINTED ANTENNAS

NEW TRENDS, TECHNIQUES AND APPLICATIONS

Editors

Debatosh Guha
Institute of Radio Physics and Electronics, University of Calcutta, India

Yahia M.M. Antar
Royal Military College, Canada

WILEY

A John Wiley and Sons, Ltd, Publication

This edition first published 2011

Registered office
John Wiley & Sons Ltd, The Atrium, Southern Gate, Chichester, West Sussex, PO19 8SQ, United Kingdom

For details of our global editorial offices, for customer services and for information about how to apply for permission to reuse the copyright material in this book please see our website at www.wiley.com.

Library of Congress Cataloging-in-Publication Data

Microstrip and printed antennas : new trends, techniques, and applications /
edited by Debatosh Guha, Yahia M.M. Antar.
p. cm.
Includes bibliographical references and index.
ISBN 978-0-470-68192-3 (cloth)
1. Microstrip antennas--Design. 2. Polarization (Electricity) I. Guha,
Debatosh. II. Antar, Yahia.
TK7871.67.M5M53 2011
621.382'4–dc22

2010022377

A catalogue record for this book is available from the British Library.

Print ISBN: 9780470681923 (hb)
ePDF ISBN: 9780470973387
oBook ISBN: 9780470973370

Set in 10/12 pts Times New Roman by Thomson Digital, Noida, India

Contents

Preface

Microstrip technology has been popular for microwave and millimeter wave applications since the 1970s and recently has taken off, with the tremendous growth in communications, wireless, as well as space-borne/airborne applications, although the concept dates back to 1952 [1]. The basic microstrip configuration is very similar to a printed circuit board (PCB) used for low frequency electronic circuits. It constitutes a low-loss thin substrate, both sides being coated with copper film. Printed transmission lines, patches, etc. are etched out on one side of the microstrip board and the other copper-clad surface is used as the ground plane. In between the ground plane and the microstrip structure, a quasi-TEM electromagnetic wave is launched and allowed to spread.

Such a structure offers some unique basic advantages such as low profile, low cost, light weight, ease of fabrication, suitability to conform on curved surface, etc. All these have made microstrip technology attractive since the early phase of its development.

Within a year of the pioneering article "Microstrip – a new transmission technology for the kilomegacycle range" appearing [1], Deschamps [2] had conceived of microstrip as "microwave antenna." But its practical application started nearly two decades later. Howell [3] and Munson [4] may be regarded as the pioneer architects of microstrip antenna engineering.

These early developments immediately attracted some potential research groups and the following studies were mainly concerned with theoretical analysis of different patch geometries and experimental verifications [5–12]. A parallel trend also developed very quickly and some researchers tried to implement conventional antennas such as dipole, wire, aperture, etc. in planar form [13–16]. They are commonly referred to as printed circuit antennas or simply *printed antennas*. Their operations and characteristics are completely different from those due to microstrip patches, although microstrip patch antennas, in many papers, are casually called printed circuit antennas. The topic *printed antenna* had acquired tremendous importance by the late 1970s and a three-day workshop held at New Mexico State University in Las Crises in October, 1979 was dedicated to *Printed Circuit Antenna Technology*.

The developments in microstrip antennas that occurred up to the late1970s were documented by Bahl and Bhartia in their famous book [17], published in 1980. The analysis and design aspects were addressed in another book by James, Hall and Wood [18], published in 1981. A contemporary article by Carver and Mink [19] discussed the fundamental aspects of microstrip antennas and this is still regarded as a good review paper for a beginner.

More activities in the area grew gradually and many applications were realized. The suitability of deploying such lightweight low profile antennas in airborne and space-borne systems initiated major developments in microstrip array technology. With the development of mobile and wireless communications, microstrip and other printed antennas attained a new focus to serve in different technology from the mobile handset to base station antennas. General information, gathered from journals, symposia and conference articles, reveals that about 50% of the whole antenna community has been active in microstrip or printed antenna practice for the past two or three decades.

The first handbook [20] was published in 1989, nearly a decade after the first book by Bahl and Bhartia [17]. Within another five years, microstrip antenna research had attained a level of maturity as is reflected in the title and topics of the microstrip antenna books published around the middle of 1990s [21–23]. The edited volume by Pozar and Shaubert [21] contains some published articles bearing the results of contemporary interests, such as bandwidth enhancement approaches, analysis and design techniques, aperture coupling and other feeding methods, active integrated antennas, conformal and phased arrays, etc. Narrow impedance bandwidth appears an inherent limitation of thc microstrip element. The research and consequent developments in bandwidth enhancement were documented in [22]. Lee and Chen [23] covered some key areas of advances reported up to 1997.

The growing need and interest in microstrip antenna designs are reflected in three design handbooks [24–26] published at close interval from 2001 to 2004. Compacting, along with bandwidth widening of printed antennas, has attracted worldwide interest to support new wireless technology since the beginning of this century and its importance was reflected in titles [27–32] which appeared between 2002 and 2007.

The book edited by Lee and Chen [23] was a timely effort to incorporate major technological developments that had occurred up to1997, under the same cover. Since then, more than a decade has passed during which many new trends, techniques and applications in planar antenna technology have been developed. For example, RFID (Radio Frequency Identification) is an ideal example to showcase the need to this day. This application needs low cost antennas, printed on paper or very thin substrate. Another example is printed antenna using unconventional and new innovations, such as using metamaterials and defected ground structures (DGSs). Replacing a large parabolic dish with a flat microstrip array with a special feeding mechanism is also a new area of activity. The design of small ultrawideband (UWB) antennas with good performance is a challenging area. Antenna for the body area network is another interesting new topic.

From our long experience in teaching and mentoring doctoral and post-doctoral students and working with practicing engineers, we certainly feel there is a need for a book that is to address more recent topics of microstrip and printed antennas. We have chosen some topics that have recently been developed or have considerably advanced during the past decade and at the same time appear to be important to the new generation of researchers, developers and application engineers. We shared the ideas with some of our colleagues and friends who are the real technical experts and potential developers in those selected topics. They fully agreed with our views, gave valuable suggestions and delivered on their promise to contribute. Our collaborative efforts have finally culminated in the present title.

As indicated by the title, the focus is on the *New Trends, Techniques and Applications* of *Microstrip and Printed Antennas*. The chapters are organized as follows: Chapters 1–4 address advances in design, analysis, and optimization techniques, Chapters 5–10 focus on some important new techniques and applications, Chapters 11 and 12 deal with engineered materials applied to printed antenna designs, and finally Chapter 13 addresses advanced methods and designs of printed leaky wave antenna.

Chapter 1 deals with numerical techniques, which are essential in analyzing and designing planar antennas of any arbitrary geometry. A brief overview of the commonly used methods are discussed and the finite difference time domain (FDTD) technique is elaborated on, with special emphasis on the recent developments that occurred after 2003. Chapter 2 presents the advances in computer aided designs (CAD) of microstrip antennas reported during 2001 and onwards. The aim of this chapter is to provide accurate closed form expressions, which can be reliably used to compute essential design parameters such as operating frequency, input impedance and matched feed-location for a given antenna involving single or multiple dielectric layers. Chapter 3 embodies the Generalized Scattering Matrix (GSM) approach to analyzing the multilayer finite printed array structures. The methodology is demonstrated through examples. Chapter 4 deals with antenna optimization techniques. Optimization in terms of performance, size and cost is discussed and the basic concept of stochastic optimization techniques is demonstrated.

Chapter 5 describes microstrip reflectarray technology, its general principle, design, operation, and applications. Microstrip's inherent demerit of narrow bandwidth is dealt with in terms of spatial and frequency dispersions and some of the techniques to suppress these factors are presented. Chapter 6 deals with Reconfigurable Microstrip Antennas, which use switches, tunable materials, or control circuitry to give additional degrees of operational freedom or to make a single element operative in multiple frequencies. A wide variety of reconfigurability is discussed. The emerging trends and directions for future research have also been indicated.

Chapter 7 describes wearable antennas for body area networks. The properties of the human body in terms of electromagnetic radiations and the performance of multiple antenna systems in presence of the human body are described. Chapter 8 presents printed wireless antennas. These include three primary configurations: microstrip patch, slot, and monopole showing multiband, wideband, or ultra wideband performances. Significant developments reported since 2000 are addressed in this chapter. Chapter 9 deals with printed antennas for RFID tags. An RFID system may be one of the following types: active, passive, or in between of these two, based on the nature of the devices used and also any of LF, HF, or UHF type based on the frequency of operations. Passive tags operating at UHF place several specialized requirements on the associated antenna structures and these are described in this chapter. Chapter 10 deals with printed antennas for ultra-wideband (UWB) applications. This incorporates the innovative technologies to minimize ground plane effects on the performance of small printed antennas.

Chapter 11 presents applications of metamaterials to planar antenna and radiative system designs. Both leaky wave and resonant metamaterial antennas are discussed with special emphasis on their recent and somewhat exotic applications. Chapter 12 deals with defected ground structures (DGS) applied to microstrip antennas. This is a recently developed topic and all the major developments that have occurred after 2002 are discussed, indicating the future scope of development. This is probably addressed here as an exclusive book chapter for the first time. Chapter 13 concludes with printed leaky wave antennas. It includes both theory and some applications based on recent advances in technology.

Each chapter is designed to cover the range from fundamental concepts to the state-of-the-art developments. We have tried to satisfy a wide cross-section of readers. A student or a researcher may consider this a guide book to understanding the strength and weaknesses of the contemporary topics. To a practicing engineer, we hope that the book will be a ready reference to many new areas of applications. To an educator, the book appears as a comprehensive review and a source of up-to-date information.

Our sincere efforts and exercise will be successful if our readers appreciate and find it useful for their respective purposes.

Debatosh Guha
Yahia M. M. Antar

References

1. D. D. Greig and H. F. Engleman, "Microstrip – a new transmission technology for the kilomegacycle range," *Proc. IRE*, vol. 40, pp. 1644–1650, 1952.
2. G. A. Deschamps, "Microstrip microwave antennas," presented at the *3rd USAF Symp. on Antennas*, 1953.
3. J. Q. Howell, "Microstrip antennas," *Dig. IEEE Int. Symp. Antennas Propagat.*, pp. 177–180, Dec. 1972.
4. R. E. Munson, "Conformal microstrip antennas and microstrip phased arrays," *IEEE Trans. Antennas Propagat.*, vol. 22, pp. 74–78, 1974.
5. T. Itoh and R. Mittra, "Analysis of microstrip disk resonator," *Arch. Elek. Ubertagung*, vol. 21, pp. 456–458, Nov. 1973.
6. T. Itoh, "Analysis of microstrip resonator," *IEEE Trans. Microwave Theory Tech.*, vol. 22, pp. 946–952, Nov. 1974.
7. A. Derneryd, "Linearly polarized microstrip antennas," *IEEE Trans. Antennas Propagat.*, vol. 24, no. 6, pp. 846–851, 1976.
8. G. Dubost, M. Nicolas and H. Havot, "Theory and applications of broadband microstrip antennas," *Proc. 6th European Microwave Conference*, pp. 275–279, 1976.
9. P. Agrawal and M. Bailey, "An analysis technique for microstrip antennas," *IEEE Trans. Antennas Propagat.*, vol. 25, no. 6, pp. 756–759, 1977.
10. W.F. Richards, Y.T. Lo and D. D. Harrison, "Improved theory for microstrip antennas," *Electronics Letters*, vol. 15, no. 2, pp. 42–44, 1979.
11. Y.T. Lo, D. Solomon and W. Richards, "Theory and experiment on microstrip antennas," *IEEE Trans. Antennas Propagat.*, vol. 27, no. 2, pp. 137–145, 1979.
12. P. Hammer, D. Van Bouchaute, D. Verschraeven and A. Van de Capelle, "A model for calculating the radiation field of microstrip antennas," *IEEE Trans. Antennas Propagat.*, vol. 27, no. 2, pp. 267–270, 1979.
13. K. Keen, "A planar log-periodic antenna," *IEEE Trans. Antennas Propagat.*, vol. 22, no. 3, pp. 489–490, 1974.
14. D.T. Shahani and Bharathi Bhat, "Network model for strip-fed cavity-backed printed slot antenna," *Electronics Letters*, vol. 14, no. 24, pp. 767–769, 1978.
15. Inam E. Rana and N. G. Alexopoulos, "On the theory of printed wire antennas," *9th European Microwave Conference*, 1979, pp. 687–691, 1979.
16. A. Mulyanto, and R. Vernon, "A V-shaped log-periodic printed-circuit antenna array for the 1 to 10 GHz frequency range," *Antennas and Propagation Society Intl. Symp.*, 1979, vol. 17, pp. 392–395.
17. I. J. Bahl and P. Bhartia, *Microstrip Antennas*, Artech House, Dedham, MA, 1980.
18. J. R. James, P. S. Hall and C. Wood, *Microstrip Antennas: Theory and Design*, Peter Peregrinus, London, 1981.
19. K. Carver and J. Mink, "Microstrip antenna technology," *IEEE Trans. Antennas Propagat.*, vol. 29, pp. 2–24, Jan. 1981.
20. J. R. James and P. S. Hall, *Handbook of Microstrip Antennas*, Peter Peregrinus, London, 1989.
21. D. M. Pozar and D. H. Schaubert, *Microstrip Antennas*, IEEE Press, New York, 1995.
22. J. F. Zürcher and F. E. Gardiol, *Broadband Patch Antennas*, Artech House, Boston, 1995.
23. K. F. Lee and W. Chen, *Advances in Microstrip and Printed Antennas*, John Wiley & Sons, Inc., New York, 1997.
24. R. Garg et al., *Microstrip Antenna Design Handbook*, Artech House, Boston, 2001.
25. R. Waterhouse, *Microstrip Patch Antennas: A Designer's Guide*, Springer, Berlin, 2003.
26. R. Bancroft, *Microstrip and Printed Antenna Design*, Noble Publishing, 2004.
27. Kin-Lu Wong, *Compact and Broadband Microstrip Antennas*, John Wiley & Sons, Inc., New York, 2002.

28. G. Kumar and K. P. Ray, *Broadband Microstrip Antennas*, Artech House, Boston, 2002.
29. Kin-Lu Wong, *Planar Antennas for Wireless Communications*, John Wiley & Sons, Inc., New York, 2003.
30. Zhi Ning Chen and Michael Yan Wah Chia, *Broadband Planar Antennas: Design and Applications*, John Wiley & Sons, Inc., New York, 2006.
31. Peter S. Hall and Yang Hao, *Antennas and Propagation for Body-Centric Wireless Communications*, Artech House, Boston, 2006.
32. Zhi Ning Chen (eds.), *Antennas for Portable Devices*, John Wiley & Sons, Inc., New York, 2007.

List of Contributors

Yahia M. M. Antar, Royal Military College, Canada
Jennifer T. Bernhard, University of Illinois at Urbana-Champaign, USA
Arun K. Bhattacharyya, Northrop Grumman Corporation, USA
Sujoy Biswas, Institute of Technology and Marine Engineering, India
Christophe Caloz, École Polytechnique, Montreal, Canada
Reza Chaharmir, Communication Research Centre Canada, Ottawa, Canada
Zhi Ning Chen, Institute for Infocomm Research, Singapore
Daniel Deavours, University of Kansas, USA
Daniel Dobkin, Enigmatics, USA
Ramesh Garg, Indian Institute of Technology, Kharagpur, India
Debatosh Guha, Institute of Radio Physics and Electronics, University of Calcutta, India
Peter S. Hall, University of Birmingham, UK
Yang Hao, Queen Mary College, University of London, UK
Samir F. Mahmoud, University of Kuwait, Kuwait
Rabindra K. Mishra, Electronic Science Department, Berhampur University, India
Xianming Qing, Institute for Infocomm Research, Singapore
Shie Ping Terence See, Institute for Infocomm Research, Singapore
Lotfollah Shafai, University of Manitoba, Canada
Jafar Shaker, Communication Research Centre Canada, Ottawa, Canada
Satish K. Sharma, San Diego State University, USA
Jawad Y. Siddiqui, Institute of Radio Physics and Electronics, University of Calcutta, India

Acknowledgments

Editing a book, like this, is a rare experience involving both liberty and responsibility. We came up with this idea in 2008 and started consulting with the experts who could be potential authors for different chapters of this book. The idea has turned into reality only due to the unstinted cooperation of the authors, who could dedicate time from their extremely busy schedules and contribute to different topics. We are grateful to all of them for their spontaneous help and support.

We would also like to express our thanks to a number of our colleagues, researchers and students who helped with many tasks throughout the process. Mr. Sujoy Biswas of Institute of Technology and Marine Engineering, India, Dr. Jawad Y. Siddiqui of the University of Calcutta, India (currently associated with the Royal Military College, Canada), Mr. Chandrakanta Kumar of the Indian Space Research Organization, and Mr. Anjan Kundu of University of Calcutta have extended their constant help and technical support throughout the whole process. We have also received help from some of our students: Sudipto Chattopadhyay of Siliguri Institute of Technology, India, Symon Podilchak of Queen's University and the Royal Military College, Canada, and Mr. David Lee of CRC in Ottawa. Dr. Somnath Mukherjee of RB Technology, USA, helped us tremendously in resolving the organization of the book. We have received constant help and support from Sarah Tilley, Anna Smart, and Genna Manaog of Wiley, which made our job easy. We are extremely grateful to all of them.

We cannot but acknowledge the ungrudging support and cooperation received from our families and from our respective Institutions: the University of Calcutta and the Royal Military College of Canada. It is always challenging to bring so many people from different parts of the world to work together on one task at the same time. We express our indebtedness to all members of this team for contributing to this volume in their different capacities.

1

Numerical Analysis Techniques

Ramesh Garg
Indian Institute of Technology, Kharagpur, India

1.1 Introduction

Microstrip and other printed antennas are constituted of, in general, patches, strips, slots, packaged semiconductor devices, radome, feed, etc. in a nonhomogeneous dielectric medium. Finite substrate and ground plane size are the norm. The dielectric used is very thin compared to the other dimensions of the antenna. The design of these antennas based on models such as transmission line model or cavity model is approximate. Besides, these designs fit regular-shaped geometries (rectangular, circular, etc.) only, whereas most of the useful antenna geometries are complex and do not conform to these restrictions [1]. The effect of surface waves, mutual coupling, finite ground plane size, anisotropic substrate, etc. is difficult to include in these types of design. The numerical techniques, on the other hand, can be used to analyze any complex antenna geometry including irregular shape, finite dielectric and ground plane size, anisotropic dielectric, radome, etc. The popular numerical techniques for antenna analysis include method of moments (MoM), finite element method (FEM), and finite difference time domain method (FDTD). MoM analysis technique, though efficient, is not versatile because of its dependence on Green's function. FEM and FDTD are the most suitable numerical analysis techniques for printed antennas. FDTD is found to be versatile because any embedded semiconductor device in the antenna can be included in the analysis at the device-field interaction level. This leads to an accurate analysis of active antennas. Maxwell's equations are solved as such in FDTD, without analytical pre-processing, unlike the other numerical techniques. Therefore, almost any antenna geometry can be analyzed. However, this technique is numerically intensive, and therefore require careful programming to reduce computation cost. We shall describe the advances in FDTD. Our reference in this respect is the classic book on FDTD by Taflove and Hagness [2].

A large number of FDTD algorithms have been developed. These can be classified as conditionally stable and unconditionally stable. The conditionally stable schemes include the original or Yee's FDTD also called FDTD (2,2), FDTD (2,4), sampling bi-orthogonal time-domain (SBTD) and their variants; and the unconditionally stable schemes include ADI

Microstrip and Printed Antennas: New Trends, Techniques and Applications. Edited by Debatosh Guha and Yahia M.M. Antar

(Alternate Direction Implicit), CN (Crank Nicolson), CNSS (Crank Nicolson Split Step), LOD (Local One-Dimensional) and their variants. The updating of fields in conditionally stable schemes does not require a solution of matrix equation as an intermediate step, and are therefore fully explicit. However, these schemes have a limit on the maximum value of the time step, which is governed by the minimum value of the space step through the Courant-Friedrich-Levy (CFL) condition.

$$c.\Delta t_{CFL} \leq \frac{1}{\sqrt{1/\Delta x^2 + 1/\Delta y^2 + 1/\Delta z^2}} \tag{1.1}$$

Due to the heterogeneous nature of the dielectric in the printed antennas, the wave velocity is less than c and may vary from cell to cell and from one frequency to another. We therefore introduce a safety margin and choose $\Delta t = (1/2)\Delta t_{CFL}$ uniformly to simplify coding and avoid instability. Defining the Courant number q as

$$q = \Delta t / \Delta t_{CFL} \tag{1.2}$$

implies that $q = 1/2$ and the wave takes $2\Delta t$ time to travel to the next node.

The value of Δt_{CFL} puts a severe computational constraint on the structures as they have fine geometrical features such as narrow strips or slots or thin dielectric sheets. Since the simulation time of an antenna is independent of space and time steps, the number of updates of fields increases linearly with the decrease in the time step. This results in an increase in processor time. The limitation on Δt_{CFL} is removed in some of the FDTD algorithms and these are therefore called unconditionally stable schemes. In these schemes one can use the same value of the time step over the whole geometry even if fine geometrical features exist without significantly affecting the accuracy of simulation results. Updating fields in unconditionally stable schemes is carried out in stages called time splitting and involves solving a set of simultaneous equations before going on to the next stage. These schemes therefore are more computationally intensive. However, their accuracy is similar to that of conditionally stable FDTD schemes.

The FDTD analysis of open region problems such as antennas necessitates the truncation of the domain to conserve computer resources. The truncation of the physical domain of the antenna is achieved through absorbing boundary conditions, either analytical ABC or material ABC. Material ABC in the form of PML can achieve a substantial truncation of domain with very low reflection. The design of PML should be compatible with the FDTD scheme employed for the rest of the antenna. A number of PML formulations are available. These are split-field and non split-field PML. Non split-field types are convenient for coding and are therefore preferred. Of the various PML formulations available now, uniaxial PML looks promising.

All the FDTD algorithms suffer from computational error, and the amount of error is related to the space and time step sizes employed. The error is quantified in the form of numerical dispersion. The goal of various FDTD schemes is to analyze multi-wavelength long complex geometries, efficiently and accurately. The complexity of the geometry may be in the form of fine geometrical dimensions, anisotropic dispersive medium, embedded packged semiconductor device, feed, mounting structure, etc. The efficient FDTD algorithms try to achieve this aim by increasing the permissible space step size without increasing dispersion, by an increase in the time step size compatible with fine geometrical features, the applicability of the algorithm to anisotropic and dispersive medium and reduced reflection from the PML medium. The presence of thin strips/slots makes uniform discretization an inefficient approach. New and efficient solutions are being tested in the form of a sub-cell approach, quasi-static approximation, etc. The treatment of PEC and PMC boundary conditions presented by irregular geometries is receiving due attention, while the interface conditions interior to the device are somewhat difficult to implement accurately. Modeling of fast variation of fields in metal, and analysis of curved geometries is being attempted. We shall now discuss the advances in FDTD analysis since 2003.

Yee's algorithm is outlined first in order to define the grid structure and the placement of electric and magnetic field components on the Yee cell. This grid will be used as a reference for other FDTD algorithms.

1.2 Standard (Yee's) FDTD Method

The FDTD method was first proposed by Yee in 1966 [3] and has been used by many investigators because of its host of advantages. However, computer memory and processing time for FDTD have to be huge to deal with the problems which can be analyzed using techniques based on the analytical pre-processing of Maxwell's equations such as MoM, mode matching, method of lines, FEM, etc. Therefore, the emphasis in the development of FDTD technique is to reduce the requirement for computer resources so that this technique can be used to analyze electrically large complex electromagnetic problems.

To determine time-varying electromagnetic fields in any linear, isotropic media with constants ε, μ, σ Maxwell's curl equations are sufficient; the curl equations are

$$\nabla \times \mathbf{H} = \sigma \mathbf{E} + \varepsilon \frac{\partial \mathbf{E}}{\partial t} \tag{1.3a}$$

$$-\nabla \times \mathbf{E} = \mu \frac{\partial \mathbf{E}}{\partial t} \tag{1.3b}$$

The partial differential equations (1.3) are solved subject to the conditions that: (i) the fields are zero at all nodes in the device at $t = 0$ except at the plane of excitation; (ii) the tangential components of E and H on the boundary of the domain of the antenna must be given for all $t > 0$. For computer implementation of Equation (1.3), the partial derivatives are implemented as finite difference approximations, and are partly responsible for the inaccuracy of the solution. For better accuracy, the central difference approximation is used in FDTD and is defined as,

$$\left.\frac{\partial \mathrm{F}}{\partial \mathrm{u}}\right|_{\mathrm{u_o}} = \left.\frac{\mathrm{F}\left(\mathrm{u_o} + \frac{\Delta \mathrm{u}}{2}\right) - \mathrm{F}\left(\mathrm{u_o} - \frac{\Delta \mathrm{u}}{2}\right)}{\Delta \mathrm{u}}\right|_{\Delta \mathrm{u} \to 0} + \mathrm{O}(\Delta \mathrm{u})^2 \tag{1.4}$$

where $O(\cdot)$ stands for *the order of*. Use of Equation (1.4) converts Equation (1.3) into the following form:

$$E_x^{n+1}(i+\tfrac{1}{2},j,k) = \left(\frac{\varepsilon-\sigma\Delta t/2}{\varepsilon+\sigma\Delta t/2}\right)E_x^n(i+\tfrac{1}{2},j,k) + \frac{\Delta t/\Delta y}{\varepsilon+\sigma\Delta t/2}\left(H_z^{n+1/2}(i+\tfrac{1}{2},j+\tfrac{1}{2},k)-H_z^{n+1/2}(i+\tfrac{1}{2},j-\tfrac{1}{2},k)\right) - \frac{\Delta t/\Delta z}{\varepsilon+\sigma\Delta t/2}\left(H_y^{n+1/2}(i+\tfrac{1}{2},j,k+\tfrac{1}{2})-H_y^{n+1/2}(i+\tfrac{1}{2},j,k-\tfrac{1}{2})\right) \quad (1.5a)$$

$$E_y^{n+1}(i,j+\tfrac{1}{2},k) = \left(\frac{\varepsilon-\sigma\Delta t/2}{\varepsilon+\sigma\Delta t/2}\right)E_y^n(i,j+\tfrac{1}{2},k) + \frac{\Delta t/\Delta z}{\varepsilon+\sigma\Delta t/2}\left(H_x^{n+1/2}(i,j+\tfrac{1}{2},k+\tfrac{1}{2})-H_x^{n+1/2}(i,j+\tfrac{1}{2},k-\tfrac{1}{2})\right) - \frac{\Delta t/\Delta x}{\varepsilon+\sigma\Delta t/2}\left(H_z^{n+1/2}(i+\tfrac{1}{2},j+\tfrac{1}{2},k)-H_z^{n+1/2}(i-\tfrac{1}{2},j+\tfrac{1}{2},k)\right) \quad (1.5b)$$

$$E_z^{n+1}(i,j,k+\tfrac{1}{2}) = \left(\frac{\varepsilon-\sigma\Delta t/2}{\varepsilon+\sigma\Delta t/2}\right)E_z^n\left(i,j,k+\tfrac{1}{2}\right) + \frac{\Delta t/\Delta x}{\varepsilon+\sigma\Delta t/2}\left(H_y^{n+1/2}(i+\tfrac{1}{2},j,k+\tfrac{1}{2})-H_y^{n+1/2}(i-\tfrac{1}{2},j,k+\tfrac{1}{2})\right) - \frac{\Delta t/\Delta y}{\varepsilon+\sigma\Delta t/2}\left(H_x^{n+1/2}(i,j+\tfrac{1}{2},k+\tfrac{1}{2})-H_x^{n+1/2}(i,j-\tfrac{1}{2},k+\tfrac{1}{2})\right) \quad (1.5c)$$

$$H_x^{n+\frac{1}{2}}(i,j+\tfrac{1}{2},k+\tfrac{1}{2}) = H_x^{n-\frac{1}{2}}(i,j+\tfrac{1}{2},k+\tfrac{1}{2}) - \frac{\Delta t}{\mu\Delta y}\left(E_z^n(i,j,k+\tfrac{1}{2})-E_z^n(i,j-1,k+\tfrac{1}{2})\right) + \frac{\Delta t}{\mu\Delta z}\left(E_y^n(i,j+\tfrac{1}{2},k)-E_y^n(i,j+\tfrac{1}{2},k-1)\right) \quad (1.5d)$$

$$H_y^{n+\frac{1}{2}}(i+\tfrac{1}{2},j,k+\tfrac{1}{2}) = H_y^{n-\frac{1}{2}}(i+\tfrac{1}{2},j,k+\tfrac{1}{2}) - \frac{\Delta t}{\mu\Delta z}\left(E_x^n(i+\tfrac{1}{2},j,k)-E_x^n(i+\tfrac{1}{2},j,k-1)\right) + \frac{\Delta t}{\mu\Delta x}\left(E_z^n(i,j,k+\tfrac{1}{2})-E_z^n(i-1,j,k+\tfrac{1}{2})\right) \quad (1.5e)$$

$$H_z^{n+\frac{1}{2}}(i+\tfrac{1}{2},j+\tfrac{1}{2},k) = H_z^{n-\frac{1}{2}}(i+\tfrac{1}{2},j+\tfrac{1}{2},k) - \frac{\Delta t}{\mu\Delta x}\left(E_y^n(i,j+\tfrac{1}{2},k)-E_y^n(i-1,j+\tfrac{1}{2},k)\right) + \frac{\Delta t}{\mu\Delta y}\left(E_x^n(i+\tfrac{1}{2},j,k)-E_x^n(i+\tfrac{1}{2},j-1,k)\right) \quad (1.5f)$$

The indices i,j, and k define the position of the field nodes, such that $x = i\Delta x,\ y = j\Delta y,\ z = k\Delta z$. The time instant is defined by $t = n\Delta t$. To implement the finite difference scheme in three dimensions, the antenna is divided into a number of cells, called Yee cells, of dimension $\Delta x\Delta y\Delta z$. One such cell is shown in Figure 1.1. Remarkably the positions of different components of E and H on the cell satisfy the differential and integral forms of Maxwell's equations. One may note from Figure 1.1 that the placements of the E and H nodes are offset in space by half a space step; it is called staggered grid. We note from Equation (1.5) that the time instants when the E and H field components are calculated are offset by half a time step, that is, components of E are calculated at $n\Delta t$ and components of H are calculated at $(n + 1/2)\Delta t$. The alternate update of E and H fields is called *leap frog* and saves computer processing time.

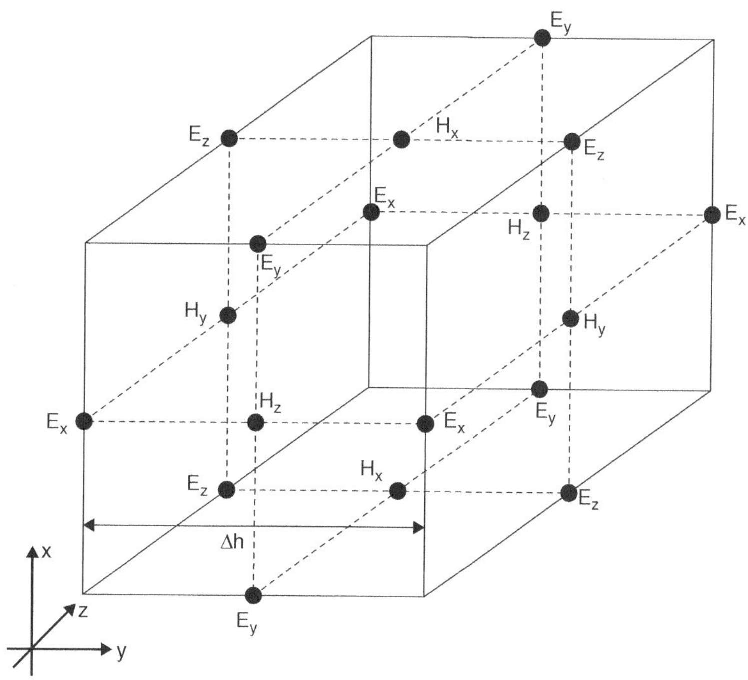

Figure 1.1 Geometry of Yee's cell used in FDTD analysis

1.3 Numerical Dispersion of FDTD and Hybrid Schemes

The finite difference form of derivative (1.4) has an error term $O(\Delta u)^2$. As a result, Equations (1.5 a–f) are second-order accurate, resulting in an approximate solution of the problem. The first sign of this approximation appears in the phase velocity v_{ph} for the numerical wave being different from that in the continuous case. This phenomenon is called numerical dispersion. The amount of dispersion depends on the wavelength, the direction of propagation in the grid, time step Δt and the discretization size Δu. The above algorithm is second-order accurate in space and time, and is therefore called FDTD(2,2). The numerical dispersion for plane wave propagation may be determined from the following expression

$$\left(\frac{\sin(\omega\Delta t/2)}{c\Delta t}\right)^2 = \left(\frac{\sin(\bar{k}\sin\theta\cos\varphi\Delta x/2)}{\Delta x}\right)^2 + \left(\frac{\sin(\bar{k}\sin\theta\sin\varphi\Delta y/2)}{\Delta y}\right)^2 + \left(\frac{\sin(\bar{k}\cos\theta\Delta z/2)}{\Delta z}\right)^2 \quad (1.6)$$

where $\bar{k}$ is the wave number for the numerical wave. The phase velocity $\bar{v} = \omega/\bar{k}$ is determined by solving Equation (1.6) as a function of discretizations $\Delta x, \Delta y, \Delta z, \Delta t$ and propagation angle θ, ϕ. The phase velocity is found to be maximum and close to the velocity of light for propagation along the diagonals and minimum for waves propagating along the axis.

1.3.1 Effect of Non-Cubic Cells on Numerical Dispersion

Devices with high aspect ratio may be analyzed by using uniform or non-uniform cell size. An alternative is to employ non-square or non-cubic cells. The influence of the aspect ratio of the unit cell on the numerical dispersion of FDTD(2,2) has been reported by Zhao [4]. It is found that the dispersion error $(c-\bar{v})/c$ increases with the increase in aspect ratio of the cell but reaches an upper limit for aspect ratios greater than 10. For N (number of cells per wavelength, $\lambda/\Delta) = 10$, the maximum dispersion error for non-cubic cells is 1.6% which decreases to 0.4% for $N = 20$, showing second-order accuracy. In general, the maximum error for non-cubic cells is about 1.5 times that of the corresponding error for cubic cells. For the non-square cells, this ratio is twice that of square cells [4]. For guidance, the minimum mesh resolution required to achieve a desired phase velocity error is plotted in Figure 1.2 for the cubic and non-cubic cells.

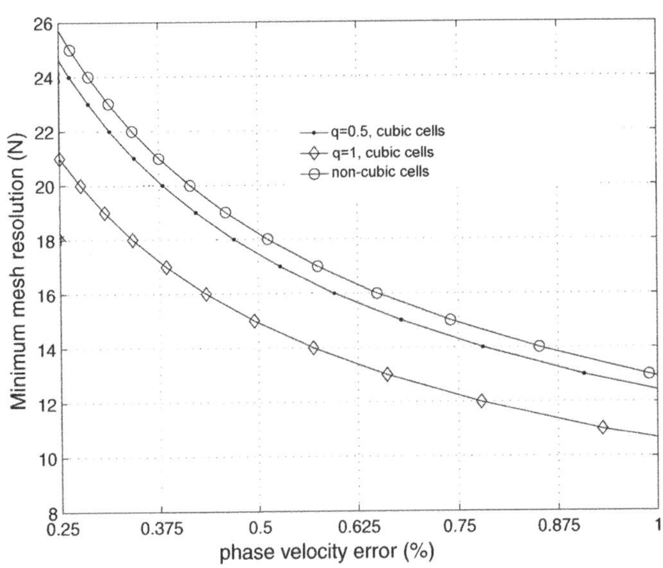

Figure 1.2 Comparison of minimum mesh resolution required for a given accuracy of phase velocity when non-cubic (with high aspect ratio) or cubic unit cells are employed. Reproduced by permission of ©2004 IEEE, Figure 8 of [4]

It may be noted from Figure 1.2 that 0.5% accuracy in phase velocity is achieved for N = 18.5, and N = 13 is needed for 1% accuracy when non-cubic cells are employed. This study shows that unit cells with very high aspect ratio may be used by sacrificing a small amount of accuracy in phase velocity. FDTD(2,2) is also employed for benchmarking other schemes.

1.3.2 Numerical Dispersion Control

The numerical dispersion can be reduced to any degree that is desired if one uses a fine enough FDTD mesh. This, however, increases the number of nodes and therefore also increases the computer memory and processor time required. An alternative way to decrease numerical dispersion is to improve upon the finite difference approximation of Equation (1.4). Higher-order finite difference schemes, also called multi-point schemes, are available to reduce the error in approximating the derivatives. The fourth-order-accurate schemes called FDTD(2,4) employ four nodal values located at $\Delta u/2$ and $3\Delta u/2$ on either side of the observation point u_0, and the space derivative is defined as [5]

$$\left.\frac{\partial F^n}{\partial u}\right|_{u_o} = \frac{9}{8}\left.\frac{F^n\left(u_o+\frac{\Delta u}{2}\right)-F^n\left(u_o-\frac{\Delta u}{2}\right)}{\Delta u}\right|_{\Delta u\to 0} - \frac{1}{24}\left.\frac{F^n\left(u_o+3\frac{\Delta u}{2}\right)-F^n\left(u_o-3\frac{\Delta u}{2}\right)}{\Delta u}\right|_{\Delta u\to 0} + O(\Delta u)^4 \qquad (1.7)$$

Another algorithm with lower dispersion called SBTD (sampling bi-orthogonal time domain) has been proposed. It is an explicit scheme with leap-frog update. It is conditionally stable wavelet-based scheme in which spatial discretization of FDTD is replaced with sampling bi-orthogonal discretization [6]. The field is expanded in wavelets or scale functions as basis functions in space domain, while the time domain expansion is in pulse functions. The coefficients of expansion of wavelets are determined by testing Maxwell's equations with the scaling functions. For the two-dimensional TM case, the expression for the fields for SBTD is of the form [6]

$$E_z^{n+1}(i,j) = E_z^n(i,j) + \frac{\Delta t}{\varepsilon\Delta x}\sum_{p=-3}^{2} c_p H_y^{n+1/2}(i+p+\tfrac{1}{2},j) - \frac{\Delta t}{\varepsilon\Delta y}\sum_{p=-3}^{2} c_p H_x^{n+1/2}(i,j+p+\tfrac{1}{2}) \qquad (1.8a)$$

$$H_y^{n+1/2}(i+\tfrac{1}{2},j) = H_y^{n-1/2}(i+\tfrac{1}{2},j) + \frac{\Delta t}{\mu\Delta x}\sum_{p=-3}^{2} c_p E_z^n(i+p+1,j) \qquad (1.8b)$$

$$H_x^{n+1/2}(i,j+\tfrac{1}{2}) = H_x^{n-1/2}(i,j+\tfrac{1}{2}) - \frac{\Delta t}{\mu\Delta y}\sum_{p=-3}^{2} c_p E_z^n(i,j+p+1) \qquad (1.8c)$$

where

$$c_0 = 1.229167 = -c_{-1}, c_1 = -0.093750 = -c_{-2}, c_2 = 0.010417 = -c_{-3} \quad (1.9)$$

The field expressions (1.8) and (1.5) are very similar. The number of terms on the RHS of (1.8) are six compared to four for the fourth-order accurate finite difference scheme (1.7), and might be responsible for lower dispersion property of SBTD. The SBTD scheme belongs to the family of multiresolution time-domain (MRTD) schemes using Cohen-Daubechies-Feauveau (CDF) wavelets [7]. The MRTD schemes simultaneously address issues of higher-order approximation of fields, multigrid structure, and accurate treatment of the interface between different media, unlike the piecemeal approach of FDTD schemes [7].

The phase velocity for the two-dimensional TM case for SBTD and FDTD(2,2) schemes are compared in Figure 1.3 [6]. The number of nodes per wavelength or spatial resolution N is 20 and $q = 0.5$. It is observed from the graph that the phase velocity for SBTD scheme is $1.001c$ independent of the direction of travel of wave. The error is also less compared to FDTD(2,2).

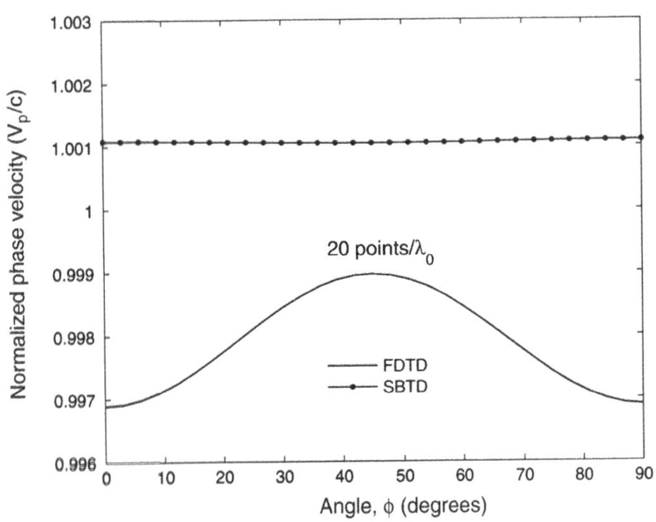

Figure 1.3 Comparison of dispersion curves for SBTD and FDTD(2,2), ($q = 0.5$). Reproduced by permission of ©2008 IEEE, Figure 1 of [6]

The normalized phase velocity for FDTD(2,2) and SBTD schemes for a cubic mesh with $N = 20$ are compared in [8] and plotted here as Figure 1.4. It is noted from Figure 1.4 that SBTD with $q = 0.5$ is isotropic and least dispersive. The combination of various spatial and temporal discretizations ($q = 0.75$) have been studied for their effect on numerical dispersion [9]. The phase velocity is plotted as a function of spatial sampling rate N in Figure 1.5 [9]. For each scheme, the phase velocity is bounded by two lines; the maximum (max) phase velocity occurs along the cell diagonal and the minimum (min) velocity occurs along the axis of the cell. It is noted from Figure 1.5 that except for FDTD(2,2), all other schemes generate fast ($>c$) waves.

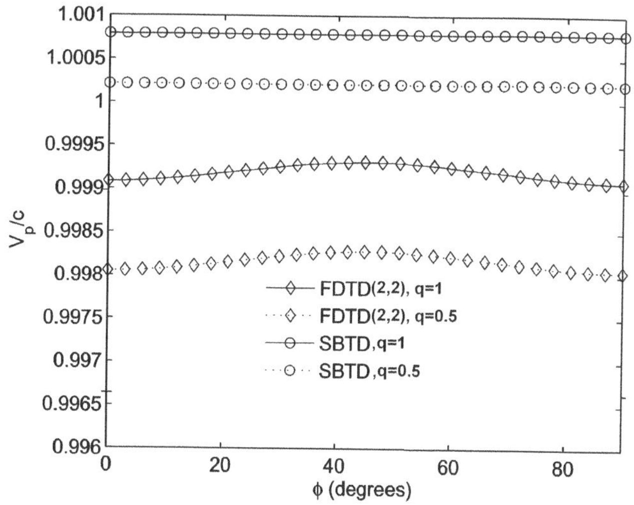

Figure 1.4 Comparison of normalized phase velocity versus azimuth angle in a cubic mesh at a spatial sampling rate of 20 points per wavelength. q is the Courant number. Reproduced by permission of ©2008 IEEE, Figure 1 of [8]

The slow and fast wave behavior of various schemes, Figure 1.5, may be exploited to reduce numerical dispersion in FDTD. For this, hybrid FDTD schemes have been proposed. The hybrid scheme based on the combination of FDTD(2,2) and FDTD(2,4) is called HFDTD(2,4), and that based on FDTD(2,2) and SBTD is called HSBTD1 [9]. Numerical dispersion produced by the hybrid schemes has been compared with non-hybrid schemes and it is found that dispersion can be minimized by properly combining the schmes with slow and fast waves [9]. The lay-out of cells for such an experiment is shown in Figure 1.6 [9]. Most of the cells are updated using higher-order schemes. The cells marked black are updated using higher-order

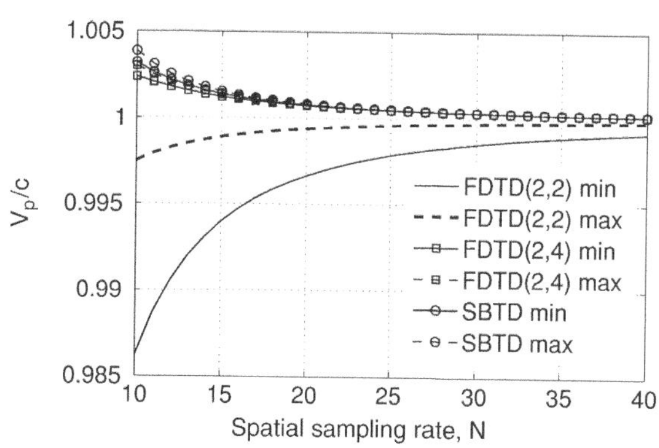

Figure 1.5 Comparison of phase velocities for SBTD and FDTD schemes as a function of spatial sampling rate N, $q = 0.75$. Reproduced by permission of ©2009 IEEE, Figure 1 of [9]

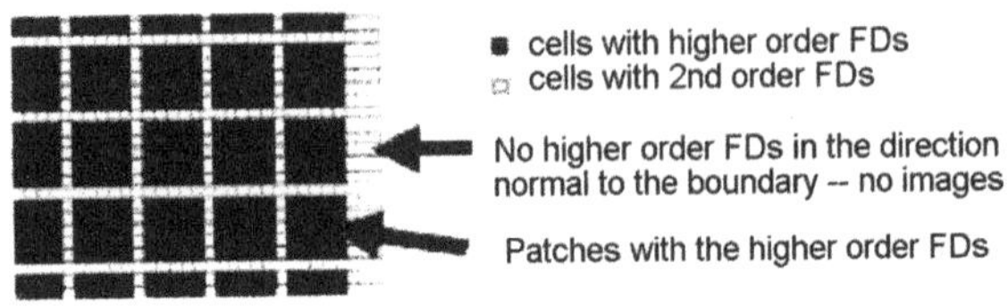

Figure 1.6 Cell pattern for the field components normal to the view. Reproduced by permission of ©2009 IEEE, Figure 2 of [9]

schemes whereas the cells marked white in each sixth row and column are updated with second-order schemes. For various schemes, the effect of spatial sampling rate or grid resolution on the error in resonant frequency of a two-dimensional cavity is compared in Figure 1.7 [9]. It is confirmed from Figure 1.7 that the hybrid schemes may be used to reduce the numerical error significantly. Further numerical experiments on a partially filled rectangular waveguide cavity confirm that the error in resonant frequency reduced by a factor of 3.1 when HFDTD(2,4) is employed; this factor increased to 22 when HSBTD1 is used. All these results are compared to standard FDTD(2,2). The spatial sampling rate used was 26.7. The processor times of the hybrid schemes are similar to those of higher-order schemes. The effects of numerical dispersion for layered, anisotropic media have been reported in [10].

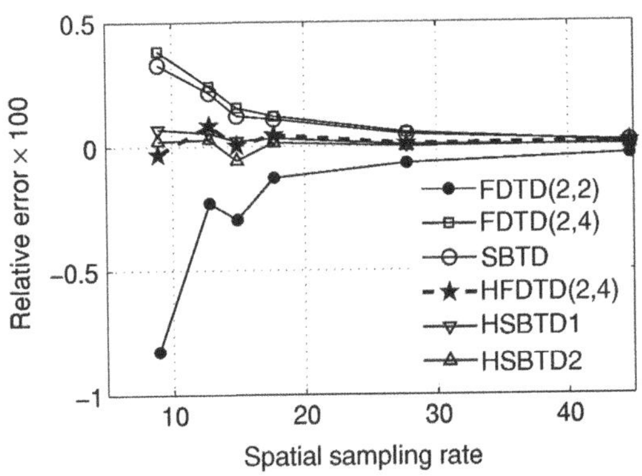

Figure 1.7 Comparison of relative error of various schemes for the resonant frequency of a two-dimensional rectangular cavity. Reproduced by permission of ©2009 IEEE, Figure 3 of [9]

One area of challenge in applying the higher-order and hybrid schemes is in the treatment of boundary conditions which are inside the computational domain, e.g. antenna conductors, feed lines, pins, dielectric interface, etc. [5]. Some of these issues for standard FDTD method are discussed in [11]. Lossy curved surface in the form of surface impedance boundary condition is modeled in [12]. The metal-semiconductor interfaces may be defined by higher-order impedance boundary conditions [13].

Numerical dispersion exhibited by the various finite difference schemes have been reviewed and expressed in the form of a general expression [8]:

$$\cos(\omega\Delta t) = \mathrm{Re}(\lambda_k) \tag{1.10}$$

where λ_k is the complex eigenvalue (not equal to unity) of the amplification matrix M. The above expression is applicable to all known conditionally and unconditionally stable algorithms; within their stability limits for conditionally stable schemes. For specific FDTD schemes, expression for λ_k in terms of discretization parameters and numerical wave number is available in [8]. For FDTD(2,2), the eigenvalues of M are obtained as $\lambda_{1,2} = 1$, $\lambda_{3,4,5,6} = 1-2d \pm 2j\sqrt{d(1-d)}$, where d is given by [8]

$$d = q_x^2 \sin^2(\phi_x/2)+q_y^2 \sin^2(\phi_y/2)+q_z^2 \sin^2(\phi_z/2) \tag{1.11a}$$

$$q_u = c\Delta t/\Delta u \quad R_u = \lambda/\Delta u \quad k_0 = 2\pi/\lambda \;\; u : x, y \;\; \text{or } z \tag{1.11b}$$

$$\phi_x = \frac{2\pi}{R_x}\frac{k_{num}}{k_0} \sin\theta \cos\phi \quad \phi_y = \frac{2\pi}{R_y}\frac{k_{num}}{k_0} \sin\theta \sin\phi \quad \phi_z = \frac{2\pi}{R_z}\frac{k_{num}}{k_0} \cos\theta \tag{1.11c}$$

1.4 Stability of Algorithms

The stability requirement of algorithm (1.5) puts an *upper limit* on time step Δt. This limit is necessary otherwise the computed field values might increase spuriously without limit as time marching continues. The reason for numerical instability is the violation of causality, that is, the minimum time Δt required for the signal to propagate from one node to the other separated by Δ is given by $\Delta t = \Delta/c$. Increasing the value of Δt beyond this value to speed up the simulation will result in instability. The upper bound on Δt is called the CFL *stability condition* or sometimes Courant limit and is given by (1.1). For the special case of cubic cell $\Delta x = \Delta y = \Delta z = \Delta$, one obtains

$$c.\Delta t_{CFL} \leq \frac{\Delta}{\sqrt{3}} \tag{1.12}$$

The stability and numerical dispersion for ADI, CNSS and CN schemes are investigated in terms of their amplification matrix M [14]. The ADI and CNSS schemes are found to have the same dispersion,. and CN and CNSS are found to be unconditionally stable. However, the unconditional stablility of ADI is contingent upon space-time discretizations, and its amplification matrix M is given by

$$||M_{ADI}|| \leq \sqrt{1+\left(\frac{c\Delta t}{min(\Delta x, \Delta y, \Delta z)}\right)^2} \tag{1.13}$$

where the matrix M is a function of wavenumber(magnitude and propagation direction), and space-time discretizations $\Delta x, \Delta y, \Delta z, \Delta t$. For a cubic cell, (1.13) reduces to

$$||M_{ADI}|| \leq \sqrt{1+\frac{q^2}{3}} \tag{1.14}$$

Another unconditionally stable FDTD scheme is LOD-FDTD. It is an efficient scheme compared to ADI, CN and CNSS schemes and is described in Section 1.5.

The FDTD algorithm may have to be run for a large number of time steps, sometimes of the order of 100,000 steps until the time domain waveform converges. It is possible to accelerate the convergence of the algorithm by using matrix-pencil (or GPOF) approach [15]. In this approach, the late time waveform can be predicted from the early time sampling data. One may be able to save about 80% of the simulation time using GPOF [16].

1.5 Absorbing Boundary Conditions

The antennas have associated open space region. The FDTD simulation of the antenna in this form will require an unlimited amount of computer memory and processing time, which is impossible to arrange. Therefore, the domain must be truncated so that the associated reflection is minimal. For this, the solution domain is divided into two regions: the interior region and the exterior region as shown in Figure 1.8. The interior region must be large enough to enclose the antenna of interest. The exterior region simulates the infinite space. It is a limited free space enclosing the interior region on one side and terminated on the other side by a perfect electric conductor. When we apply the FDTD algorithm to the interior region, it simulates wave propagation in the forward and backward directions. However, only the outward propagation in the exterior region is desired so that infinite free space conditions are simulated. Reflections are generated at the interior-exterior region interface and from the perfect electric conductor terminating the exterior region. These reflections must be suppressed to an acceptable level so that the FDTD solution is valid for all time steps. The exterior region includes the domain of absorbing boundary condition or ABC for short.

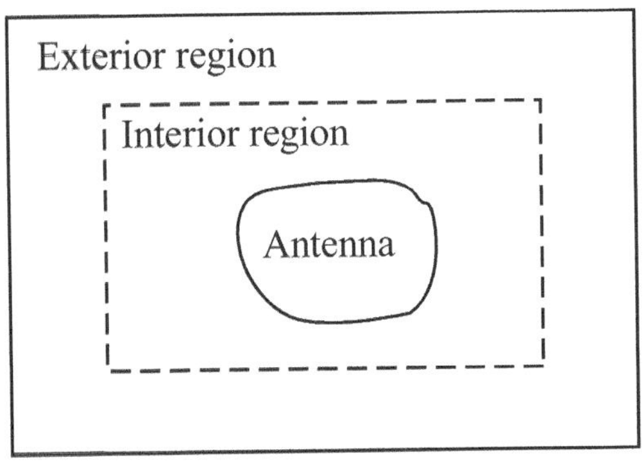

Figure 1.8 A typical truncation of the physical domain by the exterior region in FDTD algorithms

The absorbing boundary condition can be simulated in a number of ways. These are classified as analytical (or differential) ABC and material ABC. The material ABC is realized from the physical absorption of the incident wave by means of a lossy medium, whereas analytical ABC is simulated by approximating the one-way wave equation on the boundary of interior region. Whereas the analytical ABC may be able to provide upto −60 dB of reflection, the material ABC can provide better absorption with the reflection reaching an ideal limit of -80 to -120 dB from PML at the boundaries. Various types of ABCs are summarized next.

1.5.1 Analytical Absorbing Boundary Conditions

Analytical ABC is a popular technique because of its simplicity of implementation. The reflection from analytical ABC could be as low as 0.1%, i.e. -60 dB. Analysis for this absorbing boundary condition is based on the works of Enquist and Majda [17], Mur [18], Higdon [19], and Liao [20]. For a plane wave incident on a planar boundary, the wave will propagate forward (without any reflection) if the field function $F(x,y,z,t)$ satisfies one-way wave equation at the boundary. Such a boundary is therefore called the absorbing boundary. The PDE for the ABCs may be derived from the wave equation. Consider the following two-dimensional wave equation in the Cartesian coordinates

$$\frac{\partial^2 F}{\partial x^2}+\frac{\partial^2 F}{\partial y^2}-\frac{1}{c^2}\frac{\partial^2 F}{\partial t^2}=0 \tag{1.15}$$

This expression may be factored as

$$\left(\frac{\partial}{\partial x}-\sqrt{\frac{1}{c^2}\frac{\partial^2}{\partial t^2}-\frac{\partial^2}{\partial y^2}}\right)\left(\frac{\partial}{\partial x}+\sqrt{\frac{1}{c^2}\frac{\partial^2}{\partial t^2}-\frac{\partial^2}{\partial y^2}}\right)F=0 \tag{1.16}$$

Applying each of the above pseudo-operators to F gives rise to the following PDEs for the analytical ABC as

$$\left(D_x-\frac{D_t}{c}\sqrt{1-s^2}\right)F=0 \quad \text{for ABC on the left boundary} \tag{1.17a}$$

$$\left(D_x+\frac{D_t}{c}\sqrt{1-s^2}\right)F=0 \quad \text{for ABC on the right boundary} \tag{1.17b}$$

where $D_x=\partial/\partial x$, etc. and $s=cD_y/D_t$. Direct implementation of the operators with square root is not possible. The operators are therefore approximated and result in various forms of ABC depending on the approximation employed.

The first-order ABCs are obtained by a very crude approximation $\sqrt{1-s^2}\approx 1$ in (1.17a); one obtains

$$\left(\frac{\partial}{\partial x}-\frac{1}{c}\frac{\partial}{\partial t}\right)F\Big|_{x=0}=0 \tag{1.18}$$

for the left boundary at $x=0$. Similar expressions can be written down by inspection for the other boundaries also. The expression (1.18) is called the Enquist-Majda (E-M) first-order ABC. The field function F represents the electric field tangential to the boundary. For a non-planar wave, Equation (1.18) should be applied to all the components tangential to the planar boundary. It is shown that (1.18) is unconditionally stable [18]. The absorbing boundary condition (1.18) is exact only for a plane wave at normal incidence. Hence the wave will be reflected for an oblique incidence. Higher-order ABCs are suitable for non-normal incidence, and are obtained by better approximation of the square root function $\sqrt{1-s^2}$. For this, let us approximate $\sqrt{1-s^2}$ in a general form as

$$\sqrt{1-s^2} \approx \frac{d-bs^2}{1-as^2} \tag{1.19}$$

where the parameters a, b, d are optimized to define various forms of third-order ABCs such as Chebyshev, Pade and EM second-order. Use of Equation (1.19) in Equation (1.17a) gives

$$\frac{\partial^3 F}{\partial x \partial t^2} - ac^2 \frac{\partial^3 F}{\partial x \partial y^2} - \frac{d}{c}\frac{\partial^3 F}{\partial t^3} + bc\frac{\partial^3 F}{\partial t \partial y^2} = 0 \tag{1.20}$$

This expression reduces to:

(a) first-order E-M ABC for $d = 1$, $a = b = 0$;
(b) second-order E-M ABC for $d = 1$, $b = 1/2$, $a = 0$;
(c) Pade approximation or Pade ABC for $d = 1$, $b = 3/4$, $a = 1/4$;
(d) Chebyshev ABC for $d = 0.99973$, $b = 0.80864$, $a = 0.31657$.

Mur's ABCs are discretized versions of E-M ABCs, e.g. first-order Mur ABC for the boundary at $x = 0$ is obtained by discretizing Equations (1.17a) or (1.20) at $x = \Delta x/2, t = (n+1/2)\Delta t$ and is given by

$$F_0^{n+1} = F_1^{\mathbf{n}} + \left(\frac{c\Delta t - \Delta x}{c\Delta t + \Delta x}\right)\left(F_1^{n+1} - F_0^n\right) \tag{1.21}$$

Similar expressions may be obtained for other types of ABCs and for waves incident on different boundaries. One can realize about −40 dB reflection by using these ABCs. Liao's third-order ABC may be used to achieve about −50 to −60 dB reflection and is discussed next.

1.5.1.1 Liao's ABC

Liao's ABC is based on the extrapolation of fields in time and space using Newton-backward difference polynomial. Consider an outer grid boundary at $i_{\max} = M\Delta x$, then the L-th order Liao ABC may be written as [2, 21]

$$F^{n+1}(i_{\max}) = \sum_{i=1}^{L} (-1)^{l+1} C_l^L F^{n+1-l}(i_{\max} - lh) \tag{1.22}$$

where $h = \alpha c\Delta t/\Delta x$, α is the damping factor $0 \le \alpha \le 2$, and C_l^L is the Binomial coefficient defined as

$$C_l^L = \frac{L!}{l!(L-l)!} \tag{1.23}$$

It may be noted from Equation (1.22) that the field values at different time instants $(n+1-l)\Delta t$ and different locations $(i_{\max}-lh)\Delta x$ are combined judiciously to produce a near null field $F^{n+1}(i_{\max})$. This scheme is different from the one-way wave equation scheme and is found to be stable with double precision computations [2]. The choice of α and therefore h is not critical in Equation (1.22), resulting in the robustness of the scheme. The value of α may be chosen such that h is an integer and the required field locations coincide with those available in the standard FDTD scheme. Otherwise, quadratic interpolation may be employed. For the third-order ABC and h an integer, Equation (1.22) becomes

$$F_M^{n+1} = 3F_{M-1}^{n} - 3F_{M-2}^{n-1} + F_{M-3}^{n-2} \tag{1.24}$$

where F is one of the electric field components tangential to the boundary.

Interpretation of analytical ABCs in terms of surface impedance boundary condition (SIBC) is discussed in [22]. The SIBC involves tangential components of both E and H fields, whereas other analytical ABCs are applied to E field only. The SIBC type ABCs reduce the order of PDE without compromising the performance and this is discussed next.

Consider the half-space of an isotropic medium terminating the problem physical domain as shown in Figure 1.9. The interface is defined by $z=0$. Equating the tangential components of E and H at the interface gives [23]

$$k_0\sqrt{k_0^2-k_x^2-k_y^2}\,E_x = -\eta_0\left[(k_0^2-k_x^2)H_y+k_x\,k_y\,H_x\right] \tag{1.25a}$$

$$k_0\sqrt{k_0^2-k_x^2-k_y^2}\,E_y = \eta_0\left[\left(k_0^2-k_y^2\right)H_x+k_x\,k_y\,H_y\right] \tag{1.25b}$$

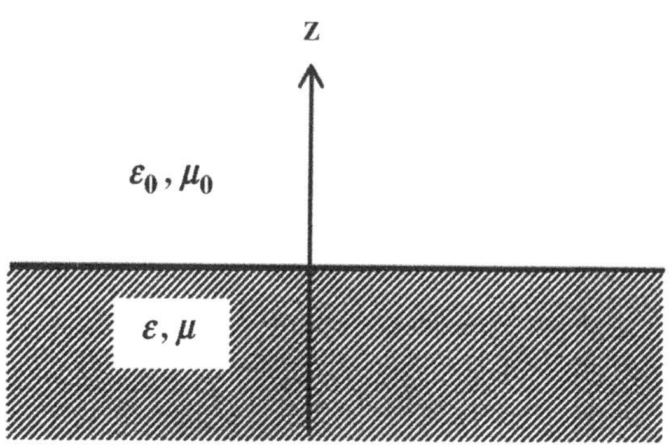

Figure 1.9 Half-space termination of physical space for deriving surface impedance boundary condition

For the two-dimensional TM_x waves $(H_x=0)$ with the fields constant along x $(k_x=0)$, Equation (1.25) reduces to

$$\sqrt{k_0^2-k_y^2}\,E_x = -\eta_0 k_0 H_y \tag{1.26}$$

Approximating the square-root factor according to Equation (1.19) gives

$$\left(dk_0^2 - bk_y^2\right)E_x = -\eta_0\left(k_0^2 - ak_y^2\right)H_y \tag{1.27}$$

The corresponding partial differential equation (PDE) for this boundary condition is obtained by using $k_0^2 = \omega^2/c^2$, $j\omega \leftrightarrow \partial/\partial t$, and $-jk_y = \partial/\partial y$ [23]

$$\frac{d}{c^2}\frac{\partial^2 E_x}{\partial t^2} - b\frac{\partial^2 E_x}{\partial y^2} = -\frac{\eta_0}{c^2}\frac{\partial^2 H_y}{\partial t^2} + \eta_0 a\frac{\partial^2 H_y}{\partial y^2} \tag{1.28}$$

It may be recast in the following form by using $\dfrac{\partial^2 H_y}{\partial t^2} = -\dfrac{c}{\eta_0}\dfrac{\partial^2 E_x}{\partial z \partial t}$

$$\frac{d}{c^2}\frac{\partial^2 E_x}{\partial t^2} - b\frac{\partial^2 E_x}{\partial y^2} = \frac{1}{c}\frac{\partial^2 E_x}{\partial z \partial t} + \eta_0 a\frac{\partial^2 H_y}{\partial y^2} \tag{1.29}$$

Discretizing it about the point $x = i\Delta x, y = \Delta y/2$ at $t = n\Delta t$ gives [23]

$$\begin{aligned}
E_x|_{i,0}^{n+1} = {} & -E_x|_{i,0}^{n-1} + \frac{2d\Delta z}{c\Delta t + d\Delta z}\left(E_x|_{i,0}^{n} + E_x|_{i,1}^{n}\right) + \frac{c\Delta t - d\Delta z}{c\Delta t + d\Delta z}\left(E_x|_{i,1}^{n+1} + E_x|_{i,0}^{n-1}\right) \\
& + \frac{b(c\Delta t)^2\Delta z}{\Delta y^2(c\Delta t + d\Delta z)}\begin{pmatrix} E_x|_{i+1,1}^{n} - 2E_x|_{i,1}^{n} + E_x|_{i-1,1}^{n} \\ +E_x|_{i+1,0}^{n} - 2E_x|_{i,0}^{n} + E_x|_{i-1,0}^{n} \end{pmatrix} \\
& + \frac{\eta_0 a(c\Delta t)^2\Delta z}{\Delta y^2(c\Delta t + d\Delta z)}\begin{pmatrix} H_y|_{i+1,1/2}^{n+1/2} - 2H_y|_{i,1/2}^{n+1/2} + H_y|_{i-1,1/2}^{n+1/2} \\ +H_y|_{i+1,1/2}^{n-1/2} - 2H_y|_{i,1/2}^{n-1/2} + H_y|_{i-1,1/2}^{n-1/2} \end{pmatrix}
\end{aligned} \tag{1.30}$$

The above expression reduces to first-order Mur ABC for $d = 1$, $a = b = 0$; second-order Mur ABC for $d = 1$, $b = {}^1\!/_2$, $a = 0$; Pade ABC for $d = 1$, $b = {}^3\!/_4$, $a = {}^1\!/_4$; and Chebyshev ABC for $d = 0.99973$, $b = 0.80864$, $a = 0.31657$.

The numerical results based on Equation (1.30) show that the second-order PDE (1.29) involving both the components of E and H have similar performance as the third-order PDE of (1.20) for one field component [22].

The performance of an ABC is determined by the power reflected by it and summed over the whole grid. It is called global error and is defined as

$$E = \sum\sum\sum e^2(i,j,k) \tag{1.31}$$

where e is the amplitude of reflected wave for an incident wave of unit amplitude. The expression (1.30) was implemented for SIBC-based Chebyshev and Pade ABC, and compared with Liao's third-order and Mur's second-order ABC on 20×200 lattice. The source was a Gaussian-like pulse of width $30\Delta t$. The global error as a function of time is compared in Figure 1.10 [22]. It may be noted that the second-order SIBC (Pade and Chebyshev) performs

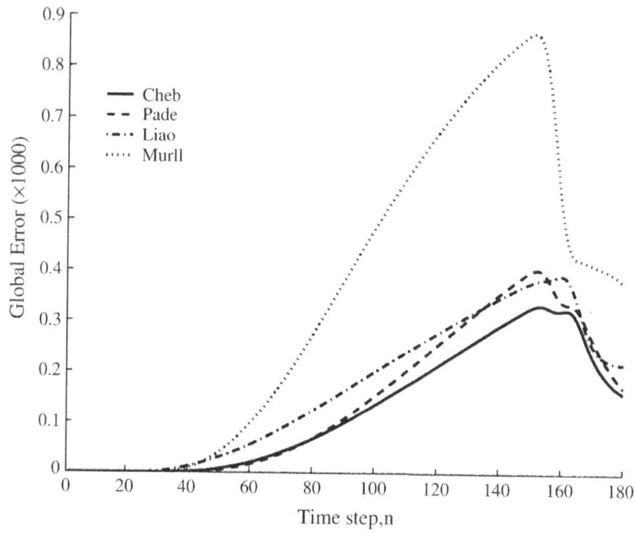

Figure 1.10 Comparison of global error as a function of time for SIBC-based Chebyshev and Pade ABCs with Liao's and Mur's second-order ABCs. Reproduced by permission of ©2003 IEEE, Figure 2 of [22]

much better than the second-order Mur's ABC. The Chebyshev approximation is better than the third-order Liao ABC. The decay in global error after about $n = 150$ is due to the fact that the source pulse went to zero some time ago. Extension of SIBC-based ABC for the three-dimensional case is also discussed in [22].

Perfectly matched layer (PML) may be employed for lower reflections and is discussed next. The performance of some of the higher-order analytical ABCs is compared with PML in [24]. The golobal error for $30 \times 30 \times 30$ geometry is compared in Figure 1.11 for higher-order analytical ABCs and PML [24]. The Gaussian-like excitation pulse is 40 Δt wide. The PML was terminated with Liao's ABC instead of electric wall. The difference between third-order E-M and Liao's fifth-order is about 20 dB. It may be observed that PML can provide an excellent termination of space domain.

1.5.2 *Material-Absorbing Boundary Conditions*

The absorbing sheets method was the first material ABC proposed and was used for FDTD computations by Hallond and Williams [25]. In this method the exterior region is filled with a lossy medium with parameters $(\mu', \varepsilon', \sigma^*, \sigma)$, where σ is the electrical conductivity and σ^* is the fictitious magnetic conductivity. The lossy medium is taken to be of finite thickness and is backed by a perfect electric conductor. The values of σ and σ^* are selected so that when a wave is incident normal to the interface separating the free space interior region and the lossy medium, there should be no reflection, which is possible only if

$$Z_0 = Z_m \tag{1.32}$$

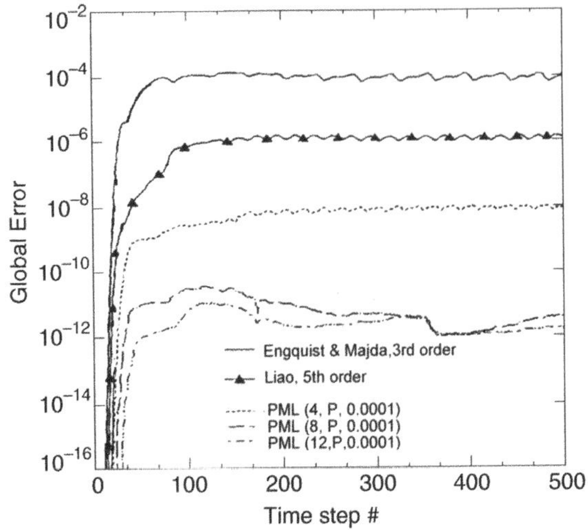

Figure 1.11 Comparison of global error as a function of time for a number of higher-order ABCs and PML. Reproduced by permission of ©1997 IEEE, Figure 1 of [24]

where Z_0 is the free space impedance defined by ($\mu_o, \varepsilon_o, \sigma^* = 0, \sigma = 0$) and Z_m is the impedance of the lossy medium dependent on ($\mu', \varepsilon', \sigma^*, \sigma$). Equation (1.32) can be expressed as

$$\sqrt{\frac{\mu_0}{\varepsilon_0}} = \sqrt{\frac{\mu' + j\sigma^*}{\varepsilon' + j\sigma}} \tag{1.33}$$

For convenience if we choose $\mu' = \mu_0$, $\varepsilon' = \varepsilon_0$ then matching condition (1.33) yields

$$\frac{\sigma^*}{\mu_0} = \frac{\sigma}{\varepsilon_0} = v \tag{1.34}$$

In practice, the conductivity of the lossy medium is increased gradually from 0 to σ_m at the thickness t of the lossy medium. The increase may be linear or parabolic, and σ is written as $\sigma(\rho)$ where ρ is the distance from the interface.

The magnitude of the reflection coefficient for a wave incident at an angle ϕ w.r.t the normal to the layers is

$$R(\phi) = [R(0)]^{\cos(\phi)} \tag{1.35}$$

with R(0) the reflection coeffficient for normal incidence and is defined as

$$R(0) = \exp\left(\frac{-2}{\varepsilon_o c}\int_0^t \sigma_n(\rho)d\rho\right) \qquad n = 0, 1, 2 \ldots \tag{1.36}$$

In general, the profile for the PML medium may be defined as

$$\sigma_n(\rho) = \sigma_m \left(\frac{\rho}{t}\right)^n \tag{1.37}$$

and the corresponding

$$R(0) = \exp\left(\frac{-2\sigma_m t}{(n+1)\varepsilon_o c}\right) \quad n = 0, 1, 2 \ldots \tag{1.38}$$

It appears from (1.38) that one may choose a sufficiently thick lossy medium to obtain any desired amount of reflection. However, the reflection at oblique incident angles is higher, and the total reflection at the interface for the absorbing sheets method is of the same order as that obtained from analytical ABCs discussed earlier.

1.5.3 Perfectly Matched Layer ABC

The perfectly matched layer (PML) formulation was proposed by Berenger [26] and represents a generalization of the absorbing sheets method to arbitrary angles of incidence. This is achieved by splitting the electric and magnetic field components in the absorbing media, and assigning different losses along different axis. The net effect of this is to create a nonphysical lossy medium in the exterior region so that the wave impedance in this medium is independent of the angle of incidence and frequency of the waves. The split-field PML makes coding difficult. However, this method is used as a benchmark for other PML formulations.

1.5.4 Uniaxial PML

Non-split field formulations in PML have been developed to simplify coding. One of these formulations is in the form of complex stretching of the Cartesian coordinates in the frequency domain [27]. The stretching variable formulation leads to convolution in the time domain, and can be avoided by the splitting of fields similar to that in Berenger's PML.

In a popular unsplit-field formulation, the PML medium employed is anisotropic with permittivity and permeability tensors, and is called Uniaxial PML, UPML for short. The general formulation for UPML is given in [28, 29]. A diagonal tensor is uniaxial if it is anisotropic along one axis only and is defined as

$$\bar{\bar{\varepsilon}} = \varepsilon_0 \bar{\bar{s}}, \quad \text{and} \quad \bar{\bar{\mu}} = \mu_0 \bar{\bar{s}} \tag{1.39}$$

where the media tensor $\bar{\bar{s}}$ for the three-dimensional PML region is defined as [30]

$$\bar{\bar{s}} = \begin{bmatrix} 1/s_x & 0 & 0 \\ 0 & s_x & 0 \\ 0 & 0 & s_x \end{bmatrix} \begin{bmatrix} s_y & 0 & 0 \\ 0 & 1/s_y & 0 \\ 0 & 0 & s_y \end{bmatrix} \begin{bmatrix} s_z & 0 & 0 \\ 0 & s_z & 0 \\ 0 & 0 & 1/s_z \end{bmatrix} = \begin{bmatrix} s_y s_z/s_x & 0 & 0 \\ 0 & s_z s_x/s_y & 0 \\ 0 & 0 & s_x s_y/s_z \end{bmatrix} \tag{1.40}$$

and s_v are defined as

$$s_v = k_v + \frac{\sigma_v}{j\omega\varepsilon} = k_v + \frac{\rho_v}{j\omega\mu} \qquad v = x, y, z \tag{1.41}$$

where σ_v and ρ_v are the UPML medium conductivity and reluctivity along the v-axis, respectively. The medium characterized as above leads to reflectionless propagation from free space to PML for both TE and TM waves, independent of angle of incidence, polarization and frequency. For exponential decay of the propagating wave in UPML, the media constants should be complex with $s_v = 1 + \frac{\sigma_v}{j\omega\varepsilon}$. The UPML is characterized by $\sigma_x(i)$ alone for layers perpendicular to x-axis, by $\sigma_y(j)$ alone for layers perpendicular to y-axis, by $\sigma_x(i)$and $\sigma_y(j)$ etc. for the edge regions, and by $\sigma_x(i)$, $\sigma_y(j)$ and $\sigma_z(k)$ for the corner regions as shown in Figure 1.12. The non-PML region or central region is characterized by $\sigma_x = \sigma_y = \sigma_z = 0$, thus simplifying coding. The values of σ_v are gradually increased as one goes deeper into the PML region.

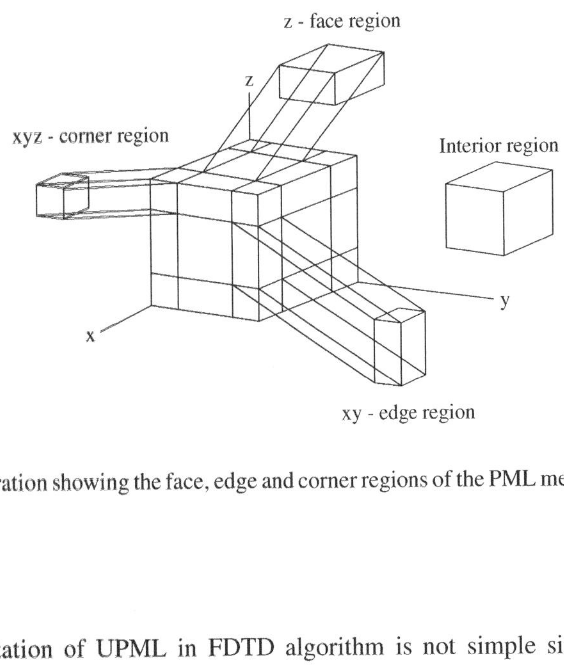

Figure 1.12 Illustration showing the face, edge and corner regions of the PML medium surrounding the interior region

The implementation of UPML in FDTD algorithm is not simple since the medium is dispersive due to the frequency dependence of permittivity and permeability tensors. Direct transformation from frequency domain to time domain involves convolution and is inefficient. Instead, a two-step time marching scheme is suggested in [2] for efficient implementation of UPML. In this scheme Maxwell's equations are formulated in terms of D and B, and E and H are derived from these. The procedure reported in [30] is D(E)-H based which can be easily implemented in most of the FDTD schemes, e.g. for the standard FDTD scheme

$$D_x^{n+1}(i+\tfrac{1}{2},j,k) = c_j^1 D_x^n(i+\tfrac{1}{2},j,k) + c_j^2 \left(\frac{H_z^{n+1/2}(i+\frac{1}{2},j+\frac{1}{2},k) - H_z^{n+1/2}(i+\frac{1}{2},j-\frac{1}{2},k)}{\Delta y} - \frac{H_y^{n+1/2}(i+\frac{1}{2},j,k+\frac{1}{2}) - H_y^{n+1/2}(i+\frac{1}{2},j,k-\frac{1}{2})}{\Delta z} \right) \tag{1.42}$$

where

$$c_j^1 = \frac{2\varepsilon k_y - \sigma_y(j)\Delta t}{2\varepsilon k_y + \sigma_y(j)\Delta t}, \quad c_j^2 = \frac{2\varepsilon\Delta t}{2\varepsilon k_y + \sigma_y(j)\Delta t} \tag{1.43}$$

$$E_x^{n+1}(i+\tfrac{1}{2},j,k) = c_k^3\, E_x^n(i+\tfrac{1}{2},j,k) + c_k^4 \left(c_{i+1/2}^5\, D_x^{n+1}(i+\tfrac{1}{2},j,k) - c_{i+1/2}^6\, D_x^n(i+\tfrac{1}{2},j,k) \right) \tag{1.44}$$

where

$$c_k^3 = \frac{2\varepsilon k_z - \sigma_z(k)\Delta t}{2\varepsilon k_z + \sigma_z(k)\Delta t}, \quad c_k^4 = \frac{1}{\varepsilon(2\varepsilon k_z + \sigma_z(k)\Delta t)},$$
$$c_{i+1/2}^5 = 2\varepsilon k_x + \sigma_x(i+1/2)\Delta t, \quad c_{i+1/2}^6 = 2\varepsilon k_x - \sigma_x(i+1/2)\Delta t \tag{1.45}$$

It is understood that the indices i, j, k within the brackets after σ_x, etc. refer to the position of the node.

Substituting for D_x^{n+1} from Equation (1.42), one can write,

$$E_x^{n+1}(i+\tfrac{1}{2},j,k) = c_k^3 E_x^n(i+\tfrac{1}{2},j,k) + \left(c_k^4 c_{i+1/2}^5 c_j^1 - c_k^4 c_{i+1/2}^6 \right) D_x^n(i+\tfrac{1}{2},j,k) + c_k^4 c_{i+1/2}^5 c_j^2 \left(\frac{H_z^{n+1/2}(i+\frac{1}{2},j+\frac{1}{2},k) - H_z^{n+1/2}(i+\frac{1}{2},j-\frac{1}{2},k)}{\Delta y} - \frac{H_y^{n+1/2}(i+\frac{1}{2},j,k+\frac{1}{2}) - H_y^{n+1/2}(i+\frac{1}{2},j,k-\frac{1}{2})}{\Delta z} \right) \tag{1.46}$$

Similar expressions may be obtained for other field components. For a two-dimensional PML, the updating equations are obtained by setting $\sigma_z = 0, k_z = 1$. The optimal value of σ_v used in the updating equations may be chosen according to that given in [31, 32].

The numerical results for FDTD(2,2), SBTD, ADI, and CN schemes employing this UPML are presented for two-dimensional TE cases in [30]. The absorbing behavior of UPML for SBTD scheme is found to be similar to that for FDTD(2,2). The UPML in ADI scheme is found to produce larger reflections compared to that in FDTD(2,2) and is shown in Figure 1.13 for the three-dimensional study [32]. The value of Δt_{CFL} used is 1.926 ps. Also, the absorbing ability of UPML deteriorates with the increase in Courant number q in Figure 1.13. The reflection error is defined as $|E_{measured}-E_{ref}|/|E_{ref\ \max}|$ where E_{ref} is the value measured by extending the dimensions outward to avoid reflections from boundaries, and $E_{ref\ \max}$ is the maximum value observed at the same point.

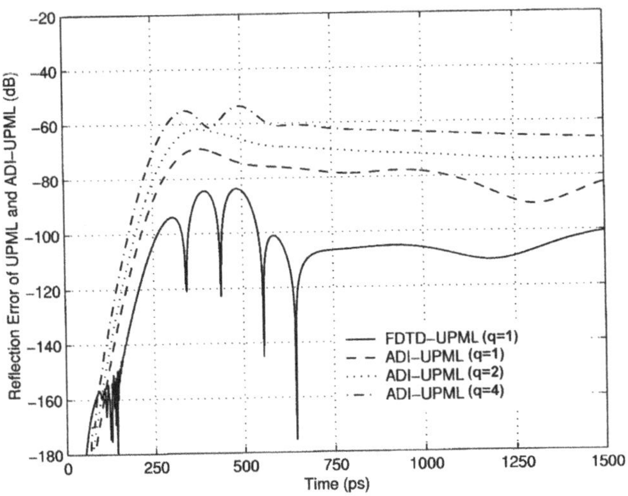

Figure 1.13 Reflection error from the FDTD-UPML and ADI-UPML as a function of time with q as a parameter. Reproduced by permission of ©2002 IEEE, Figure 1 of [32]

1.6 LOD-FDTD Algorithm

The alternating-direction implicit (ADI) FDTD algorithm was proposed by Namiki [33]. It employs the same grid as in FDTD(2,2) and is second-order accurate in both space and time. The algorithm is implicit because the update of fields is not a single step process as for FDTD(2,2) but involves solving a set of simultaneous equations as an intermediate step. The ADI algorithm has been analyzed thoroughly for its dispersion and stability properties. It has been found that the maximum permissible step size Δt in ADI-FDTD depends on the numerical errors which increase with the increase in the Courant number q. ADI exhibits splitting error which depends not only on time step size but also on the spatial derivatives of fields. Due to its implicit nature which involves solving a tridiagonal matrix equation, the number of arithmetic operations per update of fields is more compared to FDTD(2,2). CN-FDTD is more accurate than ADI-FDTD but is computationally expensive. The LOD-FDTD is simple to code, is 20% more efficient, and provides similar accuracy as ADI-FDTD for the same computer memory required [34]. We describe LOD-FDTD algorithm for three-dimensional problems next. The efficiency of LOD-FDTD can be improved to 83% by a simple transformation.

For a lossless medium the Maxwell's curl equations (1.3) reduce to

$$\nabla \times \mathbf{H} = \varepsilon \frac{\partial \mathbf{E}}{\partial \mathbf{t}} \qquad \nabla \times \mathbf{E} = -\mu \frac{\partial \mathbf{H}}{\partial \mathbf{t}} \tag{1.47}$$

These equations may be expressed in operator form as

$$\frac{\partial \mathbf{F}}{\partial t} = [A]\mathbf{F} + [B]\mathbf{F} \tag{1.48}$$

where $\mathbf{F} = [E_x, E_y, E_z, H_x, H_y, H_z]^T$ and

$$[A] = \begin{bmatrix} 0 & 0 & 0 & 0 & 0 & \frac{\partial}{\varepsilon \partial y} \\ 0 & 0 & 0 & \frac{\partial}{\varepsilon \partial z} & 0 & 0 \\ 0 & 0 & 0 & 0 & \frac{\partial}{\varepsilon \partial x} & 0 \\ 0 & \frac{\partial}{\mu \partial z} & 0 & 0 & 0 & 0 \\ 0 & 0 & \frac{\partial}{\mu \partial x} & 0 & 0 & 0 \\ \frac{\partial}{\mu \partial y} & 0 & 0 & 0 & 0 & 0 \end{bmatrix} \tag{1.49a}$$

$$[B] = \begin{bmatrix} 0 & 0 & 0 & 0 & -\frac{\partial}{\varepsilon \partial z} & 0 \\ 0 & 0 & 0 & 0 & 0 & -\frac{\partial}{\varepsilon \partial x} \\ 0 & 0 & 0 & -\frac{\partial}{\varepsilon \partial y} & 0 & 0 \\ 0 & 0 & -\frac{\partial}{\mu \partial y} & 0 & 0 & 0 \\ -\frac{\partial}{\mu \partial z} & 0 & 0 & 0 & 0 & 0 \\ 0 & -\frac{\partial}{\mu \partial x} & 0 & 0 & 0 & 0 \end{bmatrix} \tag{1.49b}$$

By applying the CN (Crank-Nicolson) algorithm

$$\partial F^{(n+1/2)}/\partial t \approx (F^{n+1}-F^{n})/\Delta t, \quad F^{n+1/2} \approx (F^{n+1}+F^{n})/2 \tag{1.50}$$

to Equation (1.48) at $t = (n+1/2)\Delta t$ we have

$$\left([I]-\frac{\Delta t}{2}[A]-\frac{\Delta t}{2}[B]\right)\mathbf{F}^{n+1} = \left([I]+\frac{\Delta t}{2}[A]+\frac{\Delta t}{2}[B]\right)\mathbf{F}^{n} \tag{1.51}$$

Equation (1.51) can be approximated as, assuming $(\Delta t)^2 AB \ll 4$,

$$\left([I]-\frac{\Delta t}{2}[A]\right)\left([I]-\frac{\Delta t}{2}[B]\right)\mathbf{F}^{n+1} = \left([I]+\frac{\Delta t}{2}[A]\right)\left([I]+\frac{\Delta t}{2}[B]\right)\mathbf{F}^{n} \tag{1.52}$$

The above expresssion may be decomposed into the following sub-steps *if the operator matrices A and B commute,*

$$\left([I]-\frac{\Delta t}{2}[A]\right)\mathbf{F}^{n+1/2} = \left([I]+\frac{\Delta t}{2}[A]\right)\mathbf{F}^{n} \tag{1.53a}$$

$$\left([I]-\frac{\Delta t}{2}[B]\right)\mathbf{F}^{n+1} = \left([I]+\frac{\Delta t}{2}[B]\right)\mathbf{F}^{n+1/2} \tag{1.53b}$$

The splitting of time step as defined above is found to suffer from non-commutivity-related error. A simple modification of the time instant for the sub-steps is proposed to reduce this error [35]. The modification is to re-define (1.53) as

$$\left([I]-\frac{\Delta t}{2}[A]\right)\mathbf{F}^{n+3/4} = \left([I]+\frac{\Delta t}{2}[A]\right)\mathbf{F}^{n+1/4} \tag{1.54a}$$

$$\left([I]-\frac{\Delta t}{2}[B]\right)\mathbf{F}^{n+5/4} = \left([I]+\frac{\Delta t}{2}[B]\right)\mathbf{F}^{n+3/4} \tag{1.54b}$$

The change-over from the fractional time steps to integer time steps for compatibility with FDTD(2,2) will be discussed at the end. Now we substitute the operator matrices and obtain the following expressions for the field components:

Sub-step 1

$$E_x^{n+3/4} = E_x^{n+1/4} + \frac{\Delta t}{2\varepsilon}\left(\frac{\partial H_z^{n+3/4}}{\partial y} + \frac{\partial H_z^{n+1/4}}{\partial y}\right), \tag{1.55a}$$

$$H_z^{n+3/4} = H_z^{n+1/4} + \frac{\Delta t}{2\mu}\left(\frac{\partial E_x^{n+3/4}}{\partial y} + \frac{\partial E_x^{n+1/4}}{\partial y}\right) \tag{1.55b}$$

Sub-step 2

$$E_x^{n+5/4} = E_x^{n+3/4} + \frac{\Delta t}{2\varepsilon}\left(\frac{\partial H_z^{n+5/4}}{\partial y} + \frac{\partial H_z^{n+3/4}}{\partial y}\right), \tag{1.56a}$$

$$H_z^{n+5/4} = H_z^{n+3/4} + \frac{\Delta t}{2\mu}\left(\frac{\partial E_x^{n+5/4}}{\partial y} + \frac{\partial E_x^{n+3/4}}{\partial y}\right) \tag{1.56b}$$

The expressions for other field components can be simply written down by inspection.

Eliminating $H_z^{n+3/4}$ from (1.55a) and (1.55b) gives

$$E_x^{n+3/4} - \frac{\Delta t^2}{4\mu\varepsilon}\frac{\partial^2 E_x^{n+3/4}}{\partial y^2} = E_x^{n+1/4} + \frac{\Delta t^2}{4\mu\varepsilon}\frac{\partial^2 E_x^{n+1/4}}{\partial y^2} + \frac{\Delta t}{\varepsilon}\frac{\partial H_z^{n+1/4}}{\partial y} \tag{1.57}$$

The finite difference approximation of the above leads to the following update [35]

$$-aE_x^{n+3/4}(i+\tfrac{1}{2},j-1,k) + (1+2a)E_x^{n+3/4}(i+\tfrac{1}{2},j,k) - aE_x^{n+3/4}(i+\tfrac{1}{2},j+1,k) =$$

$$= (1-2a)E_x^{n+1/4}(i+\tfrac{1}{2},j,k) + b\begin{pmatrix} H_z^{n+1/4}(i+\tfrac{1}{2},j+\tfrac{1}{2},k) \\ -H_z^{n+1/4}(i+\tfrac{1}{2},j-\tfrac{1}{2},k)\end{pmatrix} + a\begin{pmatrix} E_x^{n+1/4}(i+\tfrac{1}{2},j+1,k) \\ +E_x^{n+1/4}(i+\tfrac{1}{2},j-1,k)\end{pmatrix} \tag{1.58}$$

where

$$a = \frac{\Delta t^2}{4\varepsilon\mu\Delta y^2} \text{ and } b = \frac{\Delta t}{\varepsilon\Delta y}. \tag{1.59}$$

Equation (1.58) represents a matrix equation and is tri-diagonal in nature. Its solution determines $E_x^{n+3/4}$. The size of the matrix is determined by N_y, the number of nodes along the y-axis. The matrix size being large, indirect matrix solution techniques are employed. An alternate-direct implicit Douglas-Gunn algorithm has been used by Sun and Trueman [36]. Yang et al. used a biconjugate-gradient (BCG) algorithm with symmetric successive over-relaxation (SSOR) as a pre-conditioner to speed up the BCG algorithm [37].

The field $E_x^{n+3/4}$ obtained above may be used to explicitly update $H_z^{n+3/4}$ through Equation (1.55b) according to

$$H_z^{n+3/4}(i+\tfrac{1}{2},j+\tfrac{1}{2},k)$$

$$= H_z^{n+1/4}(i+\tfrac{1}{2},j+\tfrac{1}{2},k) + f\begin{pmatrix} E_x^{n+1/4}(i+\tfrac{1}{2},j+1,k) - E_x^{n+1/4}(i+\tfrac{1}{2},j,k) + \\ E_x^{n+3/4}(i+\tfrac{1}{2},j+1,k) - E_x^{n+3/4}(i+\tfrac{1}{2},j,k)\end{pmatrix} \tag{1.60}$$

where

$$f = \frac{\Delta t}{2\mu\Delta y} \tag{1.61}$$

We can similarly solve for $E_y^{n+3/4}$ and $E_z^{n+3/4}$ and update other H-field components, e.g.

$$H_x^{n+3/4}(i,j+\tfrac{1}{2},k+\tfrac{1}{2}) = H_x^{n+1/4}(i,j+\tfrac{1}{2},k+\tfrac{1}{2}) + g\begin{pmatrix} E_y^{n+1/4}(i,j+\tfrac{1}{2},k+1) - E_y^{n+1/4}(i,j+\tfrac{1}{2},k) + \\ E_y^{n+3/4}(i,j+\tfrac{1}{2},k+1) - E_y^{n+3/4}(i,j+\tfrac{1}{2},k) \end{pmatrix} \tag{1.62}$$

where

$$g = \frac{\Delta t}{2\mu\Delta z}. \tag{1.63}$$

Similarly, (1.56) may be discretized and solved to update the field components. This completes one cycle of time step updation.

The field updating procedure described above assumes that the initial field values are available at $t=(1/4)\Delta t$. Tan proposes the following input processing from $t=0$ to $t=(1/4)\Delta t$ [38]

$$\left([I] - \frac{\Delta t}{4}[B]\right)\mathbf{u}^{1/4} = \left([I] + \frac{\Delta t}{4}[B]\right)\mathbf{u}^{0} \tag{1.64}$$

Use of Equation (1.49b) for matrix B results in

$$E_x^{1/4} - \frac{\Delta t^2}{16\mu\varepsilon}\frac{\partial^2 E_x^{1/4}}{\partial z^2} = E_x^0 + \frac{\Delta t^2}{16\mu\varepsilon}\frac{\partial^2 E_x^0}{\partial z^2} - \frac{\Delta t}{2\varepsilon}\frac{\partial H_y^0}{\partial z} \tag{1.65a}$$

$$H_y^{1/4} = H_y^0 - \frac{\Delta t}{4\mu}\left(\frac{\partial E_x^0}{\partial z} + \frac{\partial E_x^{1/4}}{\partial z}\right) \tag{1.65b}$$

While the implementation of Equation (1.65a) requires the solution of a matrix equation similar to that for $E_x^{n+3/4}$ and described earlier, the update for $H_y^{1/4}$is explicit. Discretizing Equation (1.65) results in

$$-pE_x^{1/4}(i+\tfrac{1}{2},j,k-1) + (1+2p)E_x^{1/4}(i+\tfrac{1}{2},j,k) - pE_x^{1/4}(i+\tfrac{1}{2},j,k+1) =$$

$$= (1-2p)E_x^0(i+\tfrac{1}{2},j,k) + p\begin{pmatrix} E_x^0(i+\tfrac{1}{2},j,k+1) + \\ E_x^0(i+\tfrac{1}{2},j,k11) \end{pmatrix} - q\begin{pmatrix} H_y^0(i+\tfrac{1}{2},j,k+\tfrac{1}{2}) - \\ H_y^0(i+\tfrac{1}{2},j,k-\tfrac{1}{2}) \end{pmatrix} \tag{1.66a}$$

$$H_y^{1/4}(i+\tfrac{1}{2},j,k+\tfrac{1}{2}) = H_y^0(i+\tfrac{1}{2},j,k+\tfrac{1}{2})+s\begin{pmatrix} E_x^0(i+\frac{1}{2},j,k+1)-E_x^0(i+\frac{1}{2},j,k)+ \\ E_x^{1/4}(i+\frac{1}{2},j,k+1)-E_x^{1/4}(i+\frac{1}{2},j,k) \end{pmatrix} \tag{1.66b}$$

where

$$p=\frac{\Delta t^2}{16\varepsilon\mu\Delta z^2},\quad q=\frac{\Delta t}{2\varepsilon\Delta z}\ \text{and}\ s=\frac{\Delta t}{4\mu\Delta z} \tag{1.67}$$

The following transformation is recommended for output processing of data from $\mathbf{F}^{n+5/4}$ to $\mathbf{F}^{n+1}$ [38]

$$\left([I]+\frac{\Delta t}{4}[B]\right)\mathbf{F}^{n+1} = \left([I]-\frac{\Delta t}{4}[B]\right)\mathbf{F}^{n+5/4} \tag{1.68}$$

The increased efficiency of LOD-FDTD compared to ADI-FDTD is due to the reduced number of arithmetic operations on the RHS of Equation (1.53) [35]. The number of arithmetic operations for the LOD and ADI schemes are compared in Table 1.1 [[35],Table 1].

Table 1.1 Comparison between ADI and LOD-FDTD schemes for the number of arithmetical operations

Arithmetical operations		ADI	LOD
Implicit	M/D	18	18
	A/S	48	24
Explicit	M/D	12	6
	A/S	24	24
Total	M/D	30	24
	A/S	72	48

Source: Reproduced by permission of ©2007 IEEE, Table 1 of [35].

Note: *M = multiplication, D = division, A = addition and S = subtraction.

It is observed in [38] that the various implicit FDTD schemes have similar updating structures, and it is possible to improve their efficiency using a general procedure. The approach devised is to express these algorithms in a form so that the RHS of transformations e.g. (1.54) should consist of least number of terms and should be free from the matrix operator. Auxiliary variables are introduced for this purpose. For the LOD-FDTD described above we may define the auxiliary variables as [38]

$$F^{n+3/4} = H^{n+3/4} - F^{n+1/4} \quad \text{and} \quad F^{n+5/4} = H^{n+5/4} - F^{n+3/4} \tag{1.69}$$

The expressions (1.54) then transform to

$$\left(\frac{1}{2}[I]-\frac{\Delta t}{4}[A]\right)H^{n+3/4} = F^{n+1/4} \quad \left(\frac{1}{2}[I]-\frac{\Delta t}{4}[B]\right)H^{n+5/4} = F^{n+3/4} \tag{1.70}$$

This LOD algorithm is more efficient than Equation (1.54) because the RHS of Equation (1.70) is free from matrix operators. The efficiency of the modified algorithm is found to be 1.83 times that of the ADI-FDTD algorithm of Equation (1.53) [38].

The LOD-FDTD algorithm for circularly symmetric geometries has been reported. The improvement in efficiency compared to ADI-FDTD has been reported to be about 40% [39]. Extension of LOD-FDTD method to dispersive media has been reported by Shibayama et al. [40]. The improvement in efficiency is found to be about 30% compared to FDTD(2,2).

1.6.1 PML Absorbing Boundary Condition for LOD-FDTD

For the use of LOD-FDTD to antennas, the electromagnetic geometry must be terminated with reflectionless boundaries. Split field and unsplit field PML have been formulated for LOD-FDTD in [41] and [42], respectively. The reflection coefficient of the order of −75 dB has been reported for the split-field PML in the context of a two-dimensional TE case [41]. It has been observed that unsplit field non-dispersive PML gives reflection of the order of −17 dB for $q=2$, and −12 dB for $q=4$. However, complex envelope unsplit field dispersive PML gives better performance, −33 dB for $q=2$ and −25 dB for $q=4$ [42]. Convolutional PML (CPML) has been found to give better performance than other forms of PML for LOD-FDTD [43]. It is found that CPML gives lower reflection by about 16 dB compared to FDTD-PML for $q=1$. This is shown in Figure 1.14 for 8 layers of CPML. The reflection from CPML is found to decrease with increase in q number due to the increased value of $E_{ref\ \max}$ [43]. This is shown in Figure 1.15.

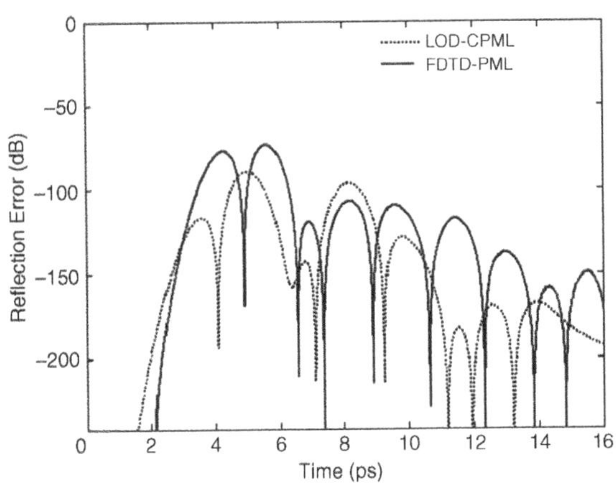

Figure 1.14 Comparison of reflection error from LOD-CPML and FDTD-PML for $q=1$. Reproduced by permission of ©2007 IEEE, Figure 2 of [43]

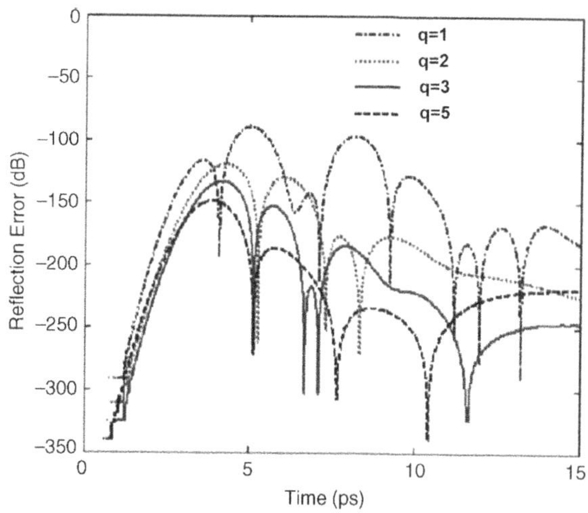

Figure 1.15 Comparison of reflection error from LOD-CPML for four values of q. Reproduced by permission of ©2007 IEEE, Figure 3 of [43]

1.7 Robustness of Printed Patch Antennas

The printed antennas are likely to get damaged when they brush against some hard object during use. When used as an RFID antenna, the environment is expected to be relatively harsh. The damage may occur to the top metallization, produce voids in dielectric subatrate, etc. In addition, the substrate may have non-uniform thickness and dielectric constant. The effect of these degradations on the resonant frequency of patch antennas has been studied analytically and verified numerically [44]. The analysis may be carried out based on perturbation technique. It has been found that the robustness of microstrip antennas is directly related to the substrate damage volume and position. The change in resonance frequency of the antenna is directly proportional to the volume and the change in permittivity of the defect. Small damage to the patch metallization alone does not affect return loss.

1.8 Thin Dielectric Approximation

The printed flat antennas on dielectric films are being used increasingly because they can be made into complicated shapes. Often the film is very thin and is less than one cell along the thickness. Use of uniform cell size is not advisable because of too much demand on computer resources. The other alternatives are non-uniform grid, or local refinement in the region, or sub-cell algorithm. Another alternative is to model the field distribution in the film region using quasi-static approximation. The geometry of a metal strip on a thin dielectric film is shown in Figure 1.16. Using the sub-cell approach for the tangential fields E_x and E_y, one needs to replace ε and σ in (1.4) by their average values defined as

$$\varepsilon_{avg} = \frac{\varepsilon_0(\Delta z - d) + \varepsilon_r d}{\Delta z} \qquad \sigma_{avg} = \frac{\sigma d}{\Delta z} \tag{1.71}$$

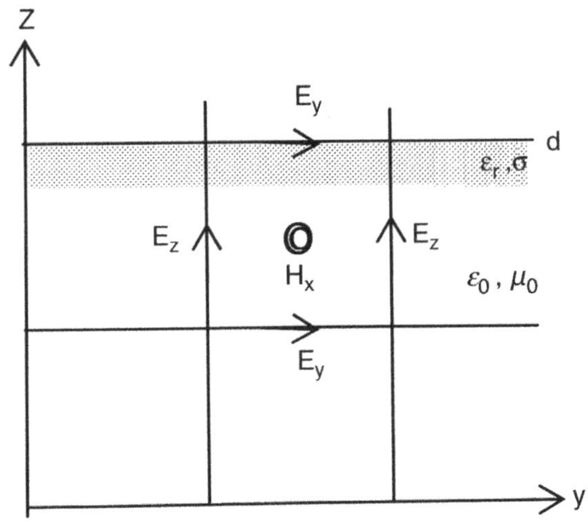

Figure 1.16 Thin dielectric film in a FDTD cell. Reproduced by permission of ©2004 IEEE [45]

where d is the film thickness, Δz is cell size along the film thickness, ε_r and σ are electrical parameters of the film. The field E_z near the strip edge is dominated by the static field, and may be included in the FDTD algorithm as described in [45, 46].

1.9 Modeling of PEC and PMC for Irregular Geometries

The PEC or PMC boundaries may not conform in places to the grid employed for the FDTD analysis of complex geometrical shapes. The modifications for such cases include conformal FDTD (C-FDTD) method [47], contour-path FDTD (CP-FDTD) method [48], sub-cell models [49], etc. These methods may require changes in existing source codes, grid, and time step. Sometimes, staircase approximation of the geometry is employed. A new method has been suggested in [50] and does not involve disturbing the original grid. The method involves extrapolation of (tangential) field values from the adjacent internal node to obtain exterior field values so that the perfect boundary condition in between is satisfied. The new method can be applied to model parallel, slanted, and curved walls in the existing FDTD grid [50]. Time step reduction is not necessary.

To illustrate this method we consider a rectangular resonator whose cross-section in the x–y plane is shown in Figure 1.17. The geometry is deliberately displaced along the y-axis so that the PEC on the LHS and RHS do not coincide but are parallel to the existing grid. Let the displacement with respect to the grid be $\xi\Delta y$, $0 < \xi < 1$. The PEC walls are now located off-grid. The field values at the LHS PEC wall at $y = \xi\Delta y$ is modeled in a linear fashion from the values at nearby node at $y = \Delta y$. Let us denote the tangential component of electric field $F(\Delta y) = f_1$. Assuming linear variation of field between adjacent layers of nodes at $y = \Delta y$ and $y = \xi\Delta y$, we can write[50]

$$F(\Delta y) = f_1 = a\Delta y + b \tag{1.72a}$$

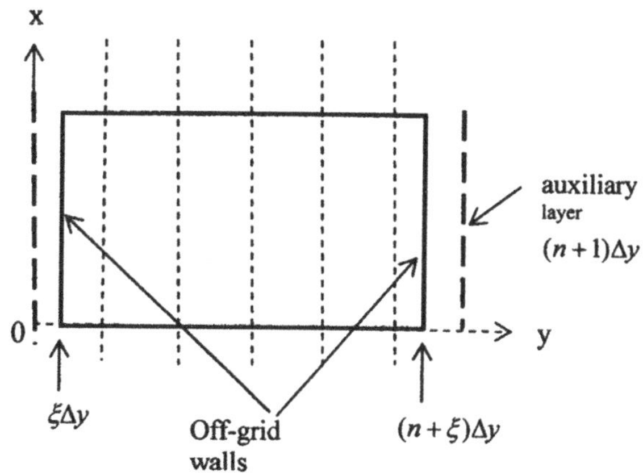

Figure 1.17 Cross-section of a rectangular cavity in the x-y plane. The cavity is offset from the grid by $\xi\Delta y$. Reproduced by permission of ©2005 IEEE, Figure 1 of [50]

$$F(\xi\Delta y) = 0 = a\xi\Delta y + b \quad \text{due to PEC} \tag{1.72b}$$

Eliminating a and b gives for the field in the vicinity of PEC boundary as

$$F(y) = \frac{f_1}{1-\xi}\left(\frac{y}{\Delta y} - \xi\right) \tag{1.73}$$

Application of (1.73) to the layer at $y=0$, left of PEC, gives

$$F(0) = f_0 = f_1\left(\frac{\xi}{\xi-1}\right), \quad \text{for } \xi \le \xi_{\max} \tag{1.74}$$

Obviously, this layer is external to the computational domain.

Similarly, the PEC boundary condition on the RHS at $y = (n+\xi)\Delta y$ can be satisfied from the known field values at $y = n\Delta y$. The resulting expression for f_{n+1} is given by

$$F((n+1)\Delta y) = f_{n+1} = \left(\frac{\xi-1}{\xi}\right) f_n \quad \text{for } \xi \ge 1-\xi_{\max} \tag{1.75}$$

The extrapolation of internal field values to the adjacent layers beyond the perfect boundaries has a limitation on the offset ξ called $\xi_{\max}$ for (1.74) and $\xi_{\max} = 1-\xi_{\max}$ for (1.75). The coefficient in these expressions approaches infinity as $\xi \to 1$ and gives rise to instability in the algorithm. The permissible value of $\xi_{\max}$ is given by [50]

$$\xi_{\max} = 0.643 + 0.584\sqrt{1-1.01q} \tag{1.76}$$

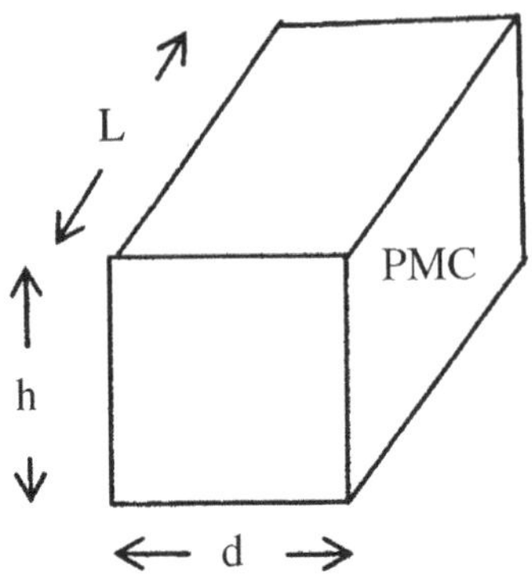

Figure 1.18 A rectangular cavity with PMC at the right-hand side

where q is the Courant number. For $0.866 \leq q \leq 0.98625$ the corresponding range of ξ_{max} is $0.85 \geq \xi_{max} \geq 0.7$. Offsets $\xi > \xi_{max}$ are also discussed in [50].

The simple approach for Dirichlet boundary condition has been extended to the off-grid Neumann boundary condition, slanted PEC walls and curved PEC walls. The above approach is verified by computing the resonant frequencies of a rectanglar cavity with PEC at one end and PMC at the other end, as shown in Figure 1.18. The cavity dimensions are: $h = 10$ mm, $d = 10$ mm, and $L = 30$ mm; the step sizes used are: $\Delta x = \Delta y = \Delta z = 1$ mm and $q = 0.88$. Comparison of the computed and analytical resonant frequencies are given in Table 1.2 for $\xi = 0.15$ [50]. The accuracy is fairly good.

Table 1.2 Relative error of the resonant frequencies of a rectangular cavity with off-grid (left) PEC and (right) PMC walls, $\xi = 0.15$

Analytical (GHz)	Calculated (GHz)	Error (%)
$f_{011} = 9.007642$	9.00277	0.054
$f_{012} = 12.491352$	12.48096	0.083
$f_{110} = 16.758909$	16.72183	0.221
$f_{111} = 17.487894$	17.45512	0.182

Source: Reproduced by permission of ©2005 IEEE, Table II of [50].

References

1. R. Garg, P. Bhartia, I. Bahl, and A. Ittipiboon, *Microstrip Antenna Design Handbook*, Artech House, Boston, 2001.
2. A. Taflove, and S.C. Hagness, *Computational Electromagnetics: The Finite-Difference Time Domain Method*, 3rd edition, Artech House, Boston, 2005.
3. K.S. Yee, "Numerical solution of initial boundary value problems involving Maxwell's equations in isotropic media," *IEEE Trans. Antennas Propagat.*, vol. AP-14, pp. 302–307, 1966.
4. A. P. Zhao, "Rigorous analysis of the influence of the aspect ratio of the Yee's unit cell on the numerical dispersion property of the 2-D and 3-D FDTD methods," *IEEE Trans. Antennas Propagat.*, vol. 52, no. 7, pp. 1633–1637, July. 2004.
5. S.V. Georgakopoulos, C.R. Birtcher, C.A. Balanis, and R.A. Renaut, "Higher-order finite-difference schemes for electromagnetic radiation, scattering, and penetration, Part I: Theory," *IEEE Antenna Propagate. Mag.*, vol. 44, pp. 134–142, Feb. 2002.

6. Z. Huang, G. Pan, and R. Diaz, "A hybrid ADI and SBTD scheme for unconditionally stable time-domain solutions of Maxwell's equations," *IEEE Trans. Packag.*, vol. 31, no. 1, pp. 219–226, 2008.
7. T. Dogaru, and L. Carin, "Multiresolution time-domain using CDF biorthogonal wavelets," *IEEE Trans. Microw. Theory Tech.*, vol. 49, no. 5, pp. 902–912, 2001.
8. S. Ogurtsov, and P. Guangwen, "An updated review of general dispersion relation for conditionally and unconditionally stable FDTD algorithms", *IEEE Trans. Antennas Propagat.*, vol. 56, no. 8, Part 2, pp. 2572–2583, Aug. 2008.
9. S. Ogurtsov, and S. V. Georgakopoulos, "FDTD schemes with minimal numerical dispersion," *IEEE Trans. Adv. Packag.*, vol. 32, no. 1, pp. 199–204, Feb. 2009.
10. C. D. Moss, F. L. Teixeira, and J. A. Kong, "Analysis and compensation of numerical dispersion in the FDTD method for layered, anisotropic media," *IEEE Trans. Antennas Propagat.*, vol. 50, no. 9, pp. 1174–1184, Sept. 2002.
11. A. P. Zhao, J. Juntunen, and A. V. Raisanen, "An efficient FDTD algorithm for the analysis of microstrip patch antennas printed on a general anisotropic dielectric substrate," *IEEE Trans. Microw. Theory Tech.*, vol. 47, no. 7, pp. 1142–1146, 1999.
12. R. M. Makinen, H. De Gersem, T. Weiland, and M. A. Kivikoski, "Modeling of lossy curved surfaces in 3-D FIT/FDTD techniques," *IEEE Trans. Antennas Propagat.*, vol. 54, no. 11, pp. 3490–3498, 2006.
13. M. K. Karkkainen, and S. A. Tretyakov, "Finite-difference time-domain model of interfaces with metals and semiconductors based on a higher order surface impedance boundary condition," *IEEE Trans. Antennas Propagat.*, vol. 51, no. 9, pp. 2448–2455, 2003.
14. S. Ogurtsov, G. Pan, and R. Diaz, "Examination, clarification, and simplification of stability and dispersion analysis for ADI-FDTD and CNSS-FDTD schemes," *IEEE Trans. Antennas Propagat.*, vol. 55, no. 12, pp. 3595–3602, Dec. 2007.
15. T. K. Sarkar, and O. Pereira, "Using the matrix pencil method to estimate the parameters of a sum of complex exponentials," *IEEE Trans. Antennas Propagat.*, vol. 37, no. 1, pp. 48–55, Feb. 1995.
16. L. Qing, X. Xiaowen, and H. Mang, "Analysis of a probe-fed cylindrical conformal microstrip patch antenna using the conformal FDTD algorithm," *IEEE 2007 Int. Sym. on Microwave, Ant., Propag. and EMC Tech. for Wireless Commun.*, pp. 876–879, 2007.
17. B. Enquist, and A. Majda, "Absorbing boundary conditions for the numerical simulation of waves," *Mathematics of Computation*, vol. 31, pp. 629–651, 1977.
18. G. Mur, "Absorbing boundary conditions for the finite difference approximation of the time domain electromagnetic field equations," *IEEE Trans., EMC-23*, pp. 377–382, 1981.
19. R. L. Higdon, "Absorbing boundary conditions for difference approximations to the multidimensional wave equation," *Math. Comput.*, vol. 47, pp. 437–459, 1986.
20. Z. P. Liao, H. L. Wong, B.-P. Yang, and Y.-F. Yuan, "A transmitting boundary for transient wave analysis," *Scientia Sinica*, Ser. A. vol. 27, pp. 1063–1076, Oct. 1984.
21. F. Costen, "Analysis and improvement of Liao ABC for FDTD," *IEEE APS-2003*, pp. 341–344, 2003.
22. M. K. Karkkainen, and S. A. Tretyakov, "A class of analytical absorbing boundary conditions originating from the exact surface impedance boundary condition," *IEEE Trans. Microwave Theory Tech.*, vol. 51, pp. 560–563, 2003.
23. S. Tretyakov, *Analytical Modeling in Applied Electromagnetics*, Artech House, Boston, 2003.
24. N. V. Kantartzis, and T. D. Tsiboukis, "A comparative study of the Berenger perfectly matched layer, the superabsorption technique and several higher-order ABC's for the FDTD algorithm in two- and three-dimensional problems," *IEEE Trans. Mag.*, vol. 33, pp. 1460–1463, March 1997.
25. R. Hallond, and J. W. Williams, "Total field versus scattered field finite difference codes: a comparative assessment," *IEEE Trans. Nuclear Science*, vol. NS-30, pp. 4583–4587, 1983.
26. J. P. Berenger, "A perfectly matched layer for the absorption of electromagnetic waves," *J. Computational Physics*, vol. 114, pp. 185–200, 1994.
27. W. C. Chew, J. Jin, E. Michielssen, and J. Song, *Fast and Efficient Algorithms in Computational Electromagnetics*, Artech House, Boston, 2001.
28. S. D. Gedney, "An anisotropic perfectly matched layer absorbing medium for the truncation of FDTD lattices," *IEEE Trans. Antennas Propagat.*, vol. AP-44, pp. 1630–1639, 1996.
29. J. A. Kong, *Electromagnetic Wave Theory*, 2nd edition, John Wiley& Sons, Ltd, New York, 1990.
30. Z. H. Huang, and G. W. Pan, "Universally applicable uniaxial perfect matched layer formulation for explicit and implicit finite difference time domain algorithms", *IET Microw. Antennas Propag.*, vol. 2, no. 7, pp. 668–676, 2008.

31. D. M. Sullivan, "An unsplit step 3-D PML for use with the FDTD method," *IEEE Microw. Guided Wave Lett.*, vol. 7, no. 7, pp. 184–186, 1997.
32. A. P. Zhao, "Uniaxial perfectly matched layer media for an unconditionally stable 3-D ADI-FDTD method," *IEEE Microw. Wireless Compon., Lett.*, vol. 12, no. 12, pp. 497–499, 2002.
33. T. Namiki, "A new FDTD algorithm based on alternating-direction implicit method," *IEEE Trans. Microwave Theory Tech.*, vol. 47, pp. 2003–2007, 1999.
34. J. Shibayama, M. Muraki, J. Yamauchi, and H. Nakano, "Efficient implicit FDTD algorithm based on locally one-dimensional scheme," *Electron. Lett.*, vol. 41, no. 19, pp. 1046–1047, 2005.
35. E. L. Tan, "Unconditionally stable LOD-FDTD method for 3-D Maxwell's equations," *IEEE Microw. Wireless Compon., Lett.*, vol. 17, no. 2, pp. 85–87, 2007.
36. G. Sun, and C. W. Trueman, "Unconditionally stable Crank-Nicolson scheme for solving two-dimensional Maxwell's equations," *Electron. Lett.*, vol. 39, pp. 595–597, April 3, 2003.
37. Y. Yang, R. S. Chen, D. X. Wang, and E. K. N. Yung, "Unconditionally stable Crank-Nicolson finite-difference time-domain method for simulation of three-dimensional microwave circuits," *IET Microw. Antennas Propag.*, vol. 4, pp. 937–942, 2007.
38. E. L. Tan, "Fundamental schemes for efficient unconditionally stable implicit finite-difference time-domain methods," *IEEE Trans. Antennas Propag.*, vol. 56, no. 1, pp. 170–177, 2008.
39. J. Shibayama, B. Murakami, J. Yamauchi, and H. Nakano, "LOD-BOR-FDTD algorithm for efficient analysis of circularly symmetric structures," *IEEE Microw. Wireless Compon., Lett.*, vol. 19, no. 2, pp. 56–58, 2009.
40. J. Shibayama, R. Takahashi, J. Yamauchi, and H. Nakano, "Frequency-dependent implementations for dispersive media," *Electron., Lett.*, vol. 42, no. 19, pp. 1084–1085, 2006.
41. V. E. Nascimento, B. H. Borges, and F. L. Teixeira, "Split-field PML implementations for the unconditionally stable LOD-FDTD method," *IEEE Microw. Wireless Comp., Lett.*, vol. 16, no. 7, pp. 398–400, 2006.
42. O. Ramadan, "Unsplit field implicit PML algorithm for complex envelope dispersive LOD-FDTD simulations," *Electron., Lett.*, vol. 43, no. 5, pp. 17–18, 2007.
43. I. Ahmed, E. Li, and K. Krohne, "Convolutional perfectly matched layer for an unconditionally stable LOD-FDTd method," *IEEE Microw. Wireless Compon., Lett.*, vol. 17, no. 12, pp. 816–818, 2007.
44. T. Olsson, J. Siden, M. Hjelm, and H.-E. Nilsson, "Robustness of printed patch antennas," *IEEE Trans. Antennas Propagat.*, vol. 55, pp. 2709–2717, Oct. 2007.
45. T. Arima, T. Uno, and M. Takahashi, "FDTD analysis of printed antenna on thin dielectric sheet including quasi-static approximation," *2004 IEEE APS Digest*, pp. 1022–1025.
46. T. Arima, T. Uno, and M. Takahashi, "Improvement of FDTD accuracy for analyzing printed antennas by using quasi-static approximation," *2003 IEEE APS Digest*, vol. 3, pp. 784–787, 2003.
47. S. Dey, and R. Mittra, "A locally conformal finite-difference time-domain (FDTD) algorithm for modeling three-dimensional perfectly conducting objects," *IEEE Microw. Guided Wave Lett.*, vol. 7, no. 9, pp. 273–275, 1997.
48. C. J. Railton, I. J. Craddock, and J. Achneider, "The analysis of general two-dimensional PEC structures using a modified CPFDTD algorithm," *IEEE Trans. Microwave Theory Tech.*, vol. 44, no. 10, pp. 1728–1733, 1996.
49. J. Anderson, M. Okoniewski, and S. S. Stuchly, "Subcell treatment of 90 degree metal corners in FDTD," *Electron., Lett.*, vol. 31, pp. 2159–2160, 1995.
50. Y. S. Rickard, and N. K. Nikolova, "Off-grid perfect boundary conditions for the FDTD method," *IEEE Trans. Microwave Theory Tech.*, vol. 53, no. 7, pp. 2274–2283, 2005.

2

Computer Aided Design of Microstrip Antennas

Debatosh Guha and Jawad Y. Siddiqui
Institute of Radio Physics and Electronics, University of Calcutta, India

2.1 Introduction

Theoretical studies of microstrip radiators started in the early 1970s [1–3] immediately after the publication of the pioneering paper by Howell in 1972 [4]. The researchers initially tried to understand the behavior of a rectangular patch in two ways: (i) as an open resonant cavity under a patch initiated by Itoh and Mittra; and (ii) as a section of a microstrip transmission line initiated by Munson [3].

In the past three decades, several books have been published and essays written on microstrip antennas [3–12], proposing rigorous analytical and numerical techniques. Analytical techniques such as full wave analysis using method of moment (MoM) are accurate, but involve rigorous mathematics and hardly provide any closed form expression. Numerical techniques such as finite difference or finite element method are equally robust and versatile for any regular or arbitrary antenna geometry, but do not provide any physical insight into the antenna mechanism, nor any closed form design equation.

The closed form equations are useful from a design aspect to approximately estimate the near-field quantities such as the frequency, around which a narrowband microstrip structure resonates and also to estimate the input impedance value which determines the matched location to feed the antenna. The cavity resonator model (CRM) is a suitable one to provide closed form expressions. It can deal with common geometries such as rectangular, circular or triangular patches and at the same time, provides a clear physical insight, which helps a beginner or an engineer to understand the antenna mechanism, and also to improvise the design, leading to improved performance.

Microstrip and Printed Antennas: New Trends, Techniques and Applications. Edited by Debatosh Guha and Yahia M.M. Antar

This chapter compiles some essential computer aided design (CAD) formulas derived using the cavity resonator model for common microstrip radiators in various configurations. The CAD available in [10, 12] have been updated with more accurate formulations developed recently. The circular, rectangular and triangular geometries are addressed in different configurations, viz. suspended substrate with variable air gap, inverted substrate, cavity enclosed and superstrate loaded [13–18]. Determination of resonant frequency is thoroughly addressed. In addition, a section has been included on the modeling of the feed reactance of a probe-fed microstrip patch [19, 20].

2.2 Microstrip Patch as Cavity Resonator

A basic microstrip radiator is a conducting patch printed on a grounded dielectric substrate. Unlike a microstrip line, a radiating patch is preferably printed on a low permittivity substrate for higher radiation efficiency. In a conventional microstrip line shown in Figure 2.1, the electromagnetic energy propagates through the region in between the printed metal strip on top and the bottom ground plane. The electric fields are vertically polarized and the pattern looks like a TEM mode when the substrate is electrically thin, that is, $h \ll \lambda$. Do we then expect similar field configuration in a truncated microstrip line having width W and length L? Indeed, a truncated microstrip line takes the shape of an isolated rectangular patch, shown in Figure 2.2, which can be looked upon as a partially open cavity having electric walls at the top and the bottom and magnetic walls surrounding its boundary. The fields then take a shape like TM^z mode and this cavity should resonate under the condition $L \approx \lambda/2$. The open boundary imposes a condition on the electric field ideally to be zero near the patch centre $x = 0$ and maximum near

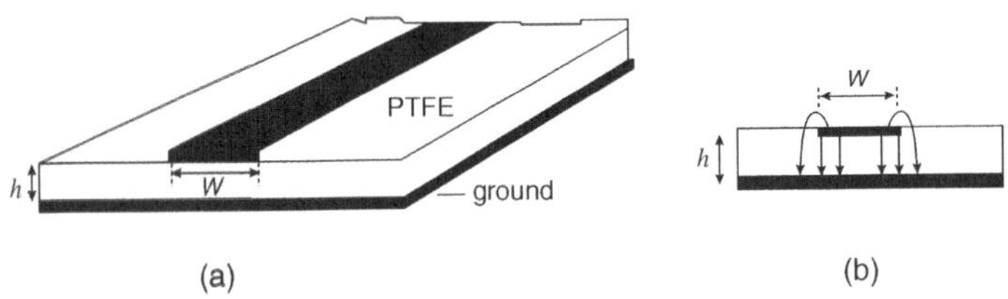

Figure 2.1 Microstrip line (a) basic geometry (b) electric field lines

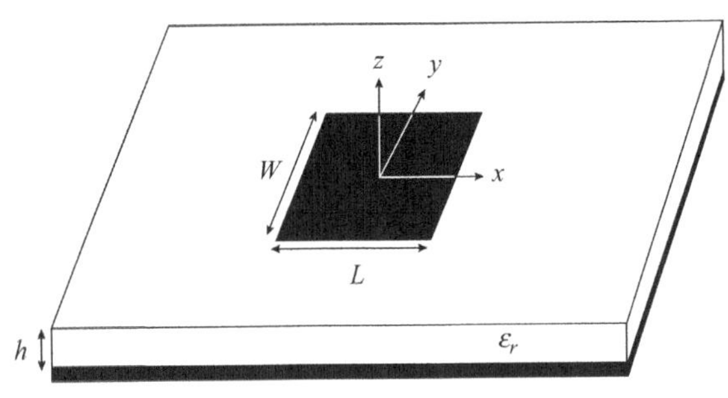

Figure 2.2 A rectangular microstrip patch with length L, and width W, printed on a grounded dielectric substrate (ε_r)

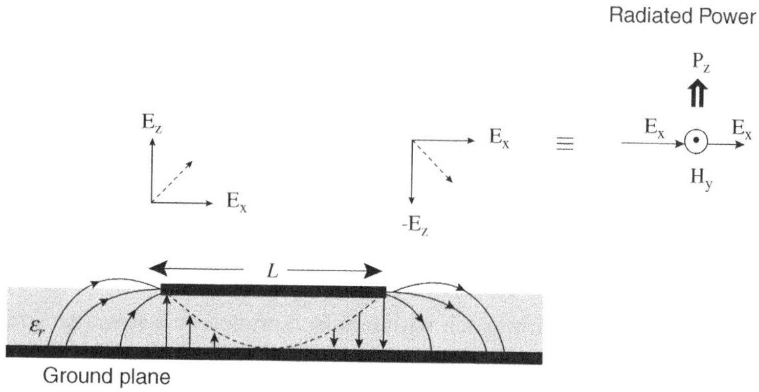

Figure 2.3 Field distribution under a rectangular microstrip patch having resonating length, *L*

the patch edges $x = \pm L/2$ with mutually opposite polarity, as shown in Figure 2.3. Once it starts resonating, the vertical electric fields near the open edges fringe around over the ground plane and the field vectors no longer remain purely vertical. The inclined electric field vectors around $x = \pm L/2$ can be split into respective horizontal and vertical components, as indicated in Figure 2.3. Looking from the top, i.e. the broad side of the patch, one will see only the horizontally polarized electric fields (E_x). The z-polarized electric fields (E_z) cancel out due to opposite polarity. So the resultant field leaking from the open microstrip resonator is E_x which is associated with orthogonal magnetic field H_y and thus radiates vertically upward following "Poynting vector" P_z. This is the simplest way to represent a microstrip patch excited with TM^z mode as an open cavity resonator radiating along the broadside of the patch. Excitation of higher modes will cause different radiation patterns.

Other common patch geometries, like circular and triangular shaped patches, have similar field distributions and can be modeled using CRM with magnetic wall along the periphery. Similar to rectangular patch, these geometries support TM^z modes. Their resonant frequencies can be expressed in terms of the geometrical dimension of the patch and the dielectric constant of the substrate. The dielectric substrate, placed on the ground plane, is considered to be electrically thin and wide enough compared to the resonant wavelength.

Let us start with the circular patch since the symmetry of this geometry helps to employ the quasi-static approach to estimate the fringing electric fields without any complexity. The simplified CAD formula derived using the quasi-static approach is then successfully extended to rectangular and triangular geometries, which are discussed in the subsequent sections.

2.3 Resonant Frequency of Circular Microstrip Patch (CMP)

The resonance of TM_{nm} mode in an open circular cavity formed by a circular microstrip patch is mathematically defined as [13]

$$f_{r,nm} = \frac{\alpha_{nm} c}{2\pi a_{eff}\sqrt{\varepsilon_{r,eff}}} \tag{2.1}$$

where, c is the velocity of light in free space and α_{nm} is the mth zero of the derivative of the Bessel function of order n. The values of this function for the fundamental and higher modes like TM_{11}, TM_{21}, TM_{01}, and TM_{31} are given as $\alpha_{11} = 1.841$, $\alpha_{21} = 3.054$, $\alpha_{01} = 3.832$, $\alpha_{31} = 4.201$, respectively. The parameters a_{eff} and $\varepsilon_{r,eff}$ are the effective parameters indicating the radius and dielectric constant which may vary from respective physical values under different situations or configurations. Those are discussed and determined in the following sections.

2.3.1 Suspended Substrate with Variable Air Gap

A coaxially fed circular microstrip patch with radius a, printed on a substrate with dielectric constant, ε_r and thickness h_2, maintaining a variable air-gap h_1 above the ground plane, is shown in Figure 2.4. The substrate commonly used for microwave printed circuits are made of low loss material namely polytetrafluoroethylene (PTFE). Details about the PTFE materials are provided in Appendix I. The excitation of the TM_{nm} modes depends on the matched location of the coaxial feed, at a distance ρ_0 from the centre of the patch. The suspended substrate with air gap was introduced by Lee et al. [21] to explore the property of mechanical tuning by varying the air gap height. Improved formulations have been reported by Guha [13], which can efficiently predict the operating frequency for electrically thick antennas where the total height h $(=h_1 + h_2)$ of the composite air/dielectric substrate is significantly thick compared to the patch dimension. The proposed formulas [13] are equally valid for electrically thin and moderate height substrates.

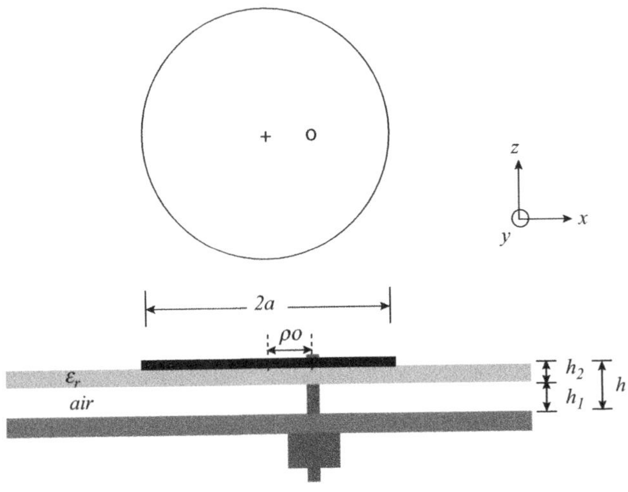

Figure 2.4 A probe-fed circular microstrip patch with variable air gap in suspended substrate configuration

A quasi-static approach has been used to develop the formulations. This means that equivalent capacitance, inductance, and resistance caused by the microstrip and associated components have been evaluated and utilized. The dielectric substrate h_2 above the air medium h_1 sandwiched between the ground plane and the metal patch resembles a parallel plate capacitor partially filled by PTFE (Figure 2.4). Its equivalent dielectric constant is, therefore, determined as

$$\varepsilon_{re} = \frac{\varepsilon_r(1 + h_1/h_2)}{(1 + \varepsilon_r h_1/h_2)} \tag{2.2}$$

This parameter is used to evaluate "dynamic dielectric constant $\varepsilon_{r,dyn}$," which was proposed by Wolff and Knoppik [22] as a function of the "static main" and "static fringing" capacitances as given below

$$\varepsilon_{r,\,dyn} = \frac{C_{dyn}(\varepsilon = \varepsilon_0\varepsilon_{re})}{C_{dyn}(\varepsilon = \varepsilon_0)} \tag{2.3}$$

where

$$C_{dyn} = C_{0,\,dyn} + C_{e,\,dyn} \tag{2.4}$$

Here, $C_{0,dyn}$ is the "dynamic main capacitance" and $C_{e,dyn}$ is the "dynamic fringing capacitance," for different modes, and they are determined from the static main and static fringing capacitances $C_{0,stat}$ and $C_{e,stat}$, respectively. The relations are given as [22]

$$C_{0,dyn} = \gamma_n C_{0,stat} \tag{2.5}$$

$$\gamma_n = \begin{matrix} 1.0 \\ 0.3525 \\ 0.2865 \\ 0.2450 \end{matrix} \quad \text{for} \quad n = \begin{matrix} 0 \\ 1 \\ 2 \\ 3 \end{matrix}$$

$$C_{e,dyn} = \frac{1}{\delta} C_{e,stat} \tag{2.6}$$

$$\begin{aligned} \delta &= 1, \text{ for } n = 0 \\ &= 2, \text{ for } n \neq 0 \end{aligned}$$

Wheeler's work [23] for deriving static fringing capacitances for "small disk," "medium disk" and "large disk" geometries are used to determine $C_{0,stat}$ and $C_{e,stat}$. The expression for the disk capacitance is given by [23]

$$C = \frac{\varepsilon_0\varepsilon_{re}\pi a^2}{h}(1 + q) \tag{2.7}$$

$$q = u + v + uv \tag{2.8}$$

$$u = (1 + \varepsilon_{re}^{-1})\frac{4}{\pi a/h} \tag{2.9}$$

$$v = s + \left(t^{-1} - 1\right)/g \tag{2.10}$$

$$s = \frac{2}{3t} \cdot \frac{\ln(p)}{8 + \pi a/h} \tag{2.11}$$

$$p = \frac{1 + 0.8(a/h)^2 + (0.31a/h)^4}{1 + 0.9a/h} \tag{2.12}$$

$$t = 0.37 + 0.63\varepsilon_{re} \tag{2.13}$$

$$g = 4 + 2.6a/h + 2.9h/a \tag{2.14}$$

Expanding Equation (2.7), the first term represents $C_{0,stat}$ and the second term is $C_{e,stat}$. Mathematically they can be expressed as

$$C_{0,stat} = \frac{\varepsilon_0 \varepsilon_{re} \pi a^2}{h} \tag{2.15}$$

$$C_{e,stat} = C_{0,stat} \cdot q \tag{2.16}$$

Here, the factor "q" accounts for the fringing electric fields around the patch and hence the resulting increment in effective area. Equation (2.7) can be rewritten as

$$C = \frac{\varepsilon_0 \varepsilon_{re} \pi}{h} \left(a\sqrt{1+q}\right)^2 \tag{2.17}$$

$$= \frac{\varepsilon_0 \varepsilon_{re} \pi}{h} a_{eff}{}^2 \tag{2.18}$$

where

$$a_{eff} = a\sqrt{(1+q)} \tag{2.19}$$

and represents the effective radius of the patch.

Although, ε_{re} and $\varepsilon_{r,dyn}$ have been deduced as two modified dielectric constant values of the medium under the patch, yet another quantity, effective dielectric constant, $\varepsilon_{r,eff}$ is derived in terms of ε_{re} and $\varepsilon_{r,dyn}$ to improve the accuracy of the formula as

$$\varepsilon_{r,eff} = \frac{4 \cdot \varepsilon_{re} \cdot \varepsilon_{r,dyn}}{\left(\sqrt{\varepsilon_{re}} + \sqrt{\varepsilon_{r,dyn}}\right)^2} \tag{2.20}$$

This is deduced to yield an average of frequencies resulting from Equation (2.1) by substituting ε_{re} and $\varepsilon_{r,dyn}$ separately in place of $\varepsilon_{r,eff}$.

Now Equation (2.1) can be used to theoretically calculate the resonant frequency of a given circular patch (a, h_1, h_2, ε_r).

Figure 2.5 shows the dependence of the fringing factor, q, on the parameter a/h. It is apparent from the curve that the fringing field varies with the equivalent dielectric constant, ε_{re} and the total height, h of the microstrip antenna and more significantly at high h values or low a/h ratio. The ratio a/h can also be termed as λ/h. A higher a/h value corresponds to electrically thin

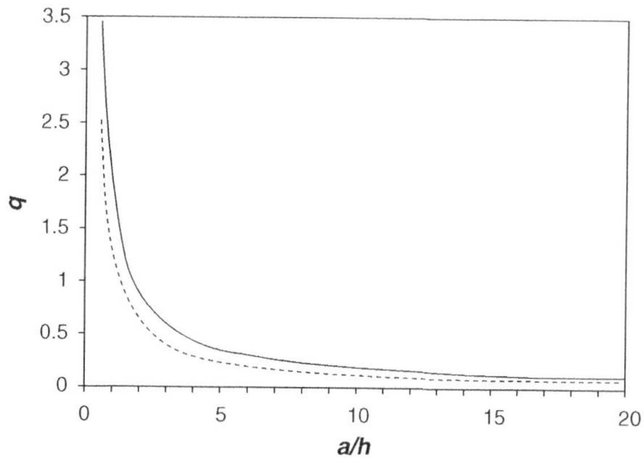

Figure 2.5 Computed values of the fringing factor q of a circular microstrip patch as a function of normalized patch radius (· · ·) $\varepsilon_{re} = 2.3$; (· · ·) $\varepsilon_{re} = 10$. Reproduced by permission of ©2001 IEEE [13]

antenna. From Figure 2.5, it is evident that "q" becomes significantly large in electrically thick antennas.

Figure 2.6 shows the theoretical values of $\varepsilon_{r,eff}$ and $\varepsilon_{r,dyn}$ as functions of the ratio a/h for two different ε_{re} values. Although the quantities $\varepsilon_{r,eff}$ and $\varepsilon_{r,dyn}$ approach close to each other for large values of a/h, they differ significantly as a/h decreases.

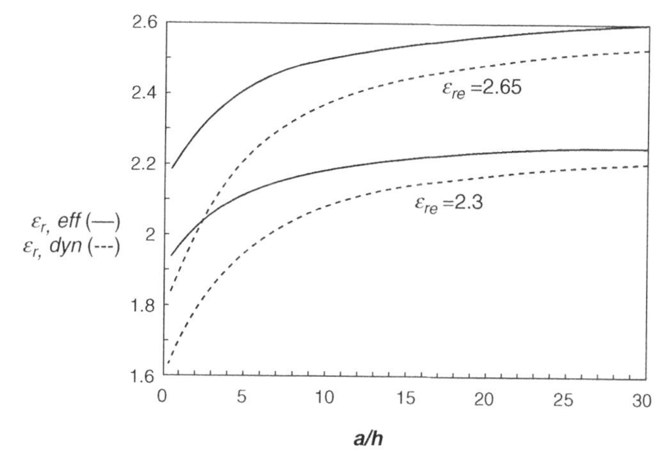

Figure 2.6 Computed values of the effective dielectric constant, $\varepsilon_{r,eff}$, and dynamic dielectric constant, $\varepsilon_{r,dyn}$, of the medium below a circular patch as a function of normalized patch radius for different equivalent dielectric constant, ε_{re}, values. Reproduced by permission of ©2001 IEEE [13]

A set of computed resonant frequencies of a circular microstrip patch with variable air gap height h_l using Equation (2.1) is compared with measured [21] and simulated results and are presented in Table 2.1. The simulations were performed using Ansoft's HFSS [24].

Table 2.1 First few resonant frequencies of a circular patch antenna: compared with measured and simulated data

Air gap height, h_1 (mm)	Mode	Measured, MHz	Computed, MHz	
			Theory	Simulation (HFSS)
	TM_{11}	1128	1131	1155
0	TM_{21}	1879	1881	1864
	TM_{31}	2596	2594	2564
	TM_{11}	1286	1275	1273
0.5	TM_{21}	2136	2120	2113
	TM_{31}	2951	2922	2906
	TM_{11}	1350	1345	1348
1.0	TM_{21}	2256	2235	2236
	TM_{31}	3106	3080	3076

Notes: $a = 50$ mm, $h_2 = 1.59$ mm, $\varepsilon_r = 2.32$)

2.3.2 *Inverted Microstrip Circular Patch (IMCP)*

The inverted configuration of a microstrip patch is shown in Figure 2.7 where the patch is printed below the substrate and separated by an air gap from the ground plane. This configuration was employed for designing active integrated antennas [25–29], where some active devices were conveniently placed in the space between the patch and the ground plane. Another distinct advantage is the reduction of surface waves. Thus it offers improved bandwidth without degradation of the radiation pattern and efficiency.

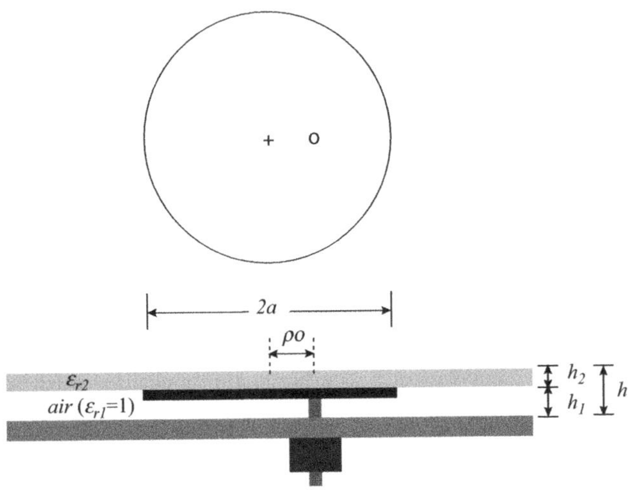

Figure 2.7 Inverted microstrip circular patch

The TM mode resonant frequencies for a circular patch in inverted configuration (Figure 2.7) was first developed in [14] as

$$f_{r,nm} = \frac{\alpha_{nm} c}{2\pi a_{eff}} \tag{2.21}$$

as a variant of (2.1) with an assumption that $\varepsilon_{r,eff} = 1$.

The effect of the dielectric medium ε_{r2} on top of the patch is accounted for in determining the fringing factor q. Effective patch radius a_{eff} derived in Equation (2.19) is modified for the inverted configuration where the fringing capacitance for "small disk" is neglected and as such the calculated q for an inverted circular patch is calculated using Equations (2.8)–(2.14) omitting the parameter s, and taking $h = h_1$. The parameter ε_{re} is replaced by an equivalent relative dielectric constant for an inverted patch, $\varepsilon_{re,i}$, introduced to account for the effect of ε_{r2} and is derived as [14]

$$\begin{aligned}\varepsilon_{re,i} &= \varepsilon_{r1} p_1 + \varepsilon_{r1}(1-p_1)^2 \times [\varepsilon_{r2}^2 p_2 p_3 + \varepsilon_{r2}\{p_2 p_4 + (p_3 + p_4)^2\}] \\ &\times [\varepsilon_{r2}^2 p_2 p_3 p_4 + \varepsilon_{r1}(\varepsilon_{r2} p_3 + p_4)(1-p_1-p_4)^2 + \varepsilon_{r2} p_4 \{p_2 p_4 + (p_3 + p_4)^2\}]^{-1}\end{aligned} \tag{2.22}$$

$$p_1 = 1 - \frac{h_1}{2w_e} \ln\left(\frac{\pi}{h_1} w_e - 1\right) - p_4 \tag{2.23}$$

$$p_2 = 1 - p_1 - p_3 - 2p_4 \tag{2.24}$$

$$p_3 = \frac{h_1 - g_1}{2w_e} \ln\left[\frac{\pi w_e}{h_1} \frac{\cos(\pi g_1/2h_1)}{\pi(0.5 + h_2/h_1) + \pi g_1/2h_1} + \sin(\pi g_1/2h_1)\right] \tag{2.25}$$

$$p_4 = \frac{h_1}{2w_e} \cdot \ln\left(\frac{\pi}{2} - \frac{h_1}{2w_e}\right) \tag{2.26}$$

$$g_1 = \frac{2h_1}{\pi} \cdot \arctan\left[\frac{\pi h_2/h_1}{(\pi/2)(w_e/h_1) - 2}\right] \tag{2.27}$$

$$\begin{aligned} w_e = \sqrt{\varepsilon_r'/\varepsilon_{re,i}} \Bigg[&\left\{1.142a + 0.882h_1 + 0.164h_1 \frac{(\varepsilon_r' - 1)}{(\varepsilon_r')^2}\right\} \\ &+ h_1 \frac{(\varepsilon_r' - 1)}{\pi \varepsilon_r'} \left\{\ln(0.94 + 1.142a/2h_1) + 1.451\right\}\Bigg] \end{aligned} \tag{2.28}$$

$$\varepsilon_r' = \frac{2\varepsilon_{re,i} - 1 + \left(1 + \frac{10h_1}{w_e}\right)^{-1/2}}{1 + \left(1 + \frac{10h_1}{w_e}\right)^{-1/2}} \tag{2.29}$$

The quantities w_e and ε'_r are determined by an iteration process starting with an approximation $\varepsilon'_r = \varepsilon_{r1}$ and $\varepsilon_{re,i} = \varepsilon'_r$. For the inverted microstrip, $\varepsilon_{r1} = 1$ and ε_{r2} is the relative permittivity of the substrate.

Figure 2.8 shows the measured [14] and computed dominant mode resonant frequencies of the inverted microstrip circular patch as a function of patch radius.

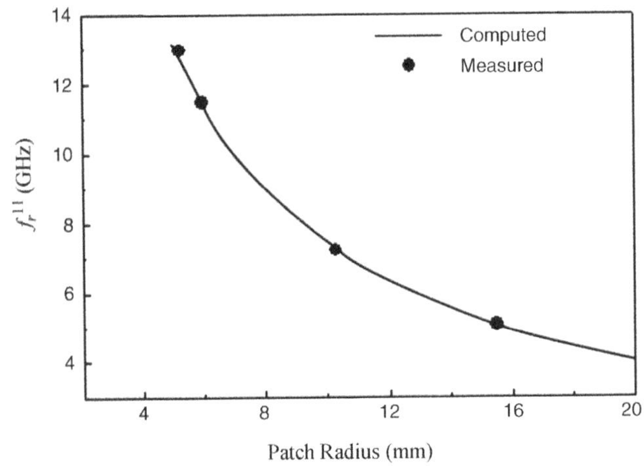

Figure 2.8 Measured and computed dominant mode resonant frequencies of an inverted microstrip circular patch as a function of patch radius ($\varepsilon_r = 2.2$, $h_1 = 1.6$ mm, $h_2 = 0.508$ mm). Reproduced by permission of ©2002 IEEE [14]

2.3.3 *IMCP Enclosed in a Cylindrical Cavity*

An inverted microstrip circular patch surrounded by a metallic cylinder was used in several active antenna designs in [25–29] and its probe-fed configuration is shown in Figure 2.9. This was first theoretically examined in [15] to determine its resonant frequency.

The proximity of the cavity wall affects the fringing fields and hence the resonant frequency. Therefore, when the radius of the cylinder r is considerably larger compared to that of a patch, it hardly influences the patch resonance. This conjecture is extended to analyze the structure. The proximity of cavity wall may be mathematically represented as $a_{eff} \approx r$ or $a_{eff} \geq r$ and under this condition, one can surmise that the cylindrical enclosure effectively turns to a circular cavity of height h_1 resonating at a frequency close to the f_r determined for an IMCP using (2.21). For low h_1 value satisfying the condition $h_1/r < 2.03$ it would then excite the TM^z_{010} mode in a circular cavity [30] resonating at

$$f_{cav} = \frac{2.4049 \times c}{2\pi r\sqrt{\varepsilon_{r,eq}}} \tag{2.30}$$

Equating the cavity resonance frequency (2.30) with (2.21) for an identical IMCP having $a_{eff} = r$ (for $a_{eff} \approx r$ or $a_{eff} \geq r$) yields an equivalent dielectric constant of the medium filling the cavity as $\varepsilon_{r,eq} \approx 1.706$. The structure now can be represented by a simple equivalent microstrip

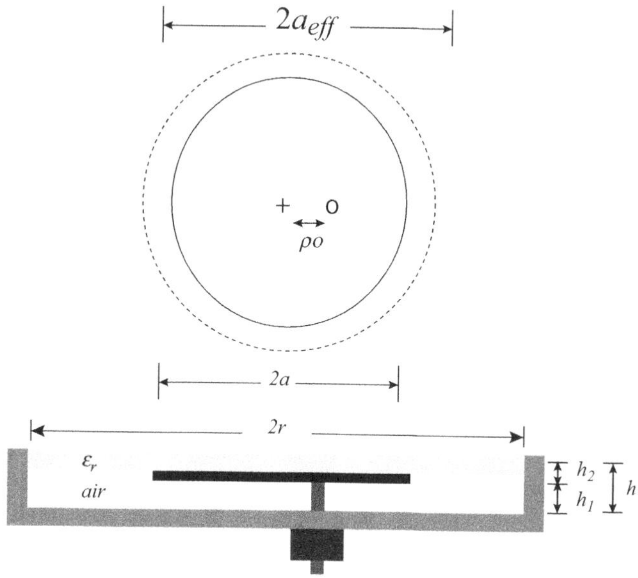

Figure 2.9 Inverted microstrip circular patch enclosed in a cylindrical cavity

configuration as was shown in Figure 2.4 with a circular patch having radius a printed on a substrate having dielectric constant $\varepsilon_{r,eq}$ (replace ε_r with $\varepsilon_{r,eq}$ in Figure 2.4) and thickness $h = h_1$. The resonant frequency of a cavity enclosed IMCP thus can be computed using (2.1).

This approach is valid for limited patch dimensions expressed as $a_{eff} \approx r$ or $a_{eff} \geq r$ and a factor is defined to represent the proximity between the cavity wall and the patch as

$$e_1 = (r - a_{eff})/h_1 \tag{2.31}$$

Another factor suggested in [15] to examine the cavity effect on fringing fields is

$$g_c = [(r - a_{eff})/h_1]/(a_{eff}/a) \tag{2.32}$$

The measured results [26] are compared with the computed values of an IMCP with and without cavity in Figure 2.10. This indicates that the cavity enclosed IMCPs with $g_c \leq 0.5$ are under the influence of the cavity and the formulation can be efficiently used to compute their dominant mode resonant frequencies.

2.3.4 Superstrate Loaded Circular Microstrip Patch (SL-CMP)

The cavity model analysis or closed form expression for a dielectric covered microstrip patch is not very common or customary. Usually full wave analysis is chosen for such structures [31–35]. Cavity model formulation developed in [16] is presented here as CAD formula for a circular patch with a dielectric superstrate. Figure 2.11 shows a cross-sectional view with a

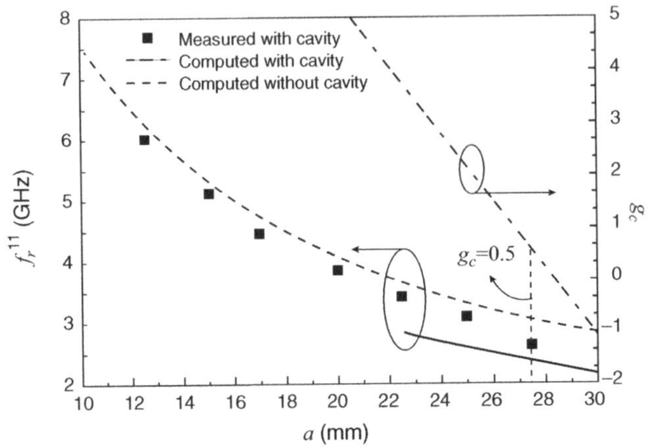

Figure 2.10 Measured and computed resonant frequencies of an inverted microstrip circular patch with and without cavity enclosure and their corresponding cavity factor g_c as a function of patch radius ($\varepsilon_r = 2.3$, $r = 30$ mm, $h_1 = 1.43$ mm, $h_2 = 1.57$ mm). Reproduced by permission of ©2004 IEEE [15]

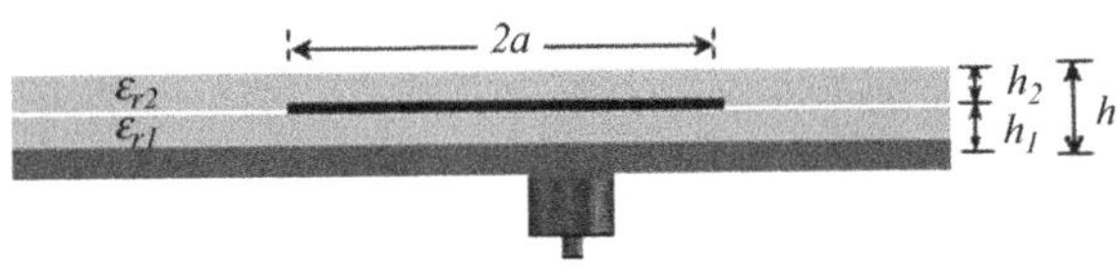

Figure 2.11 Cross-section of a coax-fed superstrate loaded circular microstrip patch

superstrate having relative permittivity ε_{r2} and height h_2 lying on top of the patch. A small air gap between the substrate and the superstrate, as also appears in Figure 2.11, is neglected in the analysis.

The dielectric layers above a microstrip patch cause the change in the fringing fields between the patch and the ground plane and the effect is incorporated in the calculation of the effective dielectric constant $\varepsilon_{r,eff}$, of the medium below the patch. The parameter $\varepsilon_{r,eff}$ can be determined from Equations (2.22) to (2.29) by substituting $\varepsilon_{re,i}$ with $\varepsilon_{r,eff}$.

The effective radius a_{eff} of the SL-CMP is calculated using Equation (2.19). The substrate-superstrate combination has an effect on the fringing factor, q and is accounted for in terms of a new equivalent dielectric constant [16]

$$\varepsilon_{re} = \varepsilon_{r1} / \varepsilon_{r,eff} \tag{2.33}$$

For such a multilayered configuration, $\varepsilon_{r,eff}$ is very sensitive to the substrate–superstrate combination and this is examined in Figure 2.12. The effect of superstrate thickness is insignificant when $\varepsilon_{r2} \approx \varepsilon_{r1}$, but that for $\varepsilon_{r2} \gg \varepsilon_{r1}$ results in a significant change in $\varepsilon_{r,eff}$ values. Similarly, ε_{re} is equally sensitive to those parameters in estimating the effective patch dimension. The measured and computed [16] resonant frequencies of a SL-CMP etched on a RT Duroid substrate and loaded with a glass epoxy superstrate are compared in Table 2.2.

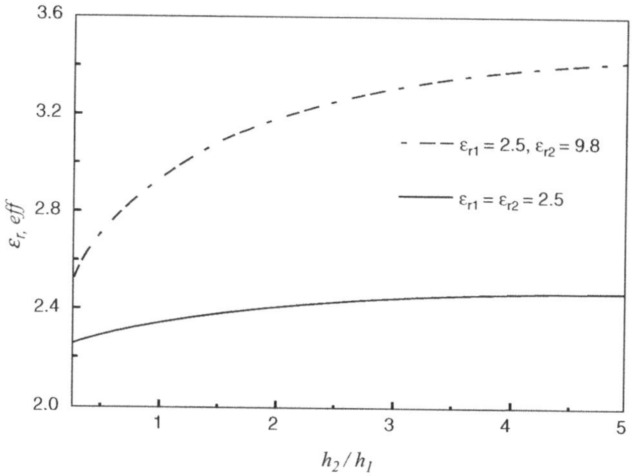

Figure 2.12 Computed effective dielectric constant, $\varepsilon_{r,eff}$ for different substrate superstrate relative permittivity combination of a superstrate loaded circular microstrip patch as a function of relative thickness of superstrate ($h_1 = 1.5875$ mm, $a/h_1 = 5$)

Table 2.2 Measured and computed values of first few resonant frequencies of a SL-CMP

Mode n, m	Resonant Frequency (GHz)	
	Measured	Computed
1,1	8.678	8.507
2,1	14.22	14.11
3,1	19.688	19.411

Notes: $a = 5.9$ mm, $h_1 = 0.508$ mm, $h_2/h_1 = 2.76$, $\varepsilon_{r1} = 2.2$, $\varepsilon_{r2} = 2.545\varepsilon_{r1}$

2.4 Resonant Frequency of Rectangular Microstrip Patch (RMP) with Variable Air Gap

The suspended substrate with variable air gap configuration studied in Section 2.3.1 is extended to the rectangular patch geometry shown in Figure 2.13. A rectangular patch having resonant length L and width W is etched on a substrate with relative permittivity ε_r and thickness h_2. Unlike a circular patch which has only one degree of freedom (radius) to control, the rectangular patch offers two degrees of freedom (length and width) to control. The excitation of the TM_{nm} modes is induced by a coaxial feed, located at x_0 and y_0 from the centre of the patch. Effective relative permittivity of the suspended substrate composite medium is represented by $\varepsilon_{r,eff}$ and its height h above the ground plane should be much less than the operating λ.

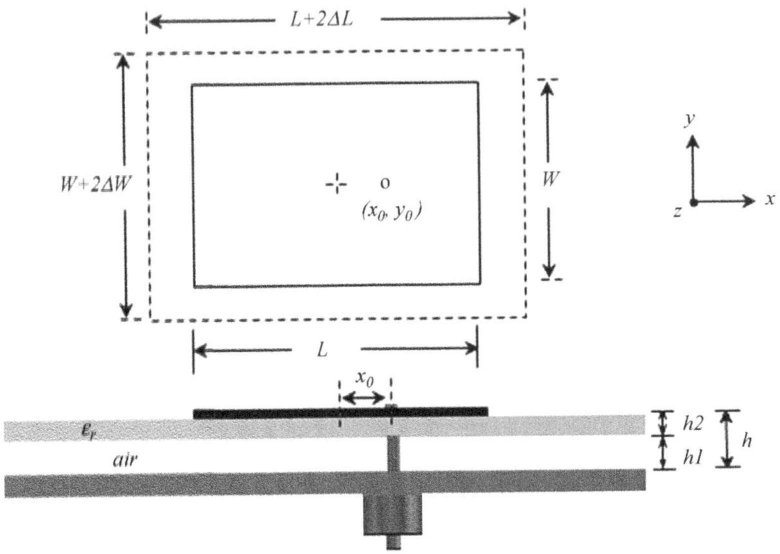

Figure 2.13 A coax-fed rectangular microstrip patch with variable air gap in suspended substrate configuration

The fringing of the electric fields at the radiating and nonradiating edges of a rectangular patch is accounted for in terms of extra linear dimensions ΔL and ΔW, respectively. The solution for the resonant frequency is then given as [17]

$$f_{r,nm} = \frac{c}{2\sqrt{\varepsilon_{r,eff}}}\left[\left(\frac{n}{L+2\Delta L}\right)^2 + \left(\frac{m}{W+2\Delta W}\right)^2\right]^{1/2} \tag{2.34}$$

An approach to estimate the quantities ΔL and ΔW is proposed using an equivalent circular patch of radius a, resonating at the same zeroth order frequency as [17]

$$f_{0,r} = \frac{c}{2L\sqrt{\varepsilon_r}} = \frac{c\alpha_{11}}{2\pi a\sqrt{\varepsilon_r}} \tag{2.35}$$

where c is the velocity of light in free space. Equation (2.35) predicts an equivalent a in terms of L. This equivalence, based on identical circumference value of both the geometries, provides the relations

$$(W+L) = \pi a \tag{2.36}$$

$$(L+2\Delta L) + (W+2\Delta W) = \pi a_{eff} \tag{2.37}$$

Solving Equations (2.35)–(2.37) yields

$$L = 1.7\,a \tag{2.38}$$

$$W = 1.44\,a \tag{2.39}$$

$$\Delta L + \Delta W = \frac{\pi[a_{eff} - 1]}{2} \quad (2.40)$$

where ΔW is empirically related to ΔL as

$$\Delta W = \Delta L(1.5 - p), \quad (2.41)$$

$$p = \frac{W}{2L} \quad (2.42)$$

The ratio *W/L*, designated as the "aspect ratio" of a rectangular patch is a significant parameter affecting both the near field and radiation properties. Equation (2.41) is valid over a wide range $2 > W/L \geq 0.5$. From Equations (2.40)–(242), we have

$$\Delta L = \frac{\pi(a_{eff} - a)}{5 - (W/L)} \quad (2.43)$$

The effective dielectric constant, i.e. $\varepsilon_{r,eff}$ in (2.34) for the rectangular patch can then be determined using (2.20) by employing an equivalent radius, *a*, for the resonating length *L*, calculated using (2.38).

Figure 2.14 shows the computed resonant frequencies of a suspended substrate RMP compared with measured [17] and simulated values obtained using HFSS for different air gap heights. The plot also reveals tunability of the resonant frequency of the patch by varying the air gap height.

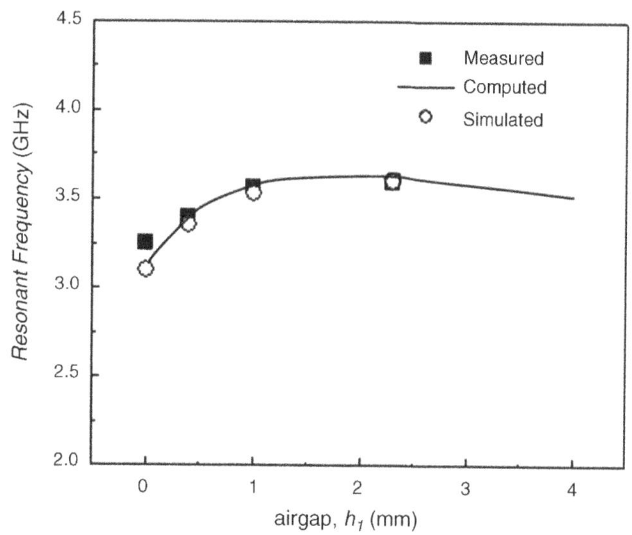

Figure 2.14 Dominant mode resonant frequencies of a suspended substrate rectangular microstrip patch as a function of varying air gap height ($L = W = 30$ mm, $\varepsilon_r = 2.33$, $h_2 = 1.575$ mm). Reproduced by permission of ©2009 John Wiley & Sons, Inc. [17]

Figure 2.15 reveals the effect of W on the resonant frequency. The formulas can accurately predict the frequency shift caused by change in W, even if the resonating length L remains unaltered.

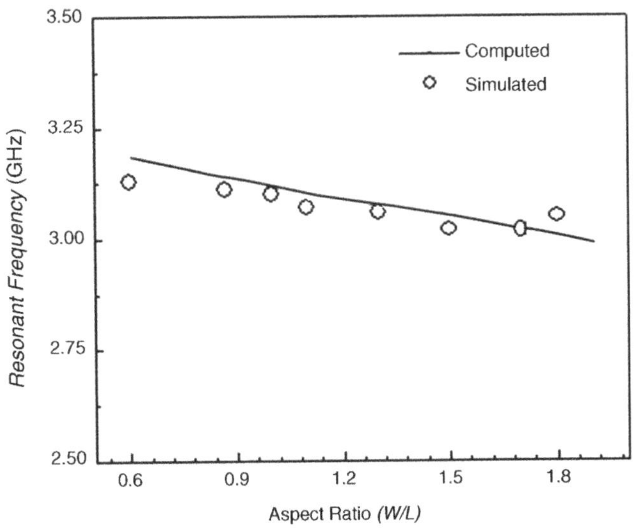

Figure 2.15 Variation of the dominant mode resonant frequency of a suspended substrate rectangular microstrip patch with varying aspect ratio W/L ($L=30$ mm, $\varepsilon_r=2.33$, $h_2=1.575$ mm, $h_l=0$). Reproduced by permission of ©2009 John Wiley & Sons, Inc. [17]

2.5 Resonant Frequency of Equilateral Triangular Microstrip Patch (ETMP) with Variable Air Gap

Compared to a circular or rectangular shape, an equilateral triangular shape is more flexible in terms of conformability. The triangular geometry can conform itself smoothly to a convex or concave ground plane which is not possible with circular or rectangle geometry. Unlike a circular or rectangular patch, a triangular patch needs three indices m, n, and l to represent the modes.

A coaxially fed equilateral triangular microstrip patch having side length, s, printed on a substrate with dielectric constant, ε_r and thickness h_2, with a variable air gap h_l is shown in Figure 2.16. Its TM_{nml} mode ($n+m+l=0$) resonances are given as [18, 36]

$$f_{r,nm} = \frac{2c}{3s_{eff}\sqrt{\varepsilon_{r,eff}}}(n^2+nm+m^2)^{1/2} \tag{2.44}$$

where c is the velocity of light in free space, $\varepsilon_{r,eff}$ is the effective dielectric constant of the medium below the patch and s_{eff} is the effective side length. This is calculated on the basis of the equivalence of the circular and triangular geometries considering identical static fringing fields working at their boundaries and is quite similar to that discussed in previous section for the rectangular patch.

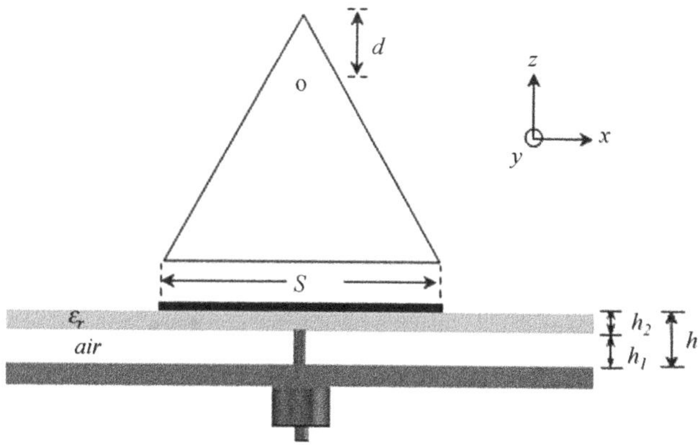

Figure 2.16 A coax-fed equilateral triangular microstrip patch with variable air gap in suspended substrate configuration

Equating the circumference of both the geometries, we obtain a relation

$$a = \frac{3}{2\pi} s \tag{2.45}$$

Once an equivalent a is determined, the equivalent effective radius, a_{eff} of a circular patch is calculated from Equations (2.8)–(2.19) and subsequently the effective side length of the triangular patch is evaluated as

$$s_{eff} = \frac{2\pi}{3} a_{eff} \tag{2.46}$$

The effective dielectric constant $\varepsilon_{r,eff}$ is again derived using Equation (2.20) with an equivalent radius a determined in Equation (2.45) and $m = n = 1$ for the dominant resonating mode.

Figure 2.17 compares the computed values using Equation (2.44) with spectral domain analysis using method of moments [37].

2.6 Input Impedance of a Microstrip Patch

A microstrip patch is considered as a resonant cavity and as such for a single isolated mode, it can be represented by an equivalent L-C-R parallel resonant circuit as shown in Figure 2.18. The input impedance is complex and involves a resistive and reactive part. These resistive and reactive components vary as a function of frequency and are symmetric around the resonant frequency. The input impedance near resonance is then given by

$$Z_{in} = R_{in} + jX_{in}$$

which can be expanded as

$$Z_{in} = \frac{R_r}{1 + Q_T^2\left(\frac{f}{f_{r,nm}} - \frac{f_{r,nm}}{f}\right)^2} + j\left[X_f - \frac{R_r Q_T\left(\frac{f}{f_{r,nm}} - \frac{f_{r,nm}}{f}\right)}{1 + Q_T^2\left(\frac{f}{f_{r,nm}} - \frac{f_{r,nm}}{f}\right)^2}\right] \tag{2.47}$$

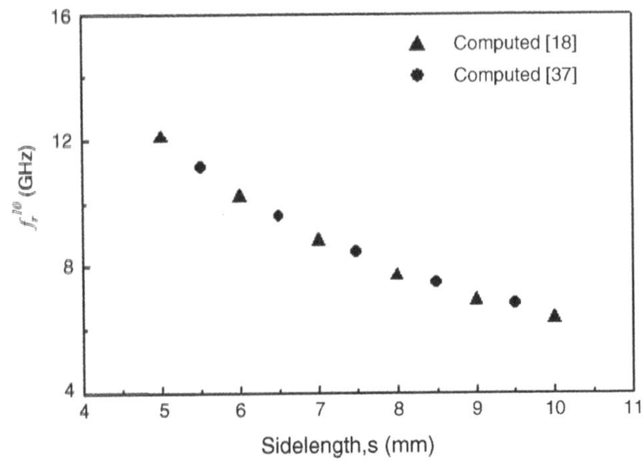

Figure 2.17 Computed resonant frequencies for the dominant mode of an equilateral triangular microstrip patch as a function of side length, s ($\varepsilon_r = 10.2$, $h_1 = 0$, $h_2 = 0.635$ mm)

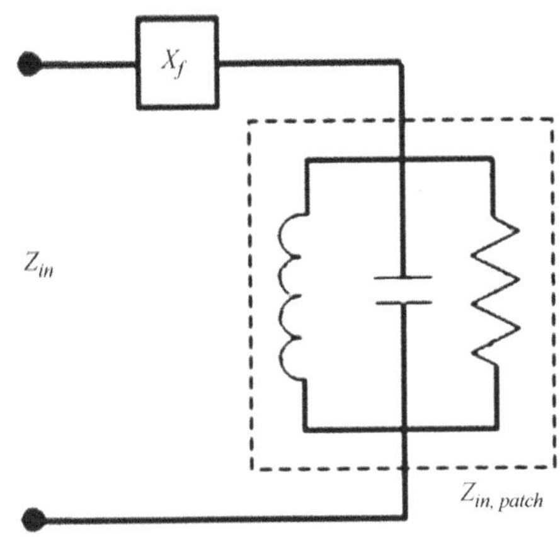

Figure 2.18 Single isolated mode equivalent circuit of a coax-fed microstrip patch

where R_r is the input resistance at resonance or the resonant resistance, Q_T is the total quality factor of the resonator and X_f is the feed reactance which accounts for the reactance contributed by the feed for probe-fed microstrip patch antennas.

The total quality factor Q_T of the resonator is given as

$$Q_T = \left[\frac{1}{Q_{rad}} + \frac{1}{Q_{con}} + \frac{1}{Q_{dlc}} + \frac{1}{Q_{sw}}\right]^{-1} \tag{2.48}$$

where Q_{rad}, Q_{con}, Q_{dlc}, and Q_{sw} account for quality factors associated with losses caused by radiation, conductivity of the patch, substrate dielectric and surface wave, respectively.

At resonance, the input impedance of the patch should be real, and hence the reactance component of Z_{in} should ideally be zero. Therefore, R_r appears to be the most significant parameter to determine appropriate feed location near resonance to match the input impedance.

2.6.1 *Input Impedance of CMP*

For a probe-fed circular patch shown in Figure 2.4, the input impedance with near resonance can be represented as a function of frequency and feed location as

$$Z_{in}(f,\rho) = R_{in}(f,\rho) + jX_{in}(f,\rho) \tag{2.49}$$

The input resistance at resonance varies with radial distance ρ from the centre of the patch as [38]

$$R_{in}(f = f_{r,nm},\rho) = R_r(\rho) = R_{edge}\frac{J_n^2(k\rho_0 \frac{a}{a_{eff}})}{J_n^2(ka)} \tag{2.50}$$

The parameter ka $(= \alpha_{nm})$ represents the mth zero of the derivative of the Bessel function of order n. For the TM_{11} mode $ka = 1.84118$.

The value of the radiation resistance at the edge, R_{edge} is determined in terms of the equivalent conductance due to the ohmic loss G_{con}, dielectric loss G_{dlc} and radiation loss G_{rad} in the magnetic wall cavity under the circular microstrip [38, 39]

$$R_{edge} = [G_{con} + G_{dlc} + G_{rad}]^{-1} \tag{2.51}$$

$$G_{con} = \frac{\pi[(ka)^2 - n^2]}{4h^2\sqrt{\sigma}(\mu_0\pi f_{r,nm})^{3/2}} \tag{2.52}$$

$$G_{dlc} = \frac{\tan\delta[(ka)^2 - n^2]}{4\mu_0 h f_{r,nm}} \tag{2.53}$$

$$G_{rad} = \frac{(ka)^2 I(X)}{960.\varepsilon_{r,eq}} \tag{2.54}$$

$$X = ka/\sqrt{\varepsilon_{r,eq}} \tag{2.55}$$

$$I(X) = \int_0^\pi \left[\{J_{n+1}(X\sin\theta) - J_{n-1}(X\sin\theta)\}^2 + \cos^2\theta\cdot\{J_{n+1}(X\cdot\sin\theta) + J_{n-1}(X\cdot\sin\theta)\}^2\cdot\sin\theta d\theta\right] \tag{2.56}$$

The function $I(X)$ can be numerically solved as [40]

$$I(X) = 2.666667378 - 1.066662519X^2 + 0.209534311X^4 - 0.019411347X^6 + 0.001044121X^8 - 0.000049747X^{10} \tag{2.57}$$

$$\varepsilon_{r,eq} = (1 + \varepsilon_{re})/2 \tag{2.58}$$

The quantity ε_{re} is the equivalent dielectric constant expressed in (2.2), tan δ is the loss tangent of the dielectric substrate, σ is the conductivity of the metal used and μ_0 is the free space permeability.

Figure 2.19 compares the measured [41] and computed input resistance at resonance for a coax-fed circular patch as a function of normalized feed location. The variation of the input resistance as a function of frequency is plotted in Figure 2.20 for a circular patch with variable air gap height.

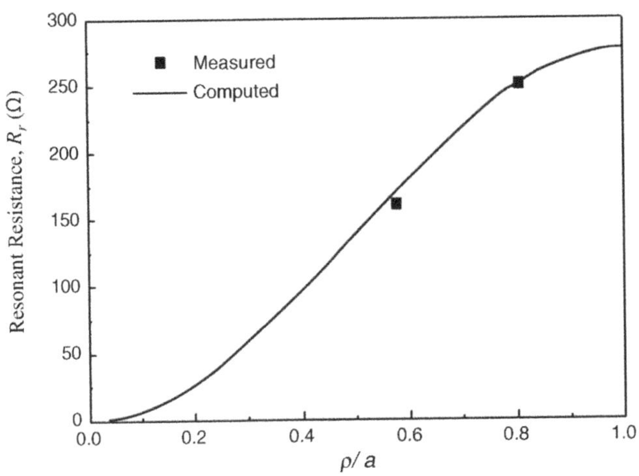

Figure 2.19 Measured and computed input resistance at resonance of a coax-fed circular microstrip patch as a function of normalized feed location ($a = 13$ mm, $h = h_2 = 4.7$ mm, $\varepsilon_r = 2.62$, *tan* $\delta = 0.001$, $f_r^{11} = 3.54$ GHz)

The Q factors, i.e. Q_{con}, Q_{rad}, Q_{dlc}, Q_{sw} may be simply calculated using simple formulas [5] as follows:

$$Q_{con} = h\sqrt{\pi f \mu_0 \sigma} \tag{2.59}$$

$$Q_{rad} = \frac{240 \cdot \left[(ka)^2 - n^2\right]}{h\mu_0 f \cdot X^2 \cdot I(X)} \tag{2.60}$$

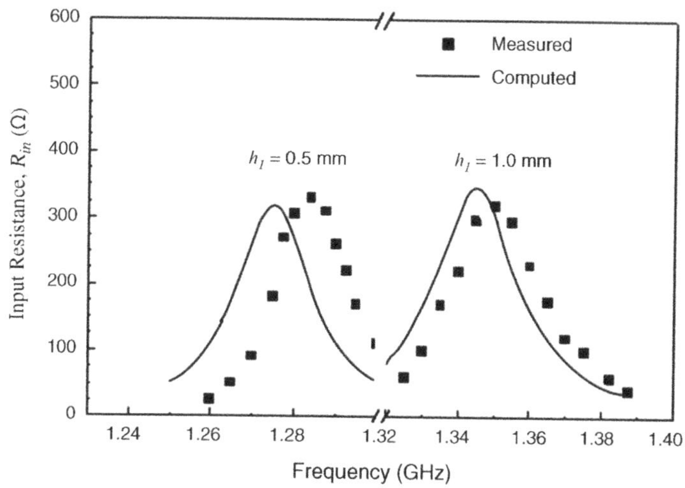

Figure 2.20 Measured and computed input resistance as a function of frequency for TM_{11} mode of a circular microstrip patch for different air gap height ($a = 50$ mm, $h_l = 1.58$ mm, $\varepsilon_r = 2.32$, $tan\,\delta = 0.0012$, $\rho/a = 0.95$)

$$Q_{dlc} = \frac{1}{\tan\delta} \tag{2.61}$$

For a thin low-dielectric constant substrate, Q_{sw}, that is, loss due to surface wave may be ignored. But for more precise calculation, Q_{sw} can be computed using expressions given in [11].

2.6.2 *Input Impedance of IMCP*

The input impedance of an IMCP is similar to (2.49) where the input resistance is calculated as (2.50), but the ratio a/a_{eff} is ignored. Equation (2.50) thus takes the form [42]

$$R_r(\rho) = R_{edge}\frac{J_n^2(k\rho_0)}{J_n^2(ka)} \tag{2.62}$$

Since here, the medium below the patch is simply air, the term *dielectric conductance* is ignored in calculating R_{edge} as

$$R_{edge} = [G_{rad} + G_{con}]^{-1} \tag{2.63}$$

$$G_{con} = \frac{\pi[(ka)^2 - n^2]}{4h_1^2\sqrt{\sigma}(\mu_0\pi f_r)^{3/2}} \tag{2.64}$$

$$G_{rad} = \frac{(ka)^2 I(X)]}{960 \cdot \varepsilon_{e,re}} \tag{2.65}$$

The function *I (X)* is expressed in Equation (2.57) where $X = ka/\sqrt{\varepsilon_{re,i}}$. Since the fringing electric field at $\rho = a$ is under the strong influence of the top dielectric medium, a new factor is introduced as $\varepsilon_{e,re} = \varepsilon_{re,i}/\varepsilon_{r2}$.

In an inverted patch, there is hardly any possibility of occurring losses due to dielectric and surface wave and as such Q_{dlc} and Q_{sw} have been neglected. The quantities Q_{con} and Q_{rad} are calculated using Equations (2.59) and (2.60) by substituting the total height h by the air gap h_1. Importantly, evaluation of $I(X)$ is conducted with the equivalent dielectric constant $\varepsilon_{re,i}$.

2.6.3 *Input Impedance of RMP*

Like Equation (2.49), the input impedance of a rectangular patch (Fig. 2.13) can be expressed as the functions of frequency and feed location (x_0,y_0) as

$$Z_{in}(f, x_0) = R_{in}(f, x_0) + jX_{in}(f, x_0) \tag{2.66}$$

Similarly, the input resistance at resonance is expressed as [43].

$$R_r(x_0) = \frac{4h}{\pi\lambda_0}\mu\eta_0 Q_T\left(\frac{L+2\Delta L}{W+2\Delta W}\right)\cos^2\left(\frac{\pi(0.5L-x_0)}{L+2\Delta L}\right) \tag{2.67}$$

where η_0 is the intrinsic impedance of free space and Q_T is expressed through (2.48).

The calculation of the total quality factor Q_T ignoring Q_{sw} is derived as [43]

$$Q_{rad} = \frac{\pi}{4G_{rad}Z_r} \tag{2.68}$$

$$\begin{aligned} G_{rad} &= \frac{W^2}{90\lambda_0^2}, \text{ for } W \leq 0.35\lambda_0 \\ &= \frac{W}{120\lambda_0} - \frac{1}{60\pi^2} \text{ for } 0.35\lambda_0 \leq W \leq 2\lambda_0 \\ &= \frac{W}{120\lambda_0}, \text{ for } 2\lambda_0 < W \end{aligned} \tag{2.69}$$

$$Z_r = \frac{120\pi\left[\frac{W}{h} + 1.393 + 0.667\ln\left(\frac{W}{h} + 1.444\right)\right]^{-1}}{\sqrt{\varepsilon_{r,n}}} \tag{2.70}$$

$$Q_{dlc} = \frac{\pi(\varepsilon_r - 1)\sqrt{\varepsilon_{r,n}}}{27.3(\varepsilon_{r,n} - 1)\sqrt{2\varepsilon_{r,n} - 1}}\frac{1}{\tan\delta} \tag{2.71}$$

$$Q_{con} = h\sqrt{\pi f \mu_0 \sigma} \tag{2.72}$$

$$\varepsilon_{r,n} = \frac{\varepsilon_{reff} + 1}{2} \tag{2.73}$$

Here, λ_0 represents the resonant wavelength in free space and $\varepsilon_{r,eff}$ is the effective dielectric constant derived in Section 2.4.

Figure 2.21 shows the variation of the input resistance at resonance as a function of feed location of a rectangular patch for three different aspect ratio $W/L = 0.7$, 1.0 and 1.5 resonating at frequencies 3.16 GHz, 3.11 GHz and 3.05 GHz, respectively.

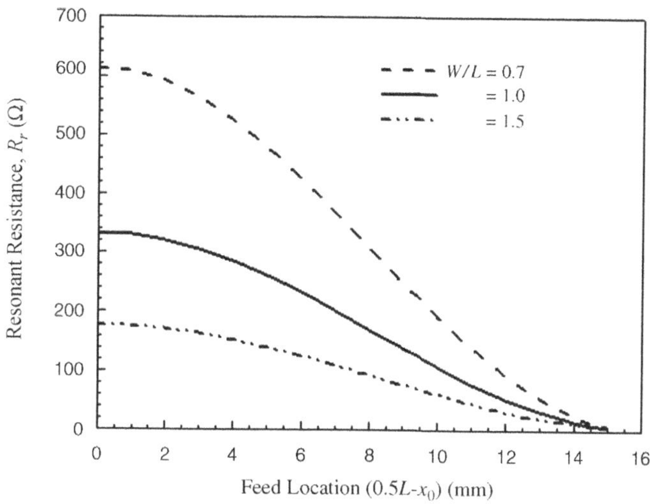

Figure 2.21 Computed input resistance at resonance of a coax-fed rectangular microstrip patch as a function of feed location for different aspect ratio W/L ($\varepsilon_r = 2.33$, $L = 30$ mm, $h_1 = 0$, $h_2 = 1.575$ mm, $\tan \delta = 0.001$)

2.6.4 Input Impedance of ETMP

An equilateral triangular microstrip patch coaxially fed at a distance d from the apex is shown in Figure 2.16. Its input impedance is expressed as [45]

$$Z(f,d) = R_{in}(f,d) + jX_{in}(f,d) \tag{2.74}$$

$$= -j\omega\mu \sum_{n=0}^{\infty}\sum_{m=n}^{\infty} \frac{4\sqrt{3}hC_{mn}}{27s^2}\left[\cos\left(\frac{2\pi ld}{\sqrt{3}s}\right)j_0\left(\frac{\pi lw}{\sqrt{3}s}\right) + \cos\left(\frac{2\pi md}{\sqrt{3}s}\right)j_0\left(\frac{\pi mw}{\sqrt{3}s}\right) + \cos\left(\frac{2\pi nd}{\sqrt{3}s}\right)j_0\left(\frac{\pi nw}{\sqrt{3}s}\right)\right]^2$$

$$\left[\frac{(\omega^2-\omega_r^2)\mu_0\varepsilon + j\delta_{eff}k^2}{\left(\omega^2-\omega_r^2\right)^2\mu_0^2\varepsilon^2 + \delta_{eff}^2 k^4}\right] \tag{2.75}$$

where, $j_0(x) = \sin(x)/x$

and

$$C_{mn} = \begin{cases} 1, \text{ if } m = n = 0 \\ 6, \text{ if } (m = 0 \text{ and } n \neq 0) \text{ or } (m \neq 0 \text{ and } n = 0) \text{ or } (m = n \neq 0) \\ 12, \text{ if } m \neq n \neq 0 \end{cases} \quad (2.76)$$

and

$$l = -m-n$$

Equation (2.75) is a double infinite series comprising terms for various modes of excitation. The parameter d is the location of the coaxial feed on the base bisector line from the apex of the triangle. The feed is modeled by a current ribbon of effective width w [44] which is several times the diameter of coax inner conductor and chosen as a best fit parameter with the measurements.

The effective loss tangent, $\delta_{eff} = 1/Q_T$ and the total quality factor Q_T are expressed as (2.48).

The total quality factor for an ETMP is derived from (2.48) neglecting Q_{sw}. Simple closed form expressions [45] are used to derive these quantities as

$$Q_{rad} = \frac{\pi}{4G_{rad}Z_{rad}} \quad (2.77)$$

$$\begin{aligned} G_{rad} &= \frac{s^2}{303\lambda_0^2} \quad for\ s \leq 0.642\lambda_0 \\ &= \frac{s}{220\lambda_0} - \frac{1}{60\pi^2} \quad for\ 0.642\lambda_0 \leq s \leq 3.67\lambda_0 \\ &= \frac{s}{220\lambda_0} \quad for\ 3.67\lambda_0 < s \end{aligned} \quad (2.78)$$

$$Z_{rad} = \frac{120\pi[((0.545s)/h)0.545s/h + 1.393 + 0.667\ln((0.545s/h) + 1.444)]^{-1}}{\sqrt{\varepsilon_{r,eff}}} \quad (2.79)$$

$$Q_{dlc} = (\frac{\pi}{19\tan\delta})\left(\frac{\varepsilon_{re}-1}{\varepsilon_{r,eff}-1}\right)\left(\frac{\sqrt{\varepsilon_{r,eff}}+1}{\sqrt{\varepsilon_{r,eff}}}\right) \quad (2.80)$$

$$Q_{con} = h\sqrt{\pi f \mu_0 \sigma} \quad (2.81)$$

2.7 Feed Reactance of a Probe-Fed Microstrip Patch

Coaxial SMA probe is the most popular candidate used to feed a microstrip patch. Its central conductor carrying electric current at microwave frequency generates considerable probe reactance X_f. This X_f acts in series with $Z_{in,patch}$, i.e., the input impedance is due to the patch only as shown in Figure 2.18 and significantly affects matching of the input impedance Z_{in} of the antenna. The effect can be visualized from the impedance locus plotted on a smith chart. High value of X_f shifts the centre of the impedance locus towards the upper part of the zero-reactance axis. For an isolated patch, a circular impedance locus should appear on the smith chart with its centre lying on the zero-reactance axis.

A simple expression was formulated by Harrington in [20] to calculate the reactance of a current carrying conductor within a parallel plate waveguide as

$$X_f(f) = \frac{\eta_0 kh}{2\pi}\left[\ln(4/kd) - 0.577\right] \tag{2.82}$$

where η is the intrinsic impedance of the free space, k is the wave number, and d is the diameter of the probe. This simple expression cannot account for accurate estimation of probe reactance as indicated in [46]. A recent formulation [19] showed that an accurate estimation of the feed reactance X_f significantly depends on the location of the probe defined as the distance ρ for a circular patch and x_0 for a rectangular patch. They proposed a new function to be multiplied with Harrington's value $X_f(f)$ and that should account for the change in feed reactance with varying location of the probe [19] as:

$$F'(x_0) = \cos^2\pi(x_0/L): \quad \text{rectangular patch} \tag{2.83}$$

$$F'(\rho) = \frac{J_n^2[ka\{1-(\rho/a)\}]}{J_n^2(ka)}: \quad \text{circular patch} \tag{2.84}$$

Thus the feed reactance can be calculated as

$$X_f(f, x_0) = X_f(f)\cdot F'(x_0) \quad \text{for rectangular patch} \tag{2.85}$$

$$X_f(f, \rho) = X_f(f).F'(\rho) \quad \text{for circular patch} \tag{2.86}$$

Figure 2.22 shows the measured and computed feed reactance, X_f with the change in the feed location under the patch for both rectangular and circular patches.

2.8 Radiation Characteristics

The CAD formulation of the far field radiation patterns of a microstrip patch using a magnetic current model for the dominant mode is derived in [12, 47]. They are capable of portraying a three-dimensional pattern at a distance r with angular position determined by θ and ϕ, shown in Figure 2.23. In Figure 2.23, the variation of electric fields over a vertical plane with $\phi = 0^0$ provides E-plane pattern and that over a vertical plane with $\phi = 90^0$ indicates the H-plane variation.

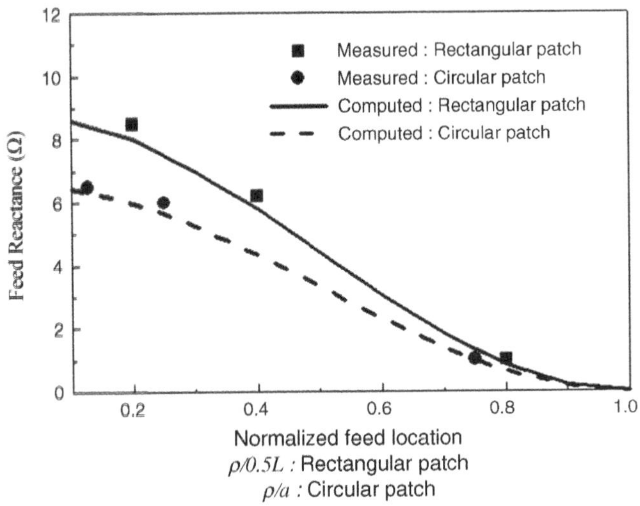

Figure 2.22 Feed reactance versus normalized feed location of probe-fed microstrip patches (rectangular patch: $L = 76.2$ mm, $\varepsilon_r = 2.53$, $d = 1.35$ mm, $h/\lambda_0 = 0.006$, $f = 1.2$ GHz; circular patch: $a = 67$ mm, $\varepsilon_r = 2.53$, $d = 1.35$ mm, $h/\lambda_0 = 0.004$, $f = 0.8$ GHz). Reproduced by permission of ©2007 IEEE [19]

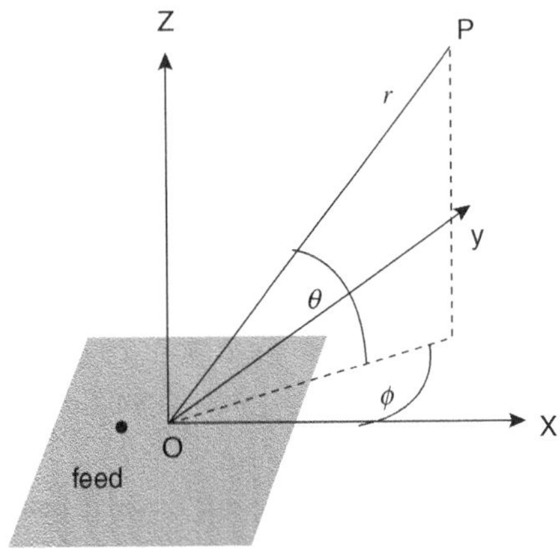

Figure 2.23 A radiating microstrip patch located at the centre of a spherical coordinate system, radiated fields being calculated at the point P.

2.8.1 Rectangular Microstrip Patch

For the truncated substrate on an infinite ground plane, the far field components due to a rectangular patch with dimensions L and W and printed on a dielectric substrate with thickness h $(=h_1 + h_2)$, are given by

$$E_\theta(r,\theta,\phi) = -2Wh\left(\frac{E_0}{\eta_0}\right)\cos\phi(1-\Gamma^{TM}(\theta))\cos\left(k_x\frac{L}{2}\right)\text{sinc}\left(k_y\frac{W}{2}\right)\tan c(k_{Z1}h) \quad (2.87)$$

$$E_\phi(r,\theta,\phi) = 2Wh\left(\frac{E_0}{\eta_0}\right)(\cos\theta\sin\phi)(1-\Gamma^{TE}(\theta))\cos\left(k_x\frac{L}{2}\right)\mathrm{sinc}\left(k_y\frac{W}{2}\right)\mathrm{tanc}(k_{Z1}h) \quad (2.88)$$

where

$$E_0 = \left(\frac{-j\omega\mu_0}{4\pi D}\right)^{e^{-jk_0D}} \quad (2.89)$$

$$k_x = k_0\sin\theta\cos\phi \quad (2.90)$$

$$k_y = k_0\sin\theta\sin\phi \quad (2.91)$$

$$k_{Z1} = k_0N_1(\theta) \quad (2.92)$$

$$N_1(\theta) = \cos\theta \quad (2.93)$$

$$\mathrm{tanc}(x) = \frac{\tan(x)}{x} \quad (2.94)$$

$$1-\Gamma^{TM}(\theta) = \frac{2}{1+j(N_1(\theta)\sec\theta)\tan(k_0hN_1(\theta))} \quad (2.95)$$

$$1-\Gamma^{TE}(\theta) = \frac{2}{1+j\left(\frac{\cos\theta}{N_1(\theta)}\right)\tan(k_0hN_1(\theta))} \quad (2.96)$$

2.8.2 *Circular Microstrip Patch*

Similarly, the far field components due to a circular patch with radius a and printed on a truncated substrate with thickness $h(=h_1 + h_2)$, are given by

$$E_\theta(r,\theta,\phi) = \left(2\pi\frac{E_0}{\eta_0}(ah)\right)\cos\phi\,\mathrm{tanc}(k_{Z1}h)J_1'(k_0a\sin\theta)(1-\Gamma^{TM}(\theta)) \quad (2.97)$$

$$E_\phi(r,\theta,\phi) = -2\pi\left(\frac{E_0}{\eta_0}(ah)\right)\sin\phi\,\mathrm{tanc}(k_{Z1}h)J_{inc}(k_0a\sin\theta)(1-\Gamma^{TE}(\theta)) \quad (2.98)$$

$$J_{inc}(x) = \frac{J_1(x)}{x} \quad (2.99)$$

2.9 Radiation Efficiency

The radiation efficiency, e_r is described as the ratio of the radiated power to the total input power provided to the antenna. In terms of the quality factors in (2.48), the radiation efficiency is given as

$$e_r = \frac{Q_T}{Q_{rad}} \quad (2.100)$$

2.10 Bandwidth

The bandwidth of a microstrip radiator is usually defined as the frequency band over which the element operates with its input voltage standing wave ratio (VSWR) remaining below a specified value (*vswr*). Mathematically it is expressed as

$$Bandwidth = \frac{vswr - 1}{\sqrt{vswr}\, Q_T} \tag{2.101}$$

Conventionally, VSWR < 2 is the accepted value for most of the antennas and thus the percent bandwidth may be determined in terms of the total quality factor Q_T as

$$\%Bandwidth = \frac{1}{\sqrt{2} Q_T} \times 100 \tag{2.102}$$

2.11 Conclusion

The formulas presented in this chapter are easy to handle using simple computer programs and estimating accurate design values for three basic patch geometries. They are highly suitable for the students for their project and classroom designs and are equally efficient for a professional design purpose.

References

1. T. Itoh and R. Mittra, "Analysis of microstrip disk resonator," *Arch. Elek. Übertagung*, vol. 21, pp. 456–458, Nov. 1973.
2. T. Itoh, "Analysis of microstrip resonator," *IEEE Trans. Microwave Theory Tech.*, vol. MTT-22 pp. 946–952, Nov. 1974.
3. R. E. Munson, "Conformal microstrip antennas and microstrip phased arrays," *IEEE Trans. Antennas Propagat.*, vol. 22, pp. 74–78, 1974.
4. J. Q. Howell, "Microstrip antennas," *Dig. IEEE Int. Symp. Antennas Propagat.*, pp. 177–180, Dec. 1972.
5. I. J. Bahl and P. Bhartia, *Microstrip Antennas*, Artech House, Dedham, MA, 1980.
6. K. Carver and J. Mink, "Microstrip antenna technology," *IEEE Trans. Antennas Propagat.*, vol. 29, pp. 2–24, Jan. 1981.
7. J. R. James, P. S. Hall and C. Wood, *Microstrip Antennas: Theory and Design*, Peter Peregrinus, London, 1981.
8. J. R. James and P. S. Hall, *Handbook of Microstrip Antennas*, Peter Peregrinus, London, 1989.
9. D. M. Pozar and D. H. Schaubert, *Microstrip Antennas*, IEEE Press, New York, 1995.
10. K. F. Lee and W. Chen, *Advances in Microstrip and Printed Antennas*, John Wiley & Sons, Ltd, New York, 1997.
11. R. Garg et al., *Microstrip Antenna Design Handbook*, Artech House, Boston, 2001.
12. D.R. Jackson, "Microstrip antennas," in J. Volakis (ed.), *Antenna Engineering Handbook*, McGraw-Hill, New York, 2007.
13. D. Guha, "Resonant frequency of circular microstrip antennas with and without air gaps," *IEEE Trans. Antennas Propagat.*, vol. 49, no. 1, pp. 55–59, Jan. 2001.
14. D. Guha and J. Y. Siddiqui, "New CAD model to calculate the resonant frequency of inverted microstrip circular patch antenna," *Microwave Opt. Technol. Lett.*, vol. 35, no. 6, pp. 434–437, Dec. 20, 2002.
15. D. Guha and J. Y. Siddiqui, "Effect of a cavity enclosure on the resonant frequency of inverted microstrip circular patch antennas," *IEEE Trans. Antennas Propagat.*, vol. 52, no. 8, pp. 2177–2180, Aug 2004.
16. D. Guha and J. Y. Siddiqui, "Resonant frequency of circular microstrip antenna covered with dielectric superstrate," *IEEE Trans. Antennas Propagat.*, vol. 51, no. 7, pp. 1649–1652, July 2003.
17. S. Chattopadhyay, M. Biswas, J. Y. Siddiqui and D. Guha, "Rectangular microstrips with variable air gap and varying aspect ratio: improved formulations and experiments," *Microwave Opt. Technol. Lett.*, vol. 51, no. 1, pp. 169–173, Jan. 2009.

18. D. Guha and J. Y. Siddiqui, "Resonant frequency of equilateral triangular microstrip antenna with and without air gaps," *IEEE Trans. Antennas Propagat.*, vol. 52, no. 8, pp. 2174–2177, Aug. 2004.
19. D. Guha, M. Biswas and J.Y. Siddiqui "Harrington's formula extended to determine accurate feed reactance of probe-fed microstrip patches," *IEEE Antennas and Wireless Propag. Lett.*, vol. 6, pp. 33–35, 2007.
20. R. F. Harrington, *Time Harmonic Electromagnetic Fields*, McGraw-Hill, New York, 1961.
21. K. F. Lee, K. Y. Ho, and J. S. Dahele, "Circular-disk microstrip antenna with an air gap," *IEEE Trans. Antennas Propagat.*, vol. 32, no. 8, pp. 880–884, Aug. 1984.
22. I. Wolff and N. Knoppik, "Rectangular and circular microstrip disk capacitors and resonators," *IEEE Trans Microwave Theory Tech.*, vol. 22, no. 10, pp. 857–864, Oct. 1974.
23. H. A. Wheeler, "A simple formula for the capacitance of a disc on dielectric on a plane," *IEEE Trans Microwave Theory Tech.*, vol. 30, no. 11, pp. 2050–2054, Nov. 1982.
24. HFSS v11, Ansoft Corporation.
25. J. A. Navarro, L. Fan, and K. Chang, "Active inverted stripline circular patch antennas for spatial power combining," *IEEE Trans Microwave Theory Tech.*, vol. 41, pp. 1856–1863, 1993.
26. J. A. Navarro, J. McSpadden, and K. Chang, AT "Experimental study of inverted microstrip for integrated antennas applications," *IEEE Antennas Propagat. Int. Symp. Seattle*, pp. 920–923, 1994.
27. J. A. Navarro, L. Fan, and K. Chang, "Novel FET integrated inverted stripline patch," *Electron. Lett.*, vol. 30, 655–657 1994.
28. R. A. Flynt, L. Fan, J. A. Navarro, and K. Chang, "Low cost and compact active integrated antenna transceiver for system applications," *IEEE Trans Microwave Theory Tech.*, vol. 44, pp. 1642–1649, 1996.
29. C. M. Montiel, L. Fan, and K. Chang, "A novel active antenna with self-mixing and wideband varactor-tuning capabilities for communication and vehicle identification applications," *IEEE Trans Microwave Theory Tech.*, vol. 44, pp. 2421–2430, 1996.
30. C. A. Balanis, *Antenna Theory: Analysis and Design*, John Wiley & Sons, New Jersey, 2005.
31. V. Losada, R. R. Boix, and M. Horno, "Resonant modes of circular microstrip patches in multilayered substrates," *IEEE Trans. Microwave Theory Tech.*, vol. 47, no. 4, pp. 488–497, April 1999.
32. F. Bouttout, F. Benabdelaziz, and A. Khellaf, "Closed-form hankel transforms for circular disk basis modes involving Chebyshev polynomials and edge condition," *Electron. Lett.*, vol. 36, no. 10, pp. 866–867, May 2000.
33. N. G. Alexopoulos and D. R. Jackson, "Fundamental superstrate (cover) effects on printed circuit antennas," *IEEE Trans. Antennas Propagat.*, vol. 32, pp. 807–816, Aug. 1984.
34. A. K. Bhattacharyya and T. Tralman, "Effects of dielectric superstrate on patch antennas," *Electron. Lett.*, vol. 24, pp. 356–358, 1988.
35. K. M. Luk, W.Y. Tam, and C. L. Yip, "Analysis of circular microstrip antennas with superstrate," *IEE Proc.*, Pt. H, vol. 136, pp. 261–262, Mar. 1989.
36. J. Helszajn and D. S. James, "Planar triangular resonators with magnetic walls," *IEEE Trans. Microwave Theory Tech.*, vol. 26, pp. 95–100, Feb. 1978.
37. A. K. Sharma and B. Bhat, "Analysis of triangular microstrip resonators," *IEEE Trans. Microwave Theory Tech.*, vol. 30, pp. 2029–2031, Nov. 1982.
38. D. Guha, Y. M. M. Antar, J. Y. Siddiqui and M. Biswas " Resonant resistance of probe- and microstrip-line-fed circular microstrip patches," *IEE Proc. Microw. Antennas Propagat.*, vol. 152, no. 6, pp. 481–484, Dec. 2005.
39. A. G. Derneryd, "Analysis of the microstrip disk antenna element," *IEEE Trans. Antennas Propagat.*, vol. 27, pp. 660–664, 1979.
40. F. Abboud, J. P. Damiano and A Papiernik, "A new model for calculating the impedance of coax-fed circular microstrip antennas with and without air gaps," *IEEE Trans. Antennas Propagat.*, vol. 38, no. 11, pp. 1882–1885, Nov. 1990.
41. M. Davidovitz, and Y.T. Lo, "Input impedance of a probe-fed circular microstrip antenna with thick substrate," *IEEE Trans. Antennas Propagat.*, vol. 34, no. 7, pp. 905–911, 1986.
42. J. Y. Siddiqui and D. Guha, "Impedance characteristics of inverted microstrip circular patch antennas," *Microwave Opt. Technol. Lett.*, vol. 39, no. 6, pp. 508–511, Dec. 20, 2003.
43. S. Chattopadhyay, M. Biswas, J. Y. Siddiqui, and D. Guha, "Input impedance of probe-fed rectangular microstrip antennas with variable air gap and varying aspect ratio" *IET Microwave, Antennas Propagat.*, vol. 3, no. 8, pp. 1151–1156, 2009.
44. K. F. Lee, K. M. Luk, and J. S. Dahele, "Characteristics of the equilateral triangular patch antenna," *IEEE Trans. Antennas Propagat.*, vol. 36, no. 11, pp. 1510–1518, Nov. 1988.

45. M. Biswas and D. Guha, "Input impedance and resonance characteristics of superstrate-loaded triangular microstrip patch," *IET Microw. Antennas Propagat.*, vol. 3, no. 1, pp. 92–98, 2009.
46. B. M. Alarjani and J. S. Dahele, "Feed reactance of rectangular microstrip patch antenna with probe feed," *Electron. Lett.*, vol. 36, pp. 388–390, 2000.
47. D. R. Jackson and J. T. Williams, "A comparison of CAD models for radiation from rectangular microstrip patches," *Intl. Journal of Microwave and Millimeter-wave Computer Aided Design*, vol. 1, no. 2, pp. 236–248, Apr. 1991.

3

Generalized Scattering Matrix Approach for Multilayer Patch Arrays

Arun K. Bhattacharyya
Northrop Grumman Corporation, USA

3.1 Introduction

Microstrip array antennas are potential candidates for airborne applications due to their low profile and light weight. For an array design it is of utmost importance that the analytical method be as accurate as possible. There are several ways to analyze microstrip array antennas including mutual coupling effects [1–6]. The most common method is the "element-by-element" approach [1–3]. This approach characterizes the array by an impedance matrix. The elements of the impedance matrix are the mutual impedances between the elements. This mutual impedance between two microstrip elements is determined using the "induced current model" which typically considers the participating elements and ignores the presence of the other array elements. For a densely populated array, this may not be a good approximation because the neighboring elements play an important role in the mutual coupling between two elements [4, 7]. Furthermore, the element-by-element approach becomes very complex for a multilayer patch array because determination of the induced currents on multilayer patches is a formidable exercise.

Another useful approach, generally known as convolution method (also called the windowing technique), is also employed to analyze finite arrays [6–10]. It is shown in [6] that a finite array can be analyzed by convolving the infinite array characteristics with the finite excitation function. Unlike the "element-by-element" approach, this approach does not determine the

Microstrip and Printed Antennas: New Trends, Techniques and Applications. Edited by Debatosh Guha and Yahia M.M. Antar

mutual impedance explicitly, but inherently incorporates the mutual coupling effects between the array elements. The drawback of this approach lies in its inability to incorporate the "edge effects" of a finite array. This is primarily due to the fact that because an edge element is not surrounded by elements in all sides, its electrical response differs from that of an inside element. However, for a practical array the effect of the edge elements on the overall array performance is minimal because the edge taper is typically large. This approach has been employed successfully by many researchers [6, 9, 10]. The element-by-element approach also does not incorporate the edge effect rigorously because generally it employs Green's function associated with an infinite dielectric layer.

This chapter presents the Generalized Scattering Matrix (GSM) approach for a finite array analysis. The GSM approach, like the element-by-element approach, essentially works under the "mutual coupling matrix" framework. However, contrary to the element-by-element approach, the GSM approach naturally incorporates the effects of all non-participating elements on the mutual coupling between two elements. Furthermore, the GSM approach is very straightforward for a multilayer patch array analysis and the effects of various patch layers on mutual coupling come naturally. It can be shown that the GSM approach and the convolution approach are analytically equivalent and the connection between two methods is established through Floquet modal analysis [11].

The aim of this chapter is to provide a general overview of the GSM approach to analyze multilayer finite printed array structures. Multilayer printed patch structures are used to enhance the bandwidth performance [5, 12, 13] of a printed array. The GSM approach essentially is a modular approach, where each layer of a multilayer structure is analyzed independently and then characterized in terms of a matrix. The matrix is called the GSM of the layer, because the reflection and transmission characteristics of the layer with respect to several incident Floquet modes are embedded within the matrix. The complete characterization of a multilayer structure is obtained by cascading the individual GSMs of the layers.

The GSM of an array essentially characterizes the periodic array that is extended to infinity in the transverse directions. In addition, the GSM is associated with an ideal Floquet excitation, defined by uniform amplitude and linear phase distributions. For a finite array and/or for a tapered excitation, the analysis involves few additional steps. In this chapter, we outline the steps with mathematical illustrations. We demonstrate that the results of an infinite array can be utilized to predict the performances of a finite array with an arbitrary excitation. The predicted result would be *exact* if we define a "finite array" as *a physically infinite array with a finite number of excited elements*. The remaining elements are non-excited, though they *must be physically present*. Such a finite array exists only in theory. A real finite array, however, has a finite number of physical elements. In many situations, radiation characteristics of a real finite array can be approximated as that of a "finite array" defined above, because the non-excited elements generally do not contribute significantly to the radiated fields, particularly in the main lobe and near-in side lobe region.

The chapter is organized as follows. Section 3.2 briefly outlines the GSM approach for an infinite multilayer array. Section 3.3 formulates the mutual coupling because the mutual coupling plays the most important role in the performance of a finite array. Section 3.4 presents the analytical procedure of a finite array employing the mutual coupling data. In Section 3.5 the methodology is demonstrated through a numerical example of a slot-fed finite patch array antenna. Important conclusions are presented in Section 3.6.

3.2 Outline of the GSM Approach

The GSM approach of a multilayer finite array involves the following steps:

1. Computation of GSM of each layer.
2. Combining GSMs of the individual layers to obtain overall GSM of the structure.
3. Mutual coupling computation between the array elements using Floquet modal theory.
4. Active element pattern computation.
5. Computations of finite array pattern and return loss of the elements.

A typical multilayer array consists of four types of basic building blocks: (1) printed elements layer; (2) dielectric layer; (3) dielectric interface; and (4) aperture (slot aperture, for instance) layer. The GSM of a printed element layer and slot aperture layer are usually obtained using Galerkin's MoM analysis [11–13]. The GSM of a dielectric layer and the interface are determined using Floquet modal analysis [11]. The GSMs of the individual layers are then combined to obtain the overall GSM of the multilayer structure. The GSM analysis and the cascading formulas are outlined in Section 3.2.1.

3.2.1 The GSM

The GSM essentially represents the input-output characteristics of an infinite array structure with respect to a set of Floquet modes. For a multilayer array structure, the GSMs of individual layers are determined and then combined together to obtain the overall GSM of the structure, as typically done in a mode-matching analysis of waveguide horns or filters.

To illustrate the GSM approach pictorially, consider a three-layer periodic array structure (patch-dielectric-patch) as shown in Figure 3.1 (a). The three-layer-structure is equivalent to five modules connected in cascade as shown in Figure 3.1 (b). Identical cell sizes and cell orientations for the periodic arrays are assumed. Also, the structure is assumed to be infinite extent along x and y directions and is under Floquet excitation (uniform amplitude and linear phase).

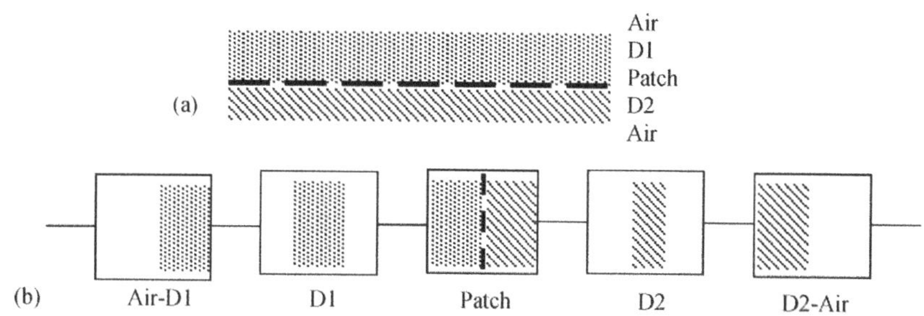

Figure 3.1 (a) A three-layer array structure; (b) modular representation of the array

The GSM of a module is defined through the relation between incident and reflected voltages as below:

$$\begin{bmatrix} a_1^- \\ a_2^- \end{bmatrix} = \begin{bmatrix} S_{11} & S_{12} \\ S_{21} & S_{22} \end{bmatrix} \begin{bmatrix} a_1^+ \\ a_2^+ \end{bmatrix} \tag{3.1}$$

In Equation (3.1), $[a_1^+]$ and $[a_2^+]$ are the incident voltage vectors with respect to the Floquet modes at the two sides (or ports) of the module and $[a_1^-]$ and $[a_2^-]$ are the corresponding reflected voltage vectors. The [S] matrix at the right-hand side of (3.1) is called the GSM of the layer. It consists of four sub-matrices, namely $[S_{11}]$, $[S_{12}]$, $[S_{21}]$ and $[S_{22}]$, respectively. The dimension of a voltage vector is equal to the number of coupling Floquet modes (coupling to the adjacent layers), which is decided by the cell size and the thickness of a layer. However, the total number of Floquet modes to be considered in obtaining the GSM is much larger because the far evanescent modes are included to account for the stored energy [12, 13].

The overall GSM of the multilayer structure is obtained by combining the individual GSMs of the layers or modules. The cascading formula for two modules A and B is given by [11]

$$[S_{11}^{AB}] = [S_{11}^A] + [S_{12}^A][I - S_{11}^B \quad S_{22}^A]^{-1}[S_{11}^B][S_{21}^A] \tag{3.2a}$$

$$[S_{12}^{AB}] = [S_{12}^A][I - S_{11}^B \quad S_{22}^A]^{-1}[S_{12}^B] \tag{3.2b}$$

$$[S_{21}^{AB}] = [S_{21}^B][I - S_{22}^A \quad S_{11}^B]^{-1}[S_{21}^A] \tag{3.2c}$$

$$[S_{22}^{AB}] = [S_{22}^B] + [S_{21}^B][I - S_{22}^A \quad S_{11}^B]^{-1}[S_{22}^A][S_{12}^B] \tag{3.2d}$$

This formula can be applied repeatedly to obtain the overall GSM of a multilayer array. The GSM cascading formulas is applicable only if the layers have identical periodicities and have identical cell orientations. The above conditions ensure that a Floquet modal vector function has an identical expression for all the layers. If the layers have different periodicities, then the process is more involved as detailed in [11, 14].

3.3 Mutual Coupling Formulation

In Section 3.2 we dealt with the characteristics of an array element under Floquet excitation. An accurate analysis of a finite array with an arbitrary excitation necessitates an estimation of the mutual coupling between the array elements. In this section we demonstrate that the mutual coupling between the elements can be estimated from the results of an infinite array under Floquet excitations.

The mutual coupling between the elements is generally quantified in terms of the following three measurable quantities:

1. mutual impedance;
2. mutual admittance;
3. scattering parameters.

The above three measurable quantities are related to each other by simple algebraic relations. In this section we will first derive the mutual impedance from Floquet impedance of an infinite array [8, 11]. We first consider a one-dimensional array. The result can be extended for a two-dimensional array.

3.3.1 Mutual Impedance

Consider an infinite array as illustrated in Figure 3.2. The elements are arranged along the x-direction with element spacing a. Suppose the elements are excited uniformly with linearly progressed phase, known as Floquet excitation. Suppose ψ is the phase difference between two adjacent elements. Then following the definition of mutual impedance, the input voltage for the 0-th element can be obtained as:

$$V_0(\psi) = \sum_{n=-\infty}^{\infty} I_n Z_{0n} \tag{3.3}$$

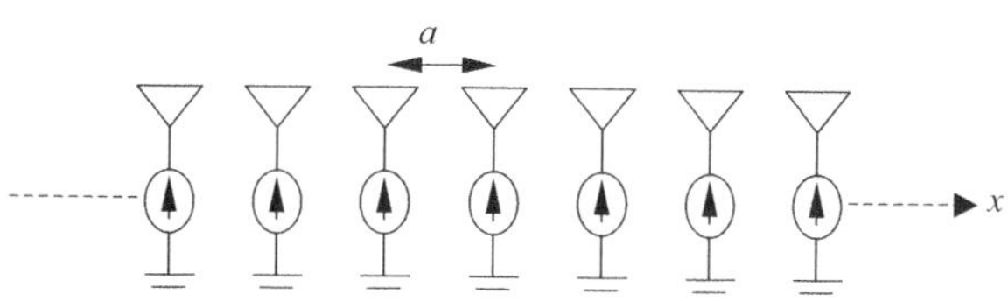

Figure 3.2 Infinite linear array

In Equation (3.3), V_0 is the input voltage for the element located at $x = 0$, I_n is the input current of the n-th element and Z_{0n} is the mutual impedance between the two elements that are located at $x = 0$ and at $x = na$, respectively. For Floquet excitations, the input currents can be expressed as:

$$I_n = I_0 \exp(-jn\psi), \tag{3.4}$$

where I_0 is the input current for the element at $x = 0$. The input impedance seen by the $n = 0$ element is

$$Z_0(\psi) = \frac{V_0(\psi)}{I_0} \tag{3.5}$$

Substituting Equations (3.3) and (3.4) in (3.5), we obtain

$$Z_0(\psi) = \sum_{n=-\infty}^{\infty} Z_{0n} \exp(-jn\psi) \tag{3.6}$$

For a Floquet excitation, the above input impedance must be equal to the Floquet impedance $Z^{FL}(\psi)$. Therefore we obtain

$$Z^{FL}(\psi) = \sum_{n=-\infty}^{\infty} Z_{0n} \exp(-jn\psi) \tag{3.7}$$

The right-hand side of Equation (3.7) is the Fourier series expansion of the Floquet impedance where the Fourier coefficients are equal to the mutual impedances. Thus, the mutual impedance between 0-th and n-th element, Z_{0n}, is readily obtained in terms of the Fourier integral as follows:

$$Z_{0n} = \frac{1}{2\pi} \int_{-\pi}^{\pi} Z^{FL}(\psi) \exp(jn\psi) d\psi \tag{3.8}$$

If the two elements are located at $x = ma$ and $x = na$, respectively, then the mutual impedance between these two elements can be expressed as

$$Z_{mn} = \frac{1}{2\pi} \int_{-\pi}^{\pi} Z^{FL}(\psi) \exp\{-j(m-n)\psi\} d\psi \tag{3.9}$$

Equation (3.9) establishes the relation between the Floquet impedance and the mutual impedance between the elements. It is important to observe that Equation (3.9) yields the mutual impedance in the array environment. Also observe that Z_{mn} and Z_{nm} are identical because $Z^{FL}(\psi) = Z^{FL}(-\psi)$ [11]. The symmetry property of $Z^{FL}(\psi)$ can be utilized to express Z_{mn} in a convenient from a computational point of view as follows:

$$Z_{mn} = \frac{1}{\pi} \int_{0}^{\pi} Z^{FL}(\psi) \cos\{(m-n)\psi\} d\psi \tag{3.10}$$

The mutual impedance deduced in Equation (3.10) includes the effects of scattering from the intermediate and surrounding elements that are open-circuited. The element-by-element approach typically ignores the scattering effects; therefore, the present formulation for mutual coupling is generally more accurate than the element-by-element approach.

It is worth pointing out that for some arrays the Floquet impedance Z^{FL} may have a finite number of singularities due to resonances of selective Floquet modes with the guided wave modes supported by the array structures. In such a situation, a singularity extraction technique [15] must be employed to compute the integral near a singular point.

The mutual admittance between the two elements in array environment can be obtained as:

$$Y_{mn} = \frac{1}{\pi} \int_{0}^{\pi} Y^{FL}(\psi) \cos\{(n-m)\psi\} d\psi \tag{3.11}$$

where $Y^{FL}(\psi)$ is the Floquet admittance, reciprocal to the Floquet impedance $Z^{FL}(\psi)$ and Y_{mn} is the mutual admittance between the m-th and the n-th elements. The distance between the two elements is $(m-n)a$. The scattering parameters between the elements also follow the similar relation. If S_{mn} represents the scattering parameter defined as the voltage received by the m-th element when the n-th element is excited with all other elements including the m-th element are matched terminated, then

$$S_{mn} = \frac{1}{\pi} \int_{0}^{\pi} \Gamma^{FL}(\psi) \cos\{(m-n)\psi\} d\psi \tag{3.12}$$

where $\Gamma^{FL}(\psi)$ is the reflection coefficient of an array element under Floquet excitation. Since $\Gamma^{FL}(\psi) = \Gamma^{FL}(-\psi)$, one can see that $S_{mn} = S_{nm}$. It should be noted that the integral for S_{mn} *does not have any singularity* because the magnitude of $\Gamma^{FL}(\psi)$ does not exceed beyond unity. Thus, from a computational point of view, a scattering matrix formulation is advantageous as compared to an impedance/admittance formulation for a finite array analysis.

The mutual coupling formulation can be extended for a two-dimensional planar array. For a rectangular lattice structure, the mutual impedance involves a two-dimensional integral with variables ψ_x and ψ_y, where ψ_x and ψ_y represent the phase difference between the adjacent elements along the x and y directions, respectively. For a triangular lattice, the formulation is slightly different (see [11, p. 141]).

3.4 Finite Array: Active Impedance and Radiation Patterns

The mutual coupling information between the elements is utilized to determine the active impedance or return loss of an element with respect to given amplitude and phase distributions. The active impedance of an element depends on the amplitude and phase distributions and the load-conditions of the non-excited elements [11, p. 146]. In the present study we will assume that the non-excited elements of a "finite array" are match terminated. This assumption is somewhat justified because a matched element has a small scattered field, thus closely resembles the absence of an element.

In such a situation the scattering matrix relation will yield the exact active input impedance solution. The relation in this situation is

$$[V^-] = [S][V^+] \tag{3.13}$$

where V^+ and V^- are the incident and reflected voltage vectors. Elements of $[S]$ are obtained using (3.12). Equation (3.13) can be utilized to obtain the complex reflection coefficients of the elements with respect to a given amplitude distribution of a finite array.

The radiation pattern of the finite array with respect to this particular condition can be obtained directly from superposition. The result becomes [11, p. 150]

$$E^{array} = E^a[P][V^+] \tag{3.14}$$

where E^a is the active element pattern, which is defined as the radiation pattern of an element in array environment while other elements are match terminated. In Equation (3.14) $[P]$ represents a row vector comprising exponential terms that appear in the array factor [11]. The vector form of the active element pattern can be obtained from a Floquet analysis and the final expression becomes [11, p. 109]

$$\vec{E}_a(\theta,\phi) = \hat{\theta}\left\{jV_{00}^{TM}\sqrt{ab/\lambda_0{}^2}\right\} - \hat{\phi}\left\{jV_{00}^{TE}\sqrt{ab/\lambda_0{}^2}\cos\theta\right\} \tag{3.15}$$

In Equation (3.15), V_{00}^{TM} and V_{00}^{TE} are the modal voltages at the array aperture for the TM_{00} and TE_{00} Floquet modes, respectively. The modal voltages are functions of scan direction (θ,ϕ). The gain can be determined by normalizing the modal voltages with respect to the incident power.

As mentioned before, the above development employs S-matrix formulation, which is rigorously valid for a "finite array" defined by an infinite array with finite number of excited elements. The remaining elements should be match-terminated. In a practical finite array only the excited elements exist. In that case, Equations (3.13) and (3.14) are not rigorously valid because the edge elements are in different environments from the inside elements. As a result, their active element patterns and mutual impedance differ from that of the inside elements. However, Equations (3.13) and (3.14) could be good approximations for tapered arrays because in that situation, errors in the edge element patterns do not alter the overall array patterns significantly, particularly in the main lobe and near-in side lobe regions.

Another important point should be mentioned here. For some applications, a finite patch array may be placed over a large ground plane. In such situations, the Y-matrix formulation is more appropriate than the S-matrix formulation, because one can approximate the ground plane region with an array of short-circuited patch elements. Then, the active impedance matrix and the array patterns will have the following expressions [11]:

$$[V] = [Y]^{-1}[I] = [Z^{sh}][I] \tag{3.16}$$

$$E^{array} = E^{a}[Y_0 + Y^{FL}(\psi')][P]\{[Y] + [Y_0]\}^{-1}[V^{+}] \tag{3.17}$$

with

Y^{FL} = Floquet admittance
$[Y]$ = admittance matrix
Y_0 = source admittance.

The elements of the admittance matrix are given in (3.11) for a linear array.

3.5 Numerical Example

To illustrate the GSM approach, we consider a multilayer finite array of slot-fed patch elements. Figure 3.3 shows the element and array structures. The numbering scheme for the array elements is also shown pictorially. We computed the Floquet return loss (return loss under Floquet excitation) versus the scan angle and this is plotted in Figure 3.4. The return loss is good near the bore-sight scan region because the array was optimized with respect to the bore-sight radiation. However, a sharp resonant spike is observed near the 39 degree scan angle along the E-plane. The resonant spike causes a complete mismatch and the array ceased radiating at this scan angle. This phenomenon is known as scan blindness. The TM_0 surface wave mode, supported by the grounded dielectric structure, is responsible for this blindness. At that scan angle, the surface wave mode has a perfect phase-match with the element phase, causing a resonance. The surface wave resonance for the D-plane scan is not present because the resonant condition is not satisfied for the cell dimensions under consideration. Generally, for the H-plane scan, the resonance does not occur because the surface wave is not excited at the first place due to a polarization mismatch between the patch mode and the surface wave mode.

Figure 3.5 shows the active element pattern cuts for the array. The patterns are normalized with respect to the incident power. The active element gain is about 6.39 dBi, which is 0.23 dB lower than that of a 100% aperture-efficient element.[1] This gain loss is due to back side

[1] Using (3.15), it can be shown that for a lossless patch the "active element aperture efficiency" for bore-sight radiation should be 100% if the element spacing is less than 1 wavelength in both planes (rectangular grid case) and the array is bore-sight matched under Floquet excitation.

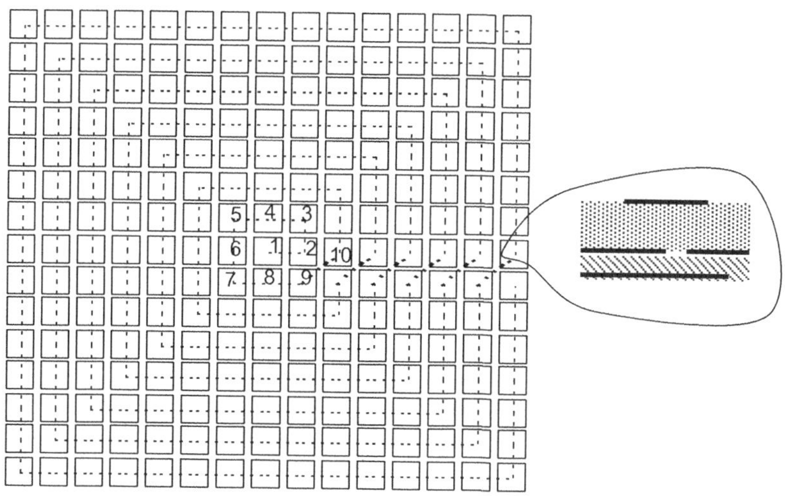

Figure 3.3 A 15 × 15 element array of slot-fed patch elements. The element numbering scheme is also shown

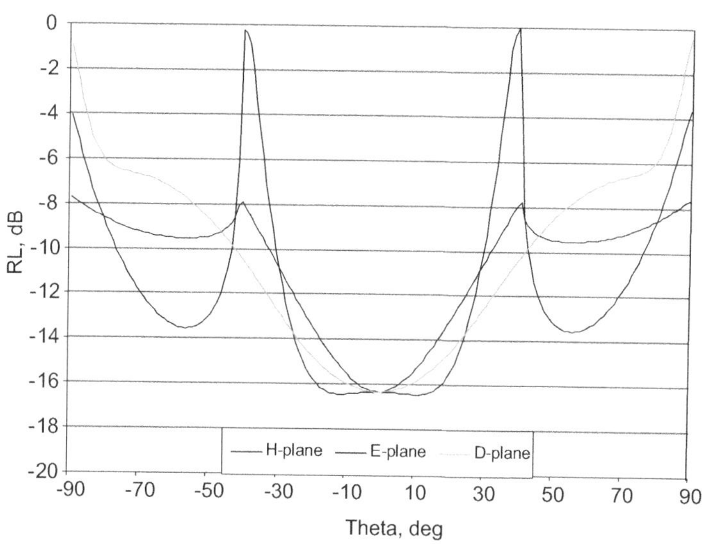

Figure 3.4 Floquet return loss of a slot-fed patch array versus scan angle. Cell size = 0.6 × 0.6, patch size = 0.24 × 0.47, slot size = 0.017 × 0.25, patch side eps = 2.53, thk. = 0.058, feed side eps = 9.8, thk = 0.026, 50 Ohms feed line. All dimensions are in wavelength in free space. Reproduced by permission of ©2006 ACES

radiation of the feed slot. The E-plane pattern has a null (blind angle) near 39 degree which is consistent with the return loss behavior. For the E and H-plane patterns, the cross-polarization components do not exist because of the symmetrical shape. The cross-polarization level is substantial at the D-plane scan, particularly near the 60-degree off-bore-sight.

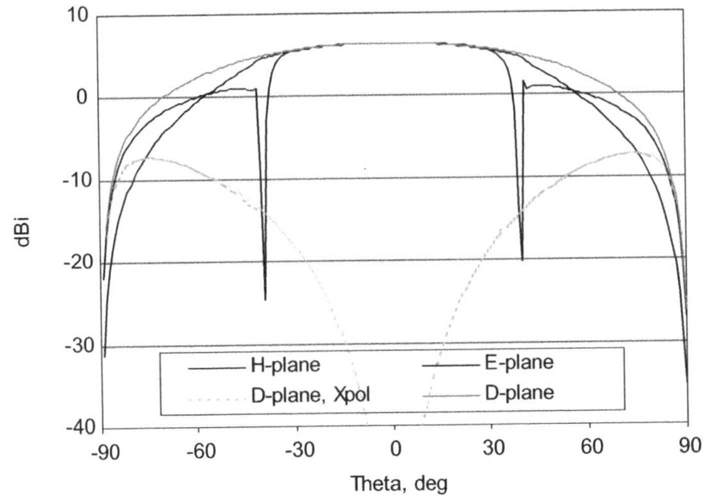

Figure 3.5 Active element pattern cuts of the array. Reproduced by permission of ©2006 ACES
Note: Element dimensions are the same as in Figure 3.4.

Figure 3.6 shows the radiation pattern of a finite array of 15 × 15 elements. Two scan angles were considered in this case. The radiation patterns of the finite array were computed using Equation (3.14). Recall that in deriving Equation (3.14) it was assumed that the non-excited elements of an infinite array are match-terminated. A Gaussian amplitude distribution with

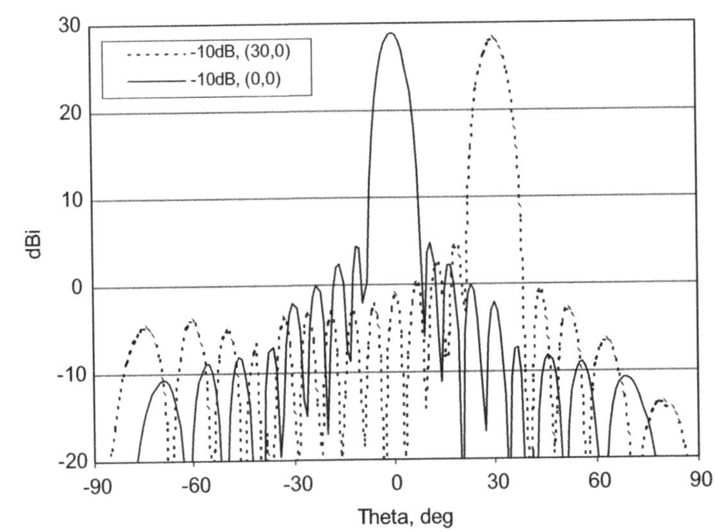

Figure 3.6 Radiation pattern of 15 × 15 element patch array. Reproduced by permission of ©2006 ACES

10 dB edge taper was assumed for both cases. For the bore-sight beam, the peak gain is about 28.95 dBi, which is about 1 dB lower than that of a uniform excitation. This 1 dB gain loss is due to the tapered distribution. The scanned beam has a peak gain of 28.06 dBi. The side lobes are 25 dB below the beam peaks for both cases.

Equation (3.12) was used to compute the mutual coupling in terms of the array scattering parameters. Figure 3.7 shows the coupling level of the array elements with respect to the center elements. The coupling with the adjacent elements is about 22 dB along both planes in this case. For a smaller element spacing this coupling is expected to be tighter. In the far region, the E-plane elements are tightly coupled than the H-plane elements.

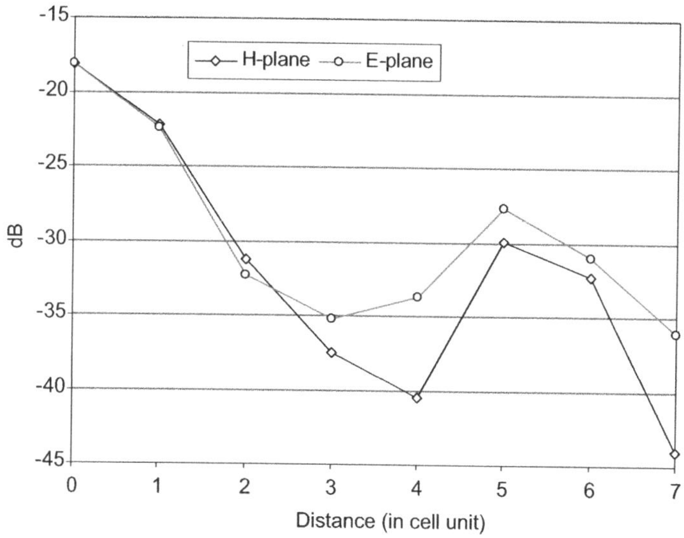

Figure 3.7 Mutual coupling between patch elements in array environment. Reproduced by permission of ©2006 ACES

Note: Patch dimensions are given in Figure 3.4.

Figure 3.8 shows the active return loss of the elements in the 15 × 15 elements patch array with uniform and tapered distributions, respectively. Three different scan angles were also considered. For the tapered array, Gaussian amplitude distributions with 10 dB edge taper in both planes were considered. For the plots, elements were numbered according to the numbering scheme depicted in Figure 3.2 (smaller numbers are associated with the elements in the central region and the bigger numbers are associated with the elements in the edge region). Four cases were considered as specified at the inset of Figure 3.8. The elements were designed to have about −16 dB bore-sight match under Floquet excitation. It is found that the active return loss for the elements in the central region does not vary significantly from one another, but the variation is quite significant for the elements near the edge. Furthermore, for the bore-sight scan, the return loss for most of the elements in the central region is about −15 dB, which is not very different from the corresponding Floquet return loss. The return losses for the elements near the edge generally differ from that of the center elements because of the edge effects.

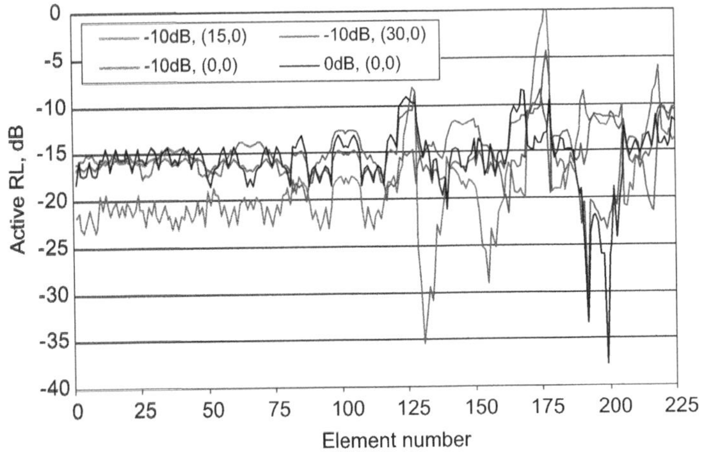

Figure 3.8 Active return loss of the 15 × 15 element patch array in Figure 3.3. Reproduced by permission of ©2006 ACES

Notes: The patch dimensions are given in Figure 3.4. The element numbering scheme is shown in Figure 3.3.

3.6 Conclusion

In this chapter we demonstrated the GSM approach to analyze a multilayer finite array. We considered a finite slot-fed patch array as an example. The mutual coupling between the elements, the active element pattern, the active return loss and the array patterns were computed and results are shown. It is found that the mutual coupling is stronger between the E-plane elements than the H-plane elements. The active return loss is substantially different for the elements near the edge, as opposed to the elements at the center region of the array.

The GSM approach is a modular approach as compared to an integrated approach. Computationally, the GSM approach for finite array analysis is much more efficient than the FEM and FDTD approaches, because the problem size of a GSM is limited to a cell only. Furthermore, the matrix size of the MoM-based GSM approach is much smaller as compared to a grid-based approach. However, a grid-based approach is much more versatile because it can be applied to non-periodic geometries also without additional complexity.

References

1. I. E. Rana and N. G. Alexopoulos, "Current distribution and input impedance of printed dipoles," *IEEE Trans., Antennas Propagat.*, vol. 29, no. 1, pp. 99–105, Jan. 1981.
2. D. M. Pozar, "Input impedance and mutual coupling of rectangular microstrip antennas," *IEEE Trans., Antennas Propagat.*, vol. 30, pp. 1191–1196, 1982.
3. A. K. Bhattacharyya and L. Shafai, "Effect of mutual coupling on the radiation pattern of phased array antennas," *IEEE APS Symp. Dig.*, pp. 891–893, 1986.
4. D. R. Jackson, W. F. Richard and A. Ali-Khan, "Series expansions for mutual coupling in microstrip patch arrays," *IEEE Trans.*, AP-37 pp. 269–274, 1989.

5. Y. Lubin and A. Hessel, "Wide-band, wide-angle microstrip stacked-patch-element phased arrays," *IEEE Trans., Antennas Propagat.*, vol. 39, no. 8, pp. 1062–1070, Aug. 1991.
6. A. Ishimaru, R. J. Coe, G. E. Miller and W. P. Green, "Finite periodic structure approach to large scanning array problems," *IEEE Trans.*, AP-33, no. 11, pp. 1213–1220, Nov. 1985.
7. A. K. Skrivervik and J. R. Mosig, "Analysis of finite phased arrays of microstrip patches," *IEEE Trans.*, AP-41, pp. 1105–1114, 1993.
8. A. K. Bhattacharyya, "Floquet modal based analysis for mutual coupling between elements in an array environment," *IEE Proc.*, MAP, vol. 144, no. 6, pp. 491–497, Dec. 1997.
9. A. J. Roscoe and R. A. Perrott, "Large finite array analysis using infinite array data," *IEEE Trans.*, AP-42, no. 7, pp. 983–992, July 1994.
10. S. K. N. Yeo and J. A. Parfitt, "Finite array analysis using iterative spatial Fourier windowing of the generalized periodic Green's function," *IEEE APS Symp. Dig.*, pp. 392–395, 1996.
11. A. K. Bhattacharyya, *Phased Array Antennas: Floquet Analysis, Synthesis, BFNs and Active Array Systems*, John Wiley & Sons, Ltd, Hoboken, NJ, 2006.
12. A. K. Bhattacharyya, "A modular approach for probe-fed and capacitively coupled multilayered patch arrays," *IEEE Trans., Antennas Propagat.*, vol. 45, no. 2, pp. 193–202, Feb. 1997.
13. A. K. Bhattacharyya, "A numerical model for multilayered microstrip phased-array antennas," *IEEE Trans., Antennas Propagat.*, vol. 44, no. 10, pp. 1386–1393, Oct. 1996.
14. A. K. Bhattacharyya, "Analysis of multilayer infinite periodic array structures with different periodicities and axes orientations," *IEEE Trans., Antennas Propagat.*, vol. 48, no. 3, pp. 357–369, Mar. 2000.
15. A. K. Bhattacharyya, *Electromagnetic Fields in Multilayered Structures: Theory and Applications*, Artech House, Boston, pp. 161–164, 1994.

4

Optimization Techniques for Planar Antennas

Rabindra K. Mishra
Electronic Science Department, Berhampur University, India

4.1 Introduction

Wireless communication has become indispensable to any technological advancement, be it biological or physical, in the twenty-first century. In all wireless systems, antenna is the front-end. The two factors influencing the design of antenna for wireless communication are: (1) performance and (2) size and cost. Planar antennas can meet most of the design criteria through proper optimization [1–3]. The optimization process often starts with a given set of specifications and an initial design. Numerical/CAD-based analyses are then used to evaluate the performance. Characteristics obtained from the analysis are compared with the desired specifications. If there is an unacceptable mismatch, then the parameters are modified in a systematic manner. The sequence of analysis, comparison with desired performance and consequent altercation of the parameters continues iteratively until optimum performance is obtained. The aim of this chapter is to illustrate the use of stochastic optimization techniques in planar antennas through examples. We will begin with the basic concepts of optimization and then move on to illustrate examples of planar antenna design using stochastic optimization techniques. The illustrative examples will include fractal antenna using RCGA and rectangular microstrip antenna using artificial neural network.

4.2 Basic Optimization Concepts

4.2.1 Cost (Fitness) Function

The optimization problem formulation usually involves minimization of a scalar objective function $E(\Theta)$, also known as error function or cost function or fitness function. Since the

Microstrip and Printed Antennas: New Trends, Techniques and Applications. Edited by Debatosh Guha and Yahia M.M. Antar

minima of a function $E(\Theta)$ correspond to the maxima of the negative of the function, i.e. $E(\Theta)$, so a proper choice of $E(\Theta)$, can convert any maximization problem to a minimization problem. $E(\Theta)$ represents the difference between the performance achieved at any stage and the desired performance. For example, in the case of a microstrip antenna, the formulation of $E(\Theta)$ may involve the specified and achieved values of the return loss in the desired band of operation.

4.2.2 Design Parameters and Space

Θ is the set of design parameters which can be modified during the optimization process. At an initial stage for microstrip antenna, they can be feeding position, length of the patch, or they could be distance between the elements in case of an array.

Usually, there are various constraints on the design parameters for a feasible solution obtained by optimization. For instance, available values of substrate dielectric constant, the minimum values of microstrip feed have practical realizable limitations on them. The elements of Θ define a space, a portion of which satisfy all the constraints and is known as the design space D. The optimization process looks for the optimum value of Θ inside D.

4.2.3 Global and Local Minima

A global minimum of $E(\Theta)$, located by a set of parameters Θ_{min}, satisfies the following relation for any feasible Θ not equal to Θ_{min}

$$E_{min} = E(\Theta_{min}) < E(\Theta) \tag{4.1}$$

Yet, finding a global minimum generally may not be guaranteed in an optimization process. However, a local minimum, which may be defined as below, will be almost always available:

$$E(\Theta_{min}) = \text{minimize } E(\Theta) \quad \text{for } \Theta \in D^{l} \tag{4.2}$$

where D^{l} is a part of D around Θ_{min}. Various remedies in this situation include restarting the optimization with another set of initial parameters, or changing to another optimization method that might be more powerful to search for the global minimum, or modifying the objective function.

4.3 Real Coded Genetic Algorithm (RCGA)

4.3.1 Genetic Algorithm

A sub-class of evolutionary computing, the Genetic Algorithm is an optimization technique based on Darwin's principle of evolution developed by John Holland in the mid-1970s. This search algorithm keeps a pool of candidate solutions. Each solution is encoded in a binary string called a chromosome with each bit being a gene. The algorithm evaluates the fitness of a solution using a selection criterion, annihilates inferior (according to the result of evaluation using the selection criteria) genes to make room for new genes and generates new chromosomes by reproduction rules for adding new genes with high fitness values into gene pool. The search process continues until the termination criteria are met. The general procedure is outlined below:

0 START: Create a random population of **n** chromosomes.

Each chromosome is a vector of genes representing a trial solution. Each gene can be a binary number, a real number or other symbols. Bit-string encoding where each gene is a binary number is the most popular approach. It develops a mapping phenotype representation, i.e. Candidate Solution (CS).

1 FITNESS: Evaluate fitness **s(x)** of each chromosome in the population.
2 NEW POPULATION

0 SELECTION: Based on **s(x)**
SELECTION CRITERIA
Two popular criteria are the Windowing and Linear Normalization:
Windowing: If s(j) is objective value of chromosome j, and C: a constant, then the fitness of chromosome j is given as:

$$s(j) = C \pm [s(j) - s(k)] \qquad where\ v(k) < v(j)\ for\ all\ j \neq k.$$

Linear Normalization: Rank objective values of chromosomes. Assign the best performed chromosome with fitness value s(best). Assign remaining j-th chromosome with fitness value

$$s(i) = s(best) - (j-k)d$$

PARENT SELECTION
Parents are selected at random with selection chances biased in relation to chromosome evaluations. This process emulates the survival-of-the-fittest mechanism in nature! In a Proportionate scheme where the growth rate of a chromosome with fitness value s(x,t) is defined as s(x,t)/S(t) where S(t) is the average fitness of the population, the following scheme is widely adopted.

Roulette Wheel Parent Selection Algorithm
Sum the fitness of all population members; named as total fitness, n.
Generate a random number between 0 and n. Return the first population member whose fitness added to the fitness of the preceding population members is greater than or equal to n.}

1 GENETIC OPERATORS: Genetic Algorithms typically use two types of operators: Crossover (Sexual Recombination), and Mutation (Asexual). Crossover is usually the primary operator with mutation serving only as a mechanism to introduce diversity in the population. However, when designing a GA to solve a problem, it is not uncommon that one will have to develop unique crossover and mutation operators that take advantage of the structure of the CSs comprising the search space.

i. RECOMBINATION: Crossover chromosomes. This greatly accelerates a search early in the evolution of a population. It leads to effective combination of schemata (sub-solutions on different chromosomes). It can be divided into two categories:

(a) Single point crossover: Given two parents, single-point crossover will generate a cut-point and recombines the first part of first parent with the second part of the second parent to create one offspring. Single-point crossover then recombines the second part of the first parent with the first part of the second parent to create a second offspring (Figure 4.1(a)).

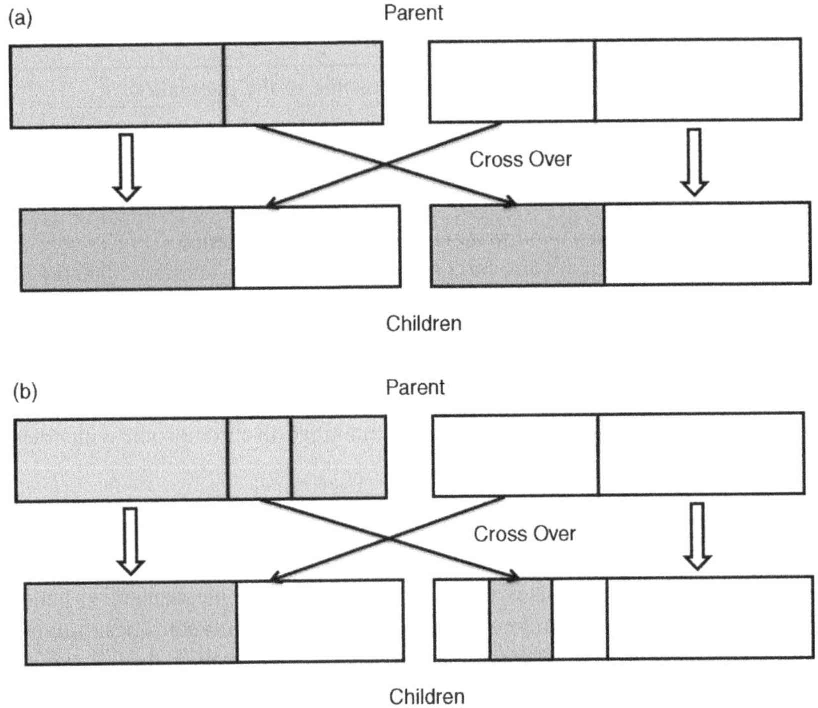

Figure 4.1 (a) Single-point crossover. (b) Multi-point crossover. Reproduced by permission of Ghatak et al. ©2009 The IET

(b) Multi-point crossover: Multi-Point crossover is very similar to single-point crossover except that multiple cut-points are generated instead of one (Figure 4.1 (b)).

ii. MUTATION: Mutate chromosomes and thereby cause movement in the search space (local or global). Restores lost information to the population. Mutation prevents the algorithm being trapped in a local minimum. In the bitstring approach, mutation will take place with a small probability. For each bit in a bit stream, a probability test is performed. If passed, then one of two methods can be used:

Method 1: That bit is flipped (0 changes to 1, and vice versa).

Method 2: Randomly generate a bit. If the randomly generated bit is different from the original bit, the original bit is flipped (Figure 4.2).

---

2 **ACCEPTATION:** Reject or accept new one.

3 **REPLACE:** Replace old with new population: the new generation. The following two strategies are common.

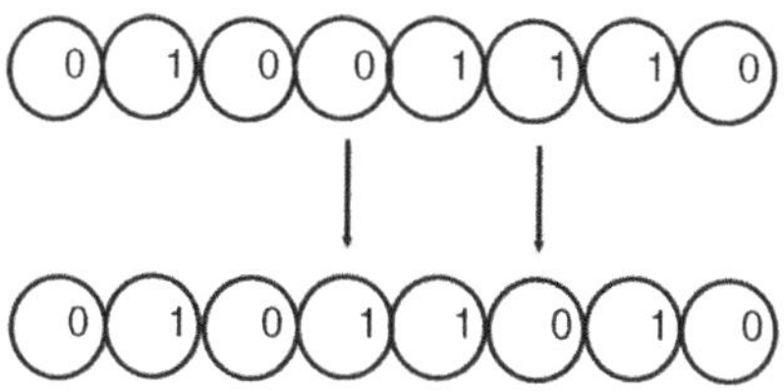

Figure 4.2 Illustration of mutation. Reproduced by permission of Ghatak et al. ©2009 The IET

Generational Replacement: Copy the best or a few of the best chromosomes into a new generation. Generate remaining new chromosome to replace current generation.
Steady State Replacement: Replace only the worst chromosomes with new chromosomes in each generation.

4 TEST: Test problem criterium
5 LOOP: Continue steps 1–4 until criterium is satisfied.

4.3.1.1 RCGA Design

Adaptation of conventional GA-based design methods inherits certain limitations. It requires the binary coding of the parameter value, which in turn decides the number of bits needed to represent the problem space. This process is time-consuming. Many variants of binary GA have been proposed to tackle such limitations [4, 5]. Due to the binary representation format, there is the common drawback of a possible wide data separation in solution space corresponding to a close separation in parameter space [6].

When the design parameters are variables in continuous domains, the optimization process eases if the differences between the *genotype* (coding) and the *phenotype* (search space) are minimal. The RCGA achieves this objective intuitively using real coded parameters, which makes the representations of the solutions very close to the natural formulation.

The parameters are coded using the chromosome mapping. A chromosome is a vector of real numbers. Its length is the vector length of the solution to the problem. Therefore, each gene represents a variable of the problem. Genetic operations are then performed on these chromosomes to generate new ones in the search space.

Chromosome Mapping
The first step in the design procedure is to identify the design parameters for optimization. The chromosomes identify an individual (*parent/child*) through Equation (4.3).

$$P^t = \{i_1^t, i_2^t, \ldots, i_n^t; s_1^t, s_2^t, \ldots, s_n^t\} \tag{4.3}$$

4.3.1.2 Genetic Operators

The efficacy of operators depends on how good a balance is made between exploration and exploitation of search space. At present, a farrago of operators is available in RCGA paradigm.

Here we consider two strategies of optimization, each with different operators. The first strategy, following Linden [7], uses *Adewuya mating* and *Gaussian mutation*. The second incorporates a new variant in GA optimization for electromagnetic by employing *Wright heuristic crossover* in conjunction with *non-uniform mutation* [6].

Non-Uniform Mutation and Gaussian Mutation

Let us assume that $P^t = \{i_1^t, i_2^t, \ldots, i_n^t; s_1^t, s_2^t, \ldots, s_n^t\}$ is the chromosome to be mutated. For the mutation operation only one chromosome is selected through the *Roulette Wheel* (RW) method. Let in the tth generation, the chromosomal value of a parameter is denoted by s_k^t, which undergoes non-uniform mutation to produce an offspring denoted by $s_k^{t'}$. Then different mutation operations shall be performed as described in the following paragraphs.

Gaussian Mutation: Let f_k be a random number drawn from a Gaussian distribution with zero mean and an adaptive variance

$$\sigma_k = \left(\frac{G-t}{G}\right)\left(\frac{s_k^t|_{Max} - s_k^t|_{Min}}{3}\right) \tag{4.4}$$

Then

$$s_k^{t'} = s_k^t + f_k \tag{4.5}$$

It can be seen that σ_k decreases when the generation counter 't' increases. Therefore, parameter tuning performed by a Gaussian mutation operator becomes finer as the generation counter 't' increases. The Gaussian mutation is very concentrated in search operation. The normal Gaussian random variable $G(0,1)$ has a density function given by Equation (4.6).

$$f_{G(0,1)}(p) = \frac{1}{\sqrt{2\pi}} e^{-\frac{p^2}{2}} \tag{4.6}$$

Yao et al. [8] had shown that the expected length of Gaussian excursion is "0.8" as proved in Equation (4.7):

$$E_{G(0,1)}(p) = E|p'_k - p_k| = 2\int_0^{+\infty} p\frac{1}{\sqrt{2\pi}} e^{-\frac{p^2}{2}} dp = \frac{2}{\sqrt{2\pi}} = 0.8 \tag{4.7}$$

Thus, in Gaussian mutation, the searching excursions are very narrow. This allows for a very rigorous search that prevents it from settling down at a strong local extreme.

Non-Uniform Mutation: Non-uniform mutation [6] provides for an exhaustive exploration of the search space at the initial generations and concentrates the search on close neighborhood regions at the later generations.

$$s_k^{t'} = \begin{cases} s_k^t + \Delta(t, Upper\_gene - s_k^t), \text{if a random number } \xi \text{ is } 0 \\ s_k^t - \Delta(t, s_k^t - Lower\_gene), \text{if a random number } \xi \text{ is } 1 \end{cases} \tag{4.8}$$

$$k \in [1, 2, \ldots, n]$$

In Equation (4.8) *upper_gene* and *lower_gene* denote respectively the upper bound and lower bound for the variable s_k. A similar formulation holds for the other chromosome that represents the pre-fractal iteration. The function $\Delta(t, y)$ is expressed as:

$$\Delta(t, y) = y.r.(1-\frac{t}{T})^b \quad (4.9)$$

In Equation (4.9), r is a uniformly distributed random number [0,1], T is the maximum number of generations and parameter b denotes the degree of non-uniformity. The value of b is usually taken as 5 [6]. It is observed from Equation (4.9) that $\Delta(t, y)$ monotonically tends to 0 as t approaches T. The expected range of parent chromosome allele, within which the child may fall after undergoing non-uniform mutation, reveals that the excursion in allele value is large at initial generations when $t \ll T$ and becomes smaller at later generations.

4.3.1.3 Heuristic Crossover and Adewuya Mating (Quadratic Crossover)

In crossover operations, the chromosomes are selected in pairs (P_a, P_b). For binary representations point crossover methods work fine, since any change in binary value results in a new continuum value. Adopting point crossover methods in RCGA practically results in no new information, since only the real values propagate from populations of one generation to the next one in different combinations.

Heuristic Crossover: A remedy in the form of Wright's heuristic crossover enhances the accuracy of the solution found by searching in the most desirable direction. For a random number "r," the child chromosome is obtained as:

$$P_a^{t+1} = P_a^t + r\left(P_b^t - P_a^t\right) \quad \text{and} \quad P_b^{t+1} = P_b^t + r\left(P_a^t - P_b^t\right) \quad (4.10)$$

Adewuya Mating: This is a single crossover operator that obtains a solution for a numerical optimization problem in the minimum number of generation with required accuracy. It produces a single individual from three different individuals by combining quadratic interpolation with heuristic extrapolation. It is defined for the maximization problem.

4.3.2 Sierpinski Gasket Fractal Microstrip Antenna Design

Antenna characteristics for the present and future wireless systems were discussed at the beginning of this chapter. They are conflicting in nature and so put antenna engineers into confusion. For example, the bandwidth is inversely proportional to antenna size while gain is directly proportional to it. So, how can one design a broadband antenna with high gain? The microstrip antennae [9], being high-Q electromagnetic structures, exhibit low bandwidth. On the other hand, fractal antennae [10] have features such as small size and multi-band characteristics. The Sierpinski gasket fractal shape is perhaps the most widely adopted geometry in later category. Empirical formulae [11] can determine the resonant frequencies of planar monopole and dipole antenna based on Sierpinski gasket geometry, Figure 4.3, with feed at the vertex. However, there are no physics-based simple closed form design-analysis formulae for such antenna, possibly due to their non-Euclidean geometry.

Thus, one constraint on simple microstrip antenna is the absence of design flexibility for multi-band operation. However, though multi-band operation is possible with a Sierpinski gasket monopole (or dipole), there is constraint on the flexibility in the choice of its feed position.

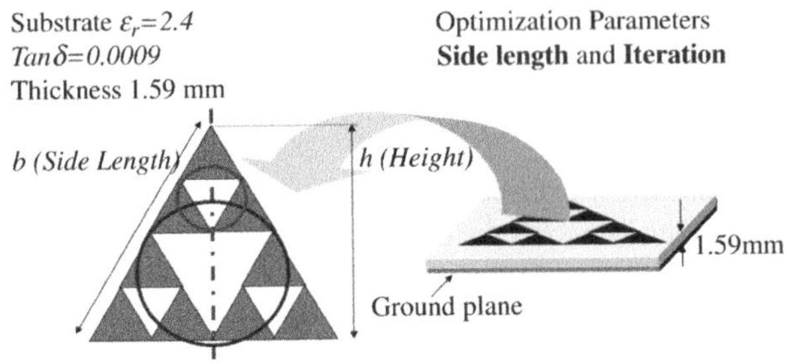

Figure 4.3 Sierpinski gasket geometry. The circles show the triangular gaps that play a major role in its behavior as antenna. Reproduced by permission of Ghatak et al. ©2009 The IET

A hybrid of microstrip and Sierpinski gasket geometry occurs as the antenna realizes a multi-band operation with open choice for feeding. It starts with the recognition that a triangular microstrip antenna is same as 0th order iterated Sierpinski gasket microstrip antenna. The patch geometry becomes non-Euclidean when the iteration number increases and then the Euclidean geometry-based antenna analysis formulations become invalid. In addition, since the structure is microstrip, the involved electromagnetism is completely different from that for the Sierpinski gasket monopole or dipole. Thus, the empirical formulae for a Sierpinski gasket monopole or dipole will also be inappropriate for such antenna.

Hence, numerical techniques [12] remain an alternative for analysis and synthesis of this proposed antenna. These methods of design optimization are time consuming and labor-intensive, and also require significant expertise and experience. Thus a design approach, using numerical techniques for the proposed antenna is extremely difficult as it involves complex electromagnetic interactions. A stochastic design approach, in recent years, has been the merger of two unconnected fields: electromagnetic optimization and GA [13], which is used for the design of such "Sierpinski Gasket Microstrip Antenna (SGMA)." This overcomes the limitations by searching the design space and automatically obtains effective design parameters to achieve performance parameters that would ordinarily not be found.

4.3.2.1 RCGS Strategy for SGMA

Designs of fractal antenna using binary GA, or its variants [14–16] are mostly meant for shape optimization of wire antennae. Here, we will discuss the strategies to implement RCGA for SGMA design optimization. Our aim is to achieve the design goal while preserving the patch shape purity.

Design of SGMA requires the determination of two important parameters, namely the side length and iteration when other parameters are fixed. They can be varied to achieve desired frequencies of operation. Hence, they are to be mapped as chromosomes in the RCGA algorithm. For example, in Equation (4.3)i_n and s_n may denote the iteration order and side length of nth individual Sierpinski gasket pre-fractal geometry of tth generation. The evolution starts by randomly generating a set of initial individuals containing side-length and iteration as two chromosomes. To generate the initial side length, the closed form formulae available for 0th order Sierpinski gasket fractal (i.e. triangular) microstrip antenna, are used. The formulae are given in Equation (4.11).

$$S = S_e - \frac{4h}{\sqrt{\varepsilon_{eff}}} \tag{4.11}$$

$$S_e = \frac{2c(m^2+mn+n^2)^{1/2}}{3f_{lmn}\sqrt{\varepsilon_{eff}}};\ l+m+n=0 \tag{4.12}$$

Fitness Function

When synthesizing a new antenna structure, an antenna design engineer normally looks for the return loss behavior. If the desired results are achieved, then one proceeds to calculate other parameters for further characterization. The fitness function is formulated to identify the desired frequencies and checking whether the return loss values are better than -10dB. This is given in Equation (4.13) which is the cost function for the RCGA algorithm.

$$S = C - RL_{f_1, f_2 \cdots f_N} \tag{4.13}$$

In Equation (4.13), C is a constant that prevents the fitness value from becoming negative and f_1, $f_2 \ldots f_N$ are the desired resonant frequencies of SGMA. The example here considers three frequencies, i.e. $N=3$. The designs with poor performances at the desired frequencies have lower fitness values and are less likely to be selected to participate in the mating process.

Design Automation in MATLAB™ with IE3D™ Link

The automation cycle starts with the synthesis strategy for SGMA design incorporating RCGA implementation in MATLAB™ program. The RCGA decides what the correct design dimensions should be for desired frequencies through iterative verifications, by linking the MATLAB™ program within IE3D™ simulation package. This provides a solution to design problems of such antenna when there are a multitude of design possibilities and completes the automation cycle. The process is depicted in Figures 4.4(a) and 4.4(b). The automation process is validated with the design of a test SGMA.

A set of an initial population of chromosomes is generated, using MATLAB, consisting of the coordinate points of the Sierpinski pre-fractal geometry using IFS [17]. To evaluate, the performance of each evolved SGMA, a MoM-based full wave electromagnetic simulation package, the IE3D™ is used [17]. The MATLAB-generated coordinates need to be formatted as per the requirement of the IE3D *geometric file format*. A separate code arranges the coordinate information in the way IE3D™ understands [18] and stores them in a file to be exported to IE3D™ for geometric information of the antenna structure during simulation.

Each evolved structure, in MATLAB, is automatically exported to IE3D™ along with a simulation file that contains the details of the supporting information, such as frequency range of simulation, mesh size, cells per wavelength, that should be used to generate the mesh and type of matrix solver to be used to accurately evaluate a single population member in the least possible time [17]. The frequency points were simulated by enabling adaptive techniques provided in the IE3D™ simulation engine. The information of return loss (RL) is extracted, with a MATLAB™ function developed to assist the evaluation process, after simulation of each entity of a population. The fitness is evaluated by a separate code written in the form of a MATLAB™ function that takes the return loss data as its input. The process is depicted in Figure 4.5.

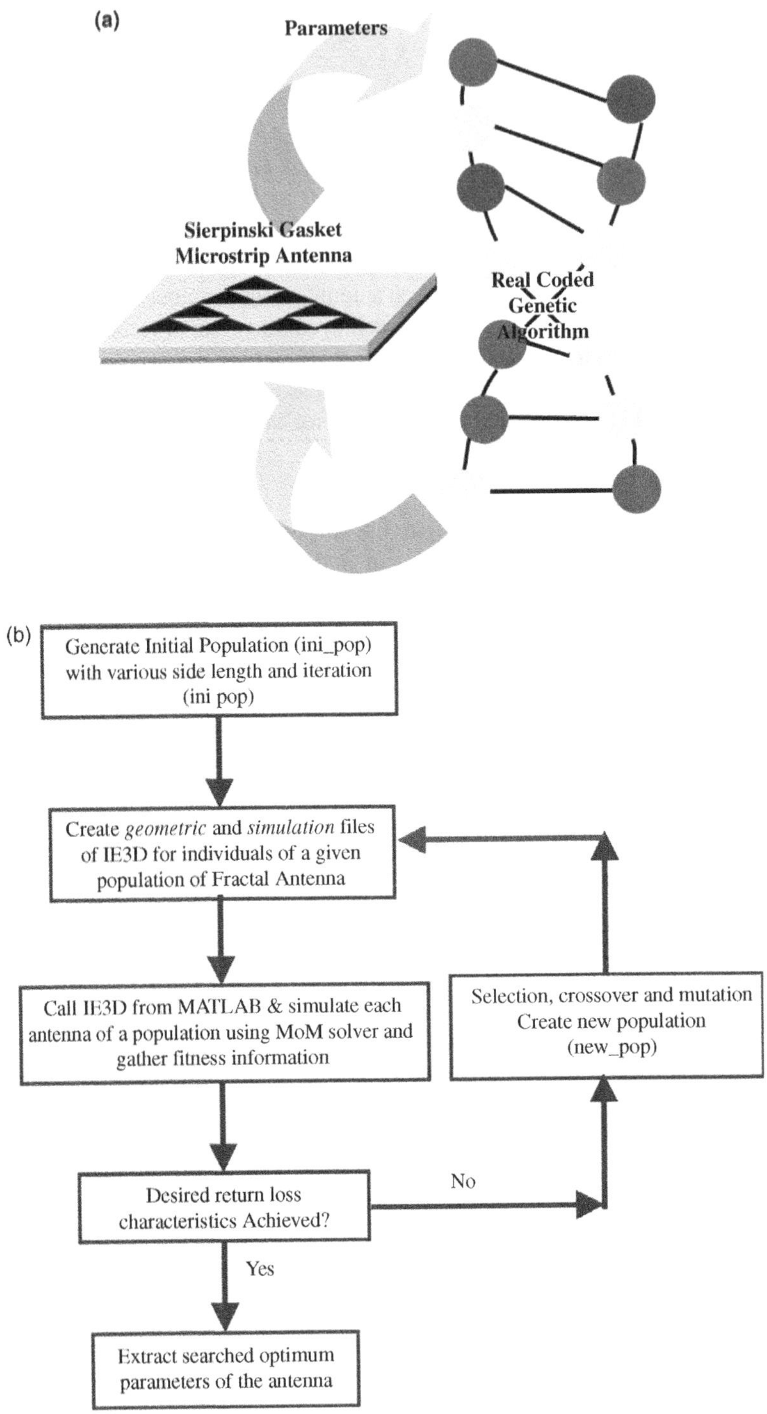

Figure 4.4 (a) RCGA optimization schematic for Sierpinski gasket antenna. (b) Process flow of the RGA optimization in conjunction with IE3D™. Reproduced by permission of Ghatak et al. ©2009 The IET

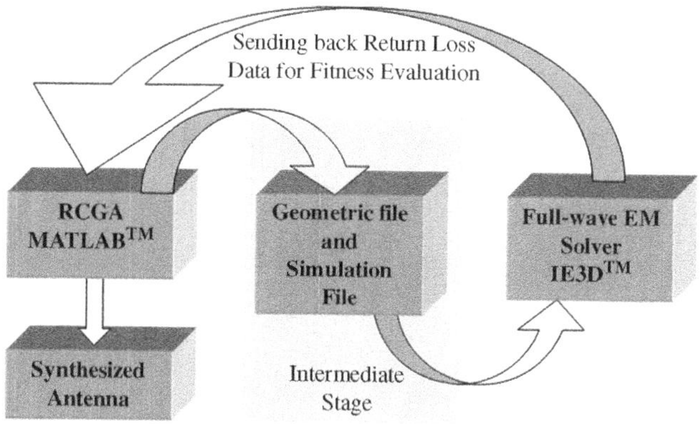

Figure 4.5 An illustration of the RCGA implementation depicting the link between IE3D™ and MATLAB™. Reproduced by permission of Ghatak et al. ©2009 The IET

Design of SGMA

Let us consider the design frequencies of 4.56, 7.51 and 11.78 GHz for the SGMA. The substrate for realizing the planar antenna has $\varepsilon_r = 2.4$ and loss tangent of 0.0009. The thickness of the substrate is 1.59 mm. The range of the allele is chosen to be 35 mm to 65 mm for the side length. Up to three pre-fractal iterations are taken. This choice of fixing the range of allele derives from the awareness of well-established design equations of an equilateral triangular microstrip patch antenna [8]. Without that knowledge, any range of positive values for side length can be given. That will put a constraint on the searching algorithm and consume an enormous amount of time in terms of the number of GA generations required for convergence. The choice of iteration is governed by the information available from work on Sierpinski gasket monopoles and dipoles reported in [19, 20]. The reason for this is the following: A fractal curve is generated after infinite iterations, but in practice that would be impossible to achieve for antenna fabrication. Thus, the number of iterations is truncated to generate pre-fractal curves. So designs until a certain stage of fractal iteration are taken that can be utilized fruitfully as antenna as far as the desired radiating frequencies are concerned.

For the Gaussian mutation, the probability of crossover, P_c is taken as 0.7 and the probability of mutation P_m as 0.01. For the non-uniform mutation, the values for P_c and P_m are chosen to be 0.7 and 0.02 respectively. The population size was taken to be 30 for both. The standard deviation in Gaussian mutation is taken to be 10% of the allele range.

The possible combinations of side length and iteration depicted as scatter plots and corresponding set of return loss plots that constitute the initial population are shown for both strategies in Figure 4.6.

The final population members give the set of desired antenna side length and the iteration is shown in Figure 4.7.

The best curves so far for both strategies are shown in Figure 4.8. The first strategy converged at 72 generations. The second strategy is found to converge a few generations earlier at 64 generations.

The values of side length and iteration for five runs of RCGA are given in Table 4.1. It is to be noted that iteration is one for all the population members. As RCGA gives a set of

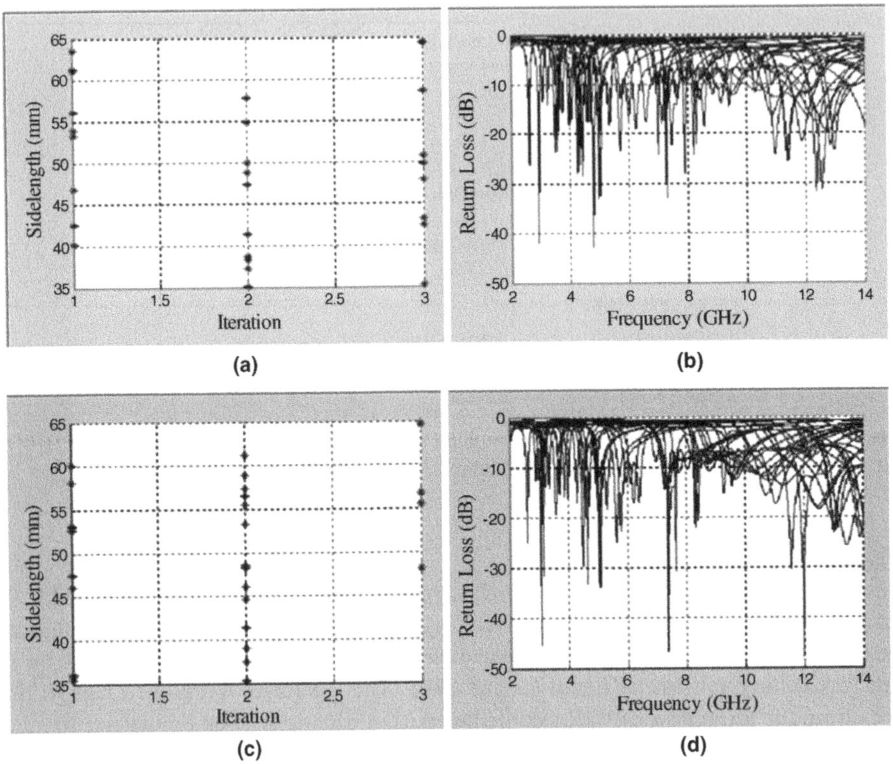

Figure 4.6 Combinations of side length and iteration depicted as scatter plots for initial population in (a) Gaussian mutation and (c) non-uniform mutation. Corresponding set of return loss plots (b) Gaussian mutation and (d) non-uniform mutation. Reproduced by permission of Ghatak et al. ©2009 The IET

values as possible solutions of side length for the design problem, so an average value in each run is tabulated. All the solution set for side length gives the same return loss plot, as the average differences between maximum and minimum value for five different runs is 0.18 mm for Gaussian mutation and that for non-uniform mutation it is 0.06 mm. For Gaussian mutation, this difference is $0.00274\lambda_1$, where λ_1 is the wavelength corresponding to the lowest resonant frequency, which is 4.56 GHz and is $0.00707\lambda_3$ that corresponds to the wavelength of the highest frequency 11.78 GHz. When non-uniform mutation is considered, the difference of 0.06 mm is $0.00091\lambda_1$ and $0.00236\lambda_2$. This small variation does not have any effect on the return loss data, which can be observed in Figure 4.7(b) and (d). The optimized antenna is further verified using a 3D electromagnetic solver CST Microwave Studio™.

A fabricated prototype, as shown in Figure 4.9, of the optimized antenna was tested using a HP8722C VNA. This prototype is of iteration 1 and the side length was 60.01 mm. A good agreement is obtained between the return loss plots evolved by RCGA and simulated in IE3D™, simulated by CST Microwave Studio™ (CST MWS) and the experiment is shown in Figure 4.10.

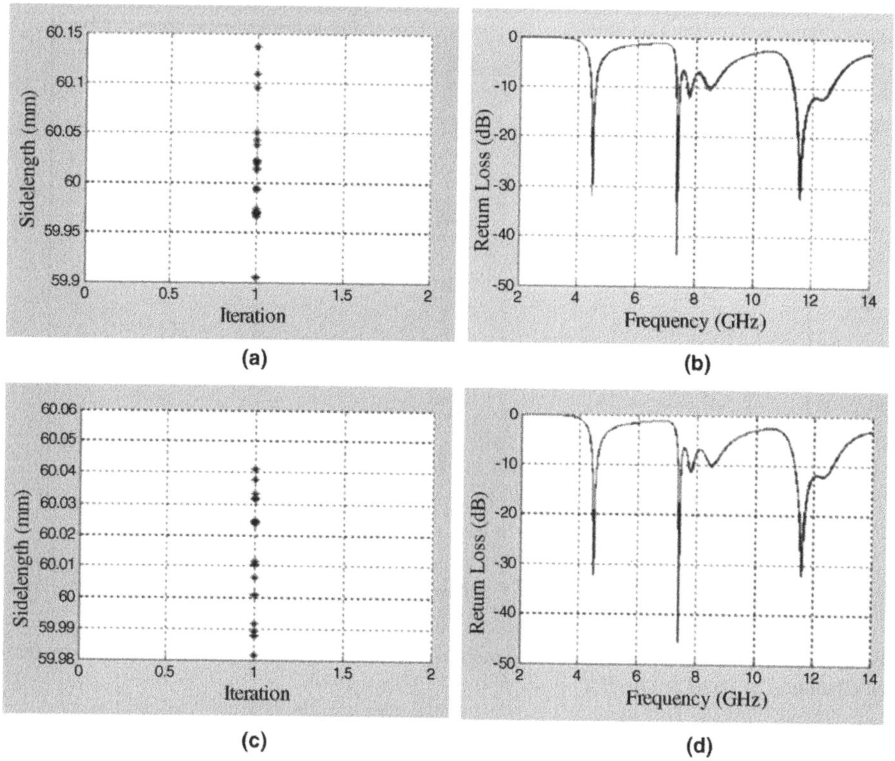

Figure 4.7 Combinations of side length and iteration depicted as scatter plots in (a) Gaussian mutation and (c) non-uniform mutation. Corresponding set of return loss plots, (b) Gaussian mutation and (d) non-uniform mutation for best-fit individuals in the final population. Reproduced by permission of Ghatak et al. ©2009 The IET

Figure 4.11 on the left-hand side, from top to bottom, shows the E_ϕ for $\theta = 90°$ at frequencies 4.56 GHz, 7.51 GHz and 11.78 GHz respectively. Sequentially for the same frequencies the E_θ for $\phi = 90°$ are shown in Figure 4.11 on the right-hand side, from top to bottom, respectively.

These simulation results are in good agreement with a tested prototype antenna. Undulations in radiation patterns are observed with increasing frequency. Thus the pattern is consistent in any given band, but changes as the frequency band changes. This reiterates the fact that the Sierpinski gasket fractal antenna, without any perturbation of the geometry, is multi-band but is not frequency independent. The design strategy can be further extended to other geometries in the fractal antenna paradigm.

4.4 Neurospectral Design of Rectangular Patch Antenna

The Spectral Domain Technique is a frequency domain numerical method with many unique features: (1) algebraic equations are solved instead of integral equations; (2) use of Galerkin's method results in a solution equivalent to that based on the variational expression, for which

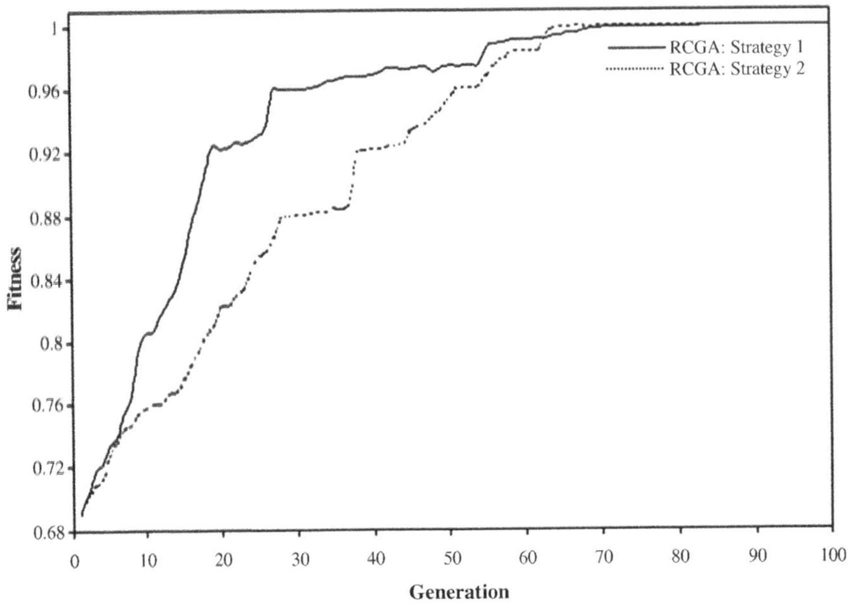

Figure 4.8 Average fitness plotted against generations for both the strategies. Reproduced by permission of Ghatak et al. ©2009 The IET

Table 4.1 Tabulated average value of side length and iteration for five runs of RCGA for both strategies

	Strategy 1		Strategy 2	
GA Run	Side length (mm)	Iteration	Side length (mm)	Iteration
1	60.0224	1	60.0115	1
2	60.0126	1	60.0137	1
3	60.0237	1	60.0026	1
4	59.9993	1	60.0034	1
5	60.0435	1	60.0108	1

Source: Reproduced by permission of Ghatak et al. ©2009 The IET

accuracy can be systematically improved by increasing the size of the matrix used; (3) numerical processing is not significantly affected due to increased structural complicacies such as the use of a multilayered substrate; and (4) unambiguous convergence of the solution and the confirmation of the mode since the physical nature of the solution is incorporated in the solution process. But its heavy computing time restricts its application in CAD tools compared to other computational electromagnetic methods. This section will discuss how an appropriately trained artificial neural network can overcome this limitation. A CAD model has two players, namely the developer and the user. The job of the developer is to create a model which can give highly accurate results with minimum computing time. There is negligible restriction on the model development time. On the other hand, the user is mostly constrained to come out with designs in a limited time period. The neurospectral technique is based on this philosophy.

Figure 4.9 Fabricated prototype of the antenna evolved using RCGA with side length 60.01mm and iteration 1. Reproduced by permission of Ghatak et al. ©2009 The IET

4.4.1 Model Development

4.4.1.1 Spectral Domain Formation

The surface electric field and the distribution of current for a rectangular microstrip antenna (Figure 4.12) are related as [11].

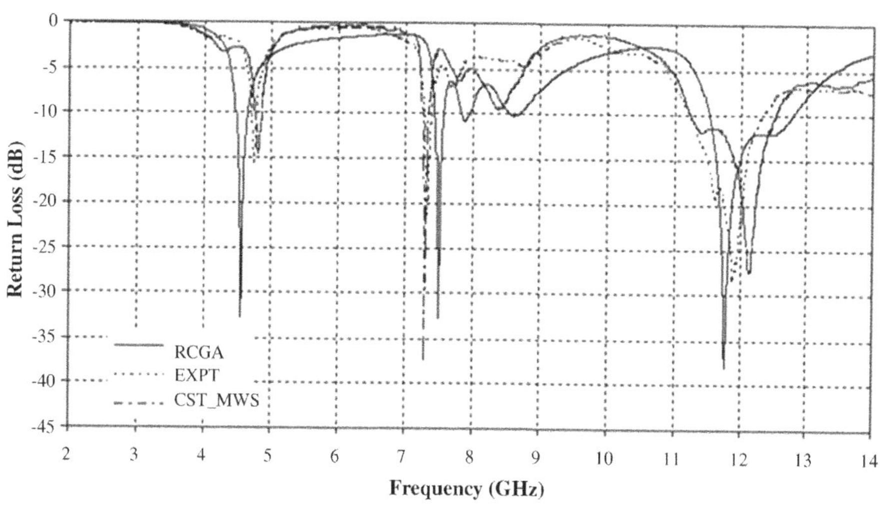

Figure 4.10 Return loss plot of the RCGA-evolved SGMA using IE3D™ compared with CST Microwave Studio™ and the experiment. Reproduced by permission of Ghatak et al. ©2009 The IET

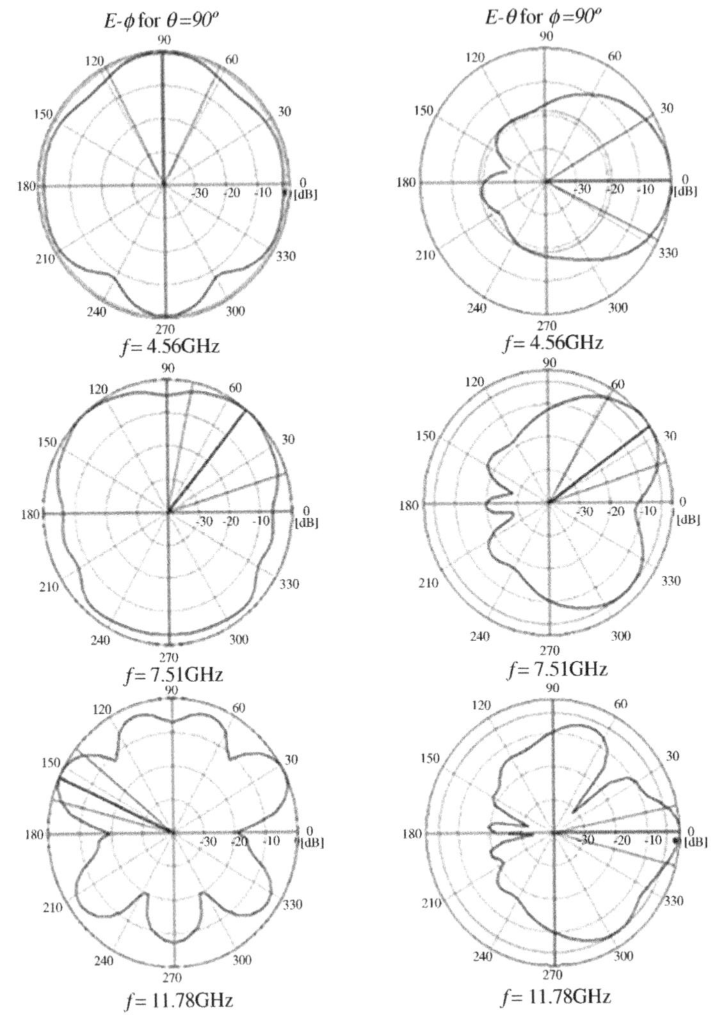

Figure 4.11 **E**-ϕ for $\theta = 90°$ at frequencies 4.56 GHz, 7.51 GHz and 11.78 GHz from top to bottom in the left hand side and **E**-θ for $\phi = 90°$ in the right hand side with the same sequence. Reproduced by permission of Ghatak et al. ©2009 The IET

$$\begin{bmatrix} \tilde{E}_x(k_x, k_y) \\ \tilde{E}_y(k_x, k_y) \end{bmatrix} = \left[\tilde{\mathbf{Z}}(k_x, k_y)\right] \begin{bmatrix} \tilde{J}_x(k_x, k_y) \\ \tilde{J}_y(k_x, k_y) \end{bmatrix} \tag{4.14}$$

$\kappa_{\zeta i} = \kappa_0^2 \varepsilon_i - (\kappa\xi^2 + \kappa\psi^2)$, $i = 1, 2$, $k_0^2 = \omega^2 \varepsilon_0 \mu_0$, $\varepsilon_1 = \varepsilon_\rho$, $\varepsilon_2 = 1$, $\eta_1 = \eta$, $\eta_2 = \infty$

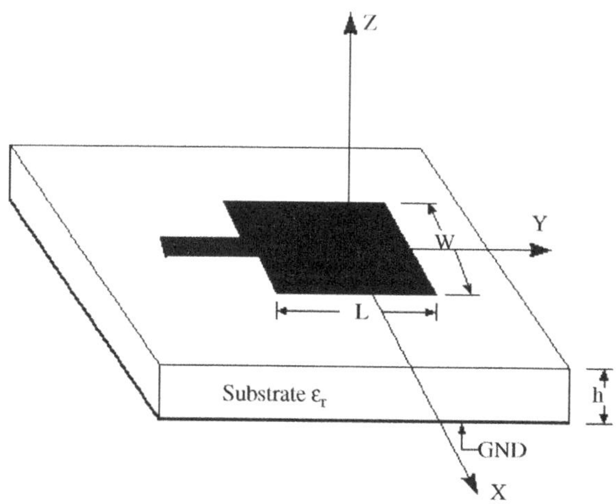

Figure 4.12 Microstripline-fed rectangular patch antenna. Reproduced by permission of Patnaik and Mishra ©2003 IEEE

Method of moment along with the modal representation for $\tilde{J}$ puts the characteristic equation in the form

$$\begin{aligned}\sum_{m=1}^{M} G_{pm}^{xx} c_m + \sum_{n=1}^{N} G_{pn}^{xy} d_n &= 0 \qquad p = 1, 2, \cdots, M \\ \sum_{m=1}^{M} G_{qm}^{yx} c_m + \sum_{n=1}^{N} G_{qn}^{yy} d_n &= 0 \qquad q = 1, 2, \cdots, N\end{aligned} \tag{4.15}$$

$$G_{jq}^{ip} = \int_{-\infty}^{\infty}\int_{-\infty}^{\infty} \tilde{J}_{ij}(k_x, k_y)\tilde{Z}_{rs}(k_x, k_y)\tilde{J}_{pq}(k_x, k_y)dk_x dk_y \tag{4.16}$$

$$(i = x,\ y;\ p = x,\ y;\ j = m,\ n;\ q = m,\ n)$$

$$\begin{aligned}\tilde{Z}_{pp}(k_x, k_y) &= \frac{k_p^2 k_{z2} k_{z1} \sin k_{z1} h}{\beta^2 T_m} + \frac{k_0^2 k_q \sin k_{z1} h}{\beta^2 T_e} \qquad p = x, y; \quad q = y, x \\ \tilde{Z}_{xy}(k_x, k_y) = \tilde{Z}_{yx}(k_x, k_y) &= \frac{k_x k_{z2} k_y \sin k_{z1} h}{\beta^2 T_m} + \frac{k_0^2 k_y k_x \sin k_{z1} h}{\beta^2 T_e} \\ T_m &= \varepsilon_r k_{z2} \cos k_{z1} h + j k_{z1} \sin k_{z1} h \\ T_e &= \varepsilon_r\, k_{z1} \cos k_{z1} h + j k_{z1} \sin k_{z1} h \\ \beta^2 &= k_x^2 + k_y^2\end{aligned} \tag{4.17}$$

The determinant of "***G***" vanishes for a non-trivial solution of Equation (4.15) at a complex frequency $f = f_r + j\, f_i$. The antenna resonant frequency is the real part of this complex frequency. Using spherical transformation, followed by a substitution of the form $H(.) = e^{(\varphi + R)}\tilde{J}_{ij}(\varphi, R)\tilde{Z}_{rs}(\varphi, R)\tilde{J}_{ij}(\varphi, R)R$, and using $r = exp\ (-R) - 1/2$ and $\varphi = \varphi - \pi$, a new function $F(L, W, h, \varepsilon_r, k_0, \varphi, r)$ given as $-e^{-\varphi+\pi}H(L, W, h, \varepsilon_r, k_0, \varphi-\pi, -ln(r-\frac{1}{2}))ln\ r$ replaces the integrand, and hence Equation (4.16) becomes

$$G_{jq}^{ip} = G_r + jG_i = \int_{-\pi}^{\pi}\int_{-1/2}^{1/2} (F_r(.) + j\,F_i(.))d\phi dr \quad (4.18)$$

$F_r(.)$ and $F_i(.)$ are respectively the real and imaginary parts of the integrand $F(.)$. Similarly, G_r and G_i are real and imaginary parts of G_{jq}^{ip}.

4.4.1.2 Artificial Neural Network Solution Technique

Using the following proposition, an artificial neural network can modify the integrand of Equation (4.18) for closed form expressions, avoiding the numerical integration of (4.18) that contains singularities in its path.

Proposition 4.1 If $f(x, y)$ is a function with singularities, at (x_i, y_i), in the ranges of x and y, and $g(x, y)$ is a continuous function such that $g(x, y) = f(x, y)$ at all non-singular points of $f(x, y)$, then $f(x, y)$ can be expressed in terms of $g(x, y)$ as

$$f(x,y) = g(x,y) + \sum_{i,j} g(x,y)\delta(x-x_i)\delta(y-y_j) \quad (4.19)$$

ANN Model for Integrand

A simple three-layer feed forward network [31], with eight input layer neurons (X_i: $i = 1, \ldots, 8$) representing, respectively, $L, W, h, \varepsilon_r, k_0, \varphi, r$ and an unity bias term can replace the integrand. The neurons in the output layer are twice the number of matrix elements in (4.16) corresponding to the real and imaginary parts of each complex matrix element. The kth output neuron corresponds to the mn^{th} matrix element through the relation $k = (m-1)N + n$ for the $(N \times N/2)$ matrix. The number of neurons, P, in the hidden layer depends on many factors, such as problem complexity, experience of the network developer, etc. Using linear transfer function $\Phi(x) = x$ for the output layer and transfer function $\psi(x) = tan\ h\ (x)$ for the hidden layer, the output at the kth neuron is

$$y_k^\mu = \phi\left(\sum_{j=1}^{n_h} v_{kj}\psi\left(\sum_{i=1}^{n_i} w_{ji}X_i^\mu\right)\right) = \phi\left(\sum_{j=1}^{n_h} v_{kj}\hbar_j\right), \quad \hbar_j = \psi\left(\sum_{i=1}^{n_i} w_{ji}X_i^\mu\right) \quad (4.20)$$

Here, μ indicates the pattern number (i.e. the data set identification number in the domain of the complete data, being used for training). v_{kj} and w_{ji} represent the weights connecting the jth-hidden layer to the kth output and the ith input, respectively. The numbers of nodes in the input, hidden and output layers are n_i, n_h, and n_o respectively.

The continuous output "y (.)" of a network is usually bounded in the range $[-1, 1]$. For the training of the network, the values of "y (.)" are found by a transformation of the continuous function "g (.)" that transforms to the integrand function at the non-singular points. The form of the transformation used to obtain the sampled values "g (.)" of the continuous function, from the network output y, is $g = |g_H|\frac{1-2y}{1+2y}$. $|g_H|$ is the highest sampled magnitude. The sampling process is described later. The Karayiannis [32] fast algorithm, using a generalized objective function, updates the ANN weights.

$$\begin{aligned} v_{kj,new} &= v_{kj,old} + \alpha\, \varepsilon_k^0(\lambda)\hbar_j \\ w_{ji,new} &= w_{ji,old} + \alpha\, \varepsilon_j^h(\lambda)x_i \end{aligned} \tag{4.21}$$

$$\varepsilon_k^0(\lambda) = \lambda\, e_k + (1-\lambda)\tanh[\eta\, e_k] \tag{4.22a}$$

$$\begin{aligned} e_k &= Y_k^\mu - y_k^\mu \\ \varepsilon_j^h(\lambda) &= (1-(\hbar_j)^2)\sum_{k=1}^{n_0}\varepsilon_{kj}^0(\lambda)v_{kj} \end{aligned} \tag{4.22b}$$

$$\lambda = exp(-\tau\ /E^2) \tag{4.23}$$

$$E = \frac{1}{2}\sum_{\mu}^{M}\sum_{k}^{n_o} {e_k}^2 \tag{4.24}$$

Y_i^μ is the value of the function being integrated. M is total number of data sets (i.e. patterns) being presented to the network. The learning rate parameter, momentum and learning rate adaptation are respectively α, η, and τ.

4.4.1.3 Closed Form Expressions for Integration

After the completion of training, Equations (4.19) and (4.20) are used in Equation (4.18) and the following closed form expressions are obtained for the integration:

$$G_{jq}^{ip} = \int_{-\pi}^{\pi}\int_{-\frac{1}{2}}^{\frac{1}{2}} g_k dX_6 dX_7 + \sum_{N} g_k(\text{Calculated at each of the N singular points}) \tag{4.25}$$

$$\int_{-\pi}^{\pi}\int_{-\frac{1}{2}}^{\frac{1}{2}} g_k dX_6 dX_7 = \frac{-0.5}{w_6 w_7}\left(\sum_j\left(\sum_{r=1}^{4} T_r\right)\right) \tag{4.26}$$

$$f_1 = \exp(w_6)\exp(6.2832w_7)\exp\left(2\sum_{j\neq 6,7} w_j X_j\right) \tag{4.26.1}$$

$$f_2 = \exp(-w_6)\exp(6.2832w_7)\exp\left(2\sum_{j\neq 6,7} w_j X_j\right) \tag{4.26.2}$$

$$f_3 = \exp(w_6)\exp(-6.2832w_7)\exp\left(2\sum_{j\neq 6,7} w_j X_j\right) \tag{4.26.3}$$

$$f_4 = \exp(-w_6)\exp(-6.2832w_7)\exp\left(2\sum_{j\neq 6,7} w_j X_j\right) \tag{4.26.4}$$

$$T_1 = 1.3\begin{pmatrix} \ln(3f_2-1)\ln(3f_2)-\ln(3f_1-1)\ln(3f_1) \\ -di,\log(3f_1)+di,\log(3f_2) \end{pmatrix} \tag{4.26.5}$$

$$T_2 = 0.2\begin{pmatrix} di,\log(f_1+1)-di,\log(f_2+1)+di,\log(-f_1)-di,\log(-f_2) \\ -\ln(-f_2-1)\ln(-f_2)+\ln(-f_1-1)\ln(-f_1)+1.885w_6w_7 \end{pmatrix} \tag{4.26.6}$$

$$T_3 = 1.3\begin{pmatrix} -\ln(3f_4-1)\ln(3f_4)+\ln(3f_3-1)\ln(3f_3) \\ +di,\log(3f_3)-di,\log(3f_4) \end{pmatrix} \tag{4.26.7}$$

$$T_4 = 0.2\begin{pmatrix} di,\log(f_4+1)-di,\log(f_3+1)+di,\log(-f_4)-di,\log(-f_3) \\ -\ln(-f_3-1)\ln(-f_3)+\ln(-f_4-1)\ln(-f_4)+1.885w_6w_7 \end{pmatrix} \tag{4.26.8}$$

Using these closed form expressions, all the elements of the matrix G are determined simultaneously at a desired frequency. Hence, the determinant of |G| is evaluated with reduced computational time and effort. With estimates of the resonant frequencies for antennae, a second ANN relates the determinant value (real and imaginary) of |G| along with the patch width, substrate thickness, permittivity and permeability and frequencies to the lengths of the antennas. The resonant frequencies for each antenna should be distributed around their estimated resonant frequencies. Once the network has adopted itself to the pattern that relates the patch dimension and substrate properties with frequencies and value of |G|, then reverse training [27, 30] will determine the patch length.

4.4.1.4 Data Generation and Pre-processing

The parameters, for generating network training data, include the frequency band being considered, the dielectric constants and thickness of the substrates available. From these parameters, the length of the antenna at the minimum and maximum frequencies can be estimated using the simple formula $L = C/(2f\sqrt{\varepsilon_r})$. This has an accuracy of less than 3%. The maximum value of L is at the lowest frequency with the lowest dielectric constant value. So we can increase this value of L by 5% to get L_{max}. Similarly, at the lowest frequency end, with highest dielectric constant, we reduce the length by 3% to get L_{min}. Now the design value of L lies in the range $L_{min} \leq L \leq L_{max}$. This range is then divided into five arbitrary parts to give L_j ($j = 1, 5$). For each combination of L_j, ε_r and h, a complex resonant frequency using Cavity model [1] is estimated. Using the real part of this frequency the patch width for that combination using the formulae of Bhal et al. [2] is estimated. Now, as in case of the length, the width W also lies in a range $W_{min} \leq W \leq W_{max}$. The band $(W_{max} - W_{min})$ is 10% of W. This range is divided into five arbitrary parts to give W_l ($l = 1, 5$).

The limits on r and φ are known to be from $-1/2$ to $+1/2$ and $-\pi$ to $+\pi$. For each set of L, W, h and ε_r, the k_o is complex since the resonant frequencies are complex. Twenty data points each for r and ϕ covering their ranges, including the end points will generally be sufficient. When choosing these points, care must be taken to ensure that singular points are properly avoided in order to form the required continuous function $g(.)$. For proper training, these data will be scaled, using $y(.) = \frac{1}{2}\frac{g_H(.) - g(.)}{g_H(.) + g(.)}$, to the range $[-1, 1]$. These will be used as the network output, which is a standard technique [34] for a continuous function approximation using ANN. Thus, $4000 \times N_{fr}$ numbers of input data sets, for N_{fr} frequency points will be obtained. Each input data set contains L, W, h, ε_r, k_0, r, φ and 1. Of these data sets 60% is be taken for training and the rest for testing. These selections are arbitrary and each set will span the complete range of data.

The data, for network output, should also be pre-processed prior to submission for the training. The complete data set can be arranged in the form of a matrix. Randomly 50% of the rows from this matrix should be taken out to form another matrix. Thus two matrices can be obtained for training. The rows of the first matrix can be interchanged, so that they are sorted in descending order of the value of the output, i.e. F (.). Similarly, the rows of the 2nd matrix will be interchanged, so that they are in the ascending order of the output. The data matrices are presented for training one after another. After the training of the first network, Equation (4.19) will give the integration results. For each selected complex frequency, the integration for all the matrix elements can be obtained in one run only. Thus the value of the determinant for each data set can easily be obtained.

4.4.2 Model Implementation

4.4.2.1 Simple Patch Antenna

Let us now discuss the network construction and reverse modeling for a simple microstrip line-fed rectangular patch antenna (Figure 4.12).

Network Model

A three-layer-feed-forward network is formed which relates the input (patch length "L" and unity bias term) and outputs (W, h, ε_r, f, $|G_r|$ and $|G_I|$). The network inputs x_I ($I = 1, 2$) correspond respectively to patch length L and a unit bias term. The network outputs y_k ($k = 1, \ldots, 6$) correspond respectively to W, h, ε_r, f, $|G_r|$ and $|G_I|$. So, once a program is written for the first network, it can be copied for the second network and necessary changes in the parameters can be made in the second program.

Data Generation

After the determinant is found for each data set, in the 1st stage, the data sets are rearranged to generate a new combination. In this combination, the inputs are L_j and 1. The outputs are W_l, h, ε_r, $|G_r|$, $|G_i|$ and f (in GHz). Here, $|G_r|$ and $|G_i|$ are the values of the determinant corresponding to frequency f. In this second phase, the aim is to find the patch length by reverse training. For each W_l, there are only five values of L, which are closely spaced. So, there is no need for pre-processing. However, it is necessary to scale (and descale) both the input and output data with respect to their highest value.

Reverse Modeling

After completion of the training for the second network, we use reverse training. In this process, the weights obtained in the previous section are kept constant throughout. The input x_1 (= L) (instead of the weights) changes itself to adapt to the desired output. (It may be noted here that in the normal training process, the connecting weights are updated to map the inputs to the outputs.) The problem here is given a set of parameters such as patch width, resonant frequency, etc., determine L so that $|G_r| = |G_i| = = 0$. The update equation for the input L, using the fast algorithm is

$$L_{new} = L_{old} + \alpha\, \varepsilon_j^h(\lambda) \sum_{j=1}^{n_h} w_{j1} \tag{4.27}$$

Figure (4.13) compares the relative errors between the neurospectral technique and other techniques defined as $100(L_{NS}-L_{OTH})/L_{NS}$, where L_{NS} is the length obtained using neurospectral method and L_{OTH} is the length obtained from the technique with which it is being compared. The antennae are on epoxy substrate ($h = 0.16$ cm and $\varepsilon_r = 4.5$) in the frequency band of 1.5–3.0 GHz with sample frequencies 0.375 GHz apart starting at 1.5 GHz. The continuous curve gives the result for the relative error between the cavity model and the neurospectral technique. It increases linearly with frequency, i.e. the substrate's electrical thickness. The circular scattered points are for the experimental results. As expected, they do not follow any regular pattern. This is due to various factors such as tolerance effects, and calibration errors (if any), etc. In all the experiments, the error is less than 4%, whereas for the cavity model the lowest value is around 5%. The triangular scattered points are for the spectral domain. The errors are less than 1.4%.

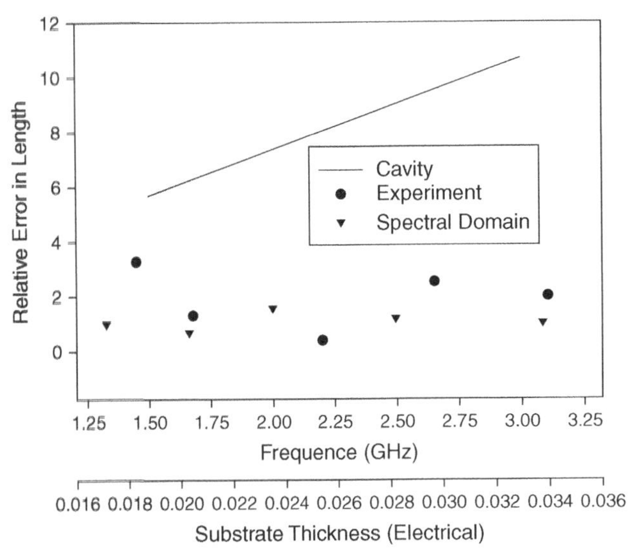

Figure 4.13 Comparison of errors in results relative to the neurospectral method for antenna on substrate 1. Reproduced by permission of Patnaik and Mishra ©2003 IEEE

Figure (4.14) compares the relative errors when a PTFE substrate of dielectric constant 2.33 and thickness 0.3175cm is used in the 2 GHz to 9 GHz band with sampling at 0.5 GHz starting at 2 GHz. The network structure remains the same as that for the first substrate. The frequency band is substantial here. So it is easy to study the validity of the method for electrically thin and thick substrates. In Figure 4.14, the error due to cavity model is nonlinearly increasing with frequency, i.e. for the substrate electrical thickness (the solid curve). An expanded view for the errors in comparisons to the HFSS (circular scattered points) and the spectral domain (triangular scattered points) is given by taking separate y axes for them.

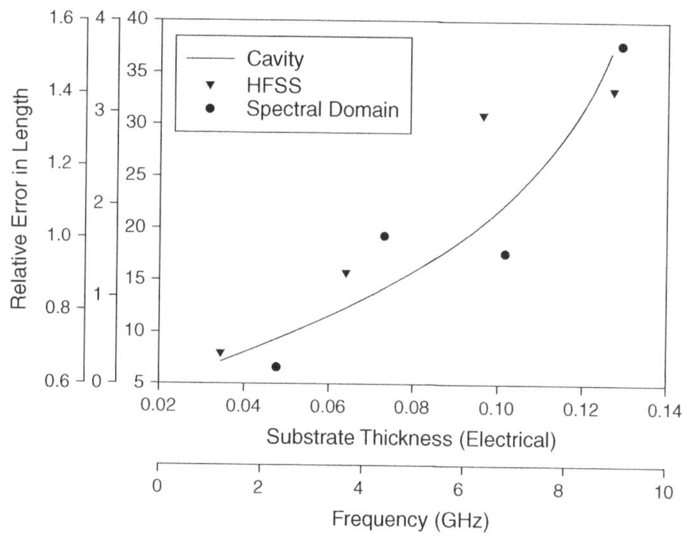

Figure 4.14 Comparison of errors in results relative to the neurospectral method for antenna on substrate 2. Reproduced by permission of Patnaik and Mishra ©2003 IEEE

For the second substrate, Figure 4.15 compares the results obtained using this method and other design methods, employing a neural network that uses a closed form formulae [26] for square patch antennas. The disagreement increases with the electrical thickness of the substrate, since the closed form formulae are normally restricted to electrically thin substrates.

4.4.2.2 Feeding Considerations

Let us consider two more feeding methods commonly used for conformal antenna, namely the coaxial probe and the aperture coupled. In both these cases, the feeding mechanism introduces loading on the resonating patch. (It may be noted that for microstripline feeding, this effect is not very prominent.) So in these cases, the resonant frequency is to be determined from the input impedance calculation. The model can easily be modified to find the feed location of the probe. It needs to add two more inputs for the x and y positions of the feed probe centers.

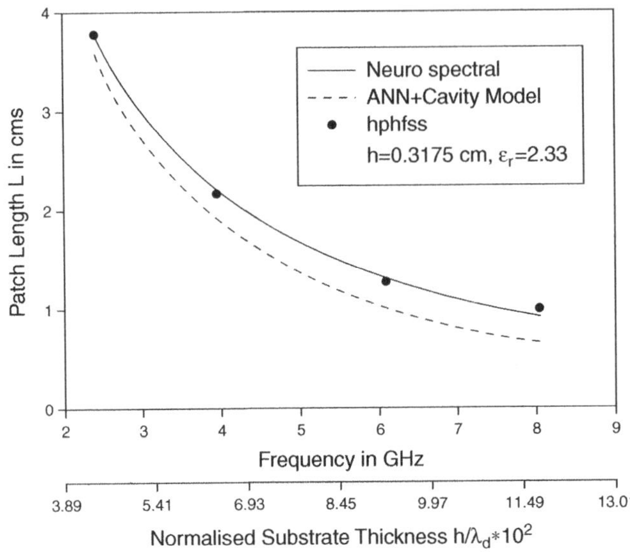

Figure 4.15 Comparison of results for square patch antenna with ANN model based on closed form formulae. Reproduced by permission of Patnaik and Mishra ©2003 IEEE

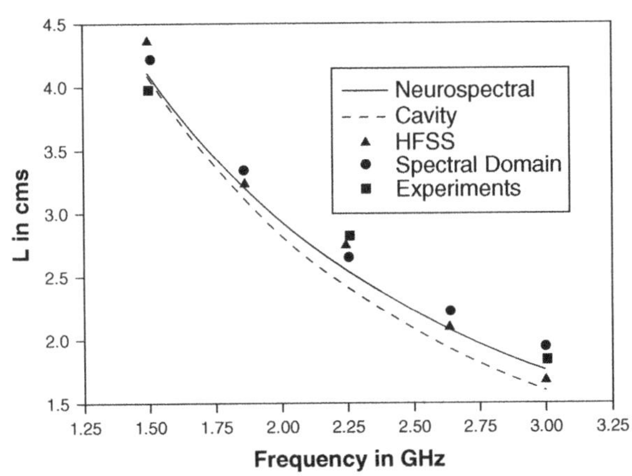

Figure 4.16 Comparison of results for probe-fed patch antenna with HFSS simulation, experiments and cavity model. Reproduced by permission of Patnaik and Mishra ©2003 IEEE

The neurospectral results, for patch lengths of probe-fed antennas on substrate 1, are compared with HFSS simulated results, cavity model results and experimental results in Figure 4.16. The neurospectral results follow the spectral domain results closely throughout. The cavity model results diverge from these results with increase in frequency. There is good matching of these results with the experimental results also.

The number of basis functions for the patch current is restricted to 1 along the x-direction and 0 along the y-direction. This can be one of the reasons for deviation from the experimental results.

Now let us consider the application of this method to the aperture-coupled antenna as given below in brief. The input impedance Z_{in} for an aperture coupled patch antenna is given [12] as

$$Z_{in} = Z - jZ_c \cot\beta\ L_s \tag{4.28}$$

In this equation:

$$Z = Z_c \frac{\Delta v^2}{Y^e + Y^a} \tag{4.29}$$

$$Y^a = [V]^t [Z]^{-1} [V] \tag{4.30}$$

The expressions for Y^e, Δv, Z and V are available in [12, (Equations 5.28, (5.30), (5.39) and (5.40)]. Z_c is the characteristic impedance of the microstrip line feeding the aperture (or slot). L_s is the stub length and β is the propagation constant in the feeding microstripline. The expressions for Y^e, Δv, Z and V are in the form of Fourier Transform Integrals. A neural network then models the integrands of these integrals and consequently closed form formulae are obtained for these integrals, as described earlier. Then another network is formed which has ten outputs and two inputs. The ten outputs are the frequency, patch width, substrate thickness of the antenna, substrate thickness of the feeding slot, dielectric constants of the substrates, the slot length, the slot width, the stub length and the reactive part of the impedance. As in the previous case, this network is trained for different frequencies. In all the cases, first the resonant frequency for a simple patch antenna is determined using the earlier networks. Then the stub length is fixed at quarter-wavelength of this frequency. The slot width is kept constant at 1mm. The slot length is always half that of the patch length. The data generation and pre-processing for this network are similar to that described earlier. The reverse modeling is applied to find a length at which the reactance value at the output becomes negligible.

Figures 4.17 and 4.18 provide, respectively, information on the variability of substrate dielectric constant and substrate thickness for both the substrates when the antenna is aperture coupled. In these figures, the substrate thickness $h = 0.16$cm and $\varepsilon_r = 4.5$ for substrate 1 and $h = 0.3175$cm and $\varepsilon_r = 2.33$ for substrate 2. First the length "L" is computed for resonant

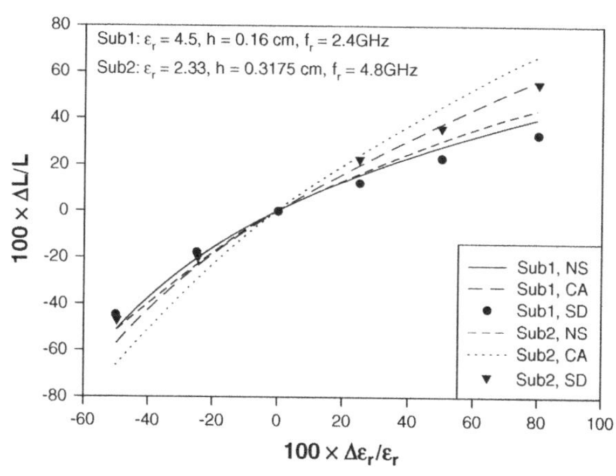

Figure 4.17 Uncertainty effects of L due to variability of substrate dielectric constant. Reproduced by permission of Patnaik and Mishra ©2003 IEEE

frequencies 2.4 GHz and 4.8 GHz corresponding respectively to substrate 1 and substrate 2. Then, keeping h constant, ε_r is changed by $\Delta\varepsilon_r$ and corresponding change in the length ΔL is shown in Figure 4.17. Similarly, change in length corresponding to thickness variations, for constant ε_r, is depicted in Figure 4.18. Thus these plots illustrate the uncertainties in patch length, for the design frequency, due to substrate variability. It is to be noted here that in both the plots, for both the substrates, the spectral domain results are closer to the neurospectral results than to the cavity model results.

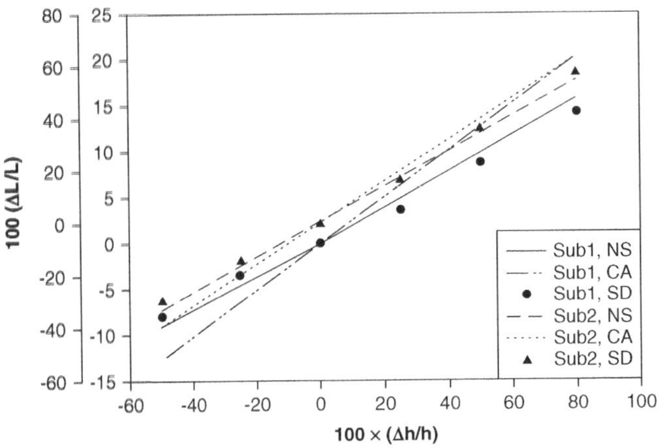

Figure 4.18 Uncertainty effects of L due to variability of substrate thickness. Reproduced by permission of Patnaik and Mishra ©2003 IEEE

4.4.2.3 Any Other Arbitrary Shape

For any arbitrary patch shape the first step in this design model is formulation of the problem in the spectral domain. The next step is evaluation of spectral integrals in closed form, using a neural network for the integrands. The procedure remains the same as that for the rectangular antenna. The last step is developing a suitable mapping between the antenna parameters. The designer has to decide what should be the input and output of the network, depending on the requirement.

4.4.2.4 Points to Note

Table 4.2 gives the parameters for the integration network in the 2nd column and the parameters for reverse training network in the 4th column, for the simple rectangular patch antenna and the square patch antenna. For the aperture-coupled antenna the network parameters are given in columns 3 and 5 of Table 4.2. In Table 4.2, the fixation of number of hidden layers is to be observed. It depends on the total number of input and output neurons. Although there is no regular pattern, but it decreases with the decrease in the total number of input and output neurons. The learning rate and the momentum decrease with an increase in the number of hidden layer neurons, whereas the learning rate adaptation increases during forward training. The pattern is opposite for the reverse training. For integration, the training times for the 1st and

Table 4.2 Network Parameters for Microstripline Fed (MF) and Aperture Coupled, i.e. Slot Fed (SF) patch antenna design stages

Network parameters	Green's Function/integration		Patch length	
	MF	SF	MF	SF
Number of Input Neurons	8	6	2	2
Number of Output Neurons	8	4	6	10
Neurons in Hidden Layers	37	17	12	28
Learning Rate (α)				
(Network training)	0.0027	0.13	0.2	0.0047
(Reverse training)	–	–	0.086	0.14
Learning rate adaptation (τ)				
(Network training)	1.8	1.4	1.2	1.6
(Reverse training)	–	–	0.89	0.68
Momentum (η)				
(Network training)	4.2	2.5	3.8	1.6
(Reverse training)	–	–	4.6	1.3
Training Tolerance	5×10^{-3}	5×10^{-3}	5×10^{-3}	5×10^{-3}

Source: Reproduced by permission of Patnaik and Mishra ©2009 IEEE

2nd substrates are approximately 1 hour 20 minutes and 1 hour 50 minutes respectively. On the second network, the forward training times are approximately 25 minutes and 40 minutes respectively. During the implementation stage, i.e. reverse training, the average time for each length calculation is almost instantaneous as evident from Table 4.3.

Table 4.3 Average time for simulation using different methods

Antenna type	HPHFSS	Spectral domain	Cavity model	Neurospectral
Microstripline fed	26 minutes	22 minutes	1 minute and 20 Sec	15 seconds
Probe fed	34 minutes	22 minutes	1 minute and 40 sec	15 seconds
Aperture coupled	2 hours and 17 minutes	1 hour and 42 minutes	5 minutes	25 seconds

Source: Reproduced by permission of Patnaik and Mishra ©2009 IEEE
Note: For the neurospectral method, only the time for reverse training is given, since the end user is concerned this time.

This method consists of three steps in two stages. In the first stage the spectral domain formulation of the problem constitutes the first step. The second step then provides an alternative to the numerical solution for this in the form of neural network. The second stage consists of the final step that uses reverse modeling for synthesis. Once this method is understood properly, it is quite easy to consider other configurations or parameters or to include different feed mechanisms. The task is simple manipulation of the network structure for the job at hand.

The forward training time, in all the stages, increases with increased number of input data sets. However, this does not affect the reverse training. This is because; in the reverse training the search starts near the vicinity of the required data set. So, it settles down to the required data set very quickly (in less than 10 iterations). Thus, it is the network developer who needs more computational resources and time. Once, a proper network is developed (i.e. the number of hidden neurons and values of the connecting weight are determined), the program can be passed onto the user (Figure 4.19). Figure 4.19 distinguishes the two stages involved. The users/clients are interested only in the implementation stage. It involves only the reverse training process to determine patch length where the results are obtained very quickly. In the neurospectral technique the time requirement is very high at the development stage, and is low in the implementation stage. For other commercial packages it is high both at development and implementation stages.

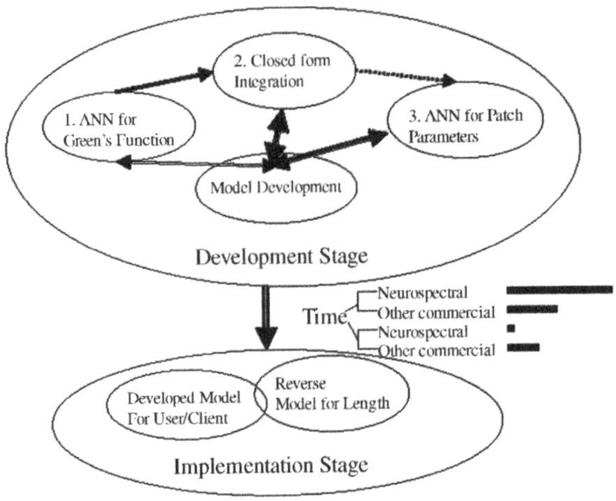

Figure 4.19 Schematic of the model development stages. Reproduced by permission of Patnaik and Mishra ©2003 IEEE

4.5 Inset-fed Patch Antenna Design Using Particle Swarm Optimization

Particle Swarm Optimization (PSO) is a social behavior simulation-inspired technique for nonlinear function optimization developed by Jim Kennedy and Russ Eberhart. It is based on the three principles of swarming theory:(i) Members of swarms steer towards the center, (ii) Members match their neighbor's velocity; and (iii) Members avoid collision. PSO is not only an optimization tool but also can represent socio-cognition of artificial agents based on the social psychology principle. It combines local searching with global searching. Therefore, it maintains a balance between exploration and exploitation of search space.

4.5.1 Explanation of PSO Terms

Like GA, the PSO is also a population-based search procedure [30]. Each individual is called a particle. The particle changes its position (i.e. state) with time in the search procedure. The particles fly around in the N-dimensional search space. Therefore, the ith particle shall have a current position ${}^{t}X_i = ({}^{t}x_i(1), {}^{t}x_i(2), \ldots, {}^{t}x_i(N))$. Its best previous position, i.e. the position from the 1st iteration to the tth iteration giving the best fitness value can be denoted as ${}^{t}P_i = ({}^{t}p_i(1), {}^{t}p_i(2), \ldots, {}^{t}p_i(N))$. Each particle is capable of changing its position based on its own experience and the experience of its neighbor during the course of the flight. Thus, the best positions encountered by a particle and its neighbor are utilized in this method. Then the best position among all particles from the 1st iteration to the tth iteration is called the global best at the tth iteration and is given as ${}^{t}G_i = ({}^{t}g_i(1), {}^{t}g_i(2), \ldots, {}^{t}g_i(N))$. Since, the position is changing continuously with iterations, the time rate of position change can be recorded as the velocity ${}^{t}V_i = ({}^{t}v_i(1), {}^{t}v_i(2), \ldots, {}^{t}v_i(N))$.

4.5.2 *Inset-fed Patch Antenna Design*

The determination of physical dimension of a simple rectangular antenna has been discussed extensively in the literature [31]. However, when an inset feed is used, the shape of the patch is perturbed. This perturbation leads to detuning of the antenna resulting in impedance mismatch. Optimization becomes important in this case. This section describes the use of PSO to compensate for the detuning and impedance mismatch.

The resonant frequency of the antenna primarily depends on its physical length, since the dielectric constant and thickness for a given substrate are constants. The inset depth perturbs this length. To compensate for this perturbation, the length of the patch and the inset depth are to be adjusted simultaneously. Similarly, the inset depth is also responsible for impedance matching. Hence, optimization of the inset depth and the patch length are important. The objective of finding a suitable inset depth and patch length is to get the desired resonant frequency and impedance matching. Thus, it is a multi-objective optimization problem.

For this design the particle position should be two-dimensional, containing the patch length (L) and the inset depth (d). So, the position of the ith particle shall be ${}^{t}X_i = ({}^{t}L_i, {}^{t}d_i(2))$. At the resonant frequency the imaginary part of the impedance shall vanish, so $Im(Z_{in}) = 0$. For acceptable matching, $S_{11} \leq -10$dB. Thus the fitness function of objective function can be formed as

$$E = |Im(Z_{in}) - 2|2 + 0.6|S_{11}\,dB + 25|2 \tag{4.31}$$

In PSO, the position of the particle is manipulated by considering its present position, its previous best position and the global best position. Figure 4.20 explains the process of manipulation. It involves two simple algebraic equations:

$${}^{t+1}V_i = {}^{t}W \cdot {}^{t}V_i + C_1 \cdot r_1 \cdot ({}^{t}P_i - {}^{t}X_i) + C_2 \cdot r_2 \cdot ({}^{t}G_i - {}^{t}X_i) \tag{4.32}$$

$${}^{t+1}X_i = {}^{t}X_i + \Delta t \cdot \tag{4.33}$$

In the above equations, C_1, C_2 are two positive constants and r_1, r_2 are two random functions in the range [0,1]. W is the inertia weight. Large inertia weight facilitates global exploration and small one facilitates local exploration. So, $\boldsymbol{W}$ is to be selected carefully. We prefer to start with large W and gradually decrease it over run. Another important parameter in PSO is the limit on the velocity. A maximum value Vmax is pre-decided, which clamps the particle velocity on each dimension. If it is too high, the particle can fly past optimal solution, whereas too low a value can result in trapping in local minima. To avoid the first condition, bounds can be fixed on the particle. Three types of bound are common in literature:(i) absorbing bound; (ii) reflecting bound; and (iii) invisible bound.

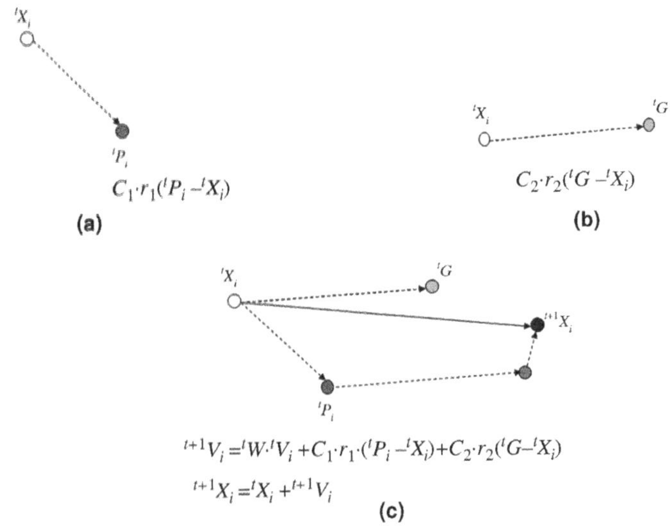

Figure 4.20 Position updating process. (a) influence of personal best position; (b) influence of global best position; (c) updated position

The following algorithm has been used for the inset-fed patch antenna. For other problems, similar algorithms can easily be developed.

Step 0: Initialize population in the search space.

Take 20 populations randomly. Fix bounds on L and D. L may vary between $\pm 10\%$ of its pre-optimized value. Lower bound on D is 0 and upper bound is half of the pre-optimized value of L.

Step 1: Evaluate fitness of each individual particle.

Use the cost function in Equation (4.31) to evaluate the fitness. Find out the fittest particle and take it as global best. In the 1st iteration, the personal best is the position of the particle in that iteration. From the 2nd iteration onward, the personal best position is to be found by comparing the fitness of the present position with the previous best personal position.

Step 2: Modify Velocity and Position using Equations (4.32) and (4.33).

If the velocity exceeds the maximum value, then update it to the maximum value (absorbing bound). Follow a similar procedure for positions.

Step 3: If the termination condition is achieved, terminate.

Fix termination condition as a fitness value of 5×10^{-4}. If any particle achieves this fitness value, then terminate. Otherwise terminate after 1000 iterations.

Step 4: Go back to step 1, if unterminated.

If terminated after the maximum number of iterations but without achieving the required fitness for any particle, then restart the process with a new random set of particles and velocities.

This algorithm was followed for a design of a 2.4 GHz inset-fed patch antenna on Ultralam® 2000 substrate having a thickness of 1.58 mm and a dielectric constant of 2.4. The width and length of the patch were found to be 33.9 mm and 41.4 mm respectively. With an inset depth of 12 mm, it was detuned to 2.32 GHz. Lower and upper bounds were fixed on the length to be 35.0 mm and 48.0 mm. The PSO terminated in 60 iterations. The optimized antenna has a length of 46.9 mm and inset depth of 13.9 mm. It resonates at 2.44 GHz and gives a S_{11} of 38 dB at the resonant frequency. The fabricated prototype is shown in Figure 4.21.

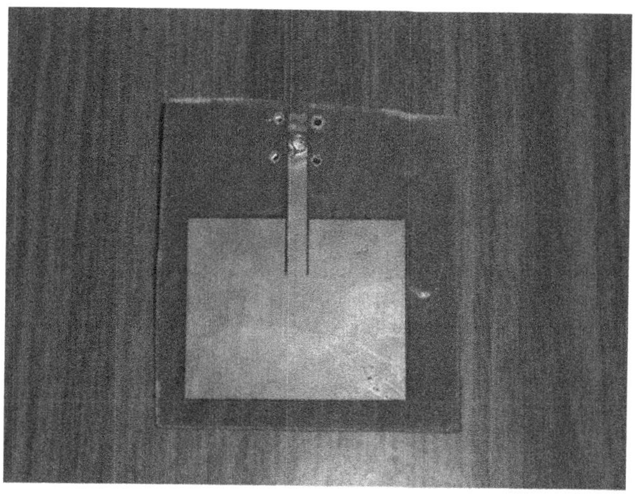

Figure 4.21 Prototype of optimized inset-fed microstrip antenna on Ultralam® 2000 substrate

4.6 Conclusion

This chapter discussed applications of three popular stochastic optimization techniques to planar antennas. In the absence of analytical formulae, the design of fractal antenna heavily depends on initial CAD simulation. The job becomes more difficult when optimization is involved. Detailed illustration of interfacing IE3D engine with MATLAB for RCGA-based optimization of Sierpinski gasket microstrip antenna has been given. Then the function approximation feature of the artificial neural network was exploited to develop an alternative analytical method in spectral domain for microstrip antennas. This was followed by an illustration of optimization of inset-fed microstrip antenna using PSO.

New stochastic optimization techniques are continuously being developed. Notable among them are Ant Colony Optimization (ACO), Bactrial Foraging Optimization (BFO), etc. Some of these will find applications in antenna engineering, depending on their complexities and performance.

References

1. G. C. Temes and D. A. Calahan, "Computer-aided network optimization, the state-of art," *Proc. IEEE*, vol. 55, pp. 1832–1863, Nov. 1967.
2. J.W. Bandler, "Optimization methods for computer-aided design," *IEEE Trans., MTT-17*, pp. 533–552, Aug. 1969.
3. K. C. Gupta, R. Garg, and R. Ghadha, *Computer-Aided Design of Microwave Circuits*, Artech House, Dedham MA, 1981.
4. Yahya Rahmat Samii and Eric Michielssen, *Electromagnetic Optimisation by Genetic Algorithms*, John Wiley and Sons Inc., New York, 1999.
5. Melanie Mitchell, *An Introduction to Genetic Algorithms*, Prentice Hall of India, New Delhi, 2004.
6. Z. Michalewicz, *Genetic Algorithms + Data Structures = Evolutionary Programming*, 3rd edition, Springer Verlag, New York, 1996.
7. Derek S. Linden, "Automated design and optimisation of wire antennas using genetic algorithms," PhD thesis, Massachusetts Institute of Technology, September 1997.
8. Xin Yao, Young Liu, and Guangming Lin, "Evolutionary programming made faster," *IEEE Transactions on Evolutionary Computation*, vol. 3, no. 2, pp. 82–102, July 1999
9. Girish Kumar and K. P. Ray, *Broadband Microstrip Antennas*, Artech House, Inc., Norwood, NJ, 2003.
10. C. T. P. Song, Peter S. Hall, and H. Ghafouri-Shiraz, "Perturbed Sierpinski multiband fractal antenna with improved feeding technique," *IEEE Transactions on Antennas and Propagation*, vol. 51, no. 5, pp. 1011–1017, May 2003.
11. F. Michielssen, J. Sajer, S. Ranjithan, and R. Mitra, "Design of lightweight, broadband microwave absorbers using genetic algorithms," *IEEE Transactions on Microwave Theory and Techniques*, vol. 41, no. 6, pp. 1024–1031, June 1993.
12. J. M. Johnson and Y. Rahmat Samii, "Genetic algorithms optimisation and its application to antenna design," *IEEE Antennas and Propagation Society Int. Symposium Digest*, pp. 326–329, June 19–24, 1994.
13. D. E. Goldberg, *Genetic Algorithms in Search Optimisation and Machine Learning*, Pearson Education, New Delhi, 2004.
14. Joshua S. Petko and Douglas H. Werner, "The evolution of optimal linear polyfractal arrays using genetic algorithm," *IEEE Transactions on Antennas and Propagation*, vol. 53, no. 11, pp. 3604–3651, November 2005.
15. Renzo Azaro, Giulia Boato, Massimo Donelli, Andrea Massa, and Edoardo Zeni, "Design of a pre-fractal monopolar antenna for 3.4–3.6 GHz Wi-max band portable devices," *IEEE Antennas and Wireless Propagation Letters*, vol. 5, pp. 116–119, 2006.
16. M. Fernandez Pantoja, F. Garcia Ruiz, A. Rubio Bretones, S. Gonzalez Garcia, R. Gomez Martin, J. M. Gonzalez Arbesu, J. Romeu, J. M. Rius, P. L. Werner, and D. H. Werner, "GA design of small thin-wire Sierpinski-type prefractal antennas," *IEEE Transactions on Antennas and Propagation*, vol. 54, no. 6, pp. 1879–1882, June 2006.
17. H. O. Peitgen, H. Jurgens, and D. Saupe, *Chaos and Fractals: New Frontiers of Science*, Springer Verlag, New York, 1992.
18. *IE3D User's Manual* (Release 9), Zeland Software Inc., March 2002.
19. C. Puente, C. B. Borau, M. N. Rodero, and J. R. Robert, "An iterative model for fractal antennas: application to the Sierpinski gasket antenna," *IEEE Transactions on Antennas and Propagation*, vol. 48, no. 5, pp. 713–719, May 2000.
20. Steven R. Best, "On the significance of self similar fractal geometry in determining the multiband behaviour of the Sierpinski gasket antenna," *IEEE Antennas and Wireless Propagation Letters*, vol. 1, pp. 22–25, 2002.
21. T. Itoh and W. Menzel, "A full-wave analysis method for microstrip structures," *IEEE Trans.*, vol. AP-29, no. 1, pp. 63–68, 1981.
22. S. Haykin, *Neural Networks: A Comprehensive Foundation*, IEEE Computer Society Press/IEEE Press, New York, 1994.

23. N. B. Karayiannis and A. N. Venetsanopoulos, "Fast learning algorithm for neural networks," *IEEE Trans. Circuits Syst. II: Analog Digital Signal Processing*, vol. 39, pp. 453–473, 1992.
24. R. K. Mishra and A. Patnaik, "Neurospectral computation for complex resonant frequency of microstrip resonators," *IEEE MGWL*, vol. 9, no. 9, pp. 351–353, Sept. 1999.
25. M. M. Vai, S. Wu, B. Li, and S. Prasad, "Reverse modelling of microwave circuits with bi-directional neural network model," *IEEE Trans.*, vol. MTT-46, no. 10, pp. 1492–1494, Oct. 1998.
26. I. J. Bhal, P. Bharatia, and S. S. Stuchly, "Design of microstrip antennas covered with dielectric layer," *IEEE Trans.*, vol. AP-30, no. 2, pp. 314–318, Mar. 1982.
27. N. Benvenuto and F. Piazza, "On the complex back propagation algorithm," *IEEE Trans. Signal Processing*, vol. 40, pp. 967–969, April 1992.
28. R. K. Mishra and A. Patnaik, "Neural network based CAD model for design of square patch antennas," *IEEE Trans.* vol. AP-46, no. 12, pp. 1890–1891, Dec. 1998.
29. D. M. Pozar "A reciprocity method of analysis for printed slot and slot-coupled microstrip antennas," *IEEE Trans*, vol. AP-34, no. 12, pp. 1439–1446, Dec. 1986.
30. X. F. Liu et al., "Design of a low-profile modified U-slot microstrip antenna using PSO based on IE3D," *Microwave and Optical Technology Letters*, vol. 49, no. 5, pp. 1111–1114, May 2007.
31. G. Kumar and K. P. Ray, *Broadband Microstrip Antenna*, Artech House, Boston, 2003.
32. Ghatak, Poddar, and R.K. Mishra, *IET-MAP*, vol. 3, 2009.
33. A. Patnaik and R.K. Mishra, *IEEE Trans.*, vol. AP-51, 2003.

5

Microstrip Reflectarray Antennas

Jafar Shaker and Reza Chaharmir
Communication Research Centre Canada, Ontario

5.1 Introduction

The concept of reflectarray was initially introduced using waveguide technology [1] which was composed of a 4×26 array of waveguides fed by a horn antenna. The complexity of implementing waveguide technology makes it impractical for higher frequency bands. Therefore, a microstrip implementation of the same concept was carried out subsequently. The microstrip approach leads to lower cost and easier fabrication procedure compared to waveguide technology. A reflectarray imitates the function of a conventional reflector by introducing the required phase shift through variation of size or geometry of planar elements [2]. Reflectarray technology can be considered a hybrid of reflector and phased-array technologies and as such it benefits from some of the strengths of these two technologies. Its similarity to phased array structures stems from its flat or at times conformal shape. The utilization of free space as the medium to focus the wave impinging onto the elements into a feed point constitutes its affinity with conventional reflector technology. Since free space serves as the power distribution medium to the elements, reflectarrays do not require the transmission line network necessary in printed phased arrays and thus avoid the associated losses. On the other hand, the fabrication technology of printed reflectarrays is simpler and more cost-effective compared to the fabrication process of the conventional reflector technology especially when it comes to complicated shaped beam reflectors realized by joining printed panels of reflectarrays [3]. Also, as will be noted in the following sections, application of printed technology leads to significant simplification of feed systems in the case of multi-polarization/band antenna systems [4–8]. In addition, there has been significant progress in new fabrication technologies in the form of integration of MEMs [9] and active devices within the antenna to carry out such functions as power combining, amplification, and adaptive beam steering [10, 11] that cannot possibly be integrated within a conventional reflector antenna. Reflectarray technology is amenable to the new advancements in fabrication technology and benefits from such advances. For instance, rather than using a complicated cluster of feeds to

Microstrip and Printed Antennas: New Trends, Techniques and Applications. Edited by Debatosh Guha and Yahia M.M. Antar

shape and reshape the radiation pattern of a conventional reflector, one can exploit MEM [9] or active elements [10] to achieve the same objective.

The main shortcoming of reflectarrays is their limited bandwidth. Although this limitation does not impose a severe restriction on their applications at higher frequency bands such as Ka-band, operation at lower frequency bands such as C-band and Ku-band can be quite challenging. The research into methods to remove this shortcoming has been concentrated on two aspects: element optimization for moderate size reflectarray (<35 dBi gain) [12, 13] and development of specific optimization routines for large (>40 dBi gain) reflectarrays [14, 15].

This chapter is organized as follows. Section 5.2 provides an overview of the general principles of design and operation of a reflectarray along with the presentation of general cell elements that have been used in the past. A quick comparison of reflectarray and reflector technologies is given in Section 5.3, noting the conventional figures of merit that are used in the evaluation of conventional reflectors as given in [16]. Then, Section 5.4 is devoted to the presentation of reflectarray antennas from the point of view of cell elements, fabrication technologies, and applications. The issue of bandwidth is dealt with afterwards through an in-depth presentation of the factors involved, namely, frequency dispersion and spatial dispersion [17]. Methods to suppress each of these two factors are presented next. An optimization technique that is developed by the authors will be described next and simulations and measurements will be given to verify the validity of the technique. Development of novel loop-based cell elements to improve the bandwidth of a moderate size reflectarray is outlined in the subsequent section.

5.2 General Review of Reflectarrays: Mathematical Formulation and General Trends

Reflectarrays are "quasi-periodic" structures. Therefore, physical concepts and mathematical tools applicable to periodic structures are indispensable in understanding the underlying physical phenomena that govern the operation of reflectarrays. On the other hand, selection of cell elements can play an important role in the performance of the reflectarray as is the case for such periodic structures as Frequency Selective Surfaces (FSS) [18]. A general review of the formulation of periodic structures that is relevant to the study of reflectarray is presented in the following. Then, general trends in the performance of the reflectarray will be presented by focusing on classical cell elements, namely, the patch or stub loaded patch.

5.2.1 *Mathematical Formulation*

Illuminating an infinite periodic structure with a plane wave leads to scattering in the form of discrete plane waves in contrast to the continuous spectrum of scattered waves from a finite structure. This is readily seen from a Green's function formulation of scattering from an infinite periodic structure given in the following as:

$$\vec{E}_{scat}(x,y,z)=\frac{1}{T_xT_y}\sum_{m=-\infty}^{+\infty}\sum_{n=-\infty}^{+\infty}\int \overleftrightarrow{G}(k_x',k_y').\vec{J}(x',y',0)e^{j(k_{xm}'(x-x')+k_{yn}'(y-y'))}e^{jk_{zmn}'z}dxdy \quad (5.1)$$

where:

$$k_{xm}{}' = \left(\frac{2\pi m}{T_x} + k_{x0}\right) \quad k_{yn}{}' = \left(\frac{2\pi n}{T_y} + k_{y0}\right) \quad k_{zmn}{}' = \sqrt{k_0^2 - k_{xm}{}'^2 - k_{yn}{}'^2} \tag{5.2}$$

and k_{x0}, k_{y0}, and k_{z0} are Cartesian components of the wavenumber of the incident plane wave. It is evident that the scattering wave is of a discrete nature and it is propagating only if k_{zmn} is a real number. The complex values of k_{zmn} represent evanescent waves. $J(x',y',0)$ is the electric current density vector on $(0,0)th$ of a typical periodic structure shown in Figure 5.1 and it can be determined using a Method of Moment (MOM) [18, 19] scheme. Equation (5.1) represents only the scattered field from the patches.

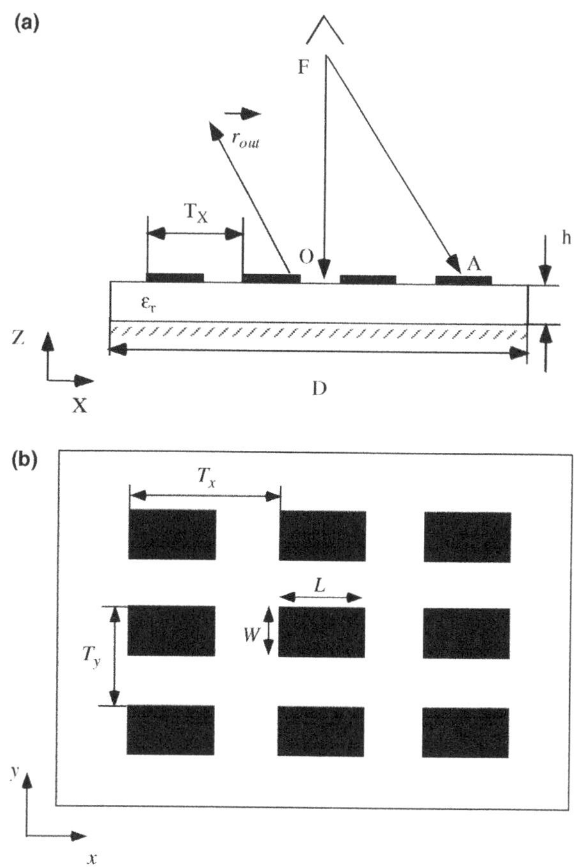

Figure 5.1 Two views of a reflectarray: (a) side view; (b) top view

Generally, collimating structures such as lens or reflector can be classified as phase front transformers. In their simplest renditions, such structures transform the spherical phase front of the feed into the planar phase front (in Tx mode) and hence comes about conventional reflector or lens. Phase front transformation in such classical realizations is carried out through geometrical shaping of the surface of the reflector or the lens. Equal optical (electrical) path from the feed to the outgoing planar phase front is the main design criterion to be followed [16]. In the absence of the shaped geometrical surface of a conventional reflector, phase front transformation can be carried out by resonant or near resonant elements etched on a flat substrate with conductor backing.

Reflectarrays are "quasi-periodic" structures composed of cell elements that are etched on a dielectric slab and the desired phase transformation at each location is achieved by adjusting the geometrical features of the cell element at that particular location (such as the size of geometrical features of the cell element or its orientation). It is to be noted that such a strategy leads to replicating the desired phase transformation in a quantized fashion. The variation of the geometrical features of the cell elements is smooth enough to warrant local approximation of the reflectarray by an infinite periodic structure. Therefore, the lattice size is maintained smaller than a half of free space wavelength in order to suppress higher-order spatial harmonics [16]. This is an important matter from thc analysis point of view in the sense that in the MoM formulation of an "infinite periodic structure" composed of similar cell elements and illuminated by a plane wave, the phase of the direct reflected wave (or (0,0) Floquet mode) is determined by the size of a specific geometric feature of the cell element. A plot showing the correspondence of the reflected phase versus the size of geometrical feature is used to implement the required phase transformation at each particular location of the reflectarray.

It is very important to note that Equation (5.2) gives only the contribution of the scattered field from the patches. Direct reflection from the dielectric substrate backed by the ground plane should be added to the scattering from patches in order to obtain the total scattered field. This is expressed in the following equation:

$$\vec{S}(\theta,\phi) = \vec{R}(\theta_{inc},\varphi_{inc}) + \int_S \overleftrightarrow{G}(k_{x0},k_{y0}).\vec{J}(x',y')e^{\,j(k_{x0}(x-x')+k_{y0}(y-y'))}dx'dy' \tag{5.3}$$

The first term represents the direct reflection from the bare dielectric substrate with conductor backing and the second term is the (0,0) Floquet mode that is generated by the patches. Direct reflection from the substrate might lead to unwanted radiation for offset-fed reflectarrays as will be seen later on.

Side and top views of a typical reflectarray are shown in Figure 5.1 along with geometrical notations that are used throughout this chapter to define the structure. A typical reflectarray cell element that was used at early stages of reflectarray research is shown in Figure 5.2 and is composed of a stub loaded patch [20]. The incident field intercepted by a particular patch is reradiated with a phase shift that is determined by that patch's stub length. The phase shift imposed by the stub can be calculated to the first degree of approximation by assuming the stub to be a microstrip transmission line [20]. Reflectarrays composed of such elements demonstrate rather high cross-polarization and narrow bandwidth.

A simple rectangular patch is another type of cell element that has been used widely in the previous literature [21, 22]. The width of the cell element is usually kept constant and the length (which is parallel to the polarization of the incoming wave) is changed to set the patch in the appropriate resonant or off-resonant condition to realize the desired phase. A typical phase versus length plot is shown in Figure 5.3. Having obtained the "design curve" shown in

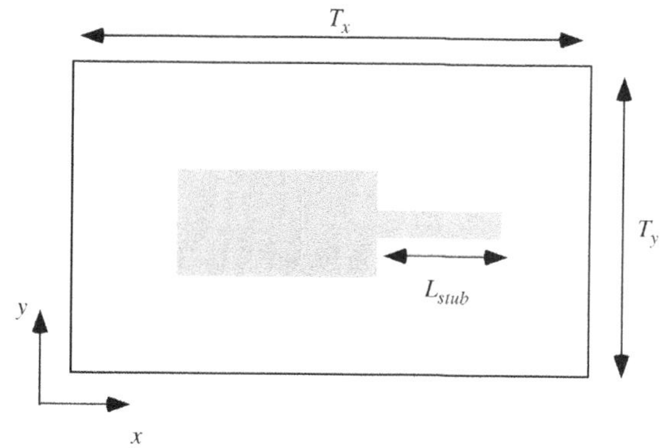

Figure 5.2 Stub-loaded patch cell element

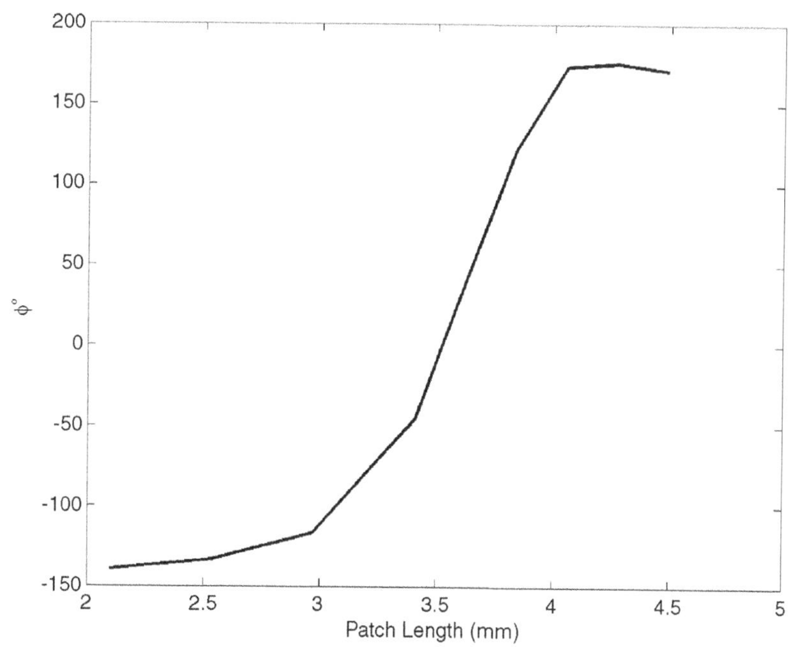

Figure 5.3 Typical phase-length design curve that is used in the design of a reflectarray. An infinite periodic structure composed of patches of 3.15 mm width. The patches were arranged in a square lattice of 5.442 mm, substrate permittivity and thickness are 2.2 and 0.20″, respectively. The structure is illuminated by normally incident plane wave and $f = 28.0$ GHz

Figure 5.3, Equation (5.4) is used to find the required phase (ϕ_{patch}) that ought to be contributed by the patch element:

$$k_0(\vec{FA} - \vec{FO}.\hat{r}_{out}) - \varphi_{patch} = 2n\pi \tag{5.4}$$

The notation of the above equation is defined in Figure 5.1a. This value of patch phase is used in Figure 5.3 to find the corresponding value for patch dimensions. It is obvious from physical considerations that the amplitude of reflection in the absence of higher order modes is unity for lossless dielectric substrate case as the presence of the ground plane blocks the transmission of power in the lower half-space.

5.2.2 *General Trends*

Considering the ample amount of research that has been carried out on reflectarrays composed of patch cell elements, we focus on this particular element to demonstrate some of the general trends such as the effect of substrate thickness, permittivity and loss, and the slope of phase versus patch length plot on the reflectarray performance. These trends are valid even for reflectarrays composed of cell elements other than the patch.

The effect of the patch width on the slope of "phase versus length" plot is shown in Figure 5.4 (a) which demonstrates a more gradual slope as the patch width is increased [21]. This is due to the fact that an increase in width leads to more efficient radiation and lower Q factor. The effect of this factor is shown in Figure 5.4 (b) by plotting across the frequency band, the gain performance of reflectarrays with narrow and thick cell elements. As can be seen, a thinner patch leads to a narrower 1-dB gain bandwidth which is defined as the relative bandwidth across which the gain of the reflectarray decreases by less than 1-dB from the maximum gain that it attains across the band. It will be shown in Section 5.5 that a more gradual slope for the phase characteristic of the cell element would indeed result in a wider bandwidth for a reflectarray of moderate size (size $< 20\lambda$).

The impact of the substrate thickness and permittivity on the slope of phase-patch length plot shown in Figure 5.5 illustrates a more gradual slope when the substrate is thicker or of lower permittivity. This is in agreement with similar observations for the bandwidth of single patch antenna if, as explained earlier, a less steep slope is understood as being representative of wider bandwidth.

Noting that a reflectarray is actually a phase transformation structure, most elements are close to resonant conditions. The dielectric loss of a periodic structure is plotted in Figure 5.6 with respect to the deviation of the length of the patch cell element from resonant conditions for different values of loss tangent for the substrate [21]. It can be seen that the highest loss occurs right at resonance which is the result of the high intensity of current flow on the cell element at the resonant condition. This implies that reduced efficiency as a result of lossy substrate is more pronounced when the cell elements display a steeper phase characteristics because of the high fraction of reflectarray cell elements that are in the near-resonant condition.

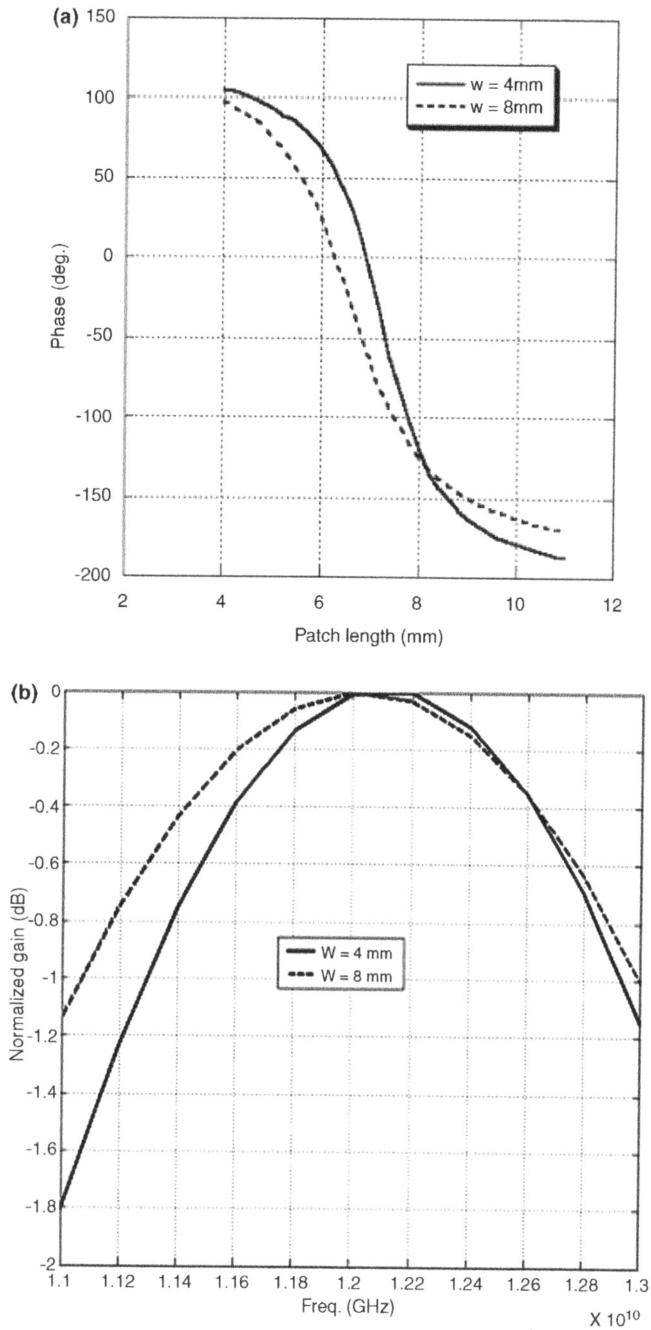

Figure 5.4 (a) Phase-length curves for patches of different widths. The structure is an infinite periodic structure composed of patches for which the lattice size is $T_x = T_y = 12.0$ mm, the substrate permittivity is 3.0, and substrate thickness is 2.0 mm. The structure is illuminated by normal incident plane wave and $f = 12.0$ GHz. (b) The effect of the width of the cell element on the gain

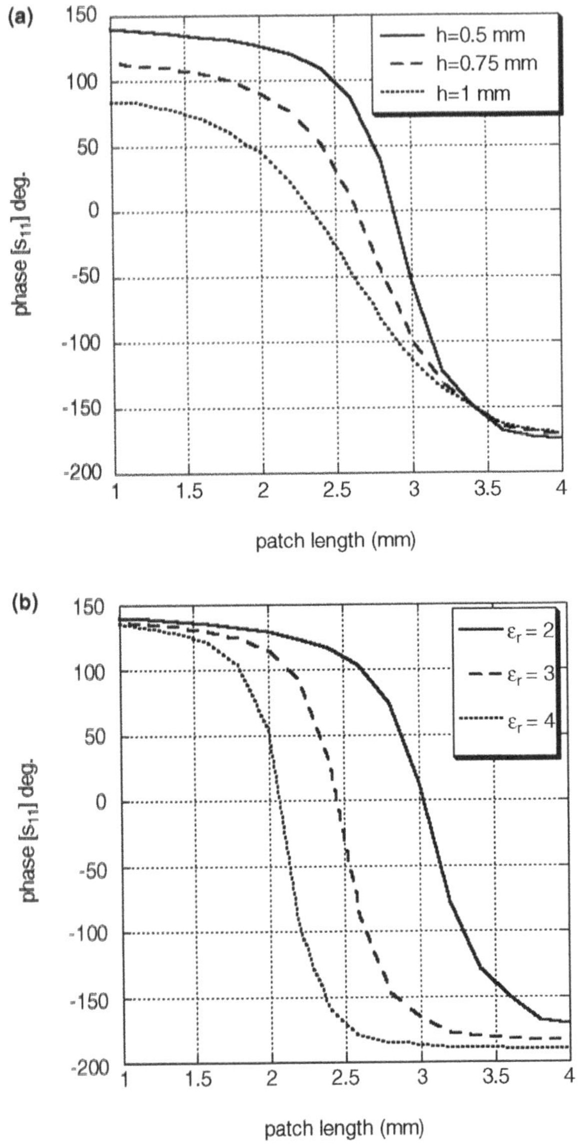

Figure 5.5 Phase-length curve for a periodic structure composed of patches illuminated by normally incidence plane wave. The lattice size is $T_x = T_y = 5.4$ mm, patch width = 3.5 mm, $f = 30.0$ GHz. The plots are for: (a) Different values of substrate thickness while $\varepsilon = 2.2$. (b) Different values of substrate permittivity while substrate thickness is 0.5 mm.

5.3 Comparison of Reflectarray and Conventional Parabolic Reflector

Euclidean geometry is the basis upon which the parabolic shape of a conventional centre-fed reflector is derived. The wideband performance of parabolic reflector is the result of the fact

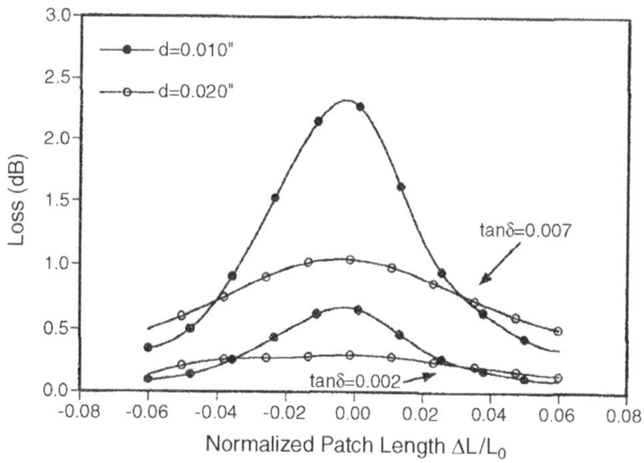

Figure 5.6 Dielectric loss for an infinite array of microstrip patches versus patch size. f = 28.0 GHz, $\varepsilon_r = 2.95$, $T_x = T_y = 0.536$ cm, $\theta_i = \phi_i = 0°$, $L_0 = W = 0.296$ cm, and substrate thickness = 0.20″. Reproduced by permission of ©1997 IEEE [21]

that light rays obey Fermat's principle when reflected from a conducting surface. The local curvature on the surface of the reflector is low enough to justify application of ray optics as a first-order approximation. Geometric Theory of Diffraction (GTD) can be used to increase the accuracy for the investigation of second-order effects such as more accurate estimation of Side Lobe Level (SLL) and cross-polarization. On the other hand, reflectarray structures achieve the same "phase transformation" by a delicate adjustment of scattering from "quasi-periodic" structures. The rather different origins of these two paradigms – the former stems from pure geometrical considerations and the latter from electromagnetic principles – can explain their different performance as gauged by conventional efficiency figures [16] such as: illumination, spill-over, cross-polarization, bandwidth, and blockage that are used to assess conventional reflectors. In the following, the performance of these two antennas is compared with respect to these efficiency figures. It is assumed that the two centre-fed broadside antennas are of the same size, focal length, and are being fed by the same feed.

5.3.1 Illumination Efficiency

Illumination efficiency is a measure of the taper of the feed illumination onto the surface of the reflector and it is derived using the following relationship [16]:

$$\eta_i = \frac{1}{\pi a^2} \frac{\left[\int_0^{2\pi} \int_0^a |E_y(\rho, \varphi)| \rho d\rho d\phi \right]^2}{\int_0^{2\pi} \int_0^a |E_y(\rho, \varphi)|^2 \rho d\rho d\phi} \tag{5.5}$$

It is to be noted that the illumination taper on the aperture determines the sidelobe level. In the case of parabolic reflector, aperture taper is solely determined by the radiation intensity of the feed along the reflector rim. However, in the case of the reflectarray, the illumination taper is a combination of the radiation pattern of the feed and also the radiation pattern of the cell element. Therefore, the edge taper in the case of a centre-fed reflectarray, composed of microstrip patches, is more often compared to its conventional counterpart. This leads to lower edge illumination in the case of the reflectarray but at the same time results in better SLL performance as a result of increased taper.

5.3.2 *Spill-over Efficiency*

This efficiency term represents the amount of power that is not intercepted by the antenna aperture and is defined as follows [16]:

$$\eta_s = \frac{\int_0^{2\pi}\int_0^{\psi/2} g(\theta,\phi)\sin\theta d\theta d\phi}{\int_0^{2\pi}\int_0^{\pi} g(\theta,\phi)\sin\theta d\theta d\phi} \tag{5.6}$$

where ψ is the subtended angle of the reflector or reflectarray and $g(\theta,\phi)$ is the radiation pattern of the feed. It should be noted that the subtended angle of the reflector might be slightly different from a reflectarray of the same size and *F/D* ratio. This might cause a slight difference between the spillover efficiency of a reflector antenna and its "equivalent" reflectarray counterpart.

5.3.3 *Polarization Efficiency*

This efficiency term represents the polarization purity of the antenna radiation and it is calculated according to the following relation [16]:

$$\eta_x = \frac{\left[\int_0^{2\pi}\int_0^{a} |E_y(\rho,\phi)|\rho d\rho d\phi\right]^2}{\int_0^{2\pi}\int_0^{a}(|E_y(\rho,\varphi)|^2 + |E_x(\rho,\varphi)|^2)\rho d\rho d\phi} \tag{5.7}$$

The above is the ratio of the power content of the desired polarization to the sum of the power content of co- and cross-polarized fields. In the case of reflectarrays, the elements generally collimate one and the same polarization while dispersing the orthogonal polarization. Therefore, even if the feed is inferior in terms of polarization specifications, the final radiation pattern can exhibit superior polarization performance. This is in contrast to the case of conventional parabolic reflectors where any degradation in the cross-polarization of the feed shows up in the cross-polarization performance of the whole antenna.

5.3.4 Phase Efficiency

Any deviation from the uniform aperture phase distribution leads to reduced efficiency. The following relationship is used to calculate phase efficiency [16]:

$$\eta_i = \frac{\left| \int_0^{2\pi} \int_0^{a} E_y(\rho, \varphi) \rho d\rho d\phi \right|^2}{\int_0^{2\pi} \int_0^{a} \left| E_y(\rho, \varphi) \right|^2 \rho d\rho d\phi} \tag{5.8}$$

As noted previously, the phase transformation is performed in a quantized fashion in the case of a reflectarray. For the case of a lattice size smaller than $\lambda/2$, the loss due to this quantization is rather minimal (<0.1 dB). However, it is to be noted that the realized phase at a particular location of the reflectarray might be different from what is desired. The deviation between desired and realized phases is caused by the infinite periodic structure assumption, which is used in the calculation of the phase as well as any error in fabrication. The former can be controlled by using a small lattice to bring about a smoother element-to-element transition in order to provide a better emulation of infinite periodic structure conditions. The fabrication error can be controlled by using more advanced lithographic and etching technologies. The most important source of phase error in the case of parabolic reflector is surface roughness which is constrained by the fabrication technology. At higher frequency bands (about Ka-band) maintaining acceptable surface roughness is quite costly and difficult to implement as compared to fabrication technology used in the case of reflectarray antennas.

Phase efficiency is the most important factor that impacts the design methodology and performance of a reflectarray. Depending on the required bandwidth and application, the appropriate cell element should be exploited to achieve the phase front transformation in the most efficient manner which is equivalent to high aperture phase efficiency.

5.3.5 Blockage Efficiency

Blockage efficiency represents the amount of energy that is blocked by the feed apparatus and it is calculated using the following relationship:

$$\eta_b = \frac{\left| \int_0^{2\pi} \int_0^{b} E_y(\rho, \varphi) \rho d\rho d\phi \right|^2}{\int_0^{2\pi} \int_0^{a} \left| E_y(\rho, \varphi) \right|^2 \rho d\rho d\phi} \tag{5.9}$$

where b is the diameter of the feed region that blocks the signal to/from the main reflector. Blockage of a portion of power by the feed apparatus of a broadside centre-fed parabolic reflector is inevitable. The larger the feed assembly, the lower is the blockage efficiency. However, a centre-fed reflectarray can be designed to shoot out its radiated beam in an off-bore-sight direction to avoid the feed assembly. Therefore, it is quite practical to have better blockage efficiency in the case of a centre-fed offset beam reflectarray as compared to the centre-fed parabolic reflector.

5.4 Cell Elements and Specific Applications: A General Survey

Reflectarray structures offer a large degree of freedom, which is the result of the large variety of cell elements that can be used. On the other hand, a number of innovative functions such as power combining, polarization rotation, beam splitting, beam shaping, etc. can be implemented by judicious selection of the cell element. Incorporation of these functions into the reflectarray system might be feasible in the case of conventional reflectors but at the expense of high complexity of the feed system or complicated shape of the reflector surface. For instance, Figure 5.7 shows the front view of a dual orthogonal linear polarization Rx/Tx Ka-band reflectarray. The Tx-band (29.5 GHz–30.0 GHz) and the Rx-band (19.5 GH–20.0 GHz) are of linear polarization and orthogonal to each other. Noting this, wide/narrow microstrip patches were selected for Tx/Rx bands, respectively. The dimensions were adjusted so that the elements of the two bands could be interlaced without touching each other. The length dimension of these two types of cell elements was orthogonal to each other to accommodate the specific polarization that each set of elements is operating at. Numerical simulations were used to ascertain low cross-coupling between the elements of the two bands. This was done by calculating the phase-length of each type of element in the absence and presence of the elements of the other band. Close agreement of the phase-length characteristics for these two simulations indicated a low level of cross-coupling between the elements of the two bands. Therefore, one can assign a dedicated focal point to each band which leads to two separate focal points as shown in Figure 5.7 [6].

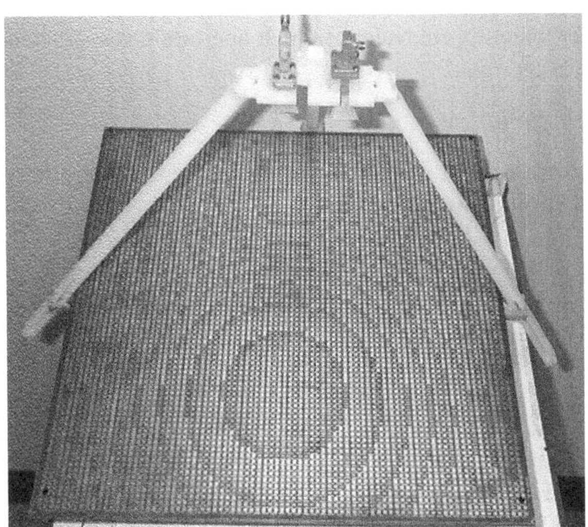

Figure 5.7 Front view of the dual-band dual-polarization bifocal reflectarray. The antenna size is 44 cm × 44 cm, F/D is 0.5, and the separation between the feeds is 5.0 cm [6]

The volume of the reflectarray can be reduced by half using the folded reflectarray technique shown in Figure 5.8 [22]. The linearly polarized signal that is radiated by feed is intercepted by the "upper reflector," reflected back towards plate "lower reflector" where its polarization is rotated by 90° while the phase front of the signal is transformed into a planar front and sent back towards the plate "upper reflector." This surface is designed to be reflective to a certain sense of linear polarization and be transparent to the orthogonal polarization. Therefore, the signal passes through the plate "upper reflector" in its second encounter. A typical radiation pattern of the folded antenna can be seen in Figure 5.8. To understand the polarization rotation mechanism, it is instructive to refer to Figure 5.8 where it is shown that polarization of the incoming wave is 45° with respect to the symmetry axis of the cell element. Therefore, the incoming signal can be decomposed into two equal components along the axis of the cell element. The orthogonal dimensions of the cell element are adjusted so that one of these components is de-phased by 180° henceforth rotating the signal by 90°.

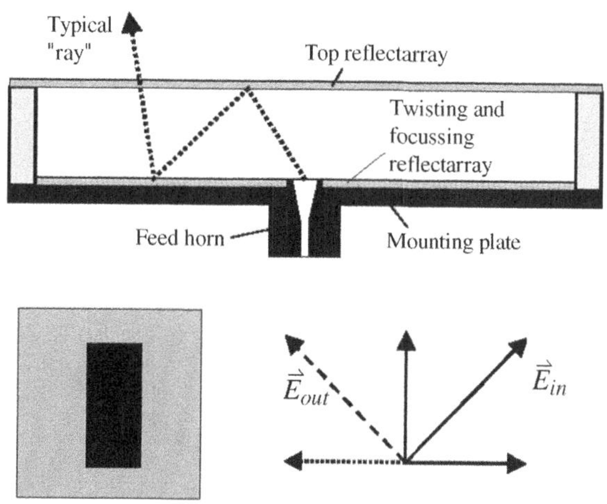

Figure 5.8 Principle of folded reflectarray antenna. Reproduced by permission of ©2005 IEEE [22]

Design of Circularly Polarized (CP) reflectors has always been a challenge because of the complexity of the design and fabrication of CP feeds. The feed system of dual band CP reflectors can be even more complicated. However, a dual band CP reflectarray can be easily implemented using Linear Polarized (LP) feeds in the case of reflectarray technology. Figure 5.9 shows the basic principle behind such an undertaking. A cross-dipole was chosen as the cell element and the incoming field is linearly polarized along the 45° line with respect to the arms of the dipole. The lengths of the arms are adjusted to establish 90° phase lead/lag that is required for the generation of RHCP/LHCP polarization and at the same time perform the phase front transformation. Figure 5.10 shows a dual band Tx/Rx reflectarray with two distinct dedicated feeds for the two bands [23]. Measured gain bandwidth and axial ratio performance of the antenna can be observed in Figure 5.11.

If a CP feed is available, one can use dedicated CP elements as shown in Figure 5.12. Each element contains two orthogonal stubs. The required phasing is imparted to the impinging CP

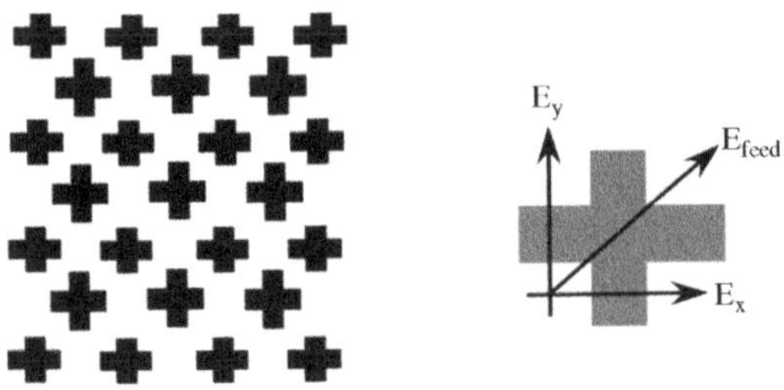

Figure 5.9 Circularly polarized reflectarray composed of cross-dipoles. The operating principle is shown on the right

Figure 5.10 Ka-band Tx/Rx reflectarray. The reflectarray utilizes a dedicated set of cross-dipoles for each of the Tx/Rx bands in order to collimate the beam [23]

field by angular rotation of the element [24, 25]. It is possible to get a rather high gain and good axial ratio performance from such reflectarrays as shown in Figure 5.13.

Multi-band operation of reflectors when the signals of the bands are of the same polarization, has always been a challenging problem in the case of conventional reflectors. The problem becomes almost intractable when distinctly different radiation footprints are required for the operating bands. The problem finds a rather simple solution in the context of reflectarray technology. It can be implemented using multilayer elements shown in Figure 5.14 [26]. The elements on the top collimate the lower band signal and the same is being performed by the lower elements within the higher frequency band.

Numerical simulations can be carried out to ensure low cross-coupling between the elements of the two bands. The radiation pattern of the reflectarray at the two bands is shown

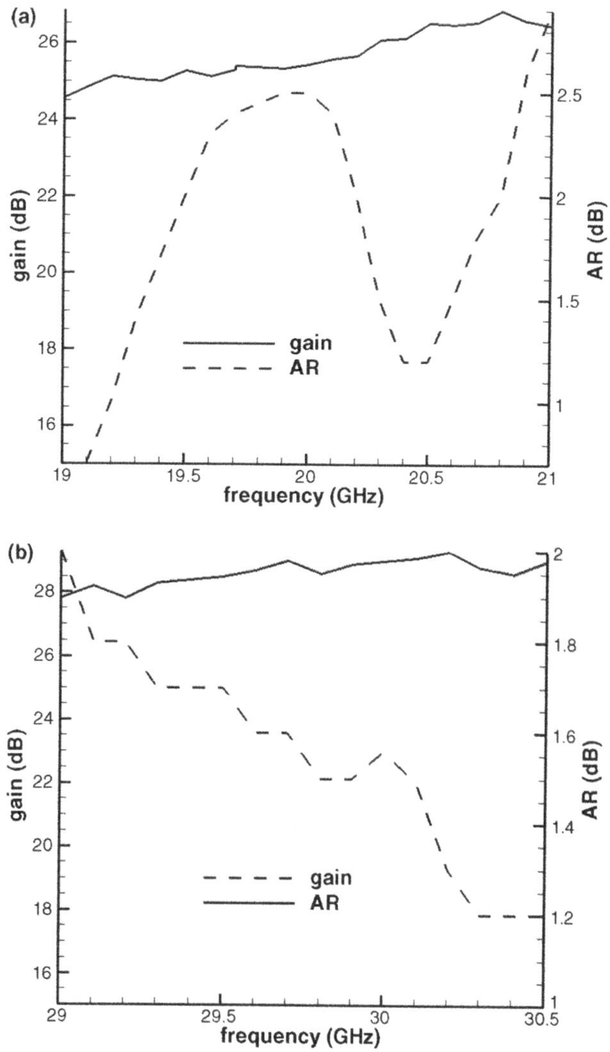

Figure 5.11 Measured gain bandwidth and axial ratio performance of the reflectarray in Tx/Rx bands: (a) Rx band; (b) Tx band [23]

in Figure 5.15. A different paradigm to achieve the same is presented in Figure 5.16, which shows a multiplicity of reflectarrays whose ground plane backing is replaced by a Frequency Selective Surfaces (FSS) [27, 28]. Each surface does the collimation in only one band. The FSS is designed to be reflective at the operating band of the reflectarray and transparent otherwise.

Multiple feeds at different locations can be utilized so that each feed is dedicated to only one reflectarray. Having maintained high isolation between the reflectarrays by judicious selections of the FSS and reflectarray cell elements, one can realize distinctly different footprints for different operating bands. Figure 5.17 shows a simple implementation of this technology in

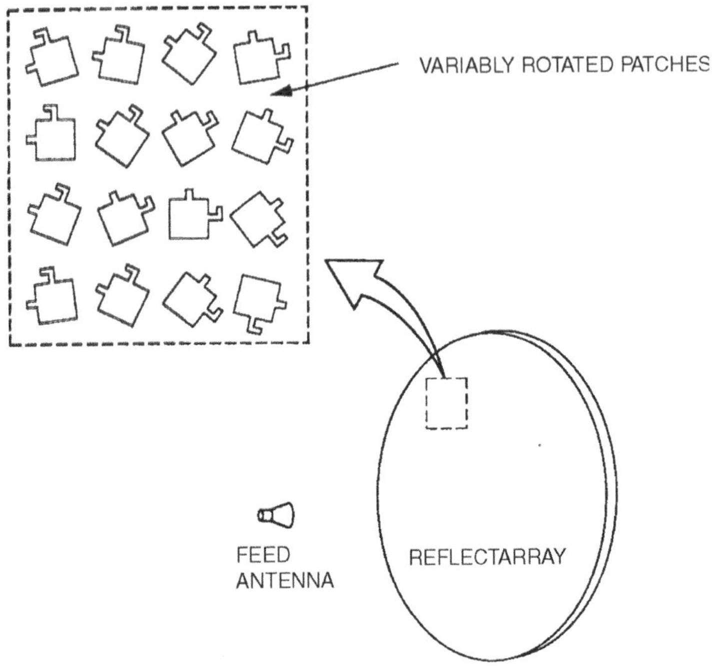

Figure 5.12 Circularly polarized microstrip reflectarray with elements having variable rotation angles. Reproduced by permission of ©1998 IEEE [24]

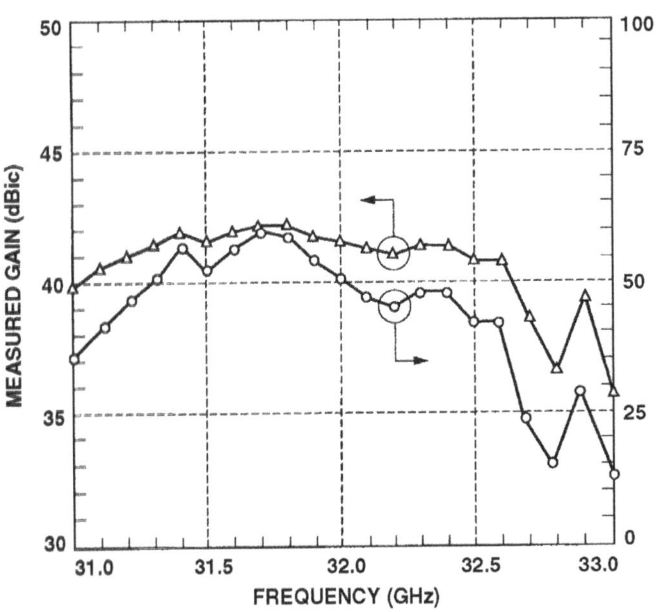

Figure 5.13 Measured bandwidth characteristics of unit 2 reflectarray with patches having variable rotation angles. The curve with small triangular markers represents the gain plot and the one with small circles represents the efficiency plot. Reproduced by permission of ©1998 IEEE [24]

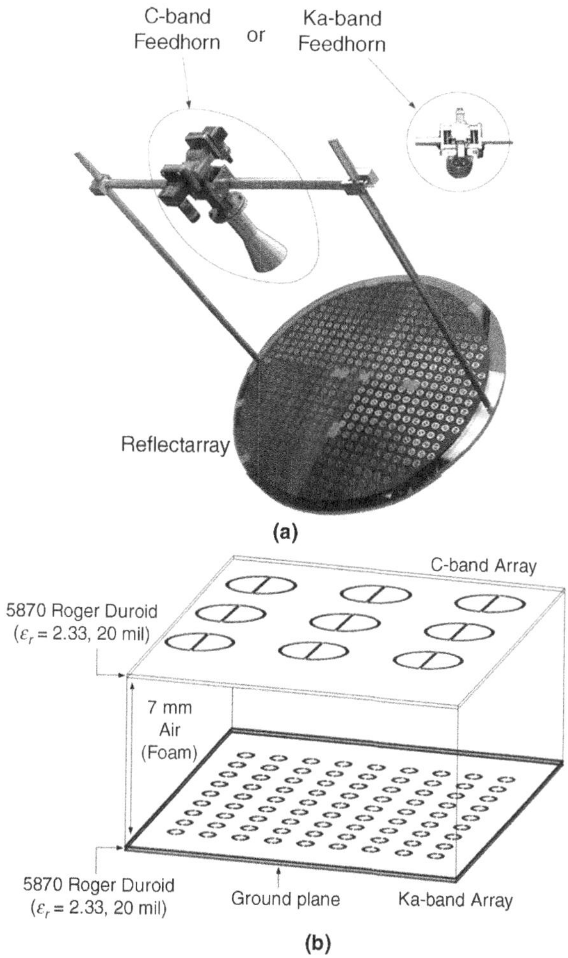

Figure 5.14 (a) Photograph of the reflectarray; (b) dual layer reflectarray topology. Reproduced by permission of ©2004 IEEE [26]

conjunction with an offset parabolic reflector. The parabolic reflector is designed to operate at Ku-band and the reflectarray with FSS backing is to operate at Ka-band. The reflectarray is reflective in the Ka-band and transparent in the Ku-band. The two reflecting surfaces can have distinct focal points.

Figure 5.18 shows the radiation patterns at the two bands which demonstrates that while the lower band signal has been collimated along the bore-sight, the upper band radiation pattern is off-bore-sight.

Novel cell elements can also be used to devise switched beam reflectarrays. The cell element type shown in Figure 5.19 was investigated thoroughly towards this end. Slots are used as loads onto patches of all the same size on the top layer [29, 30].

Figure 5.20 shows the configuration of a mechanically steerable reflectarray that utilizes this type of cell element. The top layer was composed of identical size patches and the middle and

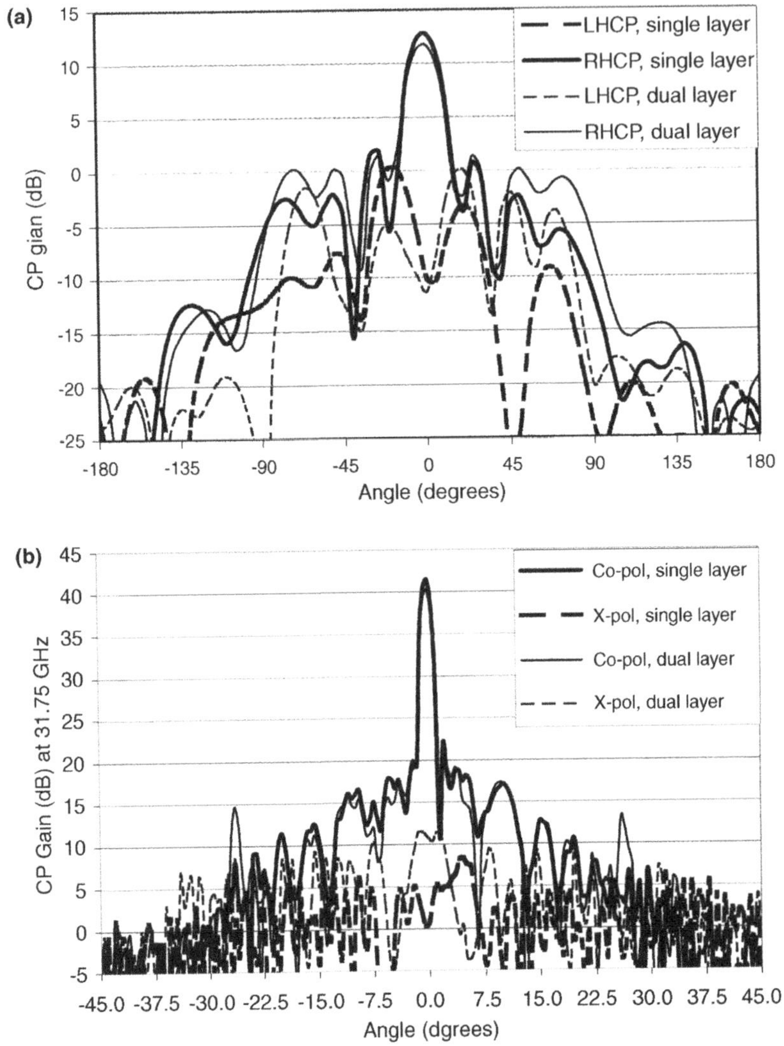

Figure 5.15 (a) Measured CP gain at 7.3 GHz for a dual layer reflectarray; (b) radiation patterns at 31.75 GHz for both the single and dual layer reflectarray. Reproduced by permission of ©2004 IEEE [26]

bottom layers were composed of two sets of slots. In the middle layer, the first set of slots (1) all had identical sizes and in the second set of slots (2) were several groups of slots of dissimilar length. There was no air gap between these two sets of slots and each group of slots of varying length corresponds to one and only one beam. The slots of identical length acted as a window through which only one set of variable slots can be seen by the patches. The slots that belonged to other sets were covered by the ground plane. Figure 5.21 shows three sets of such slots in each unit cell on layer 2 for collimating the beam at $-30°$, $0°$, and $+30°$ depending on the set of slots that are visible to the patch through the "window" in the second layer. It should be noted that

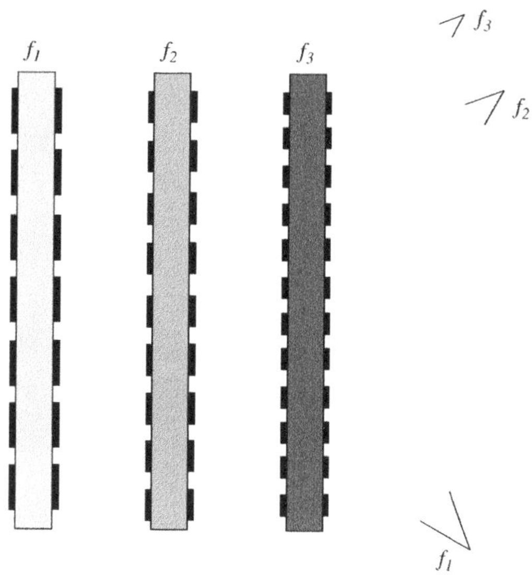

Figure 5.16 Multiple surfaces of FSS backed reflectarrays are used to operate different bands and polarizations

only the slots aligned with the uniform slots are seen by the incoming wave. The separation between the slots within the same unit cell is 1.0 mm. Therefore, upward or downward displacement of the second layer with respect to the first layer is ± 1 mm which resulted in three different beams pointing to $-30°$, $0°$, $+30°$.

One of the advantages of reflectarrays over reflectors is that they are amenable to advanced lithographic and etching technologies. For instance, Micro Electro Mechanical (MEM) techniques can used to realize innovative beam shaping or beam switching methods [9]. One implementation of this technology in the context of reflectarrays is shown in Figure 5.22 where MEM technology is used to activate cell elements in a fashion to steer the CP beam into the desired direction. This was achieved by changing the angle of rotation of the cell element by activating the appropriate MEM switch.

Laser-induced plasma technology can be used to generate plasma on a high-resistivity silicon that in turn can be used to dynamically shift the beam of a reflectarray [31]. Figure 5.23 shows a diagram of the reflectarray cell which is composed of a slot-coupled patch as the cell element. Laser illumination onto the slots can practically change slot dimensions by inducing an island of plasma in appropriate locations. The Indium Tin Oxide layer is used in the backing to decouple the optics from RF.

Coupled resonance between stacked patch elements is used in the case of a microstrip patch to improve the antenna bandwidth. The same concept can be used in the case of reflectarrays to improve the bandwidth of moderate size antenna elements [12]. The stacked patch cell element demonstrates a smooth phase variation which is the result of coupled resonance as shown in Figure 5.24. The same type of elements can be used to achieve shaped reflectarray structures [32].

Reflectarrays have been used in innovative applications such as power combining and beam splitting. Using appropriate phase transformation, the incoming field from four

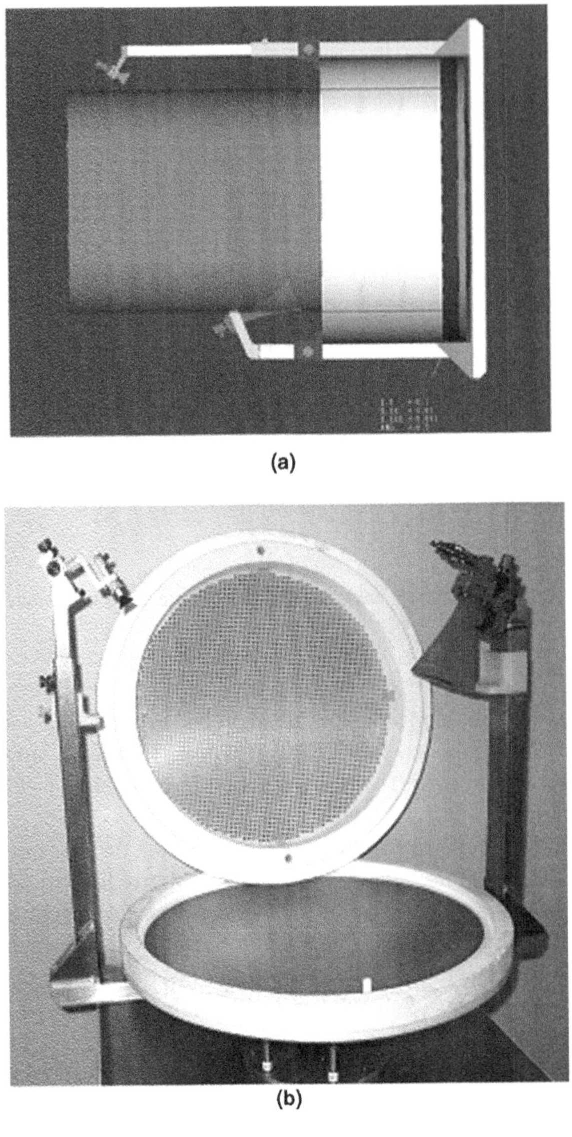

Figure 5.17 Two different views of the hybrid antenna system [27]: (a) side view of the reflectarray and the offset reflector. The reflectarray is held by a foam frame 10.0 mm away from the offset reflector rim. (b) The foam holder is used to keep the reflectarray away from the offset reflector

different feeds can be combined in one and the same beam. The top view and side views of such a reflectarray are shown in Figure 5.25. The radiation pattern of the whole system with four feeds is shown in Figure 5.25 (c) [33, 34]. A different type of phase transformation can be used to split the energy that emanates from the single feed into two radiated beams in two different directions [35].

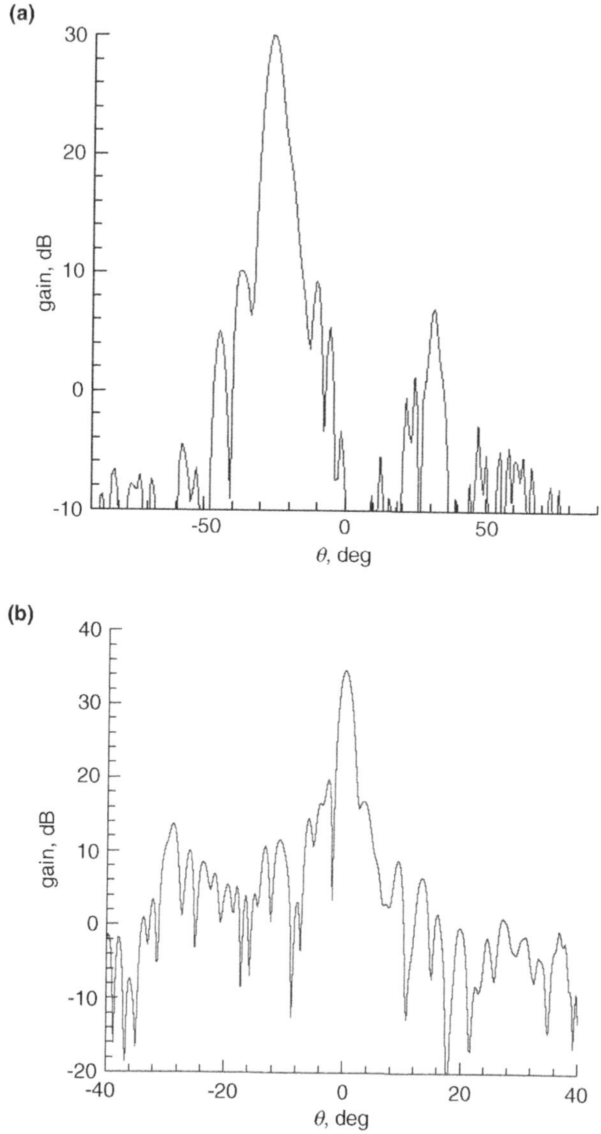

Figure 5.18 Measured radiation pattern of the FSS-backed reflectarray [27]: (a) radiation pattern in the lower band, f = 11.0 GHz; (b) radiation pattern of the hybrid reflector in the upper band (f = 27.2 GHz)

5.5 Wideband Techniques for Reflectarrays

There are two separate factors that play an important role in setting an upper bound on the bandwidth of a reflectarray antenna [17, 36]:

1. Element bandwidth
2. Time delay factor

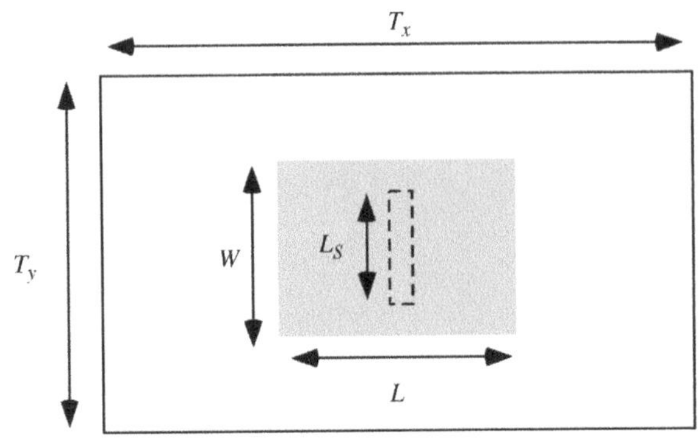

Figure 5.19 Typical slot loaded patch element. Slot length is variable to change the loading on the patch. The patch size is constant throughout the lattice

The first factor has a severe impact on the bandwidth of moderate size reflectarrays ($<20\lambda$) and $F/D > 0.5$ while the second factor becomes dominant in the case of large reflectarrays ($>30\lambda$) and $F/D > 0.5$. However, as will be seen later the second factor can adversely affect the bandwidth for $F/D < 0.25$ even for a moderate size reflectarray. In what follows, an

(a)

Figure 5.20 Mechanically steerable reflectarray [30]: (a) actual implementation; (b) the schematic diagram of the reflectarray unit cell

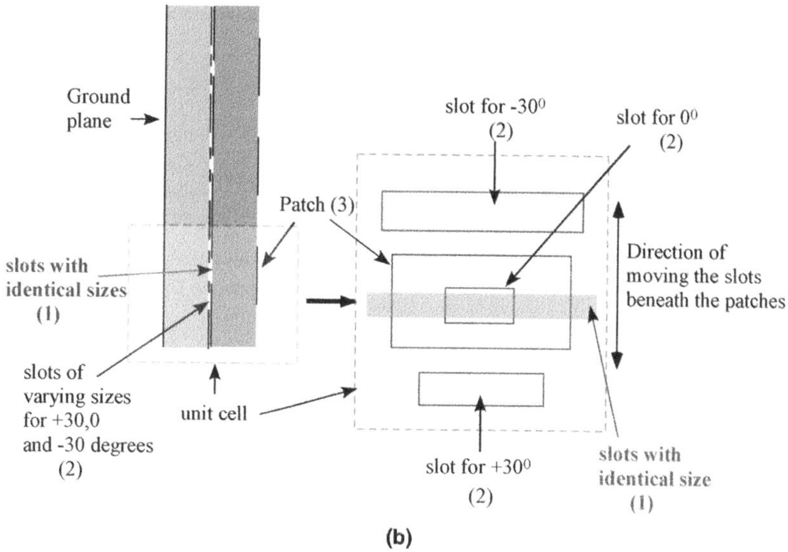

Figure 5.20 (*Continued*)

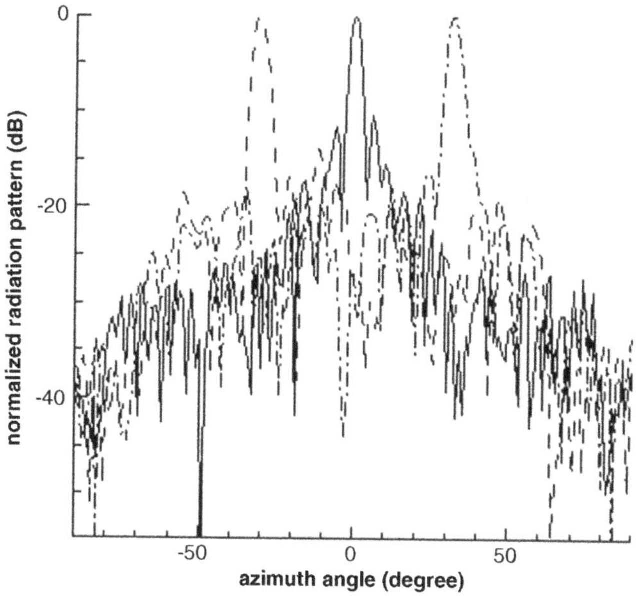

Figure 5.21 Measured radiation pattern of the reflectarray for the different positions of the moving screen [30]

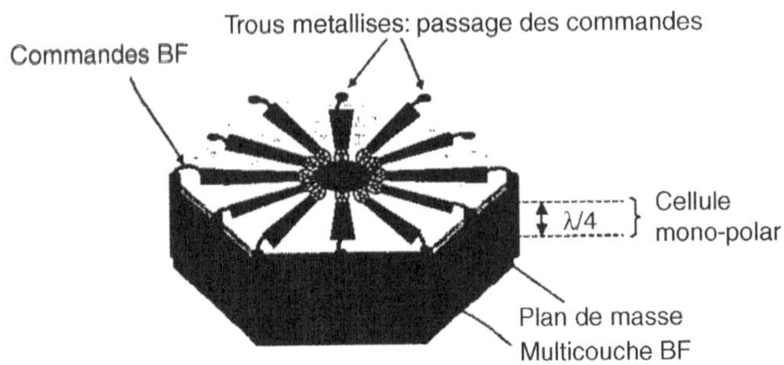

Figure 5.22 Six phase state cell used in steerable MEM reflectarray. Via holes (*trous metallisés*) are used to run bias signal (*Commandes BF*) to arms. The ground plane (*plan de masse*) is separated by a $\lambda/4$ substrate from the arms while supporting multilayer DC circuitry (*multicouche BF*) on the other side. Reproduced by permission of ©2003 IEEE [9]

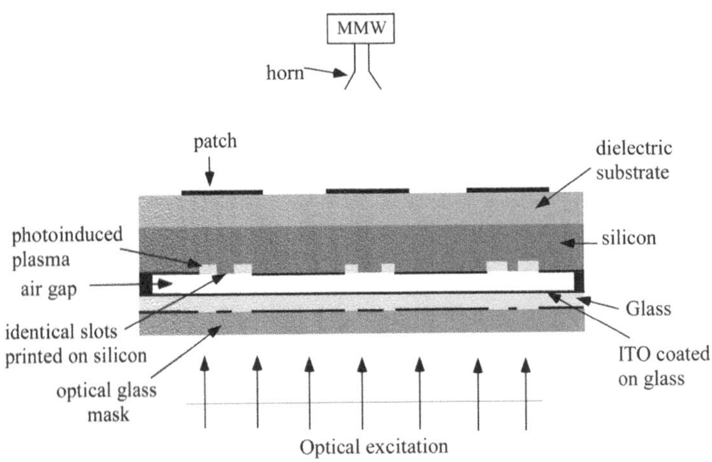

Figure 5.23 The schematic diagram of the reflectarray unit cell [31]

in-depth physical analysis will be presented to demonstrate the issue of bandwidth in the case of moderate and large size reflectarrays. Methods will be presented to improve the bandwidth, noting the physical insight attained on the factors involved in constraining the bandwidth of reflectarray. Then, results of the optimization method will be presented in theory and practice.

5.5.1 Phase Response of Reflectarrays

The phase value that is to be imparted by the elements of reflectarray in order to achieve a specific phase front transformation can be obtained using Equation (5.4). Figure 5.26 (a) shows

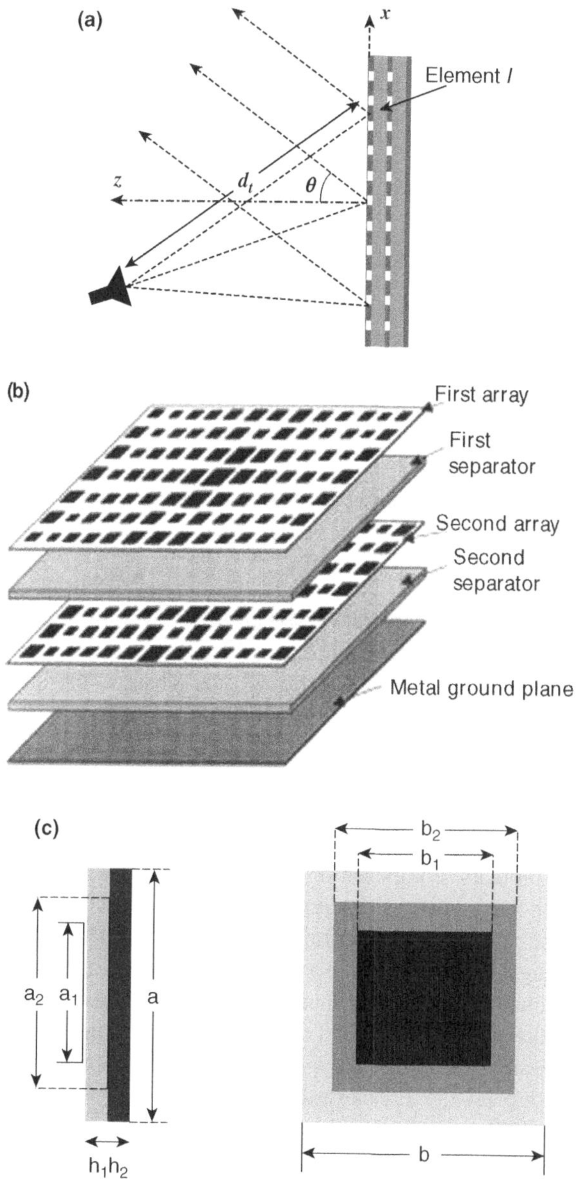

Figure 5.24 Two-layer reflectarray using patches of variable size. (a) reflectarray illuminated by a feed; (b) multilayer structure; (c) periodic cell; (d) phase versus length of the two layer stacked patches for the structure shown in (b) and (c) versus the patch side of the array closer to the ground plane: $a_1 = b_1$, $a_2 = b_2$, $a = b = 14$ mm, $h_1 = h_2 = 3.0$ mm, and $a_1 = 0.7a_2$. Reproduced by permission of ©2001 IEEE [12]

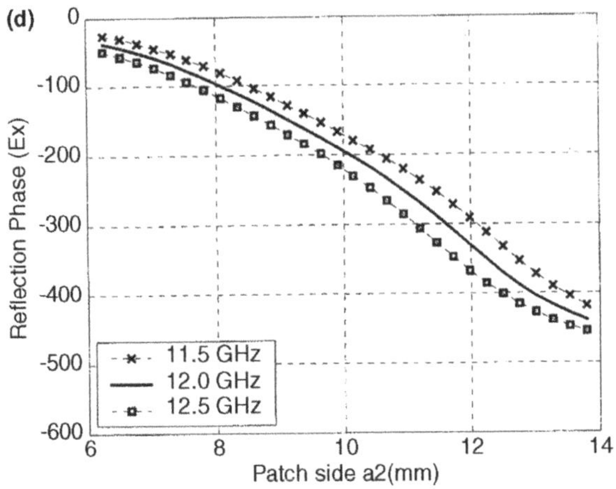

Figure 5.24 (*Continued*)

a plot of the required phase for a moderate size Ku-band reflectarray. A similar plot has been generated and shown in Figure 5.27 for a large size reflectarray in the same frequency band and the same *F/D* as the moderate size reflectarray. To demonstrate the effect of small *F/D*, a separate plot was generated for the case of a moderate size reflectarray with *F/D* = *0.25* and the result is shown in Figure 5.26 (b). The plots have been generated for the centre frequency and the edge frequencies. A close examination of these plots leads to a useful insight into understanding the cause of the limited bandwidth and for devising methods to overcome this limitation.

All of these phase plots are composed of so-called "zones" which are regions where the phase is increased smoothly until a phase jump is encountered which is the starting point of the successive adjacent zone. The first (central) zone in both cases (Figures 5.26 and 5.27) is the largest zone. Following the first zone, the zone width becomes narrower as the rim of the reflectarray is approached. This effect is more visible in the case of the large reflectarray in Figure 5.27. The proportional size of the first zone, with respect to the reflectarray size, is larger in the case of moderate size reflectarray of Figure 5.26 (a). These observations are in agreement with general knowledge on the Fresnel zone plate [37]. For the case of moderate size reflectarray, the required phase in the first zone does not change significantly as the frequency is swept across the band. On the other hand, the variation of phase is different for different zones as the frequency is swept across the band (cf. Figure 5.28). Therefore, relatively high phase efficiency can be maintained because of the comparatively large size of the first zone with respect to the size of the reflectarray. The situation is more serious in the case of a large reflectarray where the phase variation caused by frequency shift might be significant even for the first zone while the phase "jitter" for the outer zones is more severe. The difference of the desired phase values in the two band-edge frequencies ($\Delta\phi = \phi_l - \phi_u$) versus the desired phase value at the centre frequency has been plotted for moderate size and large reflectarrays in Figures 5.28 and 5.29, respectively. It is evident from these figures that $\Delta\varphi$ assumes different values for different zones.

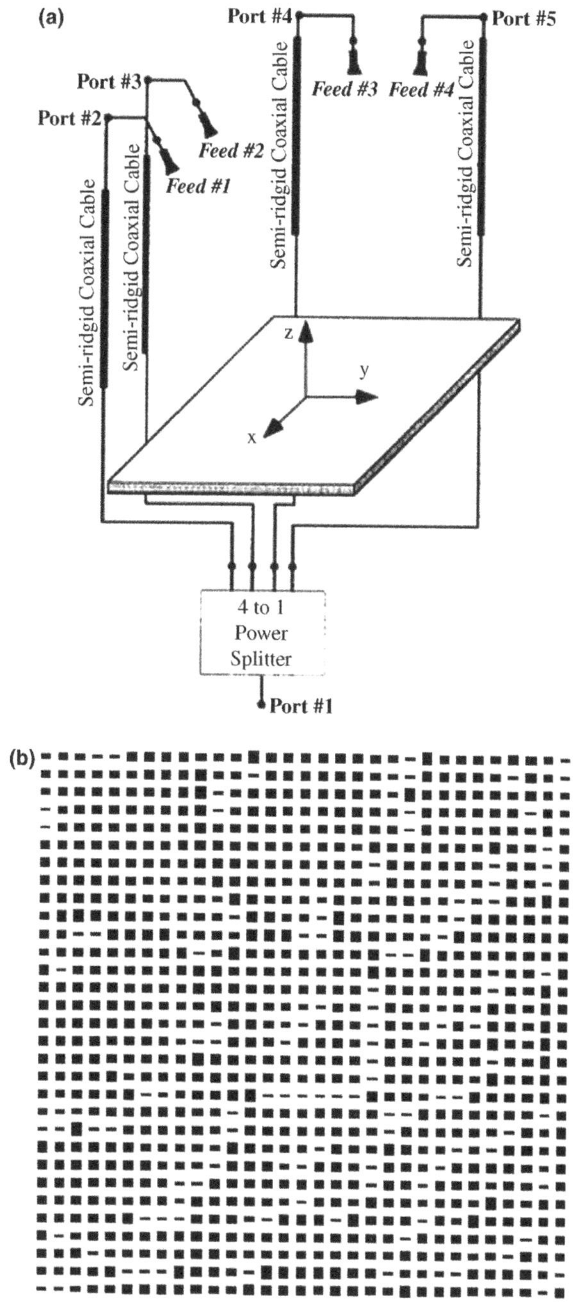

Figure 5.25 (a) Basic principle of power combining using reflectarray technology; (b) layout of the power combiner; (c) radiation pattern of the reflectarray fed simultaneously by four feeds, f = 30.0 GHz. Reproduced from ©2005 ESA [33]

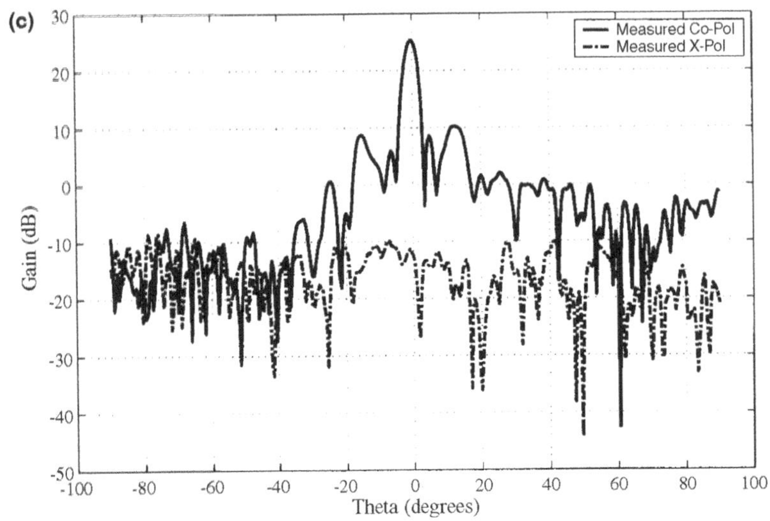

Figure 5.25 (*Continued*)

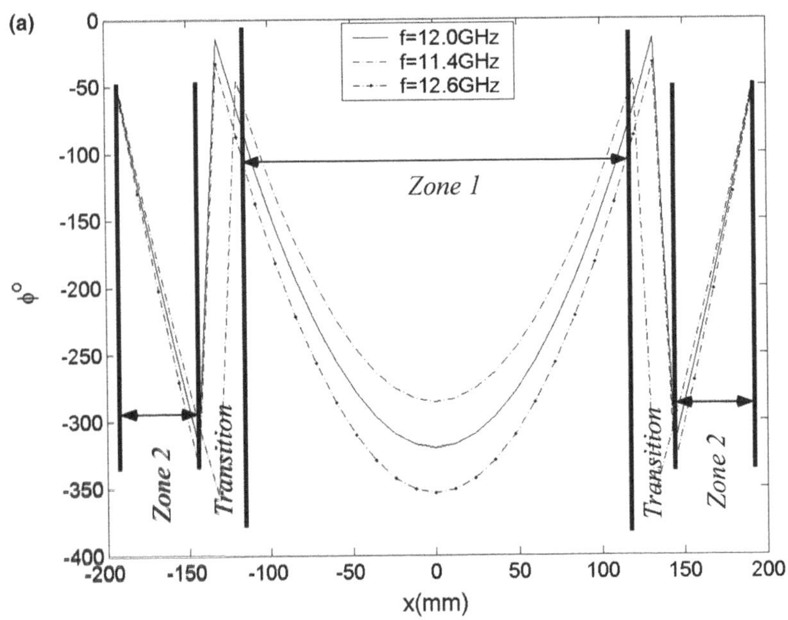

Figure 5.26 Required phase for reflectarrays of 40.0 cm × 40.0 cm size at the edge and centre frequencies: (a) $F/D = 1.0$; (b) $F/D = 0.25$

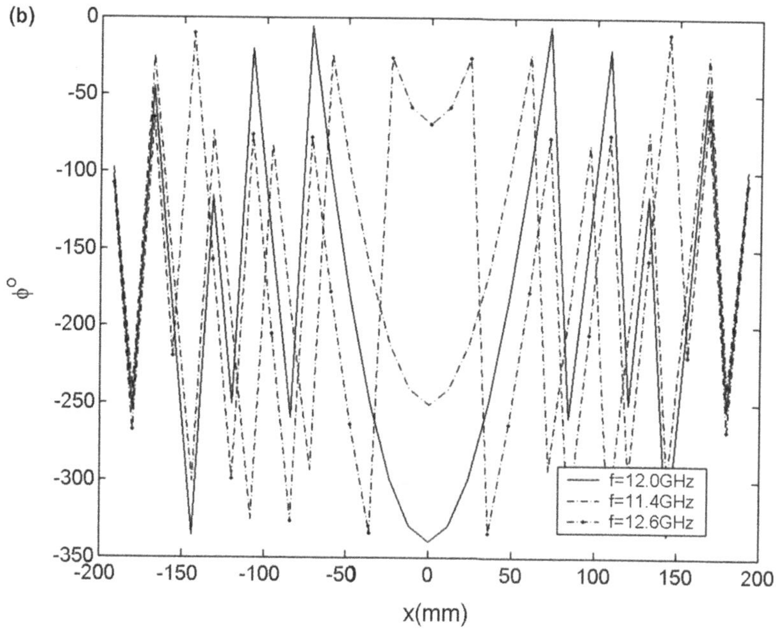

Figure 5.26 (*Continued*)

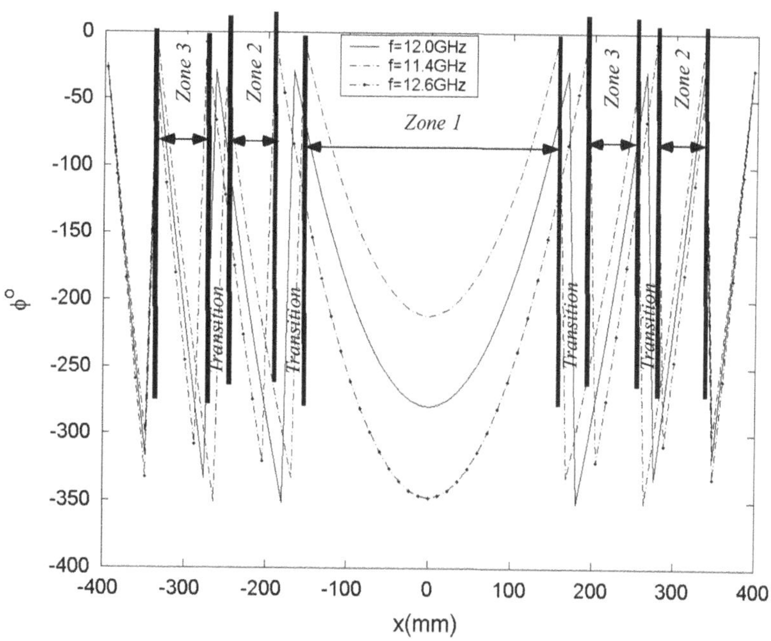

Figure 5.27 Required phase for reflectarrays of 80.0 cm × 80.0 cm size at the edge and centre frequencies

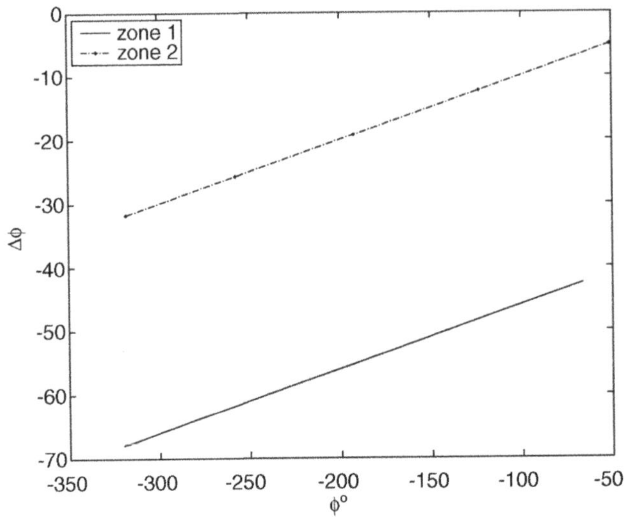

Figure 5.28 Difference of the desired phase at the edge frequencies of the moderate size reflectarray of Figure 5.26 (a) ($\Delta\phi = \phi_l - \phi_u$) versus the desired phase in the centre frequency

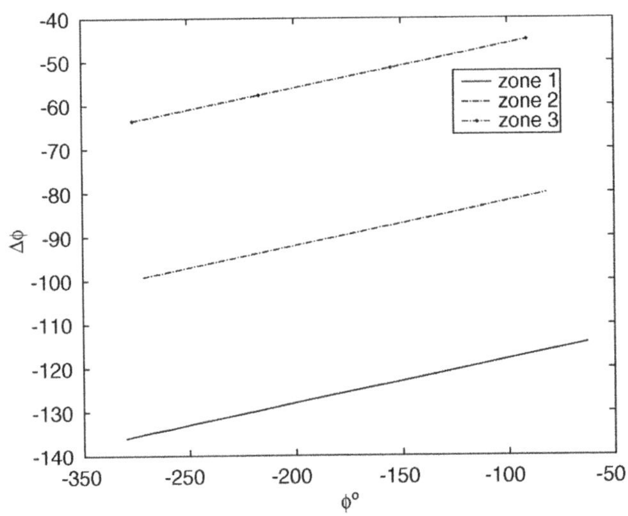

Figure 5.29 Difference of the desired phase at the edge frequencies of the large reflectarray of Figure 5.27 ($\Delta\phi = \phi_l - \phi_u$) versus the desired phase in the centre frequency

Having noted the trends that govern the required phase in the case of a moderate size reflectarray, the following conclusions are made:

1. Considering the relative size of the main zone to the size of the reflectarray, the first zone contributes significantly to the gain level as a result of its relatively large size.

2. The level of the required phase in the first zone does not change much as the frequency is swept across the band.
3. The level of $\Delta\varphi$ for different zones of the reflectarray is different.

Noting the first and second observations above, the bulk of the gain is realized by ensuring that the cell elements in the first zone have rather stable and linear phase response throughout the operating band. This latter requirement leads to wideband elements. To translate this into a specific practical design guideline, the cell element should demonstrate a smooth and gradual phase characteristic and its frequency response across the operating band of the reflectarray should resemble as closely as possible the behaviour of $\Delta\varphi$ in the first zone as shown in Figure 5.28. Therefore, it is meaningful to embark on a search for methods to synthesize cell elements with such a phase response. The stacked patch [12], which demonstrates smooth phase characteristics by exploiting of coupled resonance phenomenon, is well suited to attain this objective. A detailed account of the application of coupled resonance in synthesizing loop-based cell elements with smooth phase response will be given in the next section.

The following conclusions can be drawn from a close examination of the plot for a large reflectarray (Figure 5.29):

1. Considering the relative size of the main zone as compared to the size of the reflectarray, the first zone contribution to the whole gain is not strong as compared to a moderate size reflectarray.
2. The level of the required phase in the first zone changes quite significantly as the frequency is swept across the band.
3. The level of $\Delta\varphi$ for different zones of the reflectarray is different.

According to the first and second points above, contribution of the elements in the first zone to the gain level is not as strong as their counterpart for the case of moderate size reflectarray. Also, the elements in the centre should be capable of a wider "phase swing" throughout the frequency band. Therefore, the phase versus length characteristics should be more abrupt as compared to the similar elements in the case of moderate size reflectarray. This has implications on the impact of conductor loss on the efficiency of the reflectarray [21]. On the other hand, $\Delta\varphi$ requirements are very different for different zones as noted in Figure 5.29 and the relative size of higher order zones is larger compared to the moderate size reflectarray. This "pathological" investigation into the causes of reduced bandwidth in the case of large reflectarray might lead the attentive reader to possible solutions to overcome this restriction. For instance, the reflectarray elements ought to be selected from a pool of elements that realize the required phase at the centre frequency and at the same time are the best fit considering the phase requirements at the edge frequencies. In other words, the cell element at a particular location on the reflectarray has the dual function of realizing the desired phase in order to attain the phase transformation at the centre frequency AND edge frequencies. Having different phase requirements at these frequencies would force the designer to select from a pool of elements [38–40]. The following relationship is used as the error function to choose from a pool of elements:

$$e(m,n) = \sum_{i=l,c,u} \left|\Phi_i^{desired}(f_i)(m,n) - \Phi_i^{acieved}(f_i)(m,n)\right| \qquad (5.10)$$

Where the l, c, u refer to lower, centre, and upper frequencies, m, n refer to element indices on the reflectarray grid, and Φ is the phase shift. This error function can be used as the objective function in an optimization routine to be presented in Section 5.5.2 to design moderate and large size reflectarrays.

The issue of frequency dispersion in the case of the large reflectarray can be understood from a different perspective. Figure 5.30 shows the phase difference between the phase of the elements at the centre and the rim of the reflectarray when there is no restriction on the amount of phase that can be realized by the elements. In other words, the elements are capable of realizing phase values in excess of 360°. Wavelength is the "yardstick" that is to be used in the quantification of the desired phase. The physical size of this "yardstick" changes as the frequency is swept across the band which demonstrates the frequency dispersion. This is equivalent to different phase constraints at different frequencies in order to realize the same phase transformation. Now, if a mechanism can be contrived to account for the time delay of the signal as it travels from the feed to different points on the reflectarray, the bandwidth can be increased significantly.

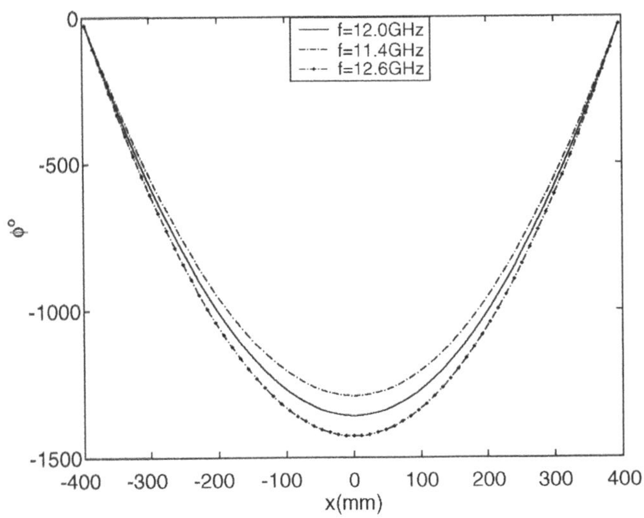

Figure 5.30 Desired phase for the large reflectarray of Figure 5.27 when there is no restriction on the amount of phase that can be attained from the cell element

Figure 5.31 shows slot coupled time delay lines that have been introduced in [41] as a measure to improve bandwidth. It can be seen that a phase shift of the order of $5{*}2\pi$ can be achieved which delays the occurrence of the first phase jump. As shown in Figure 5.32, this measure practically enlarges the size of the first zone and the same guideline as in the case of a moderate size reflectarray can be followed to improve the bandwidth performance.

5.5.2 *Verification of the Optimization Method*

Different classes of double-cross loop elements with variable length (cf. Figure 5.33) were used to design the reflectarray. Different loop elements were defined by changing the line width (d_i),

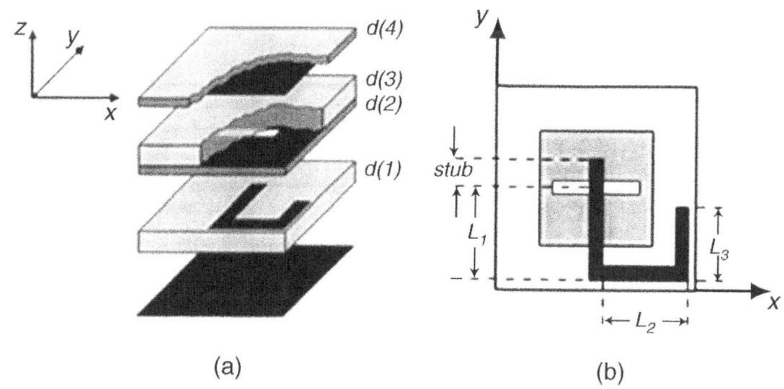

Figure 5.31 Reflectarray element based on patches aperture-coupled to delay lines: (a) expanded view; (b) top view. Reproduced by permission of ©2008 IEEE [41]

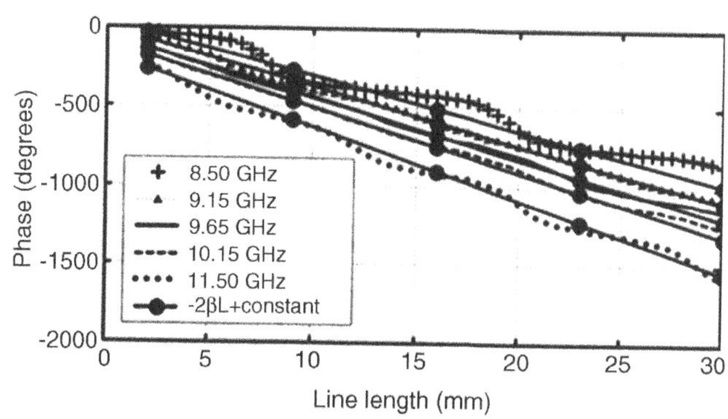

Figure 5.32 Reflection coefficient for reflectarray element as a function of line length for normal incidence. Reproduced by permission of ©2008 IEEE [41]

loops separation (g_i) and cross-width (w) (e.g. Table 5.1). These sets of loops form the search space for the selection of the elements according to the criteria outlined in the previous section. In practice, the error function introduced in Equation (5.10) is tested with all elements within the search space and the element with the lowest error function is selected as the optimum element from the point of view of bandwidth performance. This same procedure was repeated for all the elements to get the minimum frequency dispersion for all the reflectarray elements. In other words, from all the members of the search space that provide a given phase at the centre frequency, the one that minimizes the error function is chosen as the optimum element. Figure 5.34 shows the phase of reflected wave versus loop length (L) for the double-cross loop element with the line width of $d_1 = d_2 = 0.2$ mm and different loop separations (g_1). A smother phase variation can be achieved as the g_1 is increased. All the simulations were done in HFSS™ [42] using periodic boundary conditions.

A moderate size 40 cm × 40 cm offset-fed reflectarray has been designed to operate in the 10.5 GHz–13.5 GHz frequency band. The reflectarray is designed to collimate the beam at

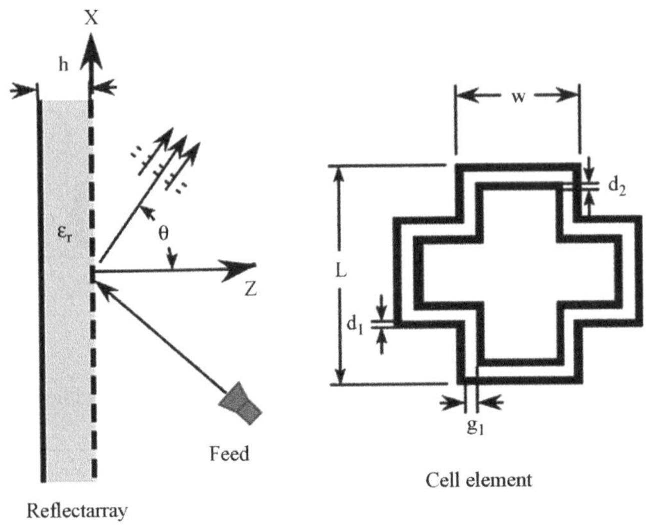

Figure 5.33 A schematic view of the reflectarray and the cell element. Reproduced from ©2009 IEEE [46]

Table 5.1 Geometrical dimensions of the cell element shown in Figure 5.33

	d_1, d_2(mm)	g_1(mm)	w (mm)
double-cross loop	0.2 -0.6 step:0.1	0.2 -0.6 step:0.1	4
double-cross loop	0.2 -0.6 step:0.1	0.2 -0.6 step:0.1	4.5

$\theta = 16^\circ$. Figures 5.35 (a) and 5.35 (b) show the calculated radiation patterns at three different frequencies for the non-optimized and optimized reflectarray when different cross-loop configurations were used in the optimization to design the reflectarray. Simple array theory was used for the calculation of the radiation pattern.

Two reflectarrays with one double cross-loop (no optimization) and 50 different cross-loops with geometrical parameters as summarized in Table 5.1 were designed and fabricated.

The measured radiation patterns for optimized and non-optimized reflectarrays are shown in Figures 5.36 (a) and 5.36 (b). The optimization was performed for 10.5 GHz, 12.0 GHz, and 13.5 GHz. As shown in Figure 5.36 (b), a very stable gain was achieved for the optimized reflectarray through the frequency band with less than 1 *dB* gain variation. The optimized reflectarray demonstrated superior radiation pattern (lower SLL and cleaner main beam) through the band as compared to the non-optimized reflectarray.

The same design procedure was applied to a large reflector of 80 cm × 80 cm. The same 50 different configurations as used for designing the previous reflectarray were utilized in the design of this antenna. The centre frequency is still the same as the previous design (f_c = 12 GHz). The calculated radiation patterns of the optimized and non-optimized reflectarrays

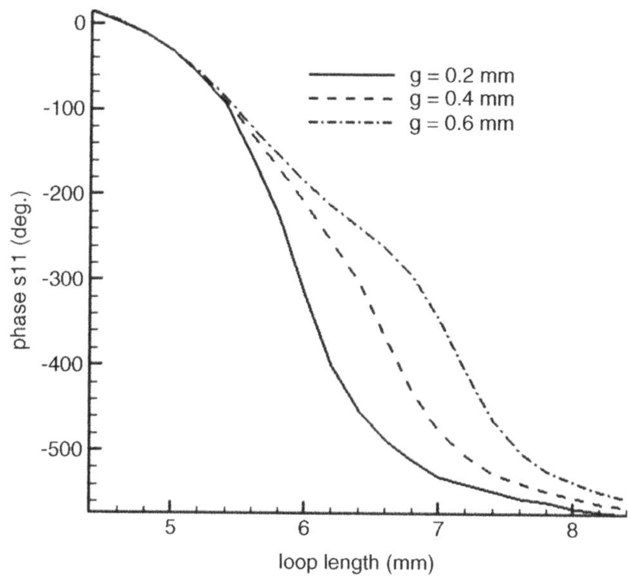

Figure 5.34 Phase of the reflected wave versus loop length for different loop separation (g_l). Reproduced from ©2009 IEEE [46]

at centre and band edge frequencies are shown in Figures 5.37 (a) and 5.37 (b). As depicted in these figures, the radiation pattern and gain of the optimized reflectarray have improved significantly compared to the non-optimized case. The gain drops by less than 0.3 dB throughout the frequency band for the optimized reflectarray.

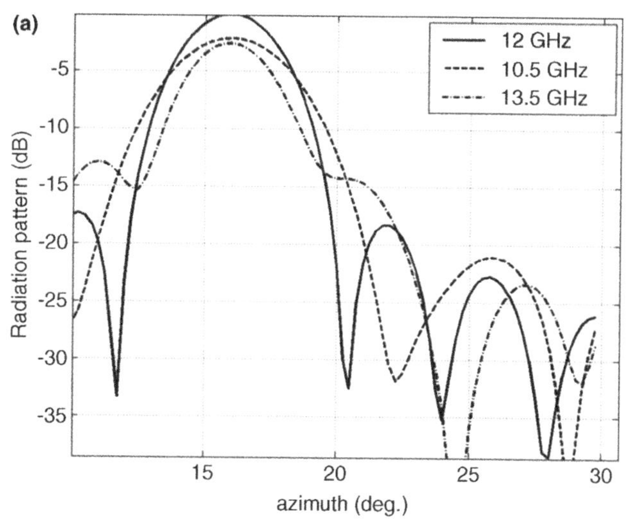

Figure 5.35 Radiation pattern of 40 cm × 40 cm reflectarray (calculation): (a) non-optimized; (b) optimized reflectarrays. Reproduced from ©2009 IEEE [46]

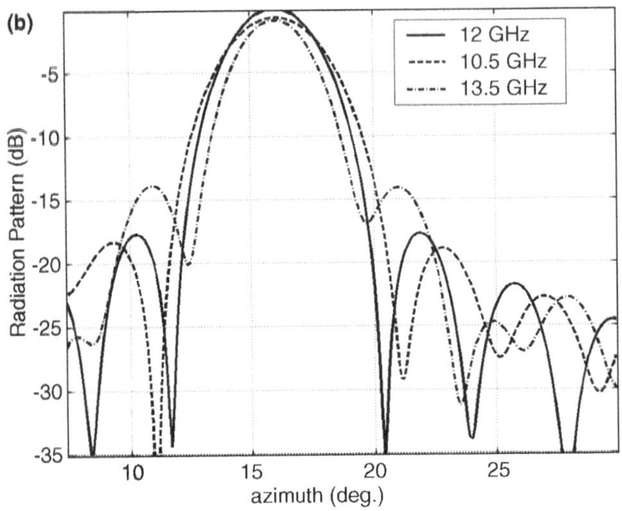

Figure 5.35 (*Continued*)

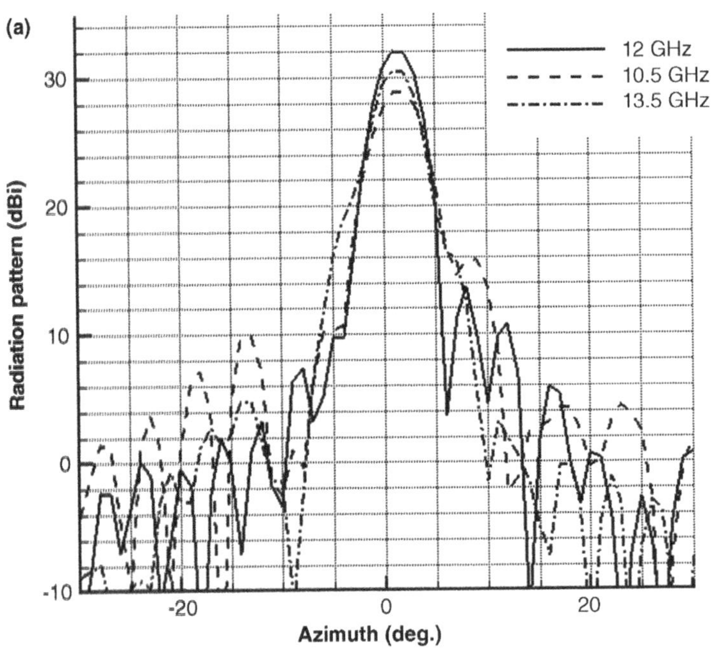

Figure 5.36 Measured radiation pattern of 40 cm × 40 cm reflectarray (measurement): (a) non-optimized; (b) optimized. Reproduced from ©2009 IEEE [46]

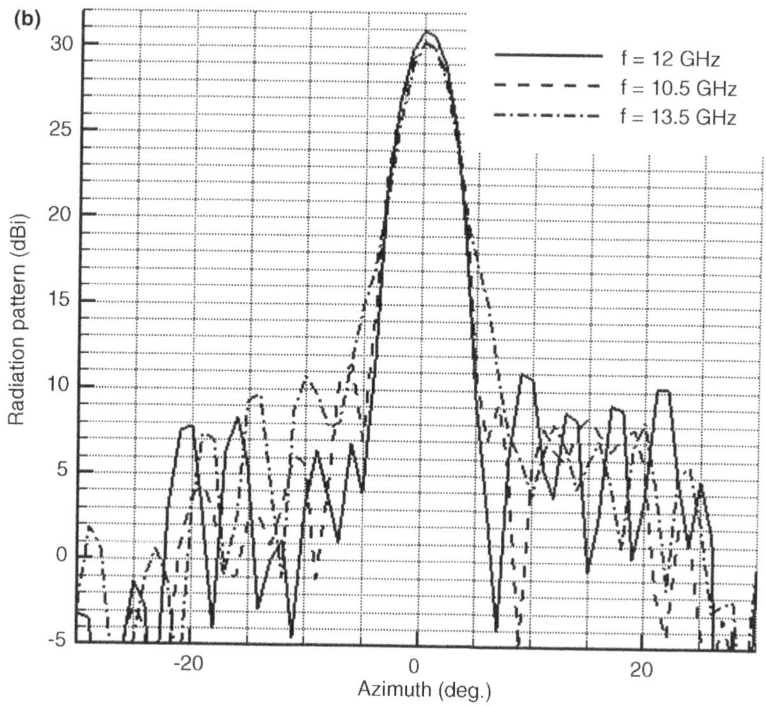

Figure 5.36 *(Continued)*

5.6 Development of Novel Loop-Based Cell Elements

The bandwidth of the cell element is the main factor that determines the bandwidth of moderate size reflectarrays as explained in the previous section. Utilization of thick substrates for single-layer reflectarray composed of patches is one of the first techniques that was used to improve the bandwidth. However, reduced attainable phase shift for such structures was synonymous with reduced gain [21]. Inspired by earlier research in the field of wideband microstrip patch, stacked patch was later suggested as the reflectarray cell element to achieve wideband operation [12]. This latter technology suffers from complexity in fabrication and reduced flexibility for multi-band or multi-polarization application. In this section, the research on realization of broadband reflectarray by focusing on using ring type cell elements is presented. The underlying physics that drives this choice is given in the next section. Subsequent sections (Sections 5.5.2 and 5.6.4) deal with different types of ring type cell elements, namely, square ring type, cross-ring type and hybrid square and cross-ring cell elements. As will be seen in the case of multi-loop structures, the same multi-resonance effect that is the basis of stacked patch design [32] is exploited in order to obtain the optimum response.

5.6.1 Motivation

Any attempt to broaden the bandwidth of the reflectarray should not solely rely on the common knowledge of the behavior of single element in isolation but should also exploit the rich

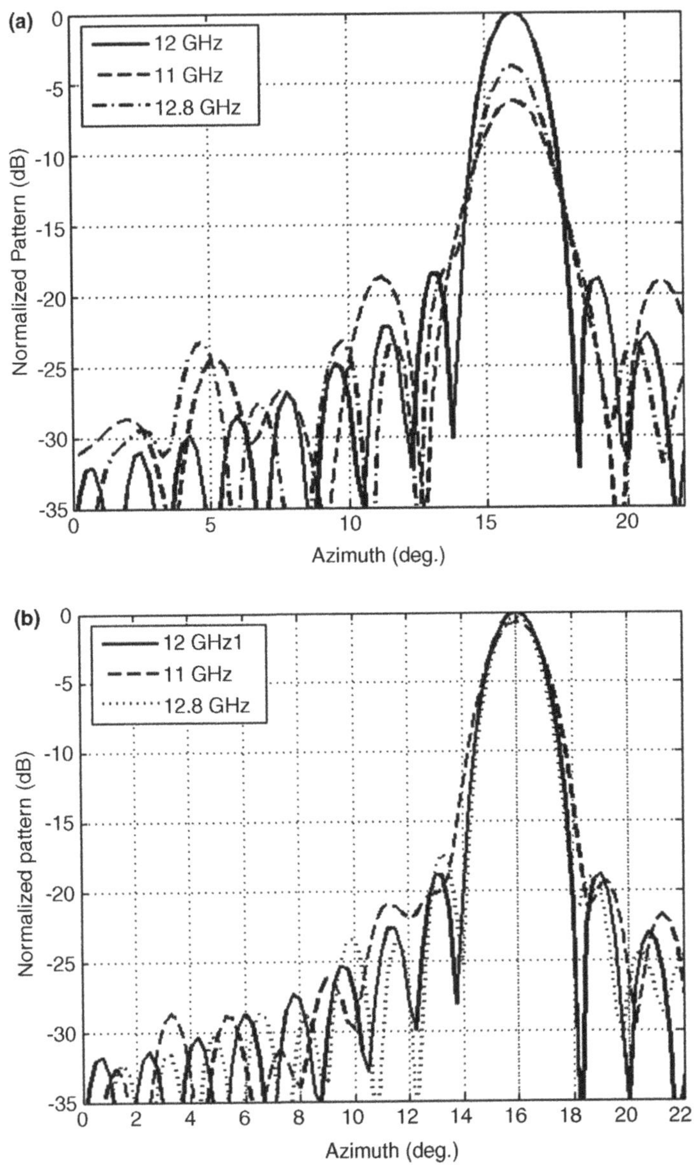

Figure 5.37 Calculated radiation pattern of 80 cm × 80 cm reflectarray: (a) non-optimized; (b) optimized. Reproduced from ©2009 IEEE [46]

literature on periodic structures [18]. Particular constraints and restrictions on the broadband operation of such extensively studied periodic structures as Frequency Selective Surfaces (FSS) are quite relevant and advantageous in the design of broadband reflectarrays. A prominent example of such an outlook is the class of cell elements known as "loop" elements in FSS nomenclature. The relatively small resonant size of this class of cell elements allows for densely packed FSSs that are superior in bandwidth performance and stability of angular response.

The small resonant size of the resonant element serves the twofold objective of exclusion of higher order modes' excitation and enhancement of the validity of the commonly used infinite periodic structure approximation that is applied in the design stage. A smaller reflectarray cell element is also more capable of capturing sharp and abrupt phase reversals on the reflectarray.

5.6.2 *Square Ring Cell Element*

To demonstrate the trends that govern the phase-length behavior of ring structures, a comparison is made in Figure 5.38 between phase characteristics of infinite periodic structures composed of a single ring cell and double ring cell elements, respectively. As can be seen, the presence of two rings leads to two resonances which is represented as two transition regions in the phase-length characteristics of this particular element.

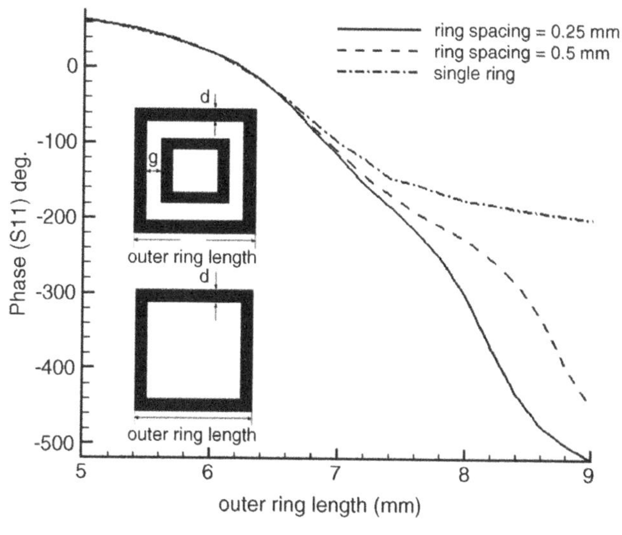

Figure 5.38 Phase-length plots for single and double square rings: Tx = Ty = 12.0 mm, d = 0.5 mm, substrate permittivity = 2.17, substrate thickness = 0.125″, f = 12.5 GHz. Reproduced from ©2008 The IET [45]

The presence of two transition regions increases the attainable phase range as compared to the single ring case. Higher slope of the phase-length curve in the second transition region of the two-ring cell element is symptomatic of the simultaneous resonance of both rings for this respective set of element dimensions. It is demonstrated in Figure 5.39 that the attainable phase range can be increased for the double ring cell element in comparison to a solid patch. A reflectarray was fabricated at Ku-band using a double ring cell element and was tested in the far-field chamber. The gain-frequency response for this particular antenna is shown in Figure 5.40, which shows 14% bandwidth for 1-dB gain reduction. The reflectarray is offset-fed with 40 cm × 40 cm dimensions. To mitigate beam squint across the frequency

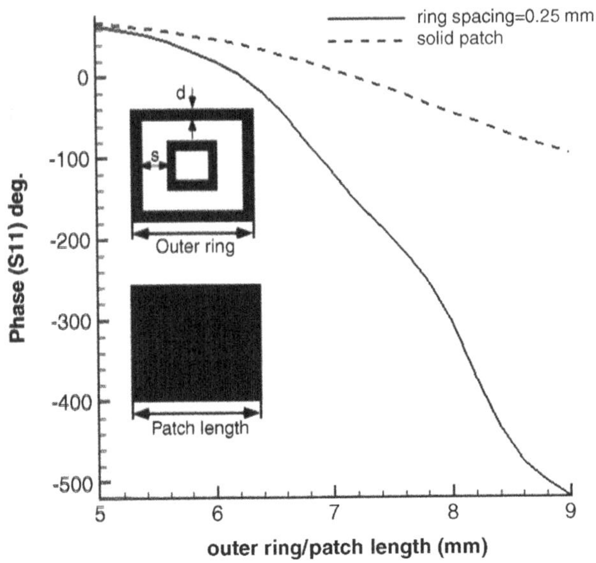

Figure 5.39 Phase-length plots for double square ring and solid square patch: Tx = Ty = 12.0 mm, d = 0.25 mm, substrate permittivity = 2.17, substrate thickness = 0.125″, f = 12.5 GHz. Reproduced from ©2008 The IET [45]

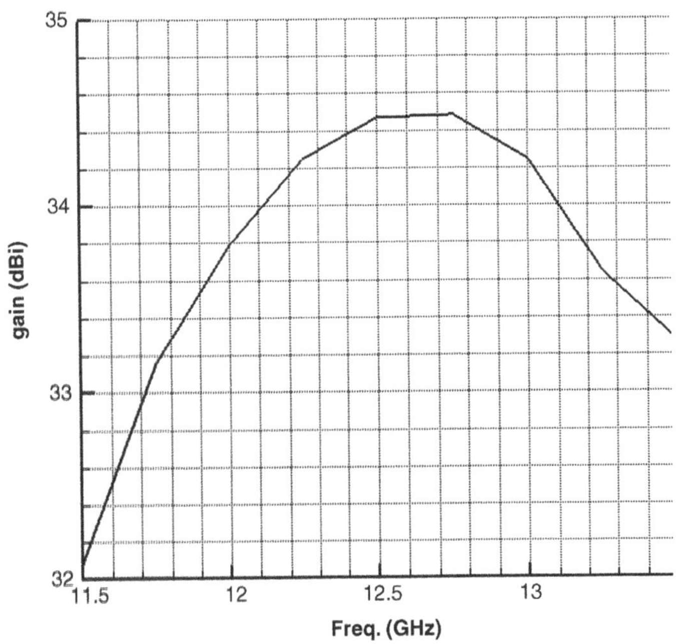

Figure 5.40 Gain-frequency plot for a reflectarray composed of double square rings. Reproduced from ©2008 The IET [45]

band, the direction of the outgoing beam was set along the direction of the direct reflected ray with respect to the ray from the focal point to the centre of the reflectarray [43]. This same practice was followed in all offset reflectarrays in this section.

5.6.3 Cross-Ring Cell Element

A cross-ring cell element has also been investigated thoroughly as the cell element for wideband reflectarray [44]. A double cross-ring cell element along with its geometrical parameters is shown in Figure 5.41. The phase versus outer ring dimension plot is shown in Figure 5.42, which again demonstrates the two transition regions that represent resonant condition of the outer and/or inner ring. An offset-fed reflectarray composed of cross-ring cell elements was designed, fabricated, and tested at Ku-band. The reflectarray dimensions was set at 40 cm × 40 cm. Gain versus frequency bandwidth of this reflectarray is shown in Figure 5.43 which shows a 16% bandwidth for 1-dB gain reduction.

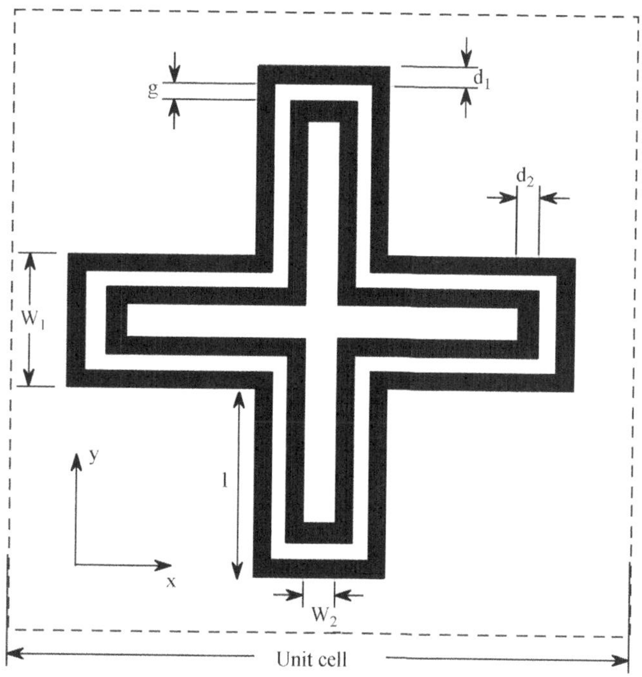

Figure 5.41 Double cross ring cell element: unit cell = 12 mm, $W_1 = 3.4$ mm, $W_2 = 0.4$ mm, $d_1 = d_2 = 0.5$ mm. Reproduced from ©2008 The IET [45]

5.6.4 Hybrid Cell Element

Following an extensive set of simulations for square ring and cross-ring cell elements, it was shown that the number of resonances is increased as more rings are added to the structure and

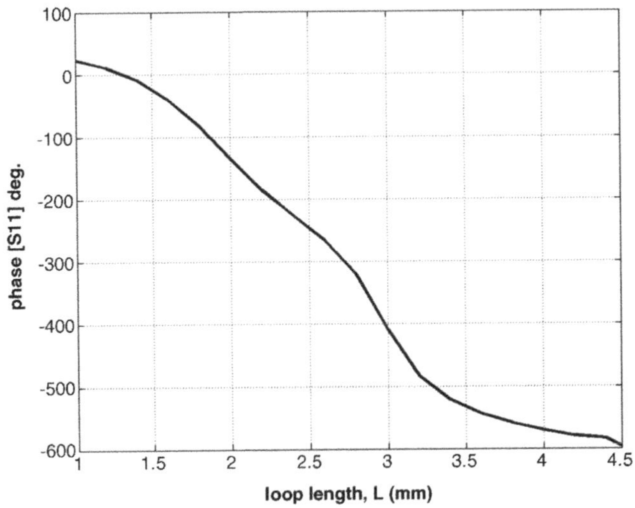

Figure 5.42 Simulated phase versus the length of cross-loop arm, $\varepsilon_r = 2.17$, h = 0.125″, cell size = 12 mm, $W_1 = 3.4$ mm, $W_2 = 0.4$ mm, $d_1 = d_2 = g = 0.5$ mm, and f = 12.5 GHz. Reproduced from ©2008 The IET [45]

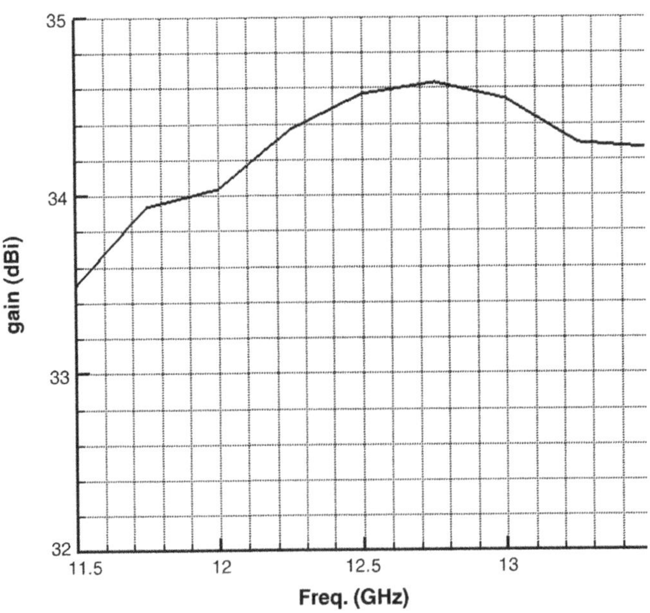

Figure 5.43 Measured gain versus frequency of the double cross-loop reflectarray: unit cell = 12 mm, $W_1 = 3.4$ mm, $W_2 = 0.4$ mm, and $d_1 = d_2 = g = 0.5$ mm. Reproduced from ©2008 The IET [45]

these resonances can be decoupled from each other by maintaining a certain relationship between the sizes of concentric rings. Also, it is important to note that simultaneous resonance of the rings results in steeper phase characteristics within the coupling region. Therefore, the rings ought to be decoupled in order to obtain a more gradual variation of the phase with respect to the ring size. A phase versus ring size plot is shown in Figure 5.44 for a cell element composed of one, two, three, and four cross-rings. It is evident that the number of resonances matches the number of rings. It was realized that in order to achieve a 360° phase shift, one has to use more than one ring.

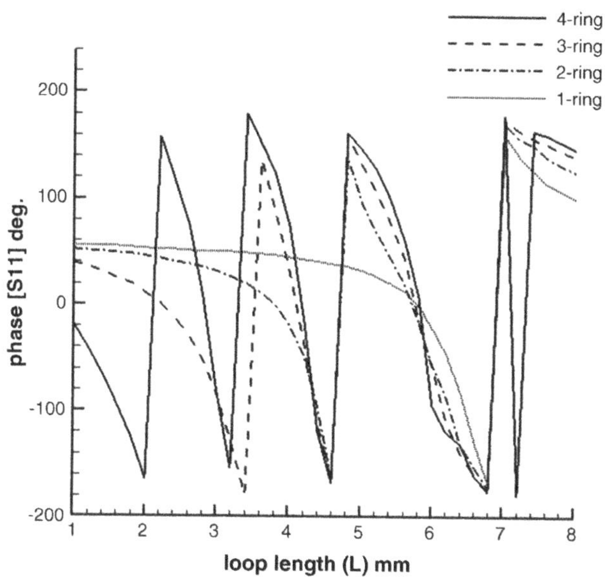

Figure 5.44 Phase-ring size characteristics for cell elements composed of one, two, and three cross-rings. Reproduced from ©2008 The IET [45]

On the other hand, the gap between the rings needs to be adjusted in order to reduce the coupling between adjacent rings of the same cell element to achieve gradual variation of phase versus ring size characteristic in all regions. These geometrical considerations led to the hybrid modified square (rectangular) and cross-ring cell elements as the optimized element to achieve the necessary phase shift range and gradual phase-ring size variation. The phase versus ring size plot of this optimized element based on cell element dimensions of [45] is shown in Figure 5.45.

An offset-fed reflectarray composed of such elements was designed, fabricated, and tested. Measured gain versus frequency plot and the radiation pattern of the antenna are shown in Figures 5.46 and 5.47. A measured 1-dB gain versus bandwidth of 24% was achieved using this novel cell element. The efficiency at the centre frequency was measured as 60%.

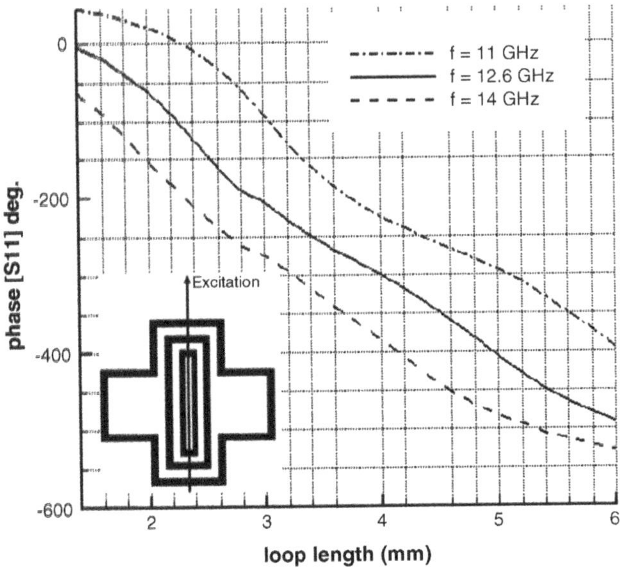

Figure 5.45 Phase-ring size characteristics for cell elements composed of one cross-ring and two rectangular-rings. Reproduced from ©2009 Antem/URSI [47]

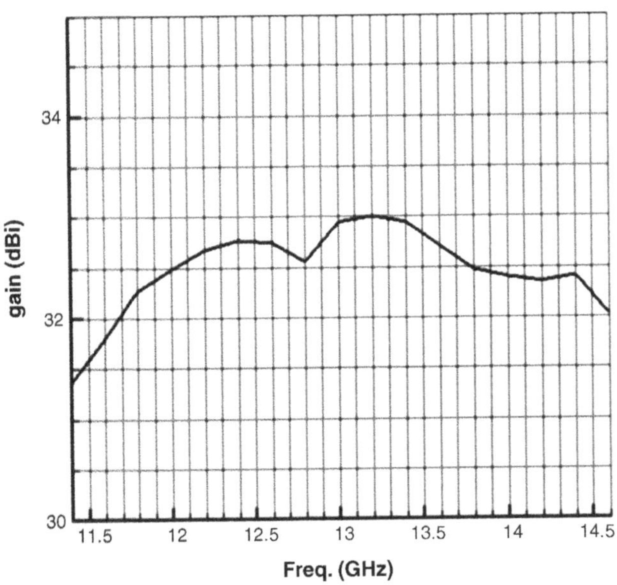

Figure 5.46 Measured gain versus frequency of the reflectarray composed of the elements depicted in Figure 5.45. Reproduced from ©2009 Antem/URSI [47]

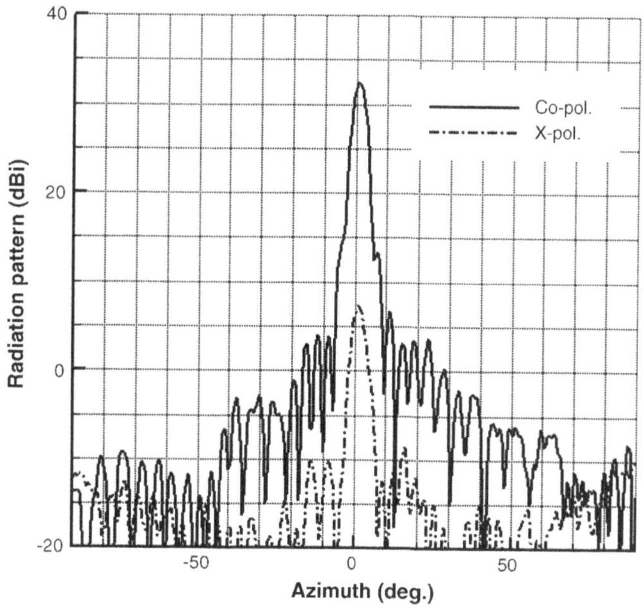

Figure 5.47 Measured radiation pattern of the reflectarray composed of cell elements described in Figure 5.45. Reproduced from ©2009 Antem/URSI [47]

5.7 Conclusion

The design guideline of a simple reflectarray was briefly presented. Reflectarray technology was compared to conventional reflector technology using the figures of merit that are often used in the evaluation of the latter. Different types of reflectarrays were presented and it was demonstrated that the feed system can be simplified for complicated multi-band multi-polarization application as compared to conventional reflector technology. Novel applications of reflectarray technology in power combining, beam splitting, and beam shaping were presented. The issue of narrow bandwidth of reflectarray was studied in detail for the cases of both moderate sized and large reflectarrays. Based on the physical insight attained from such a discussion, guidelines were developed and suggested in each case to improve the bandwidth performance.

References

1. D. G. Berry, R. G. Malech and W. A. Kennedy, "The reflectarray antenna," *IEEE Transactions on Antennas and Propagation*, vol. 6, no. 11, pp. 645–651, Nov. 1963.
2. R. E. Munson and J. Haddad, "Microstrip reflectarray antenna for satellite communication and RCS enhancement," US patent 4684952, Aug. 1987.
3. J. Huang and A. Feria, "Inflatable microstrip reflectarray antennas at X and Ka-band frequencies," *Antenna and Propagation Society International Symposium*, vol. 3, pp. 1670–1673, July 1999.
4. C. Pike, J. Shaker, M. Cuhaci, N. Jacob, S. Raut, M. Barakat and L. Shafai, "Compact Cassegrain Ka-band antennas for a briefcase satellite terminal," in *Proceedings of Ka-band Conference*, Venice, Italy, July 1998.

5. J. Shaker, C. Pike and M. Cuhaci, "Dual-band, dual-polarisation transmit-receive Cassegrain flat reflector," *Microwave and Optical Technology Letters*, vol. 24, no. 1, pp. 7–11, Jan. 5, 2000.
6. J. Shaker and M. Cuhaci,"A novel bifocal dual-band, dual orthogonal polarisation planar reflector for SatCom applications," paper presented at AP2000, Davos, Switzerland, April 2000.
7. Jafar Shaker, Aldo Petosa, Soulideth Thirakoune, Corey Pike and Michel Cuhaci, "Investigation of the performance of a reflectarray fed by an active feed block to establish link with ACTS," in *Proceedings of ANTEM 2002*, pp. 506–509, Montreal, Canada, Aug. 2002.
8. Jafar Shaker and Michel Cuhaci, "Planar reflector for LMCS applications," *Electronics Letters*, vol. 35, no. 2, pp. 103–104, Feb. 1999.
9. H. Legay, B. Pinte, M. Charrier, A. Ziaei, E. Girard and R. A. Gillard, "Steerable reflectarray antenna with MEMS controls," *IEEE International Symposium on Phased Array Systems and Technology*, pp. 494–499, Oct. 14–17, 2003.
10. S. V. Hum, M. Okoniewsky and R. J. Davies, "Realizing an electronically tunable reflectarray using varactor diode-tuned elements," *IEEE Microwave and Wireless Components Letters*, vol. 15, no. 6, pp. 422–424, June 2005.
11. M. Bilakowski, A. Robinson and H. Song, "Design, development, and testing of X-band amplifying reflectarrays," *IEEE Transactions on Antennas and Propagation*, vol. 50, no. 8, pp. 1065–1076, Aug. 2002.
12. J. Encinar, "Design of two layer printed reflectarrays using patches of variable size," *IEEE Transactions on Antennas and Propagation*, vol. 49, no. 10, pp. 1403–1410, Oct. 2001.
13. M. R. Chaharmir, J. Shaker, M. Cuhaci and A. Ittipiboon, "A broadband reflectarray antenna with double square rings," *Microwave and Optical Technology Letters*, vol. 48, no. 7, pp. 1317–1320, July 2006.
14. J. A. Encinar and J.A. Zornoza, "Broadband design of three-layer printed reflectarrays," *IEEE Transactions on Antennas and Propagation*, vol. 51, pp. 1662–1664, July 2003.
15. M. Chaharmir, J. Shaker and H. Legay,"Broadband design of a single layer large reflectarray using multi cross loop elements," paper presented at 31st ESA Antenna Workshop, Noordwijk, The Netherlands, May 18–20, 2009.
16. R. E. Collin, *Antennas and Radiowave Propagation*, McGraw-Hill, New York, 1985.
17. D. M. Pozar, "Bandwidth of reflectarrays," *Electronics Letters*, vol. 39, no. 21, pp. 1490–1491, Oct. 2003.
18. P. Munk, *Frequency Selective Surfaces: Theory and Design*, John Wiley & Sons, Inc., New York, 2000.
19. Jafar Shaker,"Analysis of multiplayer double periodic structures and their performance," Ph.D. dissertation, Department of Electrical and Computer Engineering, University of Manitoba, 1995.
20. R. D. Javor, X. D. Wu and K. Chang, "Off-fed microstrip reflectarray antenna," *Electronics Letters*, vol. 30, no. 17, pp. 1363–1365. Aug. 1994.
21. D. M. Pozar, S. D. Targonski and H. D. Syrigos, "Design of millimeter wave microstrip reflectarrays," *IEEE Transactions on Antennas and Propagation*, vol. 45, no. 2, pp. 287–295, Feb. 1997.
22. R. Leberer and W. Menzel, "A dual planar reflectarray with synthesized phase and amplitude distribution," *IEEE Transactions on Antennas and Propagation*, vol. 53, no. 11, pp. 3534–3539, Nov. 2005.
23. R. Chaharmir, J. Shaker and M. Cuhaci, "Development of dual-band circularly polarized reflectarray," *IEE Proceedings of Microwaves, Antennas, and Propagation*, vol. 153, no. 1, pp. 49–54, Feb. 2006.
24. J. Huang and R. Pogorzelski, "A Ka-band reflectarray with elements having variable rotation angles," *IEEE Transactions on Antennas and Propagation*, vol. 46, no. 5, pp. 650–656, May 1998.
25. J. Huang, "A high-gain circularly polarized Ka-band microstrip reflectarray," *Microwave and Optical Technology Letters*, vol. 14, no. 20, pp. 163–166, Feb. 1987.
26. S. C. Han, C. Rodenbeck, J. Huang and K. Chang, "A C/Ka dual frequency dual layer circularly polarized reflectarray antenna with microstrip ring elements," *IEEE Transactions on Antennas and Propagation*, vol. 52, no. 11, pp. 2871–2876, Nov. 2004.
27. Jafar Shaker and Michel Cuhaci, "Multi-band, multi-polarization reflector-reflectarray antenna with simplified feed system and mutually independent radiation patterns," *IEE Proceedings on Microwaves, Antennas, and Propagation*, vol. 152, no. 2, pp. 97–101, Apr. 2005.
28. J. Shaker, R. Chaharmir and H. Legay, "Investigation of FSS-backed reflectarray using different classes of cell elements," *IEEE Transactions on Antennas and Propagation*, vol. 56, no. 12, pp. 3700–3706, Dec. 2008.
29. M. R. Chaharmir, J. Shaker, M. Cuhaci and A. Sebak, "Reflectarray with variable slots on ground plane," *IEE Proceedings of Microwaves, Antennas and Propagation*, vol. 150, no. 6, pp. 436–439, Dec. 2003.
30. M. R. Chaharmir, J. Shaker, M. Cuhaci and A. Sebak, "Novel mechanically controlled reflectarray antenna for beam switching and beam shaping in millimetre wave applications," *Electronics Letters*, vol. 39, no. 7, pp. 591–592, 2003.

31. M. R. Chaharmir, J. Shaker, M. Cuhaci and A. Sebak, "Novel photonically-controlled reflectarray antenna," *IEEE Transactions on Antennas and Propagation*, vol. 54, no. 4, pp. 1134–1141, Apr. 2006.
32. J. A. Encinar and J. Zornoza, "Three-layer printed reflectarrays for contoured beam space applications," *IEEE Transactions on Antennas and Propagation*, vol. 52, no. 5, pp. 1138–1148, May 2004.
33. F. Arpin, J. Shaker and D. McNamara, "Power combining using multi-feed single beam reflectarray technology," in *Proceedings of 28th ESA Workshop on Satellite Technology*, Part 1, pp. 255–258, Noordweijk, The Netherlands, May 2005.
34. F. Arpin, J. Shaker and D. A. McNamara, "Multi-feed single-beam power-combining reflectarray antenna," *Electronics Letters*, vol. 40, no. 17, pp. 1035–1037, Aug. 19, 2004.
35. M. R. Chaharmir, J. Shaker, M. Cuhaci and A. Sebak, "Applications of projection method for beam shaping in reflectarray antennas," *ANTEM 2002*, pp. 489–492, St. Hubert, Quebec, 2002.
36. M. E. Bialkowski and K. H. Sayidmarie, "Bandwidth considerations for a microstrip reflectarray," *Progress in Electromagnetics Research B*, vol. 3, pp. 173–187, 2008.
37. Y. J. Guo, *Fresnel Zone Antennas*, Kluwer Academic Publishers, Boston, 2002.
38. G. Cadoret, A. Laisne, R. Gillard and H. Legay, "A new reflectarray cell using microstrip patches loaded with slots," *Microwave and Optical Technology Letters*, vol. 44, no. 3, pp. 270–272, Feb. 2005.
39. J. A. Encinar and J. Agustin Zornoza, "Broadband design of three-layer printed reflectarrays," *IEEE Transactions on Antennas and Propagation*, vol. 51, no. 7, pp. 1662–1664, July 2003.
40. R. Chaharmir, J. Shaker and H. Legay, "Broadband design of a single layer large reflectarray using multi cross loop elements," *IEEE Transactions on Antennas and Propagation*, vol. 57, no. 10, pp. 3363–3366, Oct. 2009.
41. E. Carrasco, M. Barba and J. A. Encinar, "Bandwidth improvement in large reflectarray by using true-time delay line," *IEEE Transactions on Antennas and Propagation*, vol. 56, no. 8, pp. 2496–2503, Aug. 2008.
42. HFSS, ver.11.1, Ansoft Corp., Pittsburgh, PA, USA.
43. S. D. Targonski and D. M. Pozar, "Minimization of beam squint in microstrip reflectarrays using an offset feed," *Antennas and Propagation Society International Symposium*, vol. 2, pp. 1326–1329, July 21–26, 1996.
44. R. Chaharmir, J. Shaker, M. Cuhaci and A. Ittipiboon, "A broadband reflectarray antenna with cross square rings," *Electronics Letters*, vol. 42, no. 2, pp. 65–66, 19 Jan. 2006.
45. R. Chaharmir and J. Shaker, "Broadband reflectarray with combination of cross and rectangle loop elements" *Electronics Letters*, vol. 44, no. 11, pp. 658–659, May 22, 2008.
46. R. Chaharmir and J. Shaker, "Broadband reflectarray with combination of cross and rectangle loop elements," *IEEE Transactions on Antennas and Propagation*, vol. 57, no. 10, pp. 3363–3366, Oct. 2009.
47. M.R. Chaharmir, J. Shaker, M. Cuhaci and A. Ittipiboon, "Wideband reflectarray research at the Communications Research Centre Canada," in *Proceedings of ANTEM*, Session TA1, Paper 053, 2009.

6

Reconfigurable Microstrip Antennas

Jennifer T. Bernhard
University of Illinois at Urbana-Champaign, USA

6.1 Introduction

While reconfigurable antennas have been implemented in various ways over the past 40 years, reconfigurable microstrip antennas, in particular, have existed for almost as long as the microstrip antenna itself, dating back to the early 1980s. The motivation for implementing reconfigurable properties in an antenna in general is straightforward – the acquisition of new capabilities that eliminate the need for multiple antennas and/or that provide additional degrees of operational freedom that expand system performance. Microstrip antennas are particularly good candidates for achieving reconfigurability, since their well-defined ground planes and planar structures present clear opportunities for integration of a number of popular reconfigurable mechanisms (including switches and tunable materials) and their associated control circuitry. Additionally, since most microstrip antennas operate in resonance and their operation is well modeled and well understood, an informed designer can manipulate the antenna structure and composition in different ways to achieve a variety of reconfigurable properties.

This chapter addresses recent advances in the development of reconfigurable microstrip antennas, with an emphasis not only on the antennas themselves, but also on the practical issues surrounding their implementation, including control and system-level design and performance. Fundamentally, microstrip antenna behavior is governed by two kinds of properties: substrate properties and conductor properties. Section 6.2 describes the basic concepts behind changes in substrate parameters, including tunable changes in relative permittivity and permeability, as well as some recent examples that produce useful reconfigurability. Section 6.3 addresses changes in the conductive components of microstrip antennas that can lead to a wide variety of reconfigurability. Section 6.4 discusses many of the important issues surrounding the practical

Microstrip and Printed Antennas: New Trends, Techniques and Applications. Edited by Debatosh Guha and Yahia M.M. Antar

implementation of reconfigurable microstrip antennas, including control circuitry design and trade-offs between different reconfiguration mechanisms. Finally, Section 6.5 provides some observations on emerging trends in microstrip antenna reconfigurability, as well as directions for future research and development.

6.2 Substrate Modification for Reconfigurability

The substrate properties of a microstrip antenna help determine both its operating frequency and bandwidth. While many complex microstrip radiators exist, two fundamental relationships exist between their substrate properties and their frequency characteristics. First, given identical metallic structures and substrate heights, operating frequency is inversely proportional to the relative permittivity and permeability of the substrate. Second, the bandwidth of the structure is nominally proportional to the substrate electrical height. Reconfigurable microstrip antennas based on changes in substrate properties leverage these relationships. Specifically, ferroelectric and ferrite materials that can be biased to vary their permittivity and permeability, respectively, offer the possibility of frequency tunable microstrip antennas. In the case of ferroelectric materials, an applied DC electric field determines the relative permittivity, while in the case of ferrite materials, an applied DC magnetic field determines the substrate's effective relative permeability. These changes in the material properties can be used to alter the effective electrical lengths of antennas, providing predictable and repeatable shifts in operating frequencies. An additional prospective benefit of implementing these materials is that their relative permittivities and permeabilities are large relative to commonly used substrate materials, promising significant antenna size reductions. Unfortunately, the main complications of using standard ferroelectric and ferrite bulk materials (typically with thicknesses of the order of millimeters) are their high conductivities relative to other substrates that can severely degrade the efficiency of the antennas.

A number of examples illustrate both the possibilities and limitations of tunable substrates in a variety of microstrip geometries and configurations. In order to minimize the losses of ferroelectric materials, for example, several researchers have developed ferroelectric materials in thin film form while still providing a degree of tunability [1–3]. The use of thin films, however, means that the achievable tuning range of these structures is not as large as if the material could be used as the sole substrate without losses. To date, most of the practical applications for these materials place them in tunable parasitic elements or feed structures instead of the driven antenna element due to the limitations on the achievable uniformity of the films over large lateral areas.

Several ferrite-based reconfigurable microstrip antennas also have been studied [4–6]. In [4], a frequency-tuned antenna was enabled by a variable DC magnetic field applied in the bulk ferrite substrate plane perpendicular to the resonant dimension of a rectangular patch. While the continuous tuning range was approximately 40%, the radiation patterns possess significant cross-polarization components not typically created by a patch operating in its fundamental mode. Other factors such as non-uniform bias fields [5] and multiple mode generation in the substrate [6] also preclude the use of bulk ferrite substrates. A ferrite superstrate over a circular patch antenna has also been used to produce a degree of pattern reconfigurability (approximately 15 degrees of beam tilt) when the ferrite was biased with a permanent magnet [7]. This prototype was again limited by losses, however, and did not result in an efficient far field radiator.

Future work in this particular area of reconfigurable microstrip antennas may certainly focus on the development of new materials, structures, and fabrication processes that provide high degrees of parameter tunability without incurring significant conduction loss. For example, moving to thin film ferrite technology, the authors in [8] presented a polarization-tunable microstrip antenna based on static magnetic biasing of a ferrite film. Since the traditional copolarized fields of a microstrip patch are less dominant when ferrite material is included in the structure, the authors leveraged this behavior to tune the frequency and the polarization of the radiated fields. Specifically, an applied DC magnetic bias field was used to tune the frequency of the cross-polarized fields, resulting in a variety of elliptical polarizations. Changes in feed point position and ferrite film properties could also be used to attain linear and circular polarizations (CP) [8], and losses were not as significant as if a bulk ferrite substrate were used.

More advanced approaches to changes in dielectric properties have recently been proposed by Huff et al. [9]. In these structures, microfluidic channels are fabricated in a bulk substrate beneath the patch [9]. By pumping different fluids with various dielectric properties through the channels, the operating frequency of the patch can be tuned. Since these fluids have low conductivities, they achieve low loss tuning and, in addition, there is no external biasing circuitry that could interfere with the antenna's behavior (although the microfluidic pumps still require some form of electronic control).

6.3 Conductor Modification for Reconfigurability

The dimensions and configurations of the conductors of a microstrip antenna affect the operating frequency, bandwidth, pattern, and polarization of the far fields of the antenna. Therefore, changes in these dimensions and configurations can be used to reconfigure antenna properties.

6.3.1 Frequency Reconfigurability

The most straightforward conductor modification for microstrip antennas is a change in length, which nominally results in a change in operating frequency. Many common approaches to these modifications are based on the fundamental theory of a basic microstrip patch antenna operating in its first resonant mode. These changes result in the antenna's electrical length to be approximately an effective half wavelength, which is conditioned by the permittivity and permeability of the substrate material.

In [10], for example, a continuously tuned microstrip patch antenna uses two varactor diodes, or varactors, positioned between the radiating edges of a rectangular patch and the ground plane. Using tuning capacitances between 0.4 and 2.4 pF (corresponding to applied voltages of 30 and 0 volts, respectively), the operating frequency of the antenna was tuned over a 20–30% bandwidth [10]. In this case, the additional variable capacitance present at the radiating edges of the patch served to increase the antenna's electrical length, enabling downward tuning of the operating band.

As the need for more frequency-agile systems has grown, the approaches to attain frequency reconfigurability have become more advanced. A frequency tunable microstrip patch antenna has also been demonstrated using integrated radio frequency microelectromechanical system (RF MEMS) capacitors [11]. Rather than using pre-packaged varactors, these RF MEMS capacitors are implemented on a coplanar waveguide tuning stub as shown in Figure 6.1 [11]. This approach eliminates the need for bias vias, which in turn results in minimal pattern disturbances that could arise from bias circuitry.

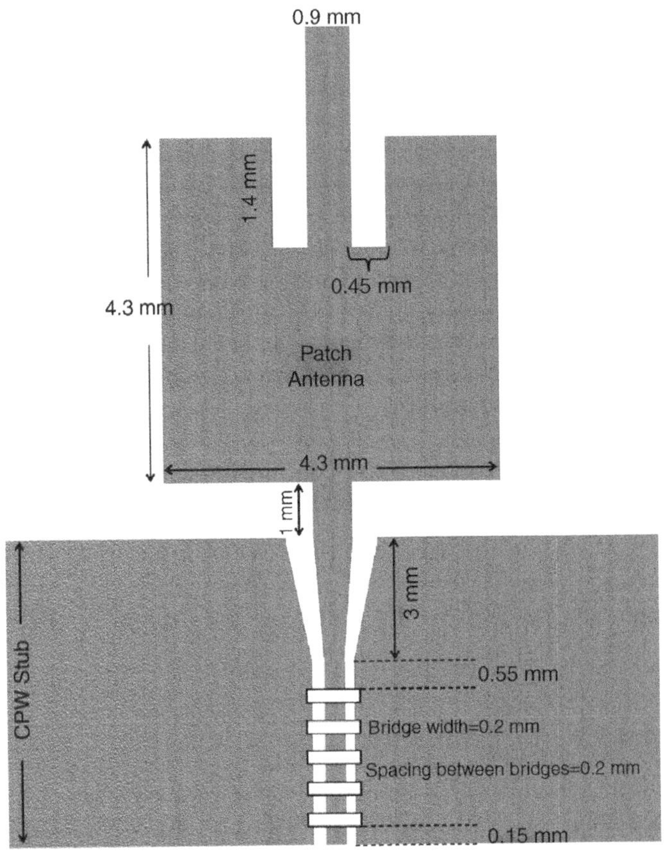

Figure 6.1 Frequency tunable microstrip patch element based on RF MEMS capacitors. Reproduced by permission of ©2007 IEEE [11]

Using the same concept but concentrating on smaller variations in reflection phase than large changes in operating frequency, the work detailed in [12] develops an electronically tunable reflectarray based on microstrip patch elements tuned with varactor diodes, shown in Figure 6.2 [12]. The complete study includes the development of a circuit model for the reflectarray element that includes the bias network and accounts for the discontinuity introduced into the patch to accommodate the varactors, as shown in Figure 6.3 [12]. An analysis of the loss incurred with the tuning varactors indicates an efficiency of roughly 40% with an achievable phase tuning range of approximately 320 degrees.

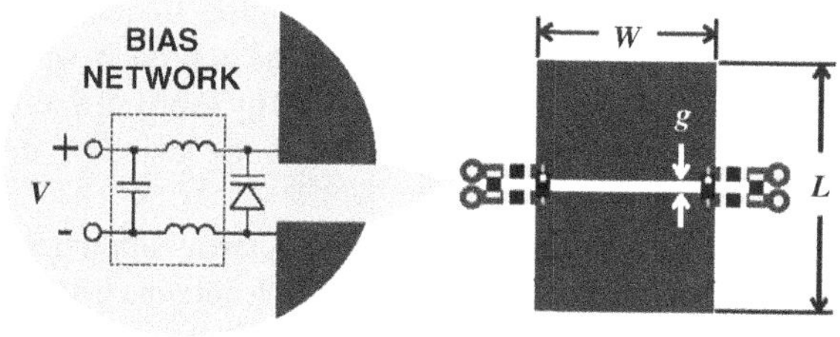

Figure 6.2 A reconfigurable microstrip reflectarray cell topology. Reproduced by permission of ©2007 IEEE [12]

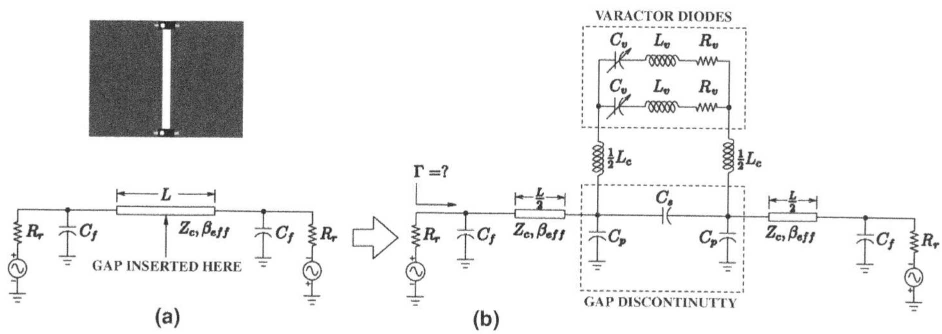

Figure 6.3 Equivalent circuit for predicting reconfigurable reflectarray cell scattering: (a) standard microstrip patch; (b) microstrip patch with varactor diode and discontinuities included. Reproduced by permission of ©2007 IEEE [12]

In addition to continuous frequency reconfigurability/tuning, design approaches to achieve switched changes in frequency have moved in a number of interesting directions. Several groups have demonstrated simple frequency switching using changes in length enabled by connection or disconnection of contiguous microstrip lines using PIN diodes [13–15].

Recognizing that the operating frequency of a microstrip antenna depends largely on the electrical length of current paths on the structure, other groups have demonstrated switched frequency reconfigurability with microstrip antennas that have non-resonant slots cut out of their footprints. One example is provided in [16], where a slot is etched with its longer dimension in a direction perpendicular to the dominant current direction of a microstrip patch's first resonance. A PIN diode placed in the slot's center can either lengthen or shorten the current paths on the patch, resulting in a lower and upper operating frequency, respectively. The spacing between these lower and upper frequencies is dictated by the length and position of the slot in the patch, and typical radiation patterns are maintained while the length of the slot does not approach the entire patch width [16]. This approach has also been extended to other patches with a variety of conductor shapes, e.g., [17], and has also been demonstrated for continuous tuning using varactors [18].

In [19], a parasitically-fed microstrip patch-based antenna is equipped with a number of switches that connect the edge of the patch to the ground plane below to enable changes in operating frequency. Optimization of the structure for frequency reconfigurability is accomplished using a genetic algorithm [19]. A similar approach was also taken in [20], where a frequency-switchable patch antenna is developed by shorting a serpentined structure to the ground plane in several places. In this case, PIN diode switches are implemented in a compact bias network [20].

Switched frequency tunability is also useful for reflectarray applications. In one embodiment, a reconfigurable reflectarray element is simple patch antenna driven by a tuned coupling slot [21]. Changes in the length of the coupling slot are controlled by RF MEMS switches, resulting in phase swings at individual elements of up to 150 degrees using four switches in various state combinations. The effects of actual RF MEMS switch properties and the effects of wire bonds are analyzed, showing that ideal connections for neither the switches nor wire bonds adequately capture the true behavior of the structures [21]. A related reflectarray structure consists of another aperture-coupled patch, but this time the variation in reflection phase is achieved through a loading of the feed transmission line with two varactor diodes, rather than through reconfiguration of the coupling slot [22]. Practical issues of loss and harmonic generation in the tuning diodes are also considered in this work.

Use of mechanical displacements instead of electrical changes in a microstrip structure can deliver significant frequency shifts, and, depending on the displacement mechanism, can provide switched or continuous frequency tuning. Additionally, with mechanical changes, the option arises to effectively change both the material and conductor configurations. However, the practical challenge with these kinds of antenna designs is the seamless integration of the actuation mechanism with the electronic components of the antenna so that one does not interfere with the operation of the other. In the realm of microstrip antennas, an early design of a mechanically tuned antenna employed a piezoelectric actuator system. This system physically varied the spacing between a rectangular microstrip antenna and a parasitic radiator of identical shape to change the antenna's operating frequency [23–25]. Figure 6.4 [25] shows a photograph of the antenna and its associated mechanical housing. The vertical movement of the

Figure 6.4 Photograph of mechanically actuated reconfigurable microstrip antenna with a movable parasitic element. Reproduced by permission of ©2001 IEEE [25]

parasitic element above the driven element delivered an effective bandwidth of about 9%, but along with these changes in frequency came changes in bandwidth and gain, which could not be decoupled from the frequency tuning in the design [25].

6.3.2 *Pattern Reconfigurability*

Pattern reconfigurability in microstrip antennas is typically achieved through the use of tuned or switched sections of microstrip line that significantly alter the currents on the structure. In particular, several microstrip structures rely on switched parasitic elements to reconfigure radiation patterns, with their foundations based on [26]. Many of these designs use this approach to change radiation patterns without significantly affecting the antennas' operating frequencies. An example developed in [27] (shown in Figure 6.5) is composed of a driven, narrow linear microstrip element combined with two parasitic elements positioned parallel to the driven element. The lengths of the parasitic elements can be changed using solid state switches [27] or varactors [28] which enable switched or continuous tilting, respectively, of the radiation pattern in a direction toward the shorter parasitic element. Figure 6.6 shows the reconfigurable antenna that uses PIN diode switches [29], and the resulting measured performance shown in Figure 6.7 [27]. Since the driven element contains no switches, the frequency response of the antenna is largely independent of the state of the parasitic elements, and the switch bias control network design explicitly minimizes any radiation disturbances [29]. Dinger [30, 31] provides other examples of steered arrays that implement tuned loads on parasitic elements based on standard microstrip patches. Often, in these more complex systems that use more than two parasitic elements, computational optimization algorithms can be used to determine the tuning reactances necessary on each parasitic element to produce a beam or null at a prescribed angle [30, 32]. However, as the driven element and the parasitic elements are more closely spaced, the frequency response of the antenna may be affected. In another example, one design incorporates a driven Z-shaped microstrip element coupled

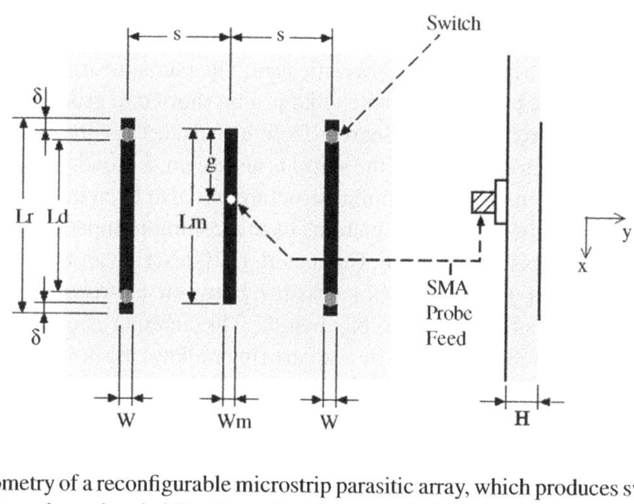

Figure 6.5 Geometry of a reconfigurable microstrip parasitic array, which produces switched beam tilts over a common impedance bandwidth. Reproduced by permission of ©2004 IEEE [27]

Figure 6.6 A reconfigurable microstrip parasitic array. Reproduced by permission of ©2005 Zhang [29]

with parasitic metallic radiators terminated with reconfigurable connections, shown in Figure 6.8 [33].

Another example of a switched reconfigurable antenna with pattern reconfigurability has been demonstrated by Huff et al. [34]. This design provides a broadside or 45 degree tilted beam over a common impedance bandwidth (of approximately 3%) shown in Figure 6.9 [34]. A simple spiral microstrip antenna in its basic structure with a length of approximately one wavelength, it delivers a broadside radiation pattern. The reconfiguration takes place using two switched connections: one that shorts the end of the spiral to ground (shown on the left of Figure 6.9 [34]) and one that opens a small gap in the spiral arm (shown on the right of Figure 6.9 [34]). With the two switches in closed and open states, respectively, the antenna operates as an open microstrip line with a parasitic arm. The parasitic arm is formed by the end section of the spiral created by the open switch that is now shorted to ground at one end. In this state, a tilted pattern (approximately 45 degrees from broadside) results. If the second switch remains open and the switch at the end of the spiral is also open, a broadside pattern at a higher frequency can also be obtained [34]. A similar structure has also been implemented to deliver switched endfire and broadside radiation patterns over a common impedance bandwidth with RF MEMS switched connections [35, 36]. Chen et al. [37] describe another microstrip-based antenna delivers switched radiation characteristics between a broadside and top-loaded monopole mode over a shared operating bandwidth. The antenna is comprised of a small, non-resonant probe-fed patch surrounded by a square ring radiator as shown in Figure 6.10 [37]. Alternate connection of two shorting straps on the square loop to the ground plane provides reconfiguration between the two operating modes (broadside and conical patterns). Other microstrip antenna designs based on spiral topologies that can also create tilted beam patterns have been studied by others [e.g., 38, 39], with beam tilting again based on parasitic effects that are easily integrated into a spiral's compact geometry.

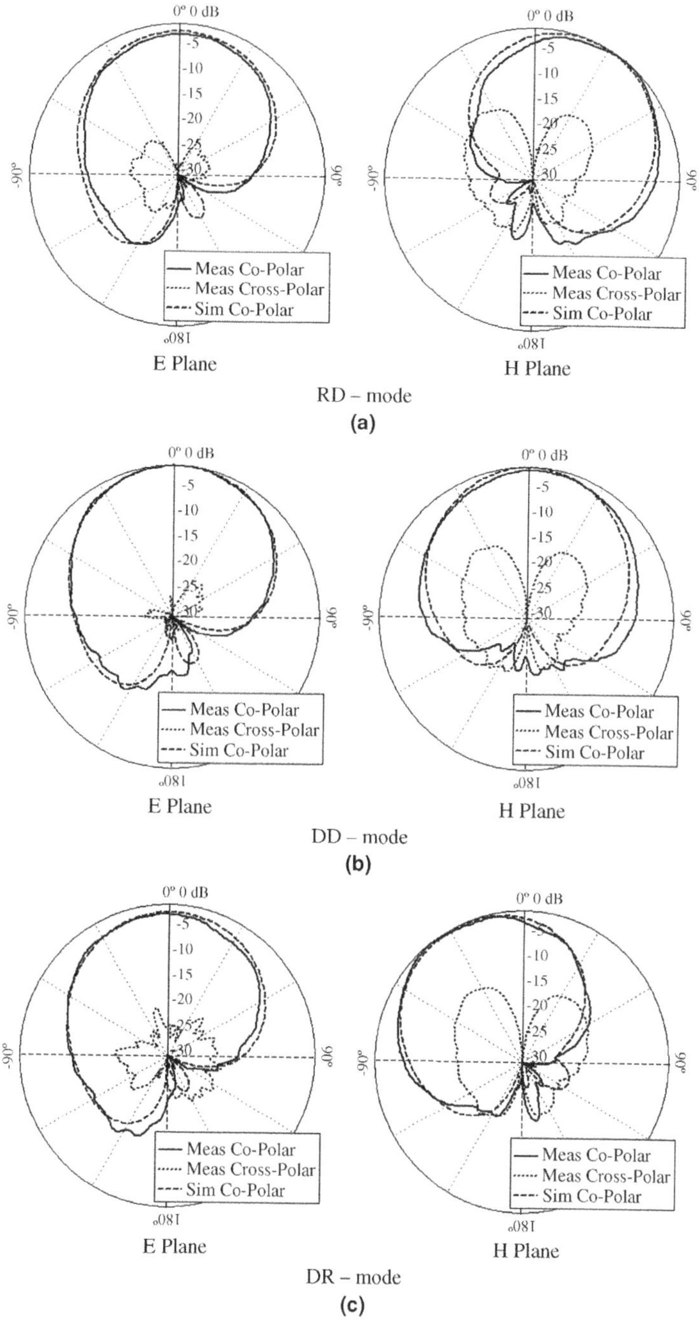

Figure 6.7 Reconfigured beam patterns from the antenna shown in Figure 6.6: (a) reflector-director (RD); (b) director-director (DD); (c) director-reflector (DR). Reproduced by permission of ©2004 IEEE [27]

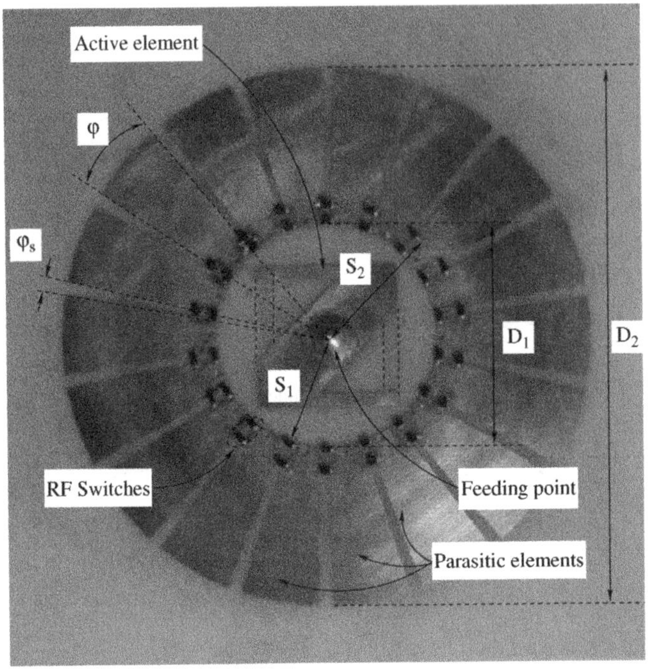

Figure 6.8 Parasitic microstrip radiator capable of reconfigurable radiation patterns. Reproduced by permission of ©2007 IEEE [33]

Finally, a compound [40] reconfigurable antenna design is described in [41], a stacked square patch configuration with two feeds supports a high-gain, circularly-polarized broadside pattern at one frequency and a low-gain omnidirectional pattern using a PIFA (planar inverted F antenna) mode at another. This particular design addresses the need for multimode operation for satellite and ground-based communication, applying principles of reconfigurability with PIN diode control to make the antenna aperture as small as possible. The authors of [41] also carefully consider the ramifications of losses incurred by the switching diodes and the interaction of the tuning structures and bias lines for each mode, including the effects of vertical feed and shorting pins used in the PIFA mode on the achievable axial ratio of the antenna's broadside mode.

6.3.3 Polarization Reconfigurability

As the concept of antenna reconfigurability has matured over the past decades, many recent developments in reconfigurable microstrip antennas have focused on polarization reconfigurability. This is due in large part to the significant increases in system capacity and throughput that are possible in complex scattering environments that can be obtained using some form of polarization diversity. Similar to the approaches taken for pattern reconfigurability, most polarization reconfigurability is enabled by the manipulation of currents on the conductors of the microstrip antenna.

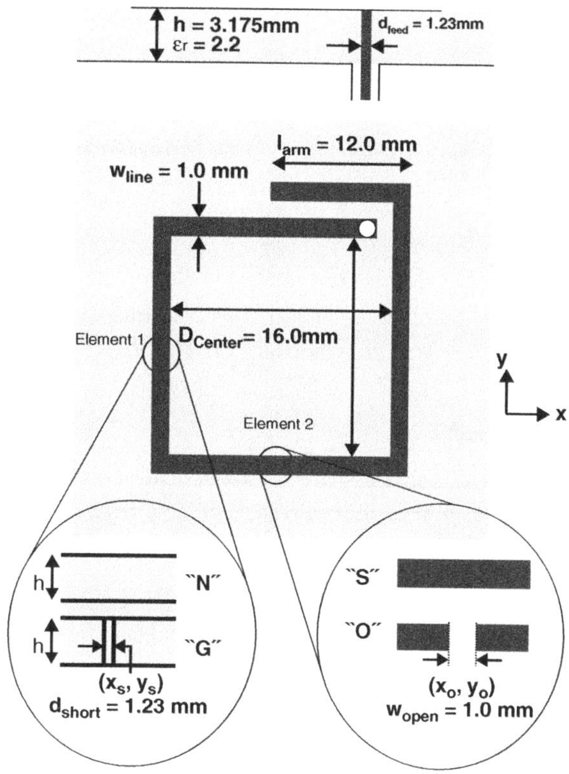

Figure 6.9 Reconfigurable square spiral microstrip antenna with two switched connections that produces both radiation and frequency reconfiguration. Reproduced by permission of ©2003 IEEE [34]

Many different kinds of microstrip antennas can be designed to reconfigure polarization. One polarization-agile antenna is the "patch antenna with switchable slots," or PASS antenna [42, 43], shown in Figure 6.11 [42]. This antenna has one or more slots cut from its top patch, similar to the one designed for frequency reconfigurability discussed in [16]. In this case, by placing two orthogonal slots in a square patch and alternating which slot is shorted with an electronic switch, the patch can radiate either left-hand circular polarization (LHCP) or right-hand circular polarization (RHCP). Similar in concept are patches with switched connections on square patch corners [44] and patches with switched excitation slots [45]. The antenna described in [46] also provides switched polarization diversity with a patch with connected slots to change the mode of resonance on the structure. The antenna can radiate in linear and circular polarization states, with a degree of shared bandwidth among all of the states [46]. Another approach for polarization reconfigurability uses a novel micro-electromechanical system (MEMS) actuator placed strategically in the footprint of a simple patch to switch between two orthogonal modes [47]. The MEMS actuator is a moveable metallic pad suspended over a conducting stub formed at the edge of the patch. Using electrostatic actuation, the metallic pad is moved downward to produce linear polarization or suspended to produce circular polarization [47]. Using a piezoelectric rather than MEMS

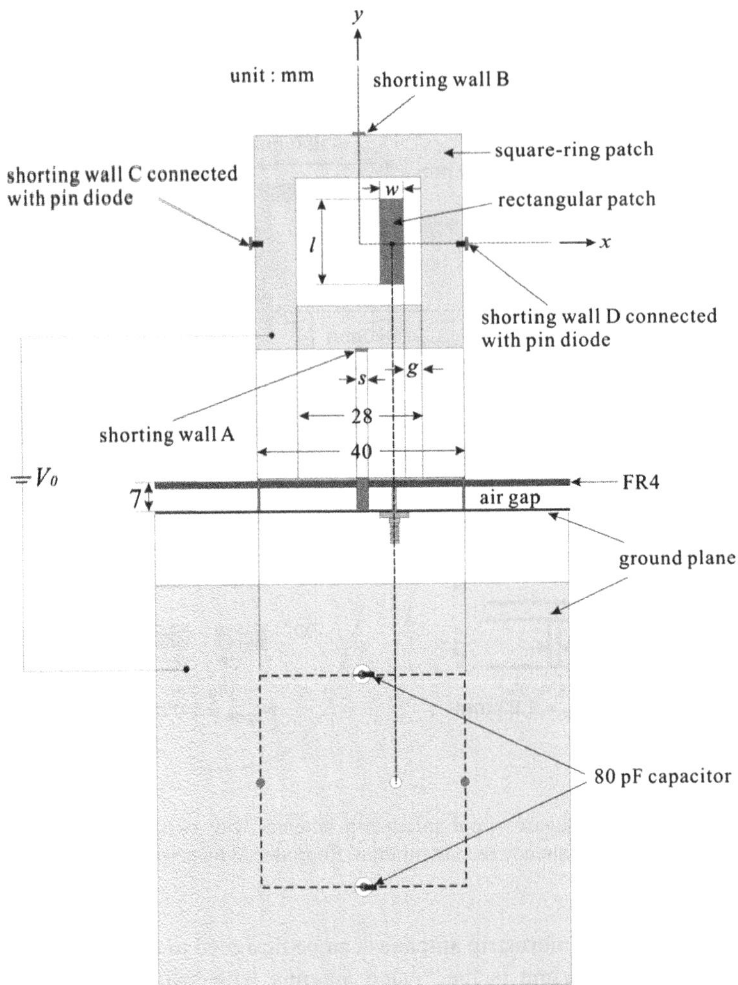

Figure 6.10 A reconfigurable square ring microstrip patch antenna for pattern diversity applications. Reproduced by permission of ©2007 IEEE [37]

transducers, a comparable structure shown in Figure 6.12 delivered the two opposing circular polarization states [48].

More recently, in [49], polarization reconfiguration has been developed to improve antenna diversity for wireless LAN applications, through the implementation of a circular patch antenna with switched shorting pins between the patch and the ground plane that support either LHCP or RHCP operation. Using a related approach to that taken in [49], the work in [50] expands the antenna functionality to include not only switching between LHCP and RHCP, but also switching between two separate frequency bands with linear polarization through the implementation of switched matching stubs. In this design, PIN diodes are used for all of the switches, which will affect the efficiency of the final antenna design.

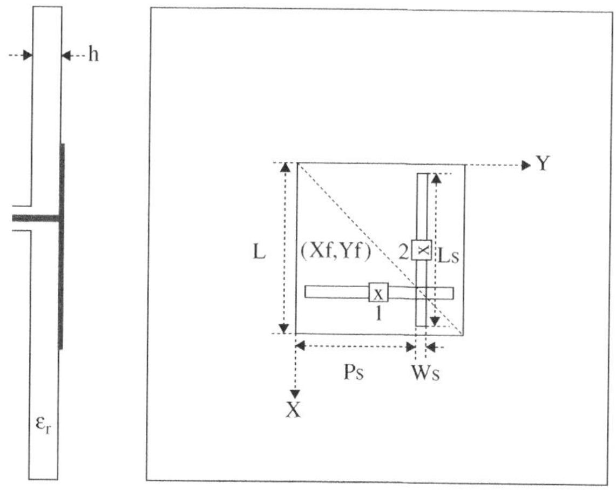

Figure 6.11 A polarization reconfigurable microstrip patch antenna based on a switchable slot. Reproduced by permission of ©2005 IEEE [42]

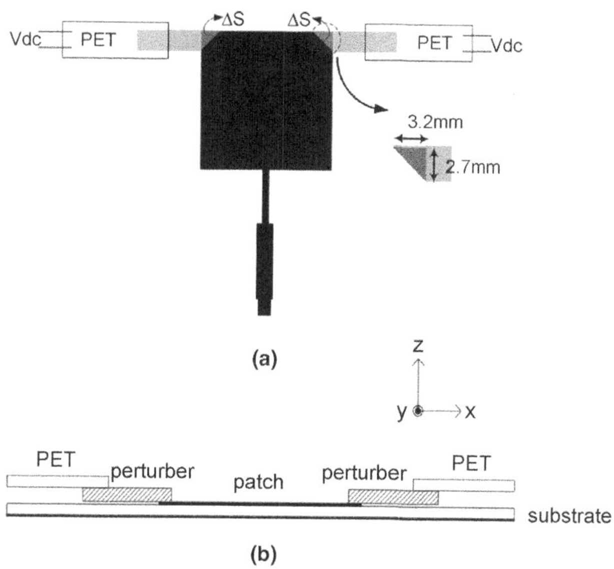

Figure 6.12 Polarization-switchable reconfigurable microstrip antenna with piezoelectric actuators. Reproduced by permission of ©2007 IEEE [48]

In [51], a polarization-reconfigurable antenna for operation at 5.8 GHz uses a coplanar waveguide (CPW) feed in order to reduce the number of metallization layers in the structure. The reconfiguration between LHCP and RHCP using a single feed line is controlled with four beam-lead PIN diodes inserted into the coupling slot below the radiating microstrip patch. The structure is shown in Figure 6.13 [51], with detail of the switched structure provided in

Figure 6.14 [51]. This antenna is an excellent example of evolution from a well-understood fixed design to a reconfigurable antenna that leverages the knowledge of its operation. Additionally, the study includes assessments of the effects of switching diode placement tolerances, dielectric layering, and diode losses and nonlinearity. Refinements and additional functionality to related structures can be found in [52].

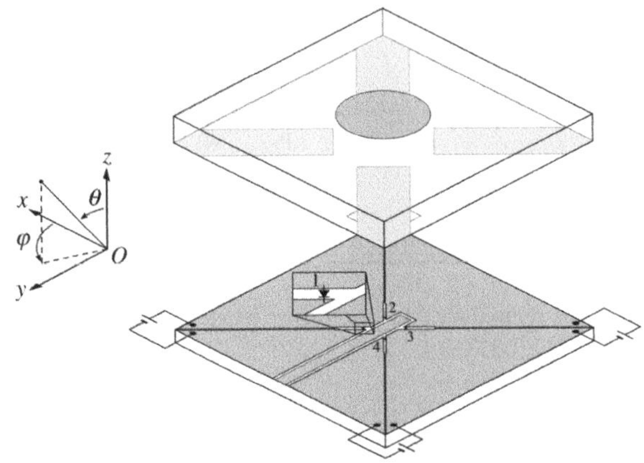

Figure 6.13 Circularly polarized reconfigurable antenna with switched feed slots. Reproduced by permission of ©2006 IEEE [51]

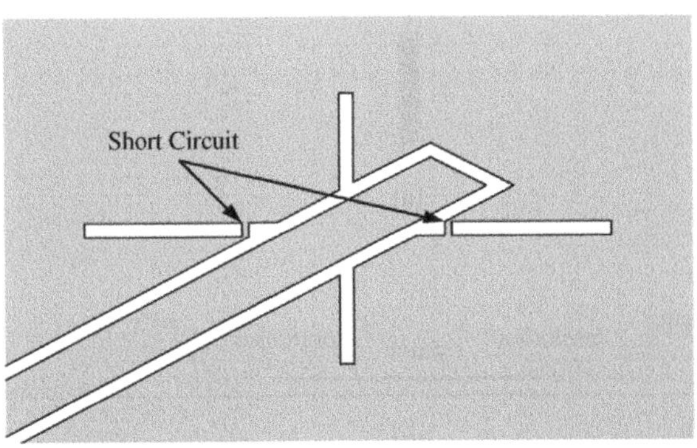

Figure 6.14 Positions of ideal shorts in the feed slots of the antenna shown in Figure 6.13. Reproduced by permission of ©2006 IEEE [51]

For the polarization reconfigurable antenna described in [53], reduction of the bias complexity and antenna size were particular goals. The novelty in the design lies with the reconfigurable feed mechanism that uses a tunable quasi-lumped coupler operated in two modes to feed two input ports of a square microstrip antenna. The antenna is illustrated in Figure 6.15 [53]. While the antenna itself is not altered, the performance and degree of reconfigurability are notable in light of the simplicity of the reconfiguration mechanism. Other researchers, including [54] and [55], have also explored the possibility of antenna reconfigurability enabled not with changes to the antenna structure, but rather changes to the port with which the antenna is excited. For example, in [54], a U-shaped coupling line is used to condition currents in a microstrip loop antenna to flow in either a clockwise or counter-clockwise manner, resulting in left-hand or right-hand circular polarization.

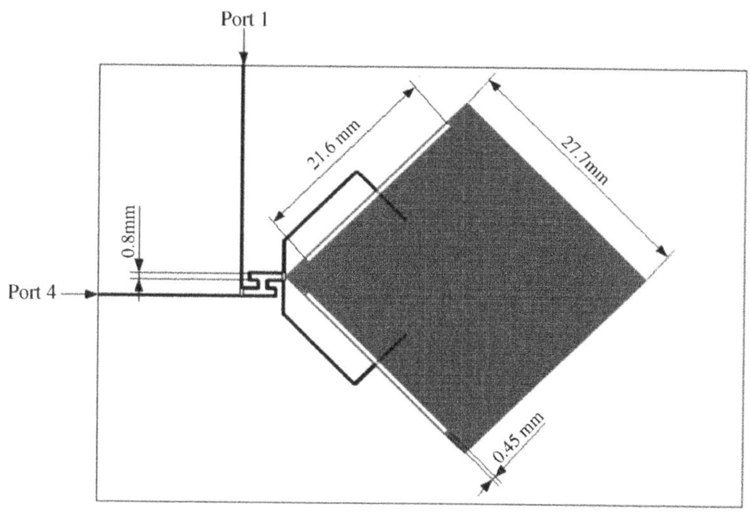

Figure 6.15 Top view of a polarization-agile antenna equipped with a reconfigurable coupler. Gray represents the signal layer of the substrate, white represents the ground plane of the upper substrate and black represents the signal layer of the lower substrate. Reproduced by permission of ©2009 IEEE [53]

6.4 Enabling Reconfigurability: Considerations for Reconfiguration Mechanisms

In earlier incarnations, reconfigurable microstrip antennas could be "demonstrated" using hard-wired connections to mimic switches or substrate substitution to mimic permittivity changes. However, as research into this area has progressed over the past decade, it has become clear that these approximations to the antenna's configuration states do not adequately capture the true behavior of the antenna once real reconfiguration mechanisms are implemented. This means that more and more of the proposed reconfigurable radiators in the open literature include both the basic antenna design as well as the practical considerations of the implemented reconfiguration mechanism. This information provides designers and researchers with a much more comprehensive and accurate assessment of both the costs and benefits of reconfigurability. This also means that there are fewer and fewer reconfigurable designs that include a large number of switches, because the inclusion of the necessary bias networks for each switch often makes the complete antenna design impractical though the hard-wired version shows promise. One can find very useful examples of approaches for selection and implementation of configuration mechanisms that can be leveraged by others designing their own reconfigurable antennas. This section describes the important issues that designers need to consider in the selection of reconfigurability mechanisms, using antennas discussed from the previous sections to illustrate cases of best practice.

Several key factors can help guide both the selection and implementation of reconfiguration mechanisms. The first of these is fundamental: isolation. In most research published to date, the reconfiguration mechanism involves some kind of DC control, whether it be for tuning of material properties or activation of electronic switches or structures. In order to have adequate isolation between the DC and RF signal paths on the antenna, care must be taken to include both DC blocking capacitors and RF chokes (typically achieved with high impedance transmission lines within the microstrip topology) in appropriate places in the structure. Holistic designs might also select particular microstrip topologies that simplify the bias networks from the outset. Several useful examples of innovative and novel approaches to achieve this DC isolation exist.

One early example of inclusion of packaged RF MEMS switches in a pattern reconfigurable antenna is provided by [36]. Since the microstrip line composing the antenna had a characteristic impedance much higher than 50 Ω, several significant changes to the antenna layout, the bias network, and even the switches themselves were required [36]. Since the ground planes of the RF MEMS switches were removed in order to decrease the impedance mismatch between the microstrip line and the switches, the required bias network also had to be specially designed to provide a higher measure of DC-RF isolation that was decreased by the removal of the switches' ground planes. Figures 6.16 and 6.17 show the antenna and its associated bias networks for the switches, respectively [36].

One approach to simplify these kinds of designs is to fabricate the antenna and any necessary switches in the same fabrication steps [38, 56]. In [38], a spiral microstrip design produced

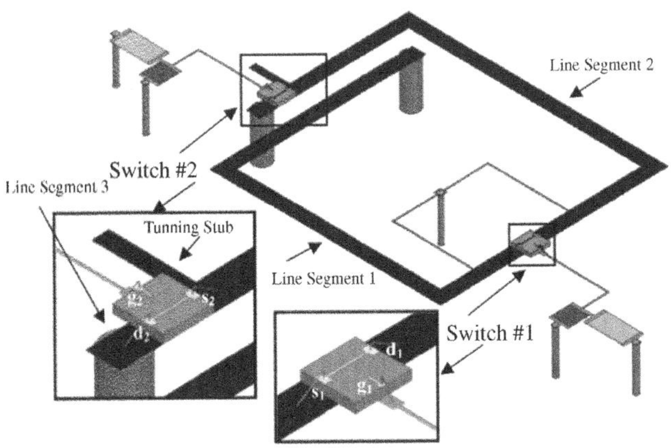

Figure 6.16 Reconfigurable square spiral microstrip antenna model using HFSS™ (Ansoft Corp., Pittsburgh, PA) that includes bias lines, vias, lumped components, and simplified switch models. Antenna ground plane not shown. Reproduced by permission of ©2006 IEEE [36]

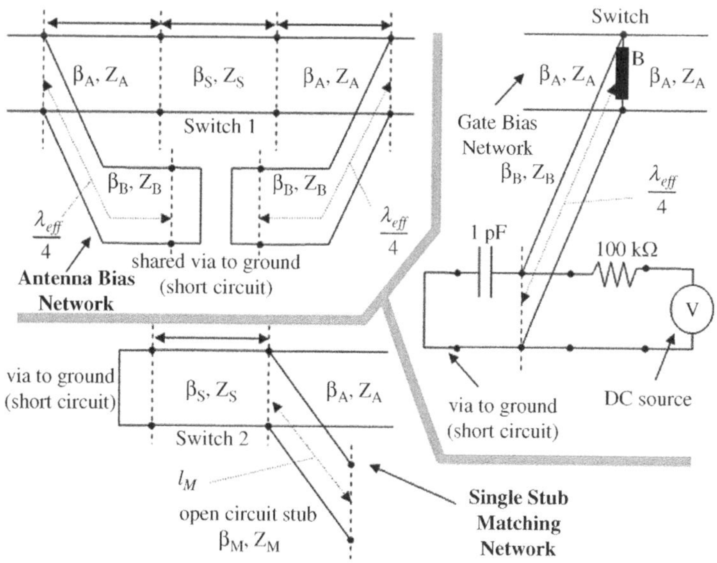

Figure 6.17 Equivalent transmission line circuits of the bias networks of the antenna shown in Figure 6.16. The subscripts A, B, M. and S reference impedances and propagation constants of the microstrip lines that compose the antenna (A), bias networks (B), matching network (M), and switch (S). Reproduced by permission of ©2006 IEEE [36]

tilted radiation patterns with actuation of a subset of directly fabricated RF MEMS switches that connect different portions of the spiral arms together. In [56], an antenna design based on that described in [36] was integrated directly with RF MEMS switches for operation at 30 GHz.

Another approach to achieve DC-RF isolation is to locate the reconfiguration devices either on parasitic elements or feeds rather than on the driven element itself. (One can make the argument that a reconfigurable feed network for an antenna does not strictly make the antenna reconfigurable, but we will not make this distinction here.) For example, in the antenna in [27], pattern reconfiguration is achieved with diode switches that change the length of parasitic elements. As discussed earlier, Rajagopalan et al. used changes in the lengths of coupling slots enabled with RF MEMS switches [21] while Aissat et al. [51] used switching diodes to reconfigure coupling slots. Additionally, Riel and Laurin [22] tuned the feed network with varactors diodes to achieve phase reconfiguration for reflectarray elements.

Several recent articles in the literature also specifically address and include the circuit behavior of the reconfigurable mechanism to capture the complete response of the antenna. Often, the switching mechanism is included in full wave simulations, captured with an equivalent circuit model [e.g., 21, 22, 27, 36, 41, 51]. This level of detail in the analysis provides insight into impedance behavior [e.g., 21, 36], losses [e.g., 22, 41, 51], efficiency [41], and when appropriate, harmonic generation arising from the use of solid state switching elements [e.g., 22, 51].

Of course, another practical design consideration is one of reconfiguration speed and the related issue of required actuation voltage. At this moment, the switching speeds of RF MEMS switches are slow (on the order of milliseconds) relative to their solid state counterparts (on the order of microseconds). Tunable/reconfigurable materials also have response times of the order of milliseconds. Additionally, RF MEMS and tunable materials usually have actuation voltages (or currents, in the case of tuned ferrites) that are at least an order of magnitude higher than those of diode and FET switches. Essentially, the properties of the switching mechanism should always align with the goals of the system in which the antenna will operate. Therefore, if slower, less frequent switching at higher control signals fits into the operational model for the antenna (for example, in satellite-based systems that typically have high voltages readily available), then selection of these kinds of switching devices makes sense. Many reconfigurable antennas intended for implementation in portable wireless devices, however, have concentrated mainly on diode or FET-based switches that require actuation voltages in the 3–5 volt range, since these operating voltages are common in these devices.

6.5 Future Trends in Reconfigurable Microstrip Antenna Research and Development

Although reconfigurable microstrip antennas have advanced well beyond their first prototypes, numerous research areas still need attention before reconfigurable microstrip antennas find their way into products and devices that can take full advantage of their flexibility. This situation is especially urgent with the emergence of software-defined radio and cognitive radio as possible responses to an increasingly crowded electromagnetic spectrum. These new applications will rely heavily on the ability of the entire wireless system to adjust to its environment, and a critical part of this ability lies in the antenna.

One of the primary areas for research that can lead to integration of reconfigurable microstrip antennas in these systems is the quantification of system performance when reconfigurability is implemented. To date, there have been a few such studies that indicate that the benefits of microstrip antenna reconfigurability can be significant, especially in the area of pattern reconfiguration. For instance, in [57], pattern reconfigurable microstrip antennas are shown to significantly increase the achievable system capacity in a multiple-input multiple-output (MIMO) scenario, and have greater effect in cases of low signal-to-noise ratios. Similar results were also achieved in [58], in which a reconfigurable leaky wave structure in microstrip is demonstrated in a MIMO system environment. In another example, pattern reconfigurable microstrip antennas integrated into a laptop computer model result in enhanced spatial coverage over typical fixed antennas, providing the capability to adjust to new operational conditions [59]. Reconfigurable microstrip antennas were also integrated into laptop models in [60], with performance evaluated with different local electromagnetic environments. More recently, Poussot et al. [61] have measured diversity gains in indoor propagation environments using a microstrip antenna that is capable of switching patterns and polarizations via a reconfigurable feed network. Of course, more significant and comprehensive studies need to be completed in this area, so that desired system performance and prescribed antenna functionality can be directly linked. In particular, in the future, a range of operating scenarios that call for specific kinds of microstrip antenna reconfigurability will be able to provide the basis for antenna development. This will result in antennas that are designed specifically for particular applications so that they provide the levels of performance enhancement required.

Another important and related area of research addresses the true implications of software-defined and cognitive radio, namely the ability to use system-level performance measures and the sensed electromagnetic environment as feedback that can be processed to determine the desired state of reconfigurable microstrip antennas. At this point, the hardware and algorithms to enable this level of functionality are critical to the advancement of more sophisticated wireless systems that society will require, but their development is not nearly as mature as the antenna technology itself.

Finally, advances in new kinds of reconfiguration mechanisms, including tunable artificial materials and microfluidically-tuned dielectrics [9], as well as refinements in existing technologies, such as microelectromagnetic actuators, are necessary to provide cost-effective devices that can be implemented in reconfigurable microstrip antennas. With these complementary research and development activities, microstrip antenna capabilities can expand beyond current performance limitations and help to fulfill the promise of next generation wireless systems.

References

1. R. R. Romanofsky, F. A. Miranda, F. W. Van Keuls, and M. D. Valerio, "Recent advances in microwave applications of thin ferroelectric films at the NASA Glenn Research Center," *Proc. Materials, Integration and Packaging Issues for High-Frequency Devices II*, vol. 833, pp. 173–181, 2004.
2. F. A. Miranda, F. W. Van Keuls, R. R. Romanofsky, C. H. Mueller, S. Alterovitz, and G. Subramanyam, "Ferroelectric thin films-based technology for frequency- and phase-agile microwave communication applications," *Integrated Ferroelectrics*, vol. 42, pp. 131–149, 2002.
3. H. Xu, N. K. Pervez, and R. A. York, "Tunable microwave integrated circuits BST thin film capacitors with device structure optimization," *Integrated Ferroelectrics*, vol. 77, pp. 27–35, 2005.
4. D. M. Pozar and V. Sanchez, "Magnetic tuning of a microstrip antenna on a ferrite substrate," *Electron. Lett.*, vol. 24, pp. 729–731, June 9, 1988.
5. R. K. Mishra, S. S. Pattnaik, and N. Das, "Tuning of microstrip antenna on ferrite substrate," *IEEE Trans. Antennas Propagat.*, vol. 41, pp. 230–233, February 1993.
6. A. D. Brown, J. L. Volakis, L. C. Kempel, and Y. Botros, "Patch antennas on ferromagnetic substrates," *IEEE Trans. Antennas Propag.*, vol. 47, pp. 26–32, January 1999.
7. A. Henderson, J. R. James, A. Fray, and G. D. Evans, "New ideas for beam scanning using magnetised ferrite," *Proc. IEE Colloquium on Electronically Scanned Antennas*, vol. 1, pp. 4/1–4/4, 21 January 1988.
8. P. Rainville and F. Harackiewicz, "Magnetic tuning of a microstrip patch antenna fabricated on a ferrite film," *IEEE Microwave Guided Wave Lett.*, vol. 2, pp. 483–485, December 1992.
9. G. H. Huff, P. Bahukudumbi, W. N. Everitt, A. Beskok, M. A. Bevan, D. Lagoudas, and Z. Ounaies, "Microfluidic reconfiguration of antennas," *Proc. 2007 Antenna Applications Symposium*, pp. 241–258, Allerton Park, Monticello, IL, Sept. 2007.
10. P. Bhartia and I. J. Bahl, "Frequency agile microstrip antennas," *Microwave J.*, vol. 25, pp. 67–70, October 1982.
11. E. Erdil, K. Topalli, M. Unlu, O. A. Civi, and T. Akin, "Frequency tunable microstrip patch antenna using RF MEMS technology," *IEEE Trans. Antennas Propagat.*, vol. 54, pp. 1193–1196, April 2007.
12. S. V. Hum, M. Okoniewski and R. J. Davies, "Modeling and design of electronically tunable reflectarrays," *IEEE Trans. Antennas Propagat.*, vol. 55, no. 8, pp. 2200–2210, August 2007.
13. E. R. Brown, "On the gain of a reconfigurable-aperture antenna," *IEEE Trans. on Antennas and Propagation*, vol. 49, pp. 1357–1362, October 2001.
14. W. H. Weedon, W. J. Payne, and G. M. Rebeiz, "MEMS-switched reconfigurable antennas," *Proc. IEEE/URSI Int. Symp on Antennas and Propagat.*, vol. 3, pp. 654–657, 2001.
15. W. Weedon, W. Payne, G. Rebeiz, J. Herd, and M. Champion, "MEMS-switched reconfigurable multi-band antenna: design and modeling," *Proc. 1999 Allerton Antenna Applications Symposium*, vol. 1, pp. 203–231, 1999.
16. F. Yang and Y. Rahmat-Samii, "Patch antenna with switchable slot (PASS): dual frequency operation," *Microwave and Optical Technology Letters*, vol. 31, pp. 165–168, November 2001.
17. B.-Z. Wang, S. Xiao, and J. Wang, "Reconfigurable patch-antenna design for wideband wireless communication systems," *IET Microw. Antennas Propagat.*, vol. 1, no. 2, pp. 414–419, April 2007.

18. S. V. Shynu, G. Augustin, C. K. Aanandan, P. Mohanan, and K. Vasudevan, "C-shaped slot loaded reconfigurable microstrip antenna," *Electron. Lett.*, vol. 42, no. 6, pp. 316–318, March 16, 2006.
19. J. L. A. Quijano and G. Vecchi, "Optimization of an innovative type of compact frequency-reconfigurable antenna," *IEEE Trans. Antennas Propagat.*, vol. 57, no. 1, pp. 9–18, January 2009.
20. A.-F. Sheta and S. F. Mahmoud, "A widely tunable compact patch antenna," *IEEE Antennas and Wireless Propagat. Lett.*, vol. 7, pp. 40–42, 2008.
21. H. Rajagopalan, Y. Rahmat-Samii, and W. A. Imbriale, "RF MEMS actuated reconfigurable reflectarray patch-slot element," *IEEE Trans. Antennas Propagat.*, vol. 56, no. 12, pp. 3689–3699, December 2008.
22. M. Riel and J.-J. Laurin, "Design of an electronically beam scanning reflectarray using aperture-coupled elements," *IEEE Trans. Antennas Propagat.*, vol. 55, no. 5, pp. 1260–1266, May 2007.
23. E. Kiely, G. Washington, and J. T. Bernhard, "Design and development of smart microstrip patch antennas," *Smart Materials and Structures*, vol. 7, no. 6, pp. 792–800, December 1998.
24. E. Kiely, G. Washington, and J. T. Bernhard, "Design, actuation, and control of active patch antennas," in *Proc. SPIE-Int. Soc. Opt. Eng.*, vol. 3328, pp. 147–155 1998.
25. J. T. Bernhard, E. Kiely, and G. Washington, "A smart mechanically-actuated two-layer electromagnetically coupled microstrip antenna with variable frequency, bandwidth, and antenna gain," *IEEE Trans. Antennas Propagat.*, vol. 49, pp. 597–601, April 2001.
26. R. F. Harrington, "Reactively controlled directive arrays," *IEEE Transactions on Antennas and Propagat.*, vol. 26, pp. 390–395, May 1978.
27. S. Zhang, G. H. Huff, J. Feng, and J. T. Bernhard, "A pattern reconfigurable microstrip parasitic array," *IEEE Trans. Antennas Propagat.*, vol. 52, pp. 2773–2776, October 2004.
28. S. Zhang, G. Huff, G. Cung, and J. T. Bernhard, "Three variations of a pattern reconfigurable microstrip parasitic array," *Microwave Opt. Technol. Lett.*, vol. 45, pp. 369–372, June 2005.
29. S. Zhang, "A pattern reconfigurable microstrip parasitic array: theory, design, and applications," Ph.D. dissertation, University of Illinois at Urbana-Champaign, 2005.
30. R. J. Dinger, "Reactively steered adaptive array using microstrip patch elements at 4 GHz," *IEEE Trans. Antennas Propagat.*, vol. 32, pp. 848–856, August 1984.
31. R. J. Dinger, "A planar version of a 4.0 GHz reactively steered adaptive array," *IEEE Trans. Antennas Propagat.*, vol. 34, pp. 427–431, March 1986.
32. R. Schlub, J. Lu, and T. Ohira, "Seven-element ground skirt monopole ESPAR antenna design from a genetic algorithm and finite element method," *IEEE Trans. Antennas Propagat.*, vol. 51, no. 11, pp. 3033–3039, Nov. 2003.
33. M. Donelli, R. Azaro, L. Fimognari, and A. Massa, "A planar electronically reconfigurable Wi-Fi band antenna based on a parasitic microstrip structure," *IEEE Antennas and Wireless Propagat. Letters*, vol. 6, pp. 623–626, 2007.
34. G. H. Huff, J. Feng, S. Zhang, and J. T. Bernhard, "A novel radiation pattern and frequency reconfigurable single turn square spiral microstrip antenna," *IEEE Microwave Wireless Comp. Lett.*, vol. 13, pp. 57–59, February 2003.
35. G. H. Huff and J. T. Bernhard, "Analysis of a radiation and frequency reconfigurable microstrip antenna," *Proc. 2004 Antenna Applications Symposium*, pp. 175–191, Sept. 2004.
36. G. H. Huff and J. T. Bernhard, "Integration of packaged RF MEMS switches with radiation pattern reconfigurable square spiral microstrip antennas," *IEEE Trans. Antennas Propagat.*, vol. 54, pp. 464–469, February 2006.
37. S.-H. Chen, J.-S. Row, and K.-L. Wong, "Reconfigurable square-ring patch antenna with pattern diversity," *IEEE Trans. Antennas Prop.*, vol. 55, pp. 472–475, February 2007.
38. C. Jung, M. Lee, G. P. Li, and F. De Flaviis, "Reconfigurable scan-beam single-arm spiral antenna integrated with RF-MEMS switches," *IEEE Trans. Antennas Propagat.*, vol. 54, pp. 455–463, February 2006.
39. A. Mehta, D. Mirshekar-Syahkal, and H. Nakano, "Beam adaptive single arm rectangular spiral antenna with switches," *IEE Proceedings-Microwaves, Antennas and Propagat.*, vol. 153, no. 1, pp. 13–18, February 2006.
40. J. T. Bernhard, *Reconfigurable Antennas*, ed. Constantine Balanis, Morgan & Claypool, San Rafael, CA, 2007.
41. M. Ali, A. T. M. Sayem, and V. K. Kunda, "A reconfigurable stacked microstrip patch antenna for satellite and terrestrial links," *IEEE Trans. Vehicular Tech.*, vol. 56, no. 2, pp 426–435, March 2007.
42. F. Yang and Y. Rahmat-Samii, "Patch antennas with switchable slots (PASS) in wireless communications: concepts, designs, and applications," *IEEE Antennas and Propagation Magazine*, vol. 47, pp. 13–29, April 2005.
43. F. Yang and Y. Rahmat-Samii, "A reconfigurable patch antenna using switchable slots for circular polarization diversity," *IEEE Microwave and Wireless Components Letters*, vol. 12, pp. 96–98, March 2002.

44. Y. J. Sung, T. U. Jang, and Y.-S. Kim, "A reconfigurable microstrip antenna for switchable polarization," *IEEE Microwave and Wireless Comp. Lett.*, vol. 14, pp. 534–536, November 2004.
45. M. Boti, L. Dussopt, and J.-M. Laheurte, "Circularly polarized antenna with switchable polarization sense," *Electron. Lett.*, vol. 36, pp. 1518–1519, August 2000.
46. Y. J. Sung, "Reconfigurable patch antenna for polarization diversity," *IEEE Trans. Antennas Propagat.*, vol. 56, no. 9, pp. 3053–3054, September 2008.
47. R. N. Simons, D. Chun, and L. P. B. Katehi, "Polarization reconfigurable patch antenna using microelectromechanical systems (MEMS) actuators," *Proc. IEEE/URSI Int. Symp. on Antennas and Propagat.*, vol. 1, pp. 6–9, 2002.
48. A S.-H. Hsu and K. Chang, "Novel reconfigurable microstrip antenna with switchable circular polarization," *IEEE Antennas and Wireless Propagat. Lett.*, vol. 6, pp. 160–162, 2007.
49. A. Khaleghi and M. Kamyab, "Reconfigurable single port antenna with circular polarization diversity," *IEEE Trans. Antennas Propagat.*, vol. 57, no. 2, pp. 555–559, February 2009.
50. B. Kim, B. Pan, S. Nikolaou, Y.-S. Kim, J. Papapolymerou, and M. M. Tentzeris, "A novel single-feed circular microstrip antenna with reconfigurable polarization capability," *IEEE Trans. Antennas Propagat.*, vol. 56, no. 3, pp. 630–638, March 2008.
51. H. Aïssat, L. Cirio, M. Grzeskowiak, J.-M. Laheurte, and O. Picon, "Reconfigurable circularly polarized antenna for short-range communication systems," *IEEE Trans. Microw. Theory Tech.*, vol. 54, no. 6, pp. 2856–2863, June 2006.
52. R.-H. Chen and J.-S. Row, "Single-fed microstrip patch antenna with switchable polarization," *IEEE Trans. Antennas Propagat.*, vol. 56, no. 4, pp. 922–926, April 2008.
53. F. Ferrero, C. Luxey, R. Staraj, G. Jacquemod, M. Yedlin, and V. Fusco, "A novel quad-polarization agile patch antenna," *IEEE Trans. Antennas Propagat.*, vol. 57, no. 5, pp 1562–1566, May 2009.
54. K.-F. Tong and J. Huang, "New proximity coupled feeding method for reconfigurable circularly polarized microstrip ring antennas," *IEEE Trans. Antennas Propagat.*, vol. 56, no. 7, pp. 1860–1866, July 2008.
55. Y.-F. Wu, C.-H. Wu, D.-Y. Lai, and F.-C. Chen, "A reconfigurable quadri-polarization diversity aperture-coupled patch antenna," *IEEE Trans. Antennas Propagat.*, vol. 55, no. 3, pp. 1009–1012, March 2007.
56. T. Z. Wojtaszek, J. T. Bernhard, G. H. Huff, D. J. Chung, and J. Papapolymerou, "Reconfigurable antennas with integrated RF MEMS switches for military MIMO applications," in *Proc. GOMACTech-2009*, Orlando, FL, March 2009.
57. J. Boerman and J. T. Bernhard, "Performance study of pattern reconfigurable antennas in MIMO communication systems," *IEEE Trans. Antennas Propagat.*, vol. 56, no. 1, pp. 231–236, Jan. 2008.
58. D. Piazza, M. D'Amico, and K. P. Dandekar, "Performance improvement of a wideband MIMO system by using a two-port RLWA," *IEEE Antennas and Wireless Propagat. Lett.*, pp. 830–834 2009.
59. T. L. Roach, G. H. Huff, and J. T. Bernhard, "A comparative study of diversity gain and spatial coverage: fixed versus reconfigurable antennas for portable devices," *Microw. Opt. Techn. Lett.*, vol. 49, no. 3, pp. 535–539, March 2007.
60. G. H. Huff, J. Feng, S. Zhang, and J. T. Bernhard, "Directional reconfigurable antennas on laptop computers: simulation, measurement, and evaluation of candidate integration positions." *IEEE Trans. Antennas Propagat.*, vol. 52, pp. 3220–3227, December 2004.
61. B. Poussot, J.-M. Laheurte, L. Cirio, O. Picon, D. Delcroix, and L. Dussopt, "Diversity measurements of a reconfigurable antenna with switched polarizations and patterns," *IEEE Trans. Antennas Propagat.*, vol. 56, no. 1, pp. 31–38, January 2008.

7

Wearable Antennas for Body Area Networks

Peter S. Hall[1] and Yang Hao[2]
[1]*University of Birmingham, UK*
[2]*Queen Mary College, University of London, UK*

7.1 Introduction

7.1.1 Overview

The increasing international activity in body area networks (BANs), wireless personal area networks (WPANs), and medical sensor networks has given rise to renewed interest in wearable antennas that are designed for operation on the human body, or the bodies of animals. The military have used portable radio equipment, incorporating antennas close to the body, for many decades, and the development of the pager and mobile phone have meant that many studies of the interaction of antennas with the body have been undertaken. However the introduction of many types of small body worn medical sensors has enhanced the potential of wireless medical sensor networks in connecting these sensors to specialists, and increased the ability of doctors to monitor their patients at a distance, thus enabling resource saving through early patient release from hospitals. The wider applications of wearable antennas have been well described in [1–6].

These developments present a number of new and exciting challenges for antenna designers and these, together with examples of current wearable antennas, form the basis of this chapter. After describing a useful classifier and pointing out the close relationship between antenna performance and radiowave propagation in this section, the properties of the human body are reviewed and some of the electromagnetic issues of sources on the body are noted. Subsequent sections describe narrowband antenna design, fabric-based antennas, ultra wideband antennas and finally the performance of multiple antenna systems on the body. Conclusions are then drawn together with some notes for future directions.

Microstrip and Printed Antennas: New Trends, Techniques and Applications. Edited by Debatosh Guha and Yahia M.M. Antar

7.1.2 Domains of Operation

While BANs, WPANs, and medical sensor networks are classifications adopted at a system level, it is useful to discuss a further classifier, at the antenna level, that gives some insight into the electromagnetic challenges that confront the antenna designer. The function of wearable antennas can be divided into three categories, namely:

1. *Communications between an antenna on the body and another off the body.* This is the conventional and most widely used function of the wearable antenna. It includes antennas for portable radio transceivers, such as military radios and mobile phones. Such antennas need to be compact so that they fit inside the transceiver and to be efficient with minimal power absorbed by the body tissue. They are usually designed to launch energy directly away from the body surface.
2. *Communications between an antenna on the body and another on the body.* In this case, the aim is to form a network on the surface of the body, which may be, for instance, a medical sensor network, or communications between a transceiver working a channel to a local base station and human interface devices, such as earpieces, microphones or visual display units. These antennas need to guide the energy along the body surface, and this involves a trade-off between energy used in communication and that dissipated in the body tissue.
3. *Communications between an antenna on the body and another in the body.* There are now many examples of medical implants that incorporate wireless communications, and the channel is primarily from the implant to an antenna on the body surface, although communications to local base stations are also used. The design of the implant antenna is very different from the wearable antennas noted above, being dominated by the surrounding high permittivity, high dissipation and very varying medium. These antennas are not normally referred to as wearable and are hence not covered in this chapter.

7.1.3 Antenna Parameters

Wearable antennas for communications to base stations off the body are generally characterized by their volume if they are to be fitted into small electronic equipment, their input match over the bands required, their efficiency and their radiation pattern. If the antenna is to be incorporated into clothing, size is less important, unless operation at VHF or UHF is needed. Determination of efficiency is problematic, and is often determined through simulation or by measurement on a partial or whole body phantom. Radiation pattern is determined by incorporating a human subject or a whole body phantom into a conventional measurement arrangement. Figure 7.1 shows a ring around the body carrying the antenna, indicating that measurement should be at the far field distance, determined by the body size rather than the antenna size, though in many cases this distance may be inconveniently large and shorter ones are used.

If the antenna is expected to be used in a heavily cluttered, multipath rich environment, then the radiation pattern is much less important, particularly for small antennas where the radiation pattern is hard to control.

If the antenna is to communicate with another one on the body then, as above, size, match and efficiency are important. However, determination of the radiation pattern is much less straightforward, as Figure 7.1 indicates. Clearly, a measurement at the far distance

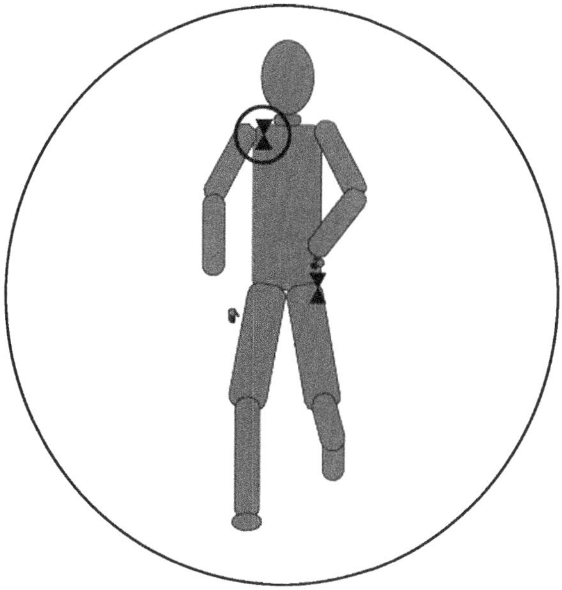

Figure 7.1 Antennas on the human body

from the body is inappropriate, as the belt-mounted receive antenna is relatively close to the shoulder-mounted transmit antenna. The small ring around the transmit antenna suggests a more appropriate distance, although there does not appear to be simple way of specifying this distance, nor of making a measurement, as part of the measurement sphere is inside the body.

7.2 Sources on the Human Body

7.2.1 Electrical Properties of the Human Body

The human body is composed of a large variety of tissue types, each having different dielectric properties, and this data is important for the design of wearable antennas. There have been a number of studies of the dielectric properties of body tissues at radio frequencies and microwaves [6–11], and a recent comprehensive one is given in [12]. Measurements cannot be made on live tissue, except in the case of skin and tongue, so that most measurements are made on human autopsy material or freshly killed animal tissue. Measurements have been made over a frequency range of 10 Hz to 20 GHz, and characterized using a 4-Cole-Cole model. A web site [13] gives a readily accessible parameter data from which the data in Figure 7.2 is obtained.

It can be seen that there is significant variation in parameters with frequency and that muscle has a considerably higher permittivity and conductivity than fat. This results in wave penetration depths that also vary with frequency. For example, for muscle at 100 MHz, the depth is 100 mm, whereas at 2 GHz it is about 30 mm. For fat, these values are about 400 mm and 200 mm, respectively. This indicates that for muscle at 2 GHz only the surface layers need

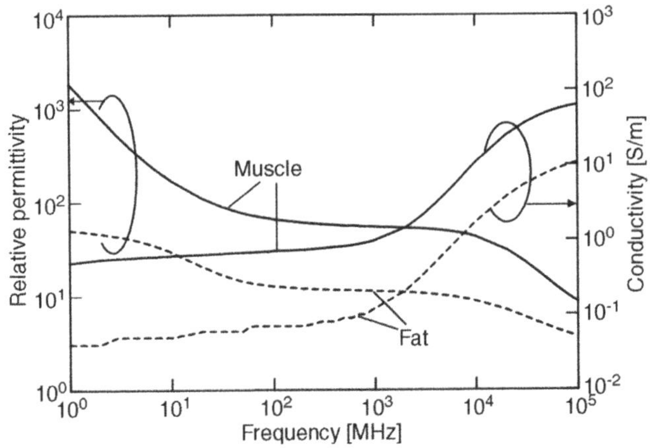

Figure 7.2 Frequency dependence of relative permittivity and conductivity of muscle and fat
Note: Solid = muscle, dotted = fat

to be taken into account when designing the antenna, whereas for fat at lower frequencies, the effect of much deeper tissue will be important.

7.2.2 Sources and Waves on the Body

The dielectric properties of the body tissues noted above can be used to find the intrinsic impedance of the material. At 2.5 GHz the impedances of muscle and fat are 50.8 + j6.1 ohms and 162.7 + j11.8 ohms respectively, assuming normal incidence. The low real part (compared to the impedance of free space) means that when an antenna is put close to the surface, the antenna impedance will be strongly mismatched. The reactive component will mean that the resonant frequency will be reduced, as it is inversely proportional to the reactance in the antenna equivalent circuit. This will occur whether the antenna is an electric source, such as a dipole, or magnetic, such as a slot. This phenomenon is observed in practice, as noted in Section 7.3. Reduction in efficiency and distortions to the radiation pattern also occur.

In the case of antennas intended for communications to another on the body, then the nature of radiowave propagation on the body should be studied. This propagation will be a mixture of surface wave, creeping wave, and space waves, with reflection, diffraction and scattering also being present. If the body is in a scattering environment such as a cluttered room, then this will also influence the received power. Studies [14] indicate that the TM surface wave mode, which has electric field normal to the body surface, is the significant transmission mode for frequencies from 1 to 5 GHz but has an upper cutoff frequency at about 10 GHz. The TE mode, which has the electric field parallel to the surface, has very high attenuation. This is borne out by comparative measurements of link loss for various antennas [5, p. 52], which show path loss lower by more than 10 dB on a channel from the belt to the chest on a moving person for a monopole antenna compared to a loop. In the case of the belt-to-wrist channel, the difference is no more than a couple of dB on average, due, it is believed, to the dominance of space wave propagation in this channel. Differences may then be accounted for by the lower efficiency of the loop, as noted below.

7.3 Narrowband Antennas

7.3.1 Performance Changes Due to Body Proximity

The changes in the performance of a typical wearable antenna due to the proximity of the body have been examined in [15]. For printed dipole and monopole type antennas, printed conformally on a substrate, detuning as the antenna is brought into contact with the body is of the order of 100 MHz at 2.45 GHz. For antennas mounted above a metal ground plane, such as a vertical monopole above a horizontal ground plane or a planar inverted F antenna, the detuning is less than this, and of course depends on the ground plane size. The position on the body is also a factor, with detuning being less on the limbs than on the trunk.

The efficiency of the antenna is also reduced when brought into proximity to the body. Measurements and simulations [16], at 2.45 GHz, in which the antenna is placed on a muscle phantom block, give the results shown in Table 7.1. In the Wheeler cap method, the phantom was placed inside the cap together with the antenna. The cap was 10 cm radius and 20 cm high. The anechoic chamber method involved conventional power and directive gain measurements, with a phantom block of 20 cm radius, 10 cm thick, having dielectric properties of relative dielectric constant of 40 and conductivity of 2.1 S/m. The simulation used the same phantom.

The radiation pattern of the antenna is deformed by the presence of the body. Figure 7.3 shows the radiation pattern of a printed dipole placed parallel to the body surface on the left side of the chest and the left ear respectively, at different distances from the body surface.

It can be seen that some pattern changes take place as the dipole approaches the body, and that different placement positions affect the direction of the main beam. Of course the patterns shown are very different from the pattern of the dipole in free space, due to the shadowing effect of the body. It should also be noted that these patterns are measured in the far field of the antenna and the body, and may thus not be the appropriate pattern to use in consideration of the on-body channel, as discussed in Section 7.2.

The detuning, pattern change and efficiency reduction that result from mounting of the antenna on the human body have been countered in a number of ways. Of course, in the patch and the PIFA antennas noted above, the size of the ground plane primarily determines the proportion of fields dissipated in the lossy tissue. If the antenna is mounted in a small electronic case, then the ground plane size is fixed. However, if the antenna is built into the clothing, then much larger ground planes can often be accommodated. The variation in resonant frequency

Table 7.1 Efficiency of antennas at 2.45 GHz in free space and on a body phantom, [16] (Monopole, 31 mm × 0.5 mm dia. on 25 mm radius × 0.4 mm thick copper ground plane. Planar inverted F antenna, top plate 25.9 mm × 14.8 mm located 9 mm above 50 mm × 40 mm ground plane)

		Wheeler cap (%)	Anechoic chamber (%)	Simulation (%)
Monopole	Free Space	96.52	90	92
	Phantom	52.41	59	63
PIFA	Free Space	93.1	97	91
	Phantom	42.1	51	56

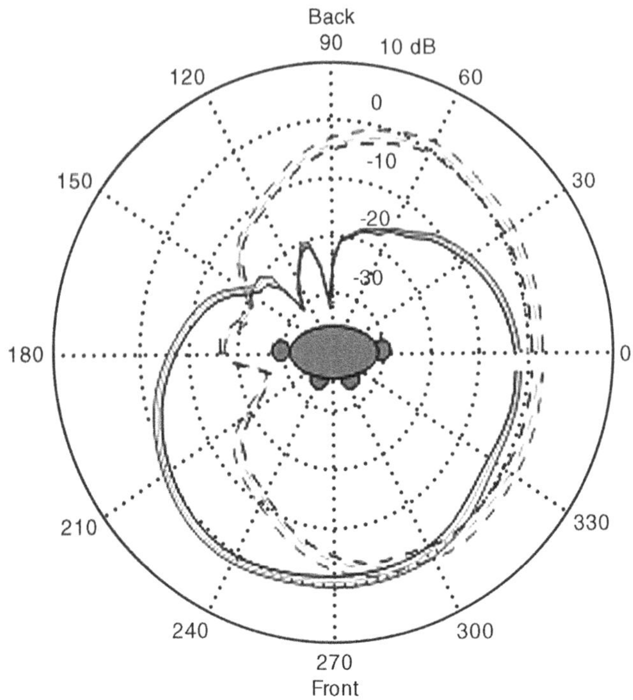

Figure 7.3 Radiation pattern of printed dipole antenna on body
Notes: Antenna on left side of chest (solid) and on left ear (dotted), freq = 2.45 GHz, pattern in azimuth plane of human body, distance from body – black = 1 mm, dark grey = 4 mm, light grey = 8 mm

and efficiency can be stabilized by the addition of thin ferrite polymer sheets to the rear surface of the antenna [17]. In experiments using PIFAs, integrated inverted F antennas, of sizes 37 × 25 mm and 55 × 20 mm respectively and coplanar wire patch antennas, designed for operation at 800 MHz, a 0.5 mm thick layer of polymer resulted in very low detuning and efficiencies similar to those found when the antennas were in contact with the human skin. Reduction of the effect of the body can also be achieved using electromagnetic bandgap (EBG) materials, as shown in Figure 7.4.

Patch antennas at 2.45 GHz [18, 19] have been shown to have very little detuning and pattern change, and high efficiency when brought close to the body. However, the higher efficiency comes at a cost of much bigger ground plane sizes than the ferrite method noted above. For example, the EBG in [18] is 80 × 100 mm.

7.3.2 Antenna Types

The performance of patch antennas on the body has been investigated. The large size of conventional patches is a concern and compact higher mode microstrip patch antennas [20] of

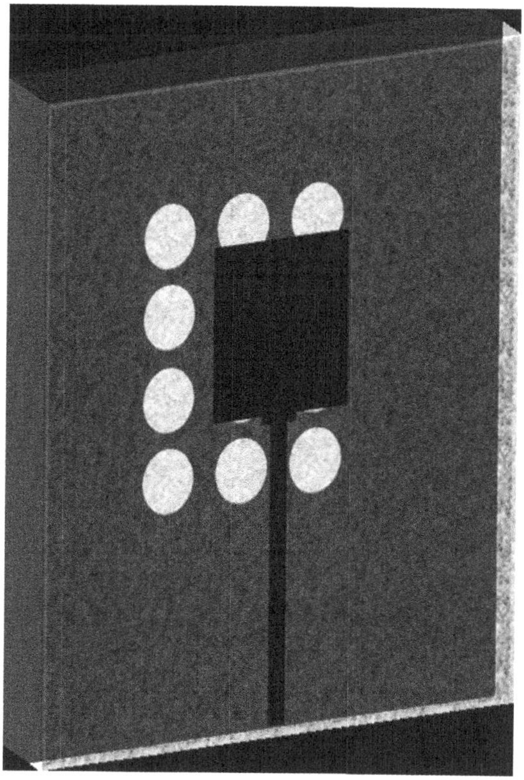

Figure 7.4 Patch antenna mounted on electromagnetic bandgap material

37 × 30 mm size, with a low profile of 5 mm and 10 mm high have been shown to have an impedance bandwidth of 6.7% and 8.6%, respectively, sufficient for the operating requirements of the 2.45-GHz industrial, scientific, and medical (ISM) band and both antennas offered 11-dB higher path gain compared to a fundamental-mode microstrip patch antenna. It was also demonstrated that a 7-dB improvement in path gain for an on-body channel could be obtained for a fundamental-mode patch through the addition of a shortening wall. Notably, the on-body antenna performance was comparable to a quarter wave monopole antenna on the same size of ground plane, indicating that the low-profile and physically more robust antenna is a promising solution for body-worn antenna applications. The same authors have also developed a switched mode patch [21] which can give either a monopolar like pattern or a patch-like pattern for operation of an on-body or off-body channel respectively. Figure 7.5 shows the patch, which is mounted with the ground plane parallel to the surface of the body. The three vias in the middle of the patch serve to reduce the size when the patch is operated in the monopolar mode. The other two are switchable with diode switches, and when short circuited force the patch to operate in a conventional mode with the main beam of the radiation pattern away from the body.

PIFA antennas have been used extensively on the body. An example described in [22] has the radiating plate, of size 50 × 50 mm, air spaced above a ground plane which is bent over the

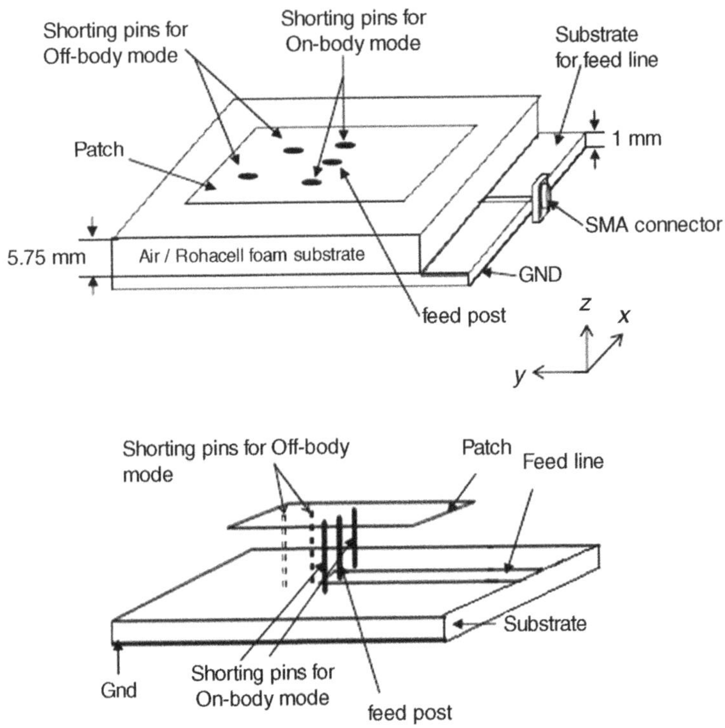

Figure 7.5 Switchable on-body patch antenna. Reproduced by permission of ©2009 IEEE [21]
Note: Switched modes are for on-body and off-body channels

shoulder of the wearer and extends 70 mm on either side of the radiating element. The antenna is aimed at application in police and fire service portable radios in the 350-MHz band. The low-profile structure, 10 mm thick, enables the antenna to be unseen when the operator wears a jacket. The antenna provides a gain of 5.5 dBd, which is 3.5 dB higher than that of an equivalent whip antenna.

V-shaped dipoles and coiled monopoles [23] have been designed at 868 MHz in the ISM band, for operation in medical sensor network applications, in contact with the body surface. Gain has not been measured, but tests inside an anechoic chamber, with the antennas working on various parts of the body and for various body orientations showed signal levels to an off-body antenna of between −54 and −63 dBm. A cavity-backed slot antenna [24] is shown in Figure 7.6 for 2.45 GHz. The overall size is ~38 × 20 × 9 mm and the antenna has been studied while mounted on the human arm and is shown to produce a vertically polarized wave along the surface of the arm, which is well supported as a surface wave.

Antennas for VHF/UHF frequencies for use on the human body include the spiral [25] and the meandered dipole [26]. The spiral, shown in Figure 7.7, is designed for operation in the 100–500 MHz range.

The input return loss is shown in Figure 7.8 and it can be seen that S_{11} is better than −5 dB from around 110 to 500 MHz and that the radiation pattern at 300 MHz is not omnidirectional

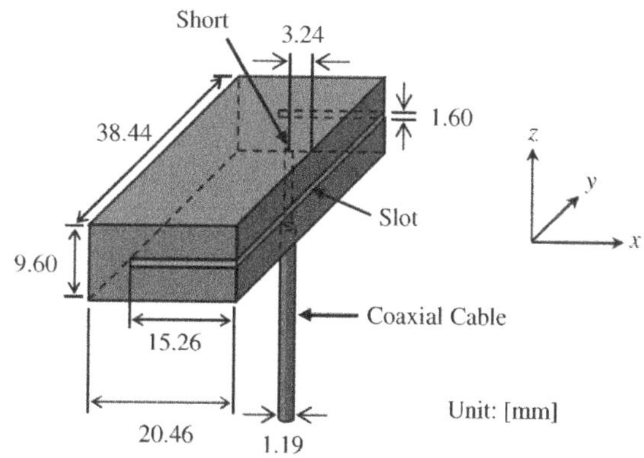

Figure 7.6 Cavity backed on-body slot antenna. Reproduced by permission of ©2008 IEEE [24]

in the azimuth plane, as desired, but shows the effect of shadowing by the body in the back direction, and this caused the gains to go down to −25 dBi.

Two antennas, front and rear, would be needed to give the required coverage in practice. [25] also describes a wideband loaded dipole for use on the body. For both antennas the spacing between the antenna and body was found to be crucial. Meandered dipoles [26] have been

Figure 7.7 Spiral body-worn antenna for VHF/UHF. Reproduced by permission of ©2008 IEEE [25]

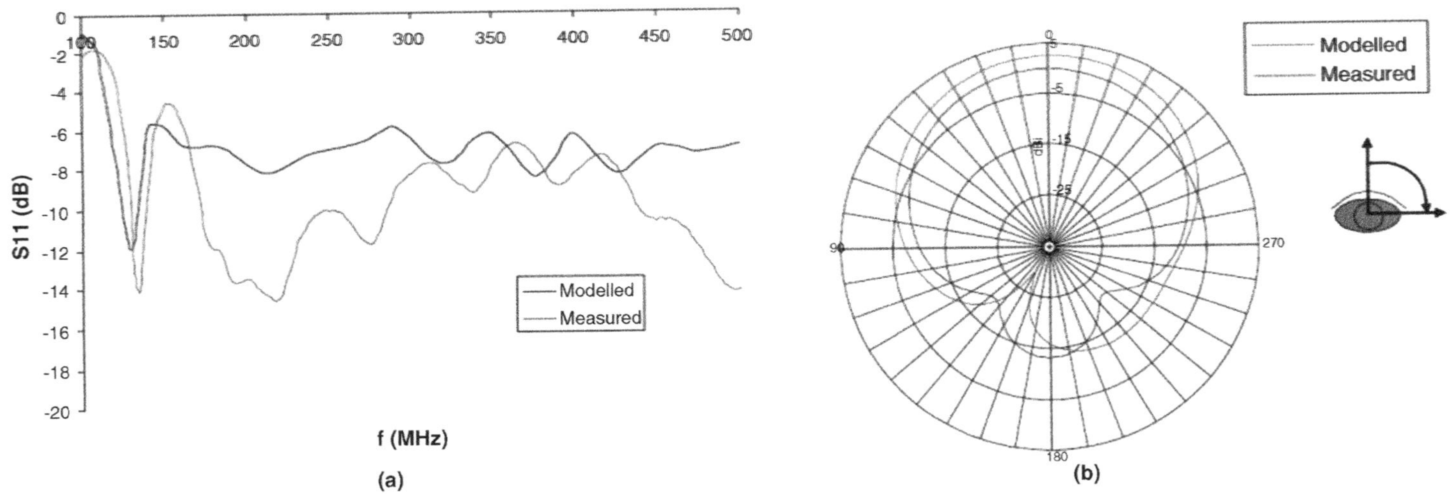

Figure 7.8 Performance of body worn spiral antenna of Figure 7.7: (a) S_{11} vs freq. in MHz; (b) radiation pattern in azimuth plane at 300 MHz. Reproduced by permission of ©2008 IEEE [25]

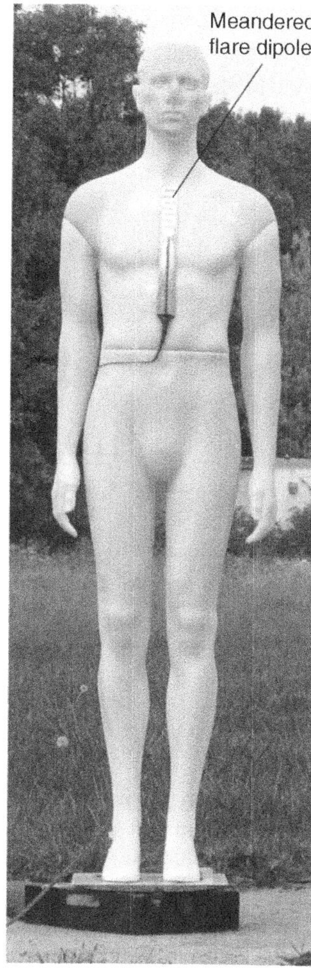

Figure 7.9 Meandered dipole for VHF/UHF on a body phantom. Reproduced by permission of ©2008 IEEE [26]

designed for the 225 MHz–370 MHz band. The antenna is printed on a flexible FR-4 substrate, 4 mils thick, enabling it to be easily mounted and conform to the body, as shown in Figure 7.9 mounted on a life-size human phantom. Radiation patterns show similar shadowing to the spiral described above and work was performed with multiple antennas to improve coverage, and this is described in Section 7.6.

Antennas have also been adapted to objects worn on the body. Clothing is the obvious case but rings and buttons are also used. The finger ring antenna [27] is shown in Figure 7.10 and consists of a PIFA on a ground plane that is wrapped around the finger. Designed for 2.45 GHz, it produces radiation that is polarized vertical to the skin surface, has a bandwidth of about 60 MHz and is slightly detuned when moved along the finger of a phantom hand. Metal buttons [28–30] have also been used as antennas, where the button post and top disc form a monopole that is fed against a circular ground plane. Designs have been described for 2.45 and 5.8 GHz [28] and the UWB band [29]. Real jeans buttons have also been investigated for use at 4.5 GHz. A belt buckle has also been used as an antenna [31] and had bandwidths of 22.8% and 9.4% at the 2.4 GHz band 5.25 GHz and gains of 2.8 dBi at 2.45 GHz and 4.5 dBi at 5.25 GHz.

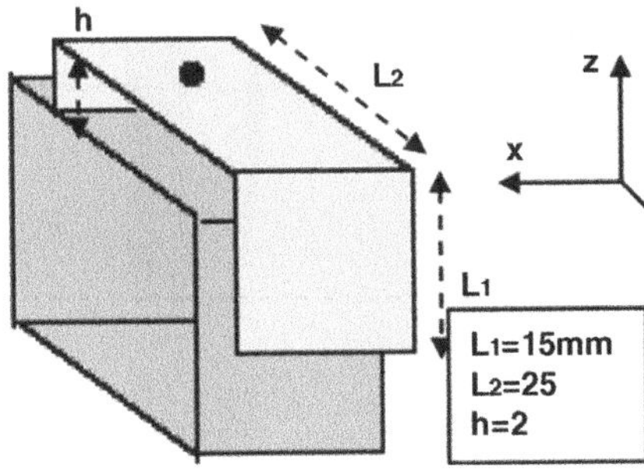

Figure 7.10 Bent inverted F antenna for finger-ring application. Reproduced by permission of ©2008 IEEE [27]

7.4 Fabric Antennas

Many of the antenna types described in the previous section have been manufactured with fabrics, and these are discussed in this section. An excellent introduction to fabric antennas is given in [32]. Fabric antennas can easily be integrated into clothing, and thus give good opportunities for integrated communications to the wearer. This may be important in special occupations, such as paramedics or firefighters, where UHF and microwave frequencies are used, and are also of interest to the defense sector, where VHF and UHF are used. The possibility of the use of large ground planes beneath the radiating element can also reduce the power dissipated in the user's body, particularly at microwave frequencies.

7.4.1 Fabric Materials

The critical design issues [32] are the attachment of the various layers and the subsequent variability of the antenna characteristics due to fabric bending and stretching. Layer attachment is a manufacturing process issue and the problems that some papers describe on delamination of the layers, can in principle, be solved using appropriate adhesives, such as adhesive sheets [33]. Fabric bending and stretching are examined in many papers described below. Typical materials used include both natural and synthetic fibers. Many different synthetic fibers are available, and each has one or more company trademark names. These include [32] with the trade names in brackets, polyamide (Nylon Cordura, Gore-Tex), acrylic (Draylon), polyester (Fleece Terylene), Para Amide (Kevlar), elastane (Lycra and polypropylene (Ulstren)). [32] describes patch-type antennas made from fleece, felt, vellux and upholstery fabric.

The effect of the quality of the conducting material is discussed in [34, 35] for antennas designed for 2.45 GHz. The conductive surfaces were made out of conductive fabrics, knitted copper and aracon fabric, and different copper tape assemblies. The results with different conductive element materials show that textile antennas can be fabricated using both conductive tape and fabrics without affecting the antenna performance. Discontinuations that are perpendicular to current flow need to be avoided and fabrics need to be densely knitted and use sufficiently conductive material. In [33], an electrical resistance of 1 Ω/□ is recommended to minimize loss of efficiency. It is noted that copper foils have this property but lack of drapability and elasticity limits their use in clothing. Antennas using a no-name nickel-plated woven fabric, with plating thickness of about 250 nm, a silver-plated knitted fabric and a silver–copper–nickel-plated woven fabric are then studied. Figure 7.11 shows details of two conducting fabrics, namely conductive, knitted P130 and woven Nora fabric.

The knitted fabric consists of entirely plated polyamide fibers. The material is flexible, which makes it ideal for integration with clothing, but does stretch which will result in antenna detuning. The woven fabric, using fibers that are plated before weaving, is less prone to stretching but less flexible than the knitted material. Table 7.2 given the properties of these materials. For the knitted fabric, simulations showed that the loss in the conductive material gives rise to a reduction of efficiency from 99% for a copper foil to 45% for the fabric.

Measurement of the average electromagnetic properties of the materials is difficult due to the flexible and non-repeatable nature of the structures, such as transmission lines, normally used in such a measurement. A method [36] which involves measuring two transmission lines of differing length has been used for fabric lines, which, when combined with the matrix-pencil approach to reduce the variations due to the materials and connections, has given good results.

Connection of the rest of the electronics to the fabric antenna has also received attention [37–39], due the inherent difficulties. The integration of RFID chips into antennas on flexible substrates is now widely demonstrated and [37] proposes similar integration of a single transceiver chip into a fabric antenna. There are connection problems in this and snap-on connectors [38] have been suggested. [39] demonstrates some of the system concepts in an integration for 17 GHz wireless BANs.

One of the most important parameters for fabric antennas is efficiency. Determination of material parameters is important as noted in the preceding paragraphs, but for an antenna on the human body efficiency is extremely difficult to determine. In many cases, only S_{11} is given, which can be misleading as material loss can contribute to a good value. A few papers report antenna gain with patterns in a limited number of pattern cuts and a few report received signal levels. Both of these allow some useful comparison.

7.4.2 Antenna Types

VHF/UHF fabric-based antennas have been demonstrated for both civilian and military application. A wide range of conductive materials were compared in [40], for use in a 0.1–1 GHz body-worn spiral, including conductive ribbon, insulated wire, conductive paint, conductive nylon, phosphor bronze mesh, embroidered conducting thread, screen printing, liquid crystal polymer, and copper-coated fabric. It was concluded that based on manufacturing and RF performance, the conducting nylon and the copper coated fabric were best. A dipole, fabricated from conducting tape and a fleece material, was designed for 100 MHz for personal

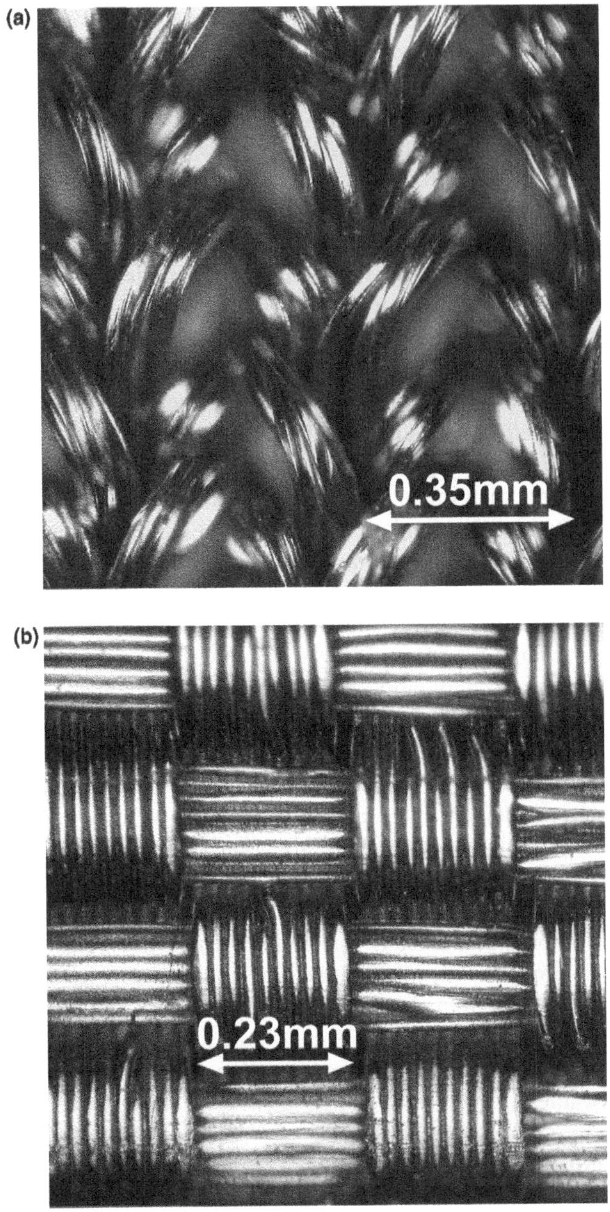

Figure 7.11 Examples of woven conductive fabrics: (a) conductive, knitted P130; (b) woven Nora fabric. Reproduced by permission of ©2004 IEEE [34]

FM radio reception [41]. The antenna was worn across the shoulders and along the arms. A −13 dBi gain was achieved with the arms held straight out, which reduced by 5 to 15 dB when the arms hung down. Conformal helmet-mounted antennas have been developed for soldier use [42]. The design was mounted on a non-conducting Kevlar helmet and used a slot geometry

Table 7.2 Properties of conductive knitted and woven fabric

	Knitted	Woven
Fibre thickness	0.55 mm ± 10%	0.15 mm
Yarn polymer	polyamide	polyamide
Yarn/thread pitch	0.35 mm	0.23 mm
Fibres per yarn	6	12
Weight	180 g/m^2	72 g/m^2
Plating	Ag	Ag
Sheet resistance	<1Ω/□	0.02 Ω/□

to get a bandwidth of 0.3–3 GHz, and used a conducting material of polyester interwoven with a nickel/copper fiber, which had a resistivity of 0.1 Ω/□. Because the material requires specialized low temperature solder, copper tape was sewn on the material in regions at which electrical connections were made. A vest antenna [43, 44] was made in the same way, using FLECTRON, which is a mixture of copper interwoven with polyester. Being larger than the helmet antenna, it achieved a lower frequency coverage, from 30–500 MHz. A body-worn fabric spiral antenna is described in [45], which has an equiangular expansion law. It used Nora, which has a nylon base and is plated with silver, copper, and nickel. Its thickness is 0.06 mm with a manufacturer's surface resistivity specification of 0.03 Ω/square. It achieved a bandwidth of 1–4 GHz for a VSWR < 4. Wearable antennas for personal digital television broadcasting reception, from 470 MHz to 770 MHz, have been demonstrated [46, 47], The slot-type antenna located in the hood of a jacket in [46] achieved 0 dBi gain and $S_{11} < -10$ dB over the band. [48] describes a fabric patch antenna, worn in the lumbar region, for military and police applications in the 380–400 MHz band.

Fabric antennas for mobile phone applications are described in [49], for wireless LAN applications in [50, 51], GPS applications in [52] and in the 2.45 GHz ISM band for rescue workers [53], for athletes [54] and for medical sensors [55]. In many of these antennas the material considerations are very similar to those at VHF/UHF and similar material choices are made. Performance is in some cases better than antennas housed in the equipment cases, as larger antennas and ground planes can be used, which has the additional benefit of significantly reducing the power dissipated in the body. The adhesive plaster antenna [55] used 8 layers of a very thin self-adhesive polyacrylate to give a substrate 1.36 mm thick. The 2.45 GHz patch antenna uses a perforated ground plane to increase the bandwidth and achieves a gain of 1.4 dBi gain with 60% efficiency when stuck to the skin.

Various fabric antenna configurations have been reported, including PIFAs [56] dual and multiple band antennas [57, 58], and printed antennas on electromagnetic bandgap materials [60, 61]. The effects of bending and crumpling of fabric antennas are discussed in some earlier noted papers and also in [62, 63].

7.5 Ultra Wideband Antennas

According to the FCC and ITU-R standard, Ultra-Wideband (UWB) antennas are referred to as antennas within the 3.1–10.6 GHz range, with a minimum bandwidth of 500 MHz, or 20% of the centre frequency and which emit a power spectral density below −41.25 dBm/MHz. UWB technology, including pulse-based systems, should be able to share spectrum with other users. Initially, UWB was intended for the efficient use of scarce radio bandwidth while enabling both high data rate personal area network (PAN) wireless connectivity and longer-range, low data rate applications as well as radar and imaging systems [64]. Recently, there have been several proposals to apply UWB technologies to Wireless Body-Area Networks (WBANs) to enable low power (below 100 μW) but high data rate applications by its control over the duty cycle. However, the main objection to the UWB is that most body-worn or implanted medical sensors only transmit/receive a limited amount of data, and, hence, will under-utilize the available bandwidth. That, therefore, is one possible reason why the operating frequency bands for UWB antenna design and on/off-body measurements are varied, ranging from 3–5 GHz [65], 3–6 GHz [66] and the full UWB band of 3.1–10.6 GHz [67], etc.

UWB antennas can be categorized as directional or non-directional and they also can be classified as electric or magnetic antennas. Most UWB antennas can be characterized using conventional antenna parameters such as return loss, bandwidth, gain, efficiency and radiation pattern, etc. In an impulse-based UWB system, which employs the unorthodox carrier-free modulation, the antenna must have its ability to preserve the pulse shape, and therefore the design needs special considerations.

In general, UWB antennas for WBAN are required to have very broad impedance bandwidth, stable and constant channel transfer response and high efficiency. Ordinary wideband antennas will cause distortion to transmitted short pulses, since they radiate various components from different parts and hence experience severe frequency-dependent changes in their phase centres, e.g., log-periodic and spiral antennas [68–70]. Any strong resonance at any frequency of the UWB antenna response causes large group delay variation and thus causes distortion in the pulse shape which in turn affects the pulse fidelity factor. Physically, UWB antennas for WBAN have to be compact, flat and integrated inside the sensor nodes and hand-held devices for health monitoring. Some antennas may be required to be made into textiles for applications for military and special forces.

Due to the presence of the human body, wearable antennas suffer from frequency-detuning, radiation pattern distortion and depolarization. A proper design of wearable UWB antennas needs to be evaluated numerically with digital human phantoms, or through on-body measurement. The human body effect on UWB antenna performance was presented by Promwong et al. in [71], where the transmitting antenna was placed off the body, while the receiving antenna on the body and two different types of antennas were used. Klemm and his colleagues [66, 72] investigated numerically time-domain characteristics of an aperture-stacked patch antenna and a textile UWB antenna in free space and on the body.

For WBAN applications, wireless sensor nodes are scattered around or in the close proximity of the body. These sensors are required to communicate with each other and with a body-worn base unit. Therefore, one of basic requirements for the wearable UWB antenna is to have omni-directional radiation parallel to the body surface, but minimal normal to the body. As is the case for narrowband wearable antennas, conventional disk-like monopoles can provide required antenna characteristics for WBAN. However, due to its bulky design, the disk-like monopole is not suited for applications with size constraints, for example, portable devices [70]. In [67], a planar inverted cone antenna (PICA) has been proposed as a reference antenna for on-body UWB channel characterization. In the following, several wearable antenna designs will be presented, including tapered slots, printed dipoles and monopoles. Antenna characteristics are shown, based on both simulation and measurement results. On-body radio channel parameters are also presented as one of important considerations for the wearable antenna system.

7.5.1 Antenna Types

The Planar Inverted Cone Antenna (PICA) is derived from the design of circular disk and volcano antennas (Figure 7.12 (a)). It is composed of a single flat element vertically mounted above a small circular ground plane with a diameter of 60 mm. The dominant antenna dimension is a quarter of the wavelength, similar to the disk cone with a conductor thickness around 0.3 mm. Both simulation and measurement results show that the antenna provides outstanding impedance and radiation pattern performance with a gain of 3–6 dBi.

Other antennas are planar types and fabricated on RT/Duroid substrate with thickness 1.524 mm and relative permittivity of 3. The planar horn-shaped self-complementary antenna (HSCA) (Figure 7.12 (b)) can be thought of as an evolution of the bow-tie antenna. Its input impedance varies slightly with frequency and therefore, a broadband impedance transformer such as a three-stage Chebyshev transformer is needed to maintain good impedance matching over the required frequency band. The proposed HSCA also exhibits an approximately constant gain of 0–2.4 dBi across the UWB band.

A printed version of planar inverted cone antenna [73–75] (Figure 7.12 (c)), consists of two elements; the top element (planar cone) and the bottom rectangular-shaped ground plane (incorporating a CPW feed). The total antenna area is $47.5 \times 50\,\text{mm}^2$ and it has good impedance matching with return loss less than 10 dB over 3–12 GHz.

A far as the design of a UWB antenna is concerned, a broadband impedance matching network is needed and it can be achieved by employing two tapered radiating slots at the end of the CPW feeding line [76] by gradually varying the feed-gap [77] and with the help of a pair of tapered radiating slots [78]. A proposed tapered slot antenna (TSA) is proposed using a similar approach to [76]; however, the difference is that in the proposed antenna, the waveguide and radiating slot are inseparable. The antenna is fabricated on RT/Duroid board (with parameters the same as those used for the PICA antenna) (Figure 7.12 (d)). The total antenna size is $27 \times 16\,\text{mm}^2$ and it has excellent impedance matching (with VSWR less than 2) across the UWB band with return loss less than 10 dB over 3–11.5 GHz.

7.5.2 Antenna On-body and Off-body Performance

All antennas seem to preserve good impedance bandwidth even when placed on the human body with slight detuning in lower frequency band for the printed ones. The conventional PICA has the least frequency shift due to the presence of the ground plane under the radiator. The phenomena of frequency-detuning of UWB wearable antennas are a little different from their narrowband counterpart as the frequency shift tends towards high frequencies (Figure 7.13). This is due to the combined effects of antenna dielectric (human body) loading and frequency-detuning of associated impedance-matching networks.

For the radiation characteristics, it has been demonstrated [67] that a conventional PICA has the most stable antenna pattern across the UWB frequency band, while the HSCA and printed PICA have deep nulls at certain directions and frequency bands. This is due to the fact that the printed antennas have a small size ground plane designated for the feed lines, which may

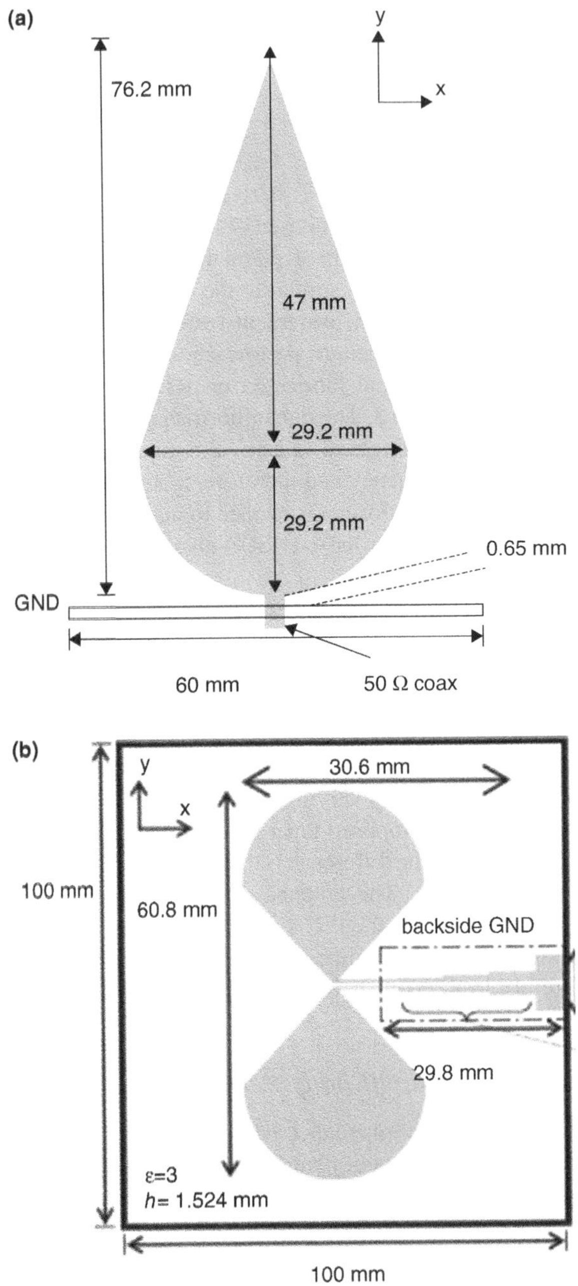

Figure 7.12 UWB antennas with dimensions in mm: (a) planar inverted cone antenna (PICA), placed perpendicular to body with ground plate on body; (b) horn-shaped self-complementary antenna (HSCA), placed parallel to the body with radiating element facing outwards from the body; (c) CPW-fed printed PICA, placed parallel to body with radiating element facing outwards from body; (d) CPW-fed tapered slot antenna (TSA), placed parallel to body with radiating element facing outwards from body

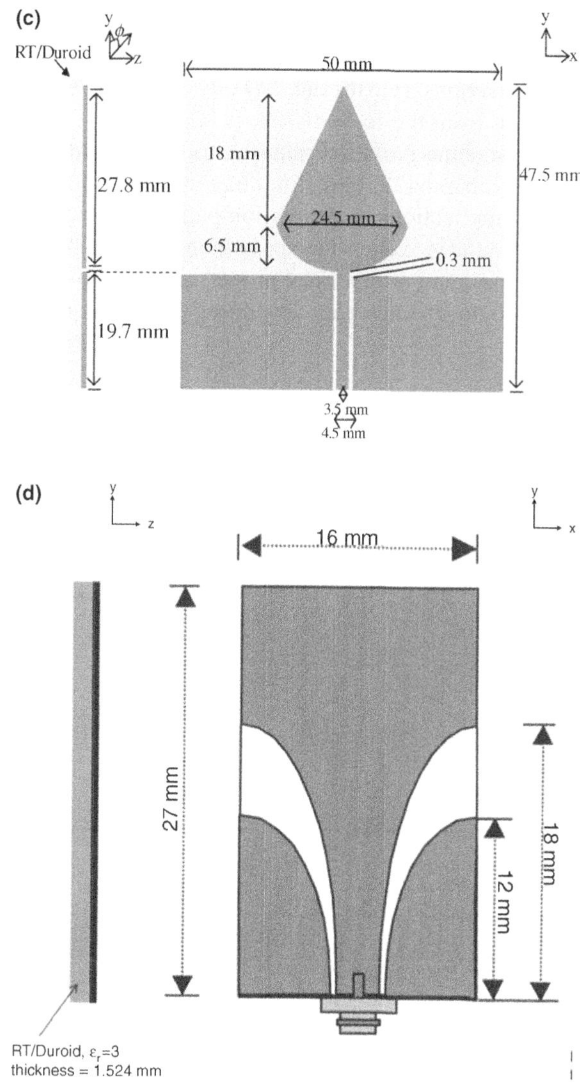

Figure 7.12 *(Continued)*

cause spurious radiation in unwanted directions. In order to understand the impact of the human body on the radiation characteristics of UWB wearable antennas, on-body antenna radiation patterns have been obtained for both printed PICA and tapered slot antenna (TSA) cases (Figure 7.14 (a)).

Radiation patterns of the two antennas are presented in three different planes at 3, 6, and 9 GHz, respectively. Both antennas exhibit omni-directional behaviour across the band except that the TSA is slightly directional at 3 GHz. The on-body radiation performance is investigated by placing the antennas on the centre of the human trunk and repeating the pattern measurements for the specified frequencies as shown in Fig. 7.14. Due to the presence of the lossy human body (with increased conductivity at high frequencies), the antenna front-to-back ratio is significantly increased to 25–30 dB. When placed on the body, the antenna directivity is increased due to wave reflections from the human body, while the total efficiency is reduced due to the wave absorption from the lossy tissue. It is interesting to note that on the x–y and y–z planes, on-body antenna radiation patterns for PICA and TSA are very similar, despite the fact that the two antennas have distinct free-space radiation characteristics. This indicates that, for WPAN applications, the radiation pattern of electrically small wearable antennas mostly depends on the size and shape of human subjects. As the human tissues are frequency-dispersive, the variation of electrical properties within the UWB frequency band contributes to the change in the antenna gain depending on the frequency of operation. The

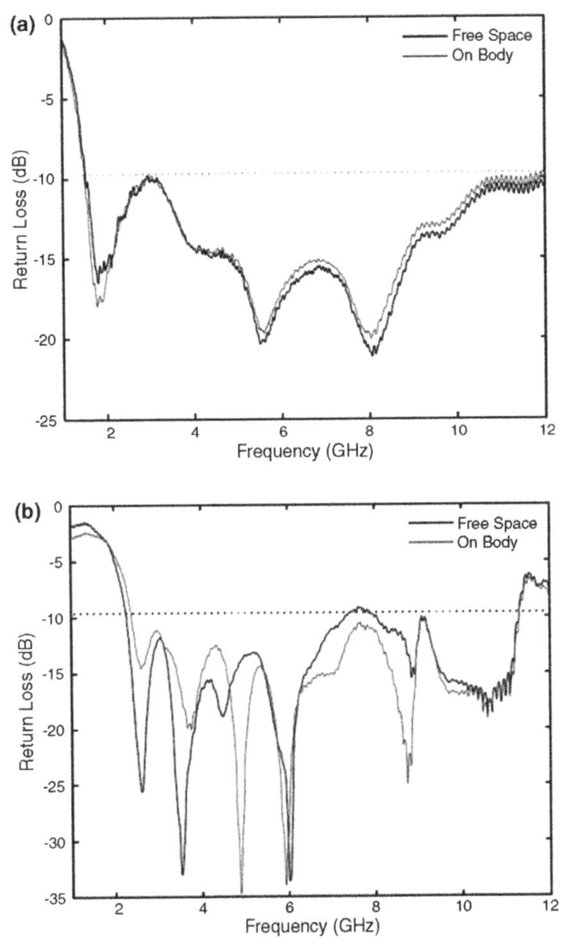

Figure 7.13 On-body and off-body return loss of the UWB wearable antennas when placed off and on the body (with antenna placed parallel to the body which means x–y plane is parallel to the antenna at 2 mm distance away from the body): (a) conventional PICA; (b) HSCA; (c) printed PICA; (d) TSA

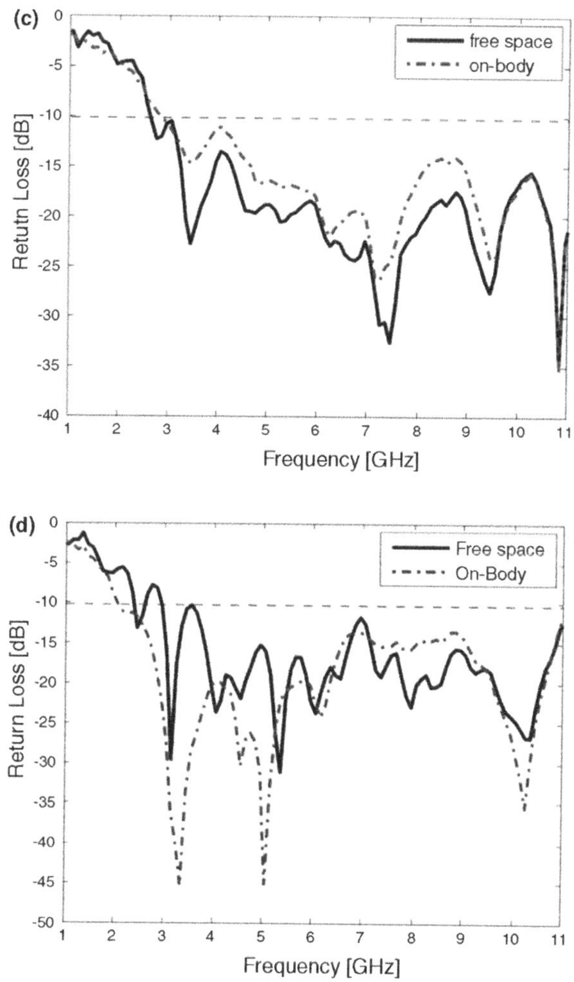

Figure 7.13 (*Continued*)

gain of the PICA ranges from 2.5–5 dBi and from 2–4 dBi for the tapered slot antenna (TSA) in free space. Overall, both PICA and TSA antennas experience, respectively, 2–3 dB and 1–4 dB increases in antenna gain for on-body scenarios (Figure 7.15). The antenna gain is obtained via numerical simulation to avoid the effects from the coax cable in practical measurement.

For WBAN applications, antenna radiation parallel to the human body surface is important as it is directly linked to the path gain between two on-body sensors. Figure 7.14 also presents the far field radiation pattern at the x–z plane, which is parallel to the body surface, when a human subject stands still above the ground. It can be seen that although the TSA has directional patterns on both the x–y and y–z planes, the antenna is reasonably omni-directional along the body surface, apart from some nulls at 3 and 9 GHz. On the other hand, the PICA has the best omni-directional radiation in free space, in the azimuth (x–y) plane, the antenna becomes more directional along the body surface when it is placed on the body, in particular it exhibits deep nulls at high frequencies, at 6 and 9 GHz, respectively. This is perhaps why the TSA demonstrates good on-body channel performance regardless of its small aperture size, compared to the PICA [79].

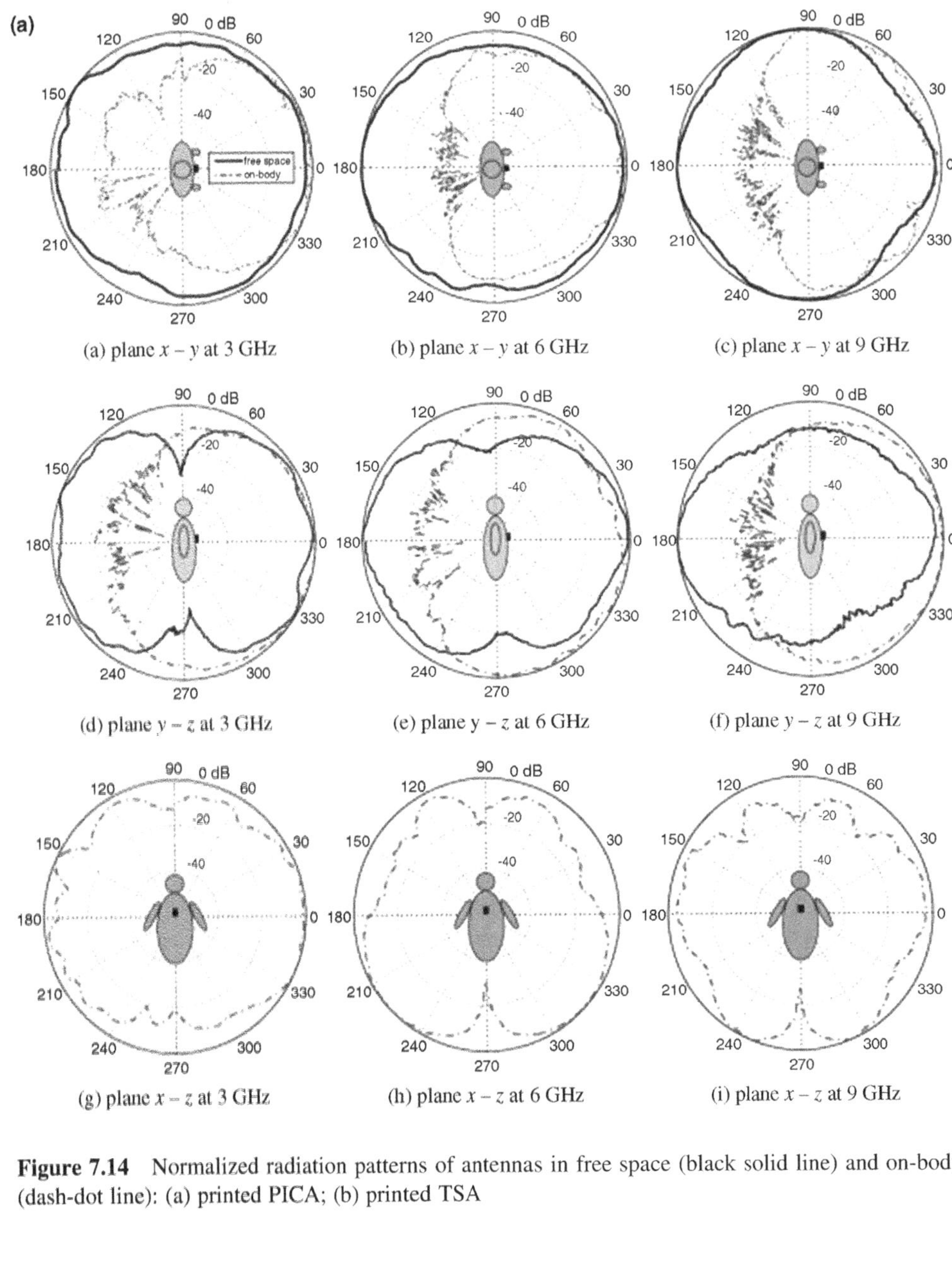

Figure 7.14 Normalized radiation patterns of antennas in free space (black solid line) and on-body (dash-dot line): (a) printed PICA; (b) printed TSA

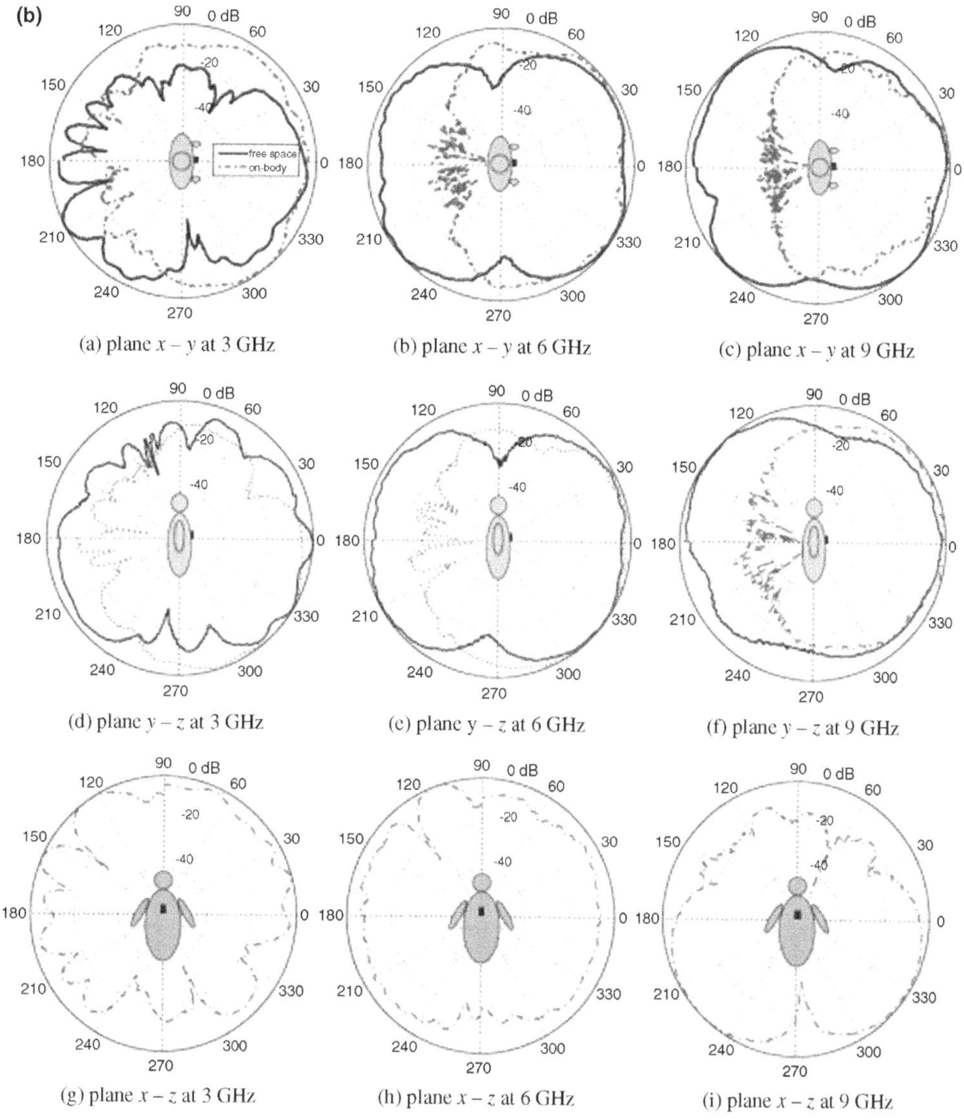

Figure 7.14 *(Continued)*

7.5.3 Antenna Fidelity and Transient Analysis

In UWB communications, specifically impulse radio systems, a correlation between the transmitted or received waveform and a template is often used to assess how the system or radio channel affects a waveform and to quantify the level of pulse distortion. A fidelity parameter involving the autocorrelation of the difference of the time domain transmitted field and a template function is described by Lamensdorf et al. [80]. In many applications, fidelity analysis

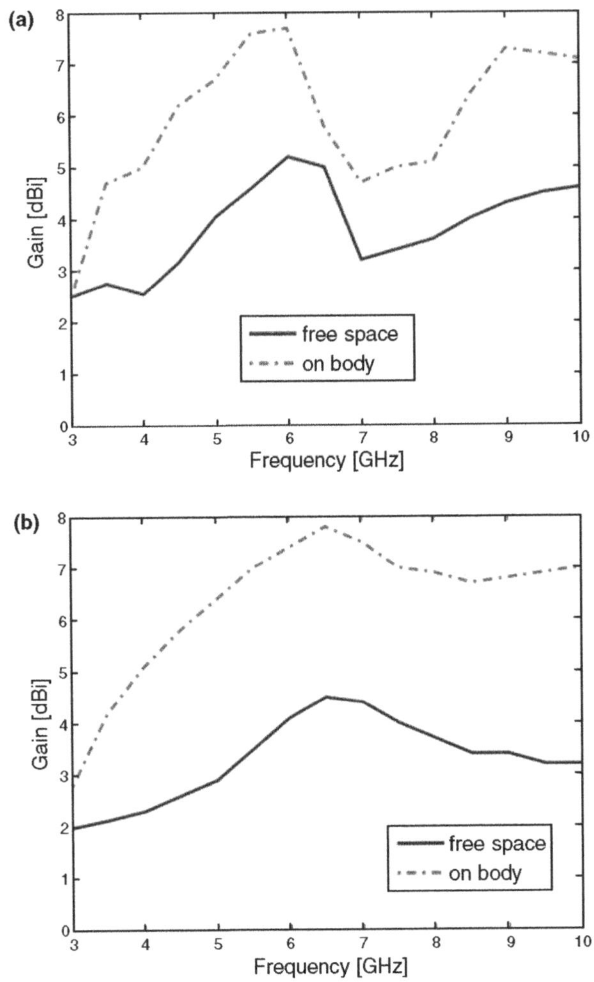

Figure 7.15 Free space and on-body gain as a function of the frequency: (a) printed PICA; (b) printed TSA

is applied to investigate the preservation of pulse shape (not amplitude) with respect to space and time [66, 80–82].

The choice of template function is not restricted to a specific reference pulse or stringent requirements except for a requirement on the separability in distance and space between the compared pulses [81].

Both PICA and TSA have been evaluated in the 3–9 GHz band for its fidelity, particularly, when the antennas are set side by side (Figure 7.16 (a)). For the PICA, the fidelity values for the measured band are 91.25% and 82.98% for 45° and 90°, respectively, with the response at 0° set as a reference; while for the TSA, the fidelity values for the measured band 3–9 GHz are 96.04% and 66.47% for 45° and 90°, respectively, which are slightly degraded compared to those for the PICA, bearing in mind that the tapered slot antenna has significant size reduction. For both

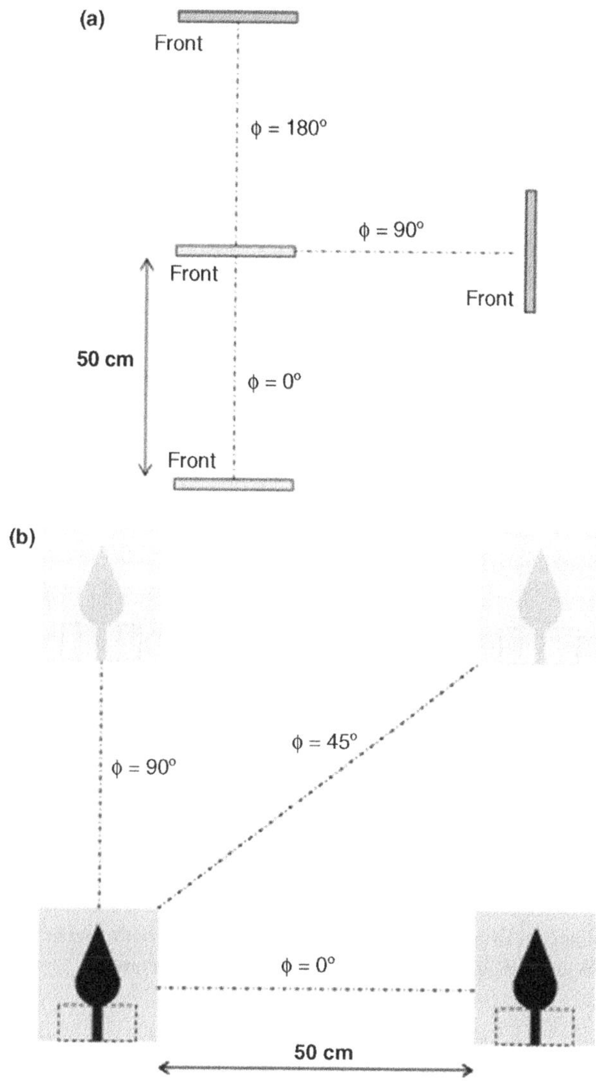

Figure 7.16 Antenna transfer functions measurement set-up with two identical antennas (PICA and tapered slot antenna) in free space with distance of 50 cm between the antennas with different orientations: (a) face to face; (b) side by side.

cases, the fidelity values are lower at 90° which is evident from the radiation patterns, where deep nulls can be seen in the corresponding direction. In addition, the PICA and TSA are tested for WPAN scenarios, where two antennas are placed face to face (Figure 7.16 (b)). For the PICA, the fidelity of the system impulse response at different directions with reference to the response at 0° (when the antennas are facing each other) is 91.7% and 95.07% for 90°, and 180°, respectively; while for TSA, the fidelity is 91.06% and 95.35% for 90° and 180°, which are comparable to the PICA case. Figure 7.17 presents a comparison between the fidelity of PICA

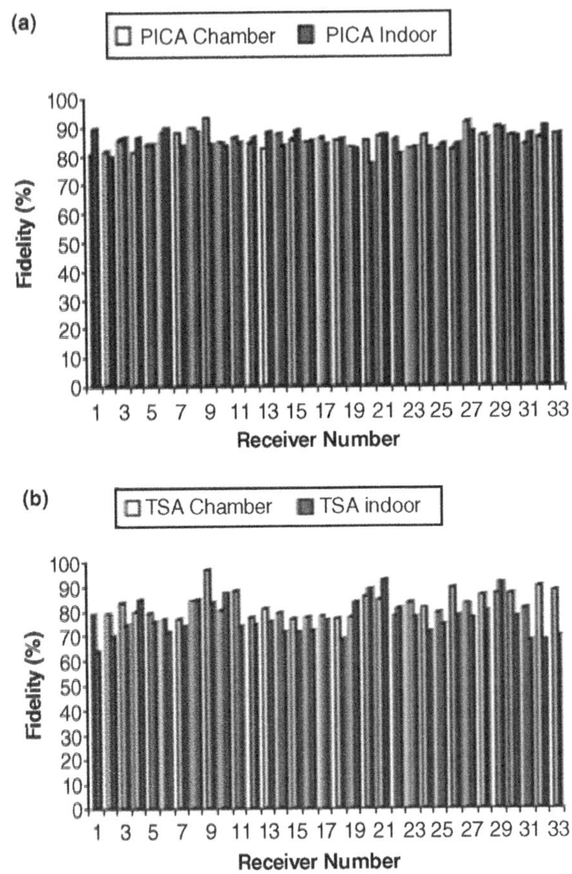

Figure 7.17 A comparison of pulse fidelity for two antennas for both anechoic chamber and indoor environment: (a) CPW-fed PICA and (b) CPW-fed tapered slot antenna

and TSA for WBAN applications in both indoor and anechoic chamber environments. The PICA seems to better preserve the pulse shape, with the mean fidelity of 85.36% in the anechoic chamber, and 84.88% in an indoor environment, while for the TSA the mean values are respectively 81.30 % and 76.41%. Those values are consistent with that calculated in free space.

Fidelity studies of commonly used antennas such as monopoles and resistively loaded dipoles presented in [66] and [80–82] have shown that a spatially averaged fidelity factor as low as 0.7 is often attained by such antennas that are widely used. In a wireless BAN study presented in [66], a value of 76–99% is derived for fidelity when numerically comparing input pulse to transmitted pulses for various antenna types. This indicates that the acceptable minimum value for fidelity is application- and environment-specific. In the study presented here, the average values obtained for the PICA and TSA (around 80%) are considered sufficient for WBAN applications.

7.6 Multiple Antenna Systems

Multiple antenna systems can be used to reduce the effect of fading through the application of diversity processing, to increase the channel capacity in a multipath environment by multiple input multiple output (MIMO) processing, or to reduce the effect of interference. All these methods have been examined for some wearable antenna scenarios, both for channels between the body to external base stations and between antennas on the body.

7.6.1 Diversity

Antenna diversity has been studied both at UHF and microwaves. At 500–600 MHz, meandered dipoles have been placed at various places on the body such as the shoulders, arms, chest and back, communicating to a distant base station in an outdoor environment [83]. Estimates were made of channel capacity using various levels of multiple antenna diversity. Capacity improvements of up to 3 bps/Hz were obtained. A study of an indoor scenario with a local base station in the same room, at 2.45 GHz [84], again with multiple antennas located in various and widely spaced positions on the body, showed that diversity gains of up to 6 dB could be obtained with selection combining. In both these cases, the channels are modified by the body movement and the multipath created by the environment. In the body-to-body case, the movement of both bodies and the environment are important. In [85], a 5.5 GHz channel sounder with a -3 dB bandwidth of 100 MHz was used, and four stacked patch antennas were mounted at the corners of a 100×50 mm printed circuit board. Distributed antenna selection diversity was shown to mitigate the body shadowing. A 5 dB diversity gain on average received power at both the mean and 10% outage levels were obtained for two distributed antennas mounted on both the transmitter and receiver person and without channel state information feedback to the transmitter.

Measurements of diversity in on-body channels have been made using closely spaced multiple antenna systems, using quarter wavelength monopoles, printed inverted F and planar inverted F antennas, at 2.45 GHz [86] and 5.8 and 10 GHz [87]. Antenna spacing was investigated and useful diversity gain was obtained for spacing as small as 10 mm for the monopoles. Gains of between 6 and 11 dB were obtained for the IFA and PIFAs depending on relative antenna orientation. A pattern switching antenna [88] consisting of five disc top-loaded monopoles, with the center monopole fed and the other four acting as parasitics, with diode switches at their base, obtained diversity gains generally lower than the space diversity arrangements in [86, 87]. However, this antenna was also used to measure signal angle of arrival in one plane, [88] in a number of on-body channels, in an anechoic environment. Small numbers of ray bundles were noticed in most cases.

7.6.2 MIMO

Electrotextile patch antennas at 2.4 GHz have been used to investigate the capacity of MIMO in body to body channels [89]. The 319% improvement in 10% outage capacity for the body-worn system over the reference system made of a full-size dipole antenna is consistent with the 288% improvement projected by theoretical modeling and the average 300% improvement found in the physical simulation of two typical indoor scenarios. Two planar inverted F antennas [90] at 2.45 GHz on a 45 × 40 mm ground plane at each end of an on-body channel in an office environment, have been measured and capacity gains close to 2 have been observed for some channels.

7.7 Conclusion

There is currently much interest in wearable antennas, due to the increased understanding of potential applications and the demands of new systems. Wearable antennas can fulfill many functions both in terms of communications from on to off the body, on-body channels and communications to implants. The common factor in design is the presence of the human body and the variability of tissue parameters around the body and of the orientation and distance of the antenna from the body. In general, the body parameters are well known, but the design of antennas that are not affected by the variations is difficult and needs further study. Similarly, while the general propagation mechanisms are known, more work is needed to understand them and to produce good analytical and statistical models.

Narrowband antenna design generally follows conventional lines, where the limited ground plane means that they are vulnerable to detuning and efficiency loss. Wider bandwidth types can overcome the first problem. The integration of antennas into clothing is now being studied by many groups, and the possibility of large ground planes significantly reduces the variability and in addition the tissue dissipation problem. The primary issues with fabric antennas are material, assembly and connection and much progress is being made. Flexing and creasing remain a difficult area.

Antennas for ultra wideband systems are easy to produce in planar form and show great potential for short range, high capacity links. However, it is much harder to achieve the required bandwidths with a radiated polarization normal to the body surface, which gives lowest path loss. The performance increase with the use of multiple antenna systems has been demonstrated, both in terms of diversity gain and MIMO capacity, although realization in practical systems remains a cost-sensitive matter.

References

1. C. Baber, J. Knight, D. Haniff and L. Cooper, "Ergonomics of wearable computers," *Mobile Networks and Applications*, vol. 4, pp. 15–21, 1999.
2. C. Baber, T. N. Arvanitis, D. J. Haniff and R. Buckley, in M.A. Sasse and C. Johnson (eds.), *A Wearable Computer for Paramedics: Studies in Model-based, User-centred and Industrial Design*, pp. 126–132. Interact'99, Amsterdam: IOS, 1998.
3. P. P. Carvey, "Technology for wireless interconnection of wearable personal electronic accessories," *IEEE Workshop on VLSI Signal Processing*, IX, pp. 13–22, Oct. 30, 1996.
4. J. Bernhard, P. Nagel, J. Hupp, W. Strauss and T. von der Grun, "BAN – body area network for wearable computing," paper presented at 9th Wireless World Research Forum Meeting, Zurich, July 2003.
5. P. S. Hall, and Y. Hao (eds), *Antennas and Propagation for Body Centric Communications Systems*, Artech House, Norwood, MA, 2006.
6. P. S. Hall, Y. Hao, Y. I. Nechayev, A. Alomainy, C. C. Constantinou, C. G. Parini, M. R. Kamarudin, T. Z. Salim, D. T. M. Hee, R. Dubrovka, A. Owadally and W. Song, "Antennas and propagation for on-body communication systems," *IEEE Ant. and Prop. Magazine*, vol. 49, no. 3, pp. 41–58, June 2007.
7. C. H. Durney, H. Massoudi and M. F. Iskander, *Radiofrequency Radiation Dosimetry Handbook*, Brooks Air Force Base- USAFSAM-TR-85-73, 1986.

8. L. A. Geddes and L. E. Barker, "The specific resistance of biological material: a compendium of data for the biomedical engineer and physiologist," *Medical and Biological Engineering*, vol. 5, pp. 271–293, 1967.
9. M. A. Stuchly and S. S. Stuchly, "Dielectric properties of biological substances: tabulated," *Journal of Microwave Power*, vol. 15, no. 1, pp. 19–26, 1980.
10. K. R. Foster and H. P. Schwan, "Dielectric properties of tissues and biological materials: a critical review," *Critical Reviews in Biomedical Engineering*, vol. 17, no. 1, pp. 25–104, 1982.
11. F. A. Duck, *Physical Properties of Tissue: A Comprehensive Reference Book*, Academic Press, Harcourt Brace Jovanovich, Publishers, New York, 1990.
12. C. Gabriel, *Compilation of the Dielectric Properties of Body Tissues at RF and Microwave Frequencies*, Brooks Air Force, Technical Report, AL/OE-TR-1996-0037, 1996.
13. Institute for Applied Physics, Italian National Research, Council, *An Internet Resource for the Calculation of the Dielectric Properties of Body Tissues*, http://niremf.ifac.cnr.it/tissprop/.
14. R. G. Vaughan, A. Lea, P. Hui and J. Ollikainen, "Theory of propagation for direct on-body wireless sensor communications," in *IET Seminar on Antennas and Propagation for Body Centric Wireless Communications*, London, April 20, 2009.
15. A. Alomainy, Y. Hao and D. M. Davenport, "Parametric study of wearable antennas with varying distances from the body and different on-body positions," in *IET Seminar on Antennas and Propagation for Body-Centric Wireless Communications*, pp. 84–89, April 24–25, 2007.
16. T. Z. Salim and P. S. Hall, "Efficiency measurement of antennas for on-body communications," *Microwave and Optical Technology Letters*, vol. 48, no. 11, pp. 2256–2259, Nov. 2006.
17. T. Alves, R. Augustine, M. Grzeskowiak, B. Poussot, D. Delcroix, S. Protat, J-M. Laheurte and P. Queffelec, "BAN antenna design using ferrite polymer composite," 3rd European Conference on Antennas and Propagation, Berlin, Germany, March 23–27, 2009.
18. P. Salonen, M. Keskilammi and L. Sydanheimo, "A low-cost 2.45 GHz photonic band-gap patch antenna for wearable systems," in *Eleventh IEE International Conference on Antennas and Propagation* (IEE Conf. Publ. No. 480), 2001.
19. S. Zhu and R. Langley, "Dual-band wearable antennas over EBG substrate," *Electronics Letters*, vol. 43, no. 3, pp. 141–142, 2007.
20. G.A. Conway and W.G. Scanlon, "Antennas for over-body-surface communication at 2.45 GHz," *IEEE Transactions on Antennas and Propagation*, vol. 57, no. 4, Part 1, pp. 844–855, April. 2009.
21. R. Anupam, R. Chandran, G. A. Conway and W. G. Scanlon, "Pattern switching compact patch antenna for on-body and off-body communications at 2.45 GHz," 3rd European Conference on Antennas and Propagation, Berlin, Germany, March 23–27, 2009.
22. K. Ogawa, T. Uwano and M. Takahashi, "A shoulder-mounted planar antenna for mobile radio applications," *IEEE Transactions on Vehicular Technology*, vol. 49, no. 3, May 2000.
23. N. Noury, P. Barralon and D. Flammarion, "Preliminary results on the study of smart wearable antennas," in *27th Annual International Conference of the Engineering in Medicine and Biology Society, 2005*, pp. 3814–3817, Jan. 17–18, 2006.
24. N. Haga, M. Takahashi, K. Saito and K. Ito, "A cavity-backed slot antenna for on-body ban devices," *IEEE International Workshop on Antenna Technology*, IWAT2008, Chiba, Japan, 2008.
25. J. C. G. Matthew, B. Pirollo, A. Tyler and G. Pettitt, "Body wearable antennas for UHF/VHF," in *Antennas and Propagation Conference, 2008, LAPC 2008*, pp. 357–360, Loughborough, March 17–18, 2008.
26. D. Psychoudakis, C.-C. Chen and J. L. Volakis, "Wearable UHF antenna for squad area networks (SAN)," *IEEE Antennas and Propagation Int. Symposium*, 2008. AP-S 2008, pp. 1–4, July 5–11, 2008.
27. Y. Shimizu and N. Kuga, "A finger-ring antenna for BAN and PAN applications," *IEEE International Symposium on Antennas and Propagation*, ISAP2008, Taipei, Taiwan 2008.
28. B. Sanz-Izquierdo, F. Huang and J. C. Batchelor, "Dual band button antennas for wearable applications," *IEEE International Workshop on Antenna Technology, IWAT06*, pp. 132–135, March 6–8, 2006.
29. B. Sanz-Izquierdo, J. C. Batchelor and M. Sobhy, "UWB wearable button antenna," *European Conference on Antennas and Propagation, 2006*, EuCAP 2006, pp. 1–4, Nov. 6–10, 2006.
30. J. M. Floch, F. Queudet and E. Fourn, "Radio-electric characterizations of jeans buttons," *European Conference on Antennas and Propagation EuCAP 2007*, pp. 1–4, Nov. 11–16, 2007.
31. B. Sanz-Izquierdo and J. C. Batchelor, "A dual band belt antenna," *IEEE International Workshop on Antenna Technology*, IWAT2008, Chiba, Japan, 2008.

32. P. Salonen and Y. Rahmat-Samii, "Wearable antennas: advances in design, characterisation and application," in P. S. Hall and Y. Hao (eds.), *Antennas and Propagation for Body Centric Communications Systems*, Artech House, Norwood, MA, 2006.
33. I. Locher, M. Klemm, T. Kirstein and G. Troster, "Design and characterization of purely textile patch antennas," *IEEE Trans Advanced Packaging*, vol. 29, pp. 777–788, Nov. 2006.
34. P. Salonen, Y. Rahmat-Samii, H. Hurme and M. Kivikoski, "Effect of conductive material on wearable antenna performance: a case study of WLAN antennas," *IEEE Antennas and Propagation International Symposium*, pp. 455–458, June 20–25, 2004.
35. Y. Ouyang, E. Karayianni and W. J. Chappell, "Effect of fabric patterns on electrotextile patch antennas," *IEEE Antennas and Propagation Society International Symposium*, vol. 2B, pp. 246–249, vol. 2B, July 3–8, 2005.
36. F. Declercq, H. Rogier and C. Hertleer, "Permittivity and loss tangent characterization for garment antennas based on a new matrix-pencil two-line method," *IEEE Transactions on Antennas and Propagation*, vol. 56, no. 8, Part 2, pp. 2548–2554, Aug. 2008.
37. Y. Ouyang and W. J. Chappell, "Electro-textiles for body worn integrated silicon solutions," *Topical Meeting on Silicon Monolithic Integrated Circuits in RF Systems*, pp. 107–110, Jan. 2007.
38. I. Belov, M. Chedid and P. Leisner, "Investigation of snap-on feeding arrangements for a wearable UHF textile patch antenna," *Proc. Ambience 08 Smart Textiles – Technology and Design*, Boras, Sweden, June 2–3, 2008.
39. A. Rydberg, P. van Engen, S. Cheng, R. van Doremalen, M. Sanduleanu, K. Hjort, W. De Raedt, T. Fritzsch and P. Hallbjörner, "Body surface backed flexible antennas and 3d SI-level integrated wireless sensor nodes for 17 GHz wireless body area networks," *IET Seminar on Antennas and Propagation for Body Centric Wireless Communications*, London, April 20, 2009.
40. J. C. G. Matthews and G. Pettitt, "Development of flexible, wearable antennas," *3rd European Conference on Antennas and Propagation*, Berlin, March 23–27, 2009.
41. T. Kellomaki, J. Heikkinen and M. Kivikoski, "Wearable antennas for FM reception," *1st European Conference on Antennas and Propagation, EuCAP*, pp. 1–6, Nov. 6–10, 2006.
42. J. Lebaric and A.-T. Tan, "Ultra-wideband conformal helmet antenna," *IEEE Asia Pacific Microwave Conference*, pp. 1477–1481, 2000.
43. R. Abramo, R. Adams, F. V. Canez, H. Price and P. Haglind, "Fabrication and testing of the COMVIN vest antenna," *IEEE MILCOM 2000*, vol. 2, pp. 595–598, 2000.
44. J. E. Lebaric, R. W. Adler and M. E. Limbert, "Ultra-wideband, zero visual signature RF vest antenna for man-portable radios," *IEEE MILCOM 2001*, vol. 2, pp. 1291–1294, 2001.
45. J. A. Dobbins, A. W. Chu, P. W. Fink, T. F. Kennedy, G. Y. Lin, M. A. Khayat and R. C. Scully, "Fabric equiangular spiral antenna," *IEEE Antennas and Propagation Society International Symposium*, pp. 2113–2116, July 9–14, 2006.
46. E. Isogai, Y. Okano, A. Kuramoto, T. Harada, N. Tamaki, S. Yamamoto and T. Taura, "Research of wideband wearable antenna integrated on the clothing," *IEEE International Workshop on Antenna Technology*, IWAT2008, Chiba, Japan, 2008.
47. K. Furuya, Y. Taira, H. Iwasaki, S. Yamamoto, N. Tamaki, T. Harada and A. Kuramoto, "Wide band wearable antenna for DTV reception," *IEEE Antennas and Propagation Society International Symposium*, pp. 1–4, July 5-11, 2008.
48. C. Cibin, P. Leuchtmann, M. Gimersky, R. Vahldieck and S. Moscibroda,"A flexible wearable antenna," *IEEE Antennas and Propagation Society International Symp.*, pp. 3589–3592, 20–25 June 2004.
49. P. J. Massey, "Mobile phone antennas integrated within clothing," *IEE 11th International Conference on Antennas and Propagation (ICAP 2001)*, vol. 1, no. 480, pp. 344–347, Manchester, 2001.
50. B. Sanz-Izquierdo and J. C. Batchelor, "WLAN jacket mounted antenna," *International Workshop on Antenna Technology: Small and Smart Antennas Metamaterials and Applications, 2007*, IWAT '07. pp. 57–60, March 21–23, 2007.
51. P. Salonen and H. Hurme, "A novel fabric WLAN antenna for wearable applications," *2003 IEEE Antennas and Propagation Society International Symposium*, vol. 2, pp. 700–703, Columbus, Ohio, USA, 2003.
52. P. Salonen and H. Hurme, "Modeling of a fabric GPS antenna for smart clothing," in M. H. Hamza (ed.) *Proceedings of the IASTED International Conference; Modeling and Simulation*, pp. 18–23, Palm Springs, CA, February 24–26, 2003.
53. L. Vallozzi, H. Rogier and C. Hertleer, "A textile patch antenna with dual polarization for rescue workers' garments," *3rd European Conference on Antennas and Propagation*, Berlin, Germany, March 23–27, 2009.

54. A. Galehdar and D.V. Thiel, "Flexible, light-weight antenna at 2.4GHz for athlete clothing," *IEEE Antennas and Propagation International Symposium*, pp. 416, June 9–15, 2007.
55. T. Kellomaki, W. G. Whittow, J. Heikkinen and L. Kettunen, "2.4 GHz plaster antennas for health monitoring," *3rd European Conference on Antennas and Propagation*, Berlin, March 23–27, 2009.
56. P. Salonen, L. Sydänheimo, M. Keskilammi and M. Kivikoski, "A small planar inverted-F antenna for wearable applications," *Third International Symposium on Wearable Computers*, pp. 95–100, San Francisco, Oct. 1999.
57. P. Salonen and J. Rantanen, "A dual-band and wide-band antenna on flexible substrate for smart clothing," *27th Annual Conf. of IEEE Industrial Electronics Soc (IECON 2001)*, Denver, CO, 2001.
58. T. Thalmann, Z. Popovic, B. M. Notaros and J. R. Mosig, "Investigation and design of a multi-band wearable antenna," *3rd European Conf. on Antennas and Propagation*, Berlin, Germany, March 23–27, 2009.
59. C. Hertleer, A. Tronquo, H. Rogier, L. Vallozzi and L. Van Langenhove, "Aperture-coupled patch antenna for integration into wearable textile systems," *IEEE Antennas and Wireless Propagation Letters*, vol. 6, pp. 392–395, 2007.
60. P. Salonen, M. Keskilammi and L. Sydänheimo, "A low-cost 2.45 GHz photonic band-gap patch antenna for wearable systems," *IEE 11th International Conference on Antennas and Propagation (ICAP 2001)*, vol. 2, no. 480, pp. 719–723, Manchester, 2001.
61. S. Bashir, M. Hosseini, R. M. Edwards, M. I. Khattak and L. Ma, "Bicep mounted low profile wearable antenna based on a non-uniform EBG ground plane – flexible EBG inverted-l (FEBGIL) antenna," *Loughborough Antennas and Propagation Conference. LAPC 2008*, pp. 333–336, March 2008.
62. Q. Bai and R. Langley, "Wearable EBG antenna bending," *3rd European Conference on Antennas and Propagation*, Berlin, Germany, March 23–27, 2009.
63. Q. Bai and R. Langley, "Effect of bending and crumpling on textile antennas," *IET Seminar on Antennas and Propagation for Body Centric Wireless Comms*, London, April 20, 2009.
64. M. Z. Win and R. A. Scholtz, "Impulse radio: how it works," *IEEE Communications Letters*, vol. 2, no. 2, pp. 36–38, Feb. 1998.
65. A. Fort, J. Ryckaert, C. Desset, P. De Doncker, P. Wambacq and L. Van Biesen, "Ultra wideband channel model for communication around the human body," *IEEE J. Sel. Areas Commun.*, vol. 24, no. 4, pp. 927–933, Apr. 2006.
66. M. Klemm, I. Z. Kovacs, F. G. Pedersen and G. Troester, "Novel small size directional antenna for UWB WBAN/WPAN applications," *IEEE Trans. Antennas Propag*, vol. 53, no. 12, pp. 3884–3896, Dec. 2005.
67. A. Alomainy, Y. Hao, C. G. Parini and P. S. Hall, "Comparison between two different antennas for UWB on-body propagation measurements," *IEEE Antennas Wireless Propag. Lett*, vol. 4, no. 1, pp. 31–34, 2005.
68. W. Soergel, C. Waldschmidt and W. Wiesbeck, "Antenna characterisation for ultra wideband communications," in *Proc. 2003 Int. Workshop UltraWideband Syst. (IWUWBS)*, Oulu, Finland, CD-ROM. Jun. 2003.
69. T. Yang, S.-Y. Suh, R. Nealy, W. A. Davis and W. L. Stutzman, "Compact antennas for UWB applications," *IEEE Conf. Ultra Wideband Syst. Technol*, pp. 205–208, Nov. 2003.
70. Z. N. Chen, "Novel bi-arm rolled monopole for UWB applications," *IEEE Trans. Antennas Propag*, vol. 53, no. 2, pp. 672–677, Feb. 2005.
71. S. Promwong, W. Hachitani, G. S. Ching and J.-I. Takada, "Characterization of ultra-wideband antenna with human body," in *Proc. Int. Symp. Commun. Inf. Technol.* ISCIT'04, pp. 1213–1217, Sapporo, Japan, Oct. 2004.
72. M. Klemm and G. Troester, "Textile UWB antennas for wireless body area networks," *IEEE Trans. Antennas Propag*, vol. 54, no. 11, pp. 3192–3197, Nov. 2006.
73. S.-Y. Suh, W. Stutzman, W. Davis, A. Waltho and J. Schiffer, "A novel CPW-fed disc antenna," in *Proc. IEEE Antennas Propag. Soc. Symp*, vol. 3, pp. 2919–2922, Jun. 2004.
74. S.-Y. Suh, W. L. Stutzman and W. A. Davis, "A new ultra-wideband printed monopole antenna: the planar inverted cone antenna (PICA)," *IEEE Trans Ant Prop*, vol. 52, no. 5, pp. 1361–1364, May 2004.
75. A. Alomainy and Y. Hao, "Radio channel models for UWB body centric networks with compact planar antenna," *Proc. IEEE AP-S Int. Symp. Ant Prop*, pp. 2173–2176, Albuquerque, NM, July 9–14, 2006.
76. T.-G. Ma and C.-H. Tseng, "An ultra-wideband coplanar waveguide-fed tapered ring slot antenna," *IEEE Trans. Antenna Propag*, vol. 54, no. 4, pp. 1105–1110, Apr. 2006.
77. J. Liang, L. Guo, C. C. Chiau, X. Chen and C. G. Parini, "Study of CPW-fed circular disc monopole antenna for ultra wideband application," *IEE Proc. Microw. Antennas Propag*, vol. 152, no. 6, pp. 520–526, Nov. 2005.
78. H. G. Schantz and M. Barnes, "The COTAB UWB magnetic slot antenna," *Proc. IEEE AP-S Int. Symp*, vol. 4, pp. 104–107, Boston, MA, Jul. 2001.

79. A. Alomainy, A. Sani, A. Rahman, J. G. Santas and Y. Hao, "Transient characteristics of wearable antennas and radio propagation channels for ultrawideband body-centric wireless communications," *IEEE Trans. Antennas Propag*, vol. 57, no. 4, pp. 875–884, April 2009.
80. D. Lamensdorf and L. Susman, "Baseband-pulse antenna techniques," *IEEE Antennas Propag. Mag*, vol. 36, no. 1, pp. 20–30, Feb. 1994.
81. J. S. McLean, H. Foltz and R. Sutton, "Pattern descriptors for UWB antennas," *IEEE Trans. Antennas Propag*, vol. 53, no. 1, pp. 553–559, Jan. 2005.
82. T. Dissanayake and K. P. Esselle, "Correlation-based pattern stability analysis and a figure of merit for UWB antennas," *IEEE Trans. Antennas Propag*, vol. 54, no. 11, pp. 3184–3191, Nov. 2006.
83. D. Psychoudakis, G. YLee, C.-C. Chen and J. L. Volakis, "Estimating diversity for body-worn antennas," *3rd European Conference on Antennas and Propagation*, Berlin, Germany, March 23–27, 2009.
84. S.L. Cotton and W. G. Scanlon, "Indoor channel characterisation for a wearable antenna array at 868 MHz," *IEEE Wireless Comms and Networking Conference*, WCNC 2006, pp. 1783–1785, April 2006.
85. Y. Wang, I. B. Bonev, J. Ø. Nielsen, I. Z. Kovacs and G. F. Pedersen, "Characterization of the indoor multi-antenna body-to-body radio channel," *IEEE Trans on Antennas and Propagation, Special Issue on Body-Centric Wireless Communications*, vol. 57, no. 4, pp. 972–979, April 2009.
86. I. Khan, P. S. Hall, A. A. Serra, A. R. Guraliuc and P. Nepa, "Diversity performance analysis for on-body communication channels at 2.45 GHz," *IEEE Trans Antennas and Propagation*, vol. 57, no. 4, pp. 956–963, April 2009.
87. I. Khan and P. S. Hall, "Multiple antenna reception at 5.8 and 10 GHz for body-centric wireless communication channels," *IEEE Trans on Ant Prop*, vol. 57, no. 1, pp. 248–255, Jan. 2009.
88. M. R. Kamarudin, Y. I. Nechayev and P. S. Hall, "On-body diversity and angle-of-arrival measurement using a pattern switching antenna," *IEEE Trans on Antennas and Propagation, Special Issue on Body-Centric Wireless Communications*, vol. 57, no. 4, pp. 964–971, Apr. 2009.
89. Y. Ouyang, D. J. Love and W. J. Chappell, "Body-worn distributed MIMO system," *IEEE Trans on Vehicular Technology*, vol. 58, no. 4, pp. 1752–1765, May 2009.
90. I. Khan and P. S. Hall, "Experimental evaluation of MIMO capacity and correlation for narrowband body-centric wireless channels," *IEEE Trans on Antennas and Propagation*, vol. 58, no. 1, pp. 195–202, Jan. 2010.

8

Printed Antennas for Wireless Communications

Satish K. Sharma[1] and Lotfollah Shafai[2]
[1]*San Diego State University, USA*
[2]*University of Manitoba, Canada*

8.1 Introduction

The aim of this chapter is to introduce and discuss some significant developments in improving the performance of the printed antennas such as the microstrip patch and slot type of antennas, that have been reported since 2000, in general for various wideband wireless communication applications. Initially, we will discuss the microstrip patch antennas, followed by the microstrip slot antennas and finally the microstrip planar monopole antennas. In all cases, emphasis will be placed on the wide-bandwidth or multiband performance, extending to the Ultra-wide bandwidth (UWB) performance. There are numerous text books, handbooks, and research papers which cover these antenna topics in details [1–6], however, here the effort is to include some of the recent developments on these antennas. Chapter 2 of this book discussed a microstrip patch antennas in detail, hence it is not repeated here again.

8.2 Broadband Microstrip Patch Antennas

A broadband patch antenna is defined by its impedance bandwidth, i.e. a frequency band in which the reflection coefficient magnitude is better than $S_{11} = -10\,\text{dB}$ or $\Gamma = 0.333$. A conventional patch antenna, as mentioned earlier, shows a narrow band performance, having a fractional bandwidth 2–3%. A broadband patch antenna is normally defined one which can provide a bandwidth of the order of 10% and more, either by using a single layer patch design or by stacked patch designs. Regardless of the design, there are fundamentally two methods of broadbanding: changing the substrate parameters or using coupled resonators. In the former case, one can increase the patch impedance bandwidth by increasing the substrate height, or lowering its relative permittivity. This method is not used in most designs, as the substrate

Microstrip and Printed Antennas: New Trends, Techniques and Applications. Edited by Debatosh Guha and Yahia M.M. Antar

parameters are often decided by the cost or weight, and the fact that commercial substrates have standard dimensions and parameters. The latter method is easier to implement, as one can build the patch parameters into the design process. Since the patch itself is a resonator, one needs to couple it to another resonator, which can be another patch or a slot. Slots are normally cut on the ground plane or the patch surface. This method is preferred, as it results in a single layer low profile design, and will be discussed in this chapter. Stacked and slot coupled patches are discussed in the existing literature.

Recently, the use of impedance surfaces, in lieu of a perfectly conducting electric ground plane, has been receiving increasing attention. Conceptually, a magnetic conductor such as the ground plane will enhance the radiation efficiency of the patch. Studies have shown that such a ground plane will also enhance the impedance bandwidth and gain. Bandwidth in excess of 20% and gain in order of 10 dBi are obtained with a single patch antenna, which are remarkable improvements over those of a patch antenna over a conventional perfect electric ground plane. It should be noted that the focus of discussion in this chapter is on the single linear polarized antennas.

8.2.1 Single Layer Broadband Patch Antennas

The single layer microstrip patch antennas suffer from narrow impedance bandwidth. This can be improved significantly by employing slots on the patch in a controlled fashion which consequently provides the coupled resonator structures such as the U-slot loaded patch, E-shape patch and Ψ-shape patch. Both E-shape and U-slot loaded single layer rectangular microstrip patch antennas have shown 2 : 1 VSWR impedance bandwidths of 30–35%, on electrically thick substrate materials. The geometry of the E-shape patch antennas and its impedance match performance are shown in Figures 8.1(a) and (b). It can be seen that both the simulated and measured results provide a bandwidth of 32.2%. This antenna has dual resonant frequencies at 2.12 GHz and 2.66 GHz, respectively. The antenna frequency band is from 2.05–2.84 GHz. This antenna is discussed in [7, 8]. The measured E-plane and H-plane radiation patterns of the antenna are also shown in Figures 8.1(c) and (d), respectively, for both 1.9 GHz and 2.4 GHz, two wireless communication frequencies. In the E plane, the 3-dB beamwidth is 42 degrees at 2.4 GHz and 63 degrees at 1.9 GHz. The peak cross-polarization level at 1.9 GHz is −15 dB, which is higher than −25 dB at 2.4 GHz. This is due to different current distributions at these two frequencies. Similarly, in the H-plane, the radiation patterns at 1.9 GHz and 2.4 GHz are almost similar and the 3-dB beamwidth is 60 degrees at both frequencies, and the peak cross-polarization is −7 dB at 50 degrees. This high cross-polarization is generated by the leaky radiation of the slots, and the excitation of the next mode, which is orthogonal to the dominant mode.

As mentioned earlier, a U-slot loaded rectangular patch antenna was also reported in [9, 10], both on foam and conventional microwave substrate material. The microwave substrate (dielectric constant $\varepsilon_r = 2.33$) based U-slot loaded patch antenna geometry is shown in Figure 8.2(a) and selected design parameters for three antenna designs namely Antenna A, B, and C are shown in Figure 8.2(b). The antennas B and C provided an impedance bandwidth in the range of 25% and 27%, respectively, slightly less than that reported with the foam substrate antenna (30%) [9], but all showed good acceptable directive radiation patterns. The detailed design procedure/guidelines for designing a coaxial probe fed U-slot loaded micro-

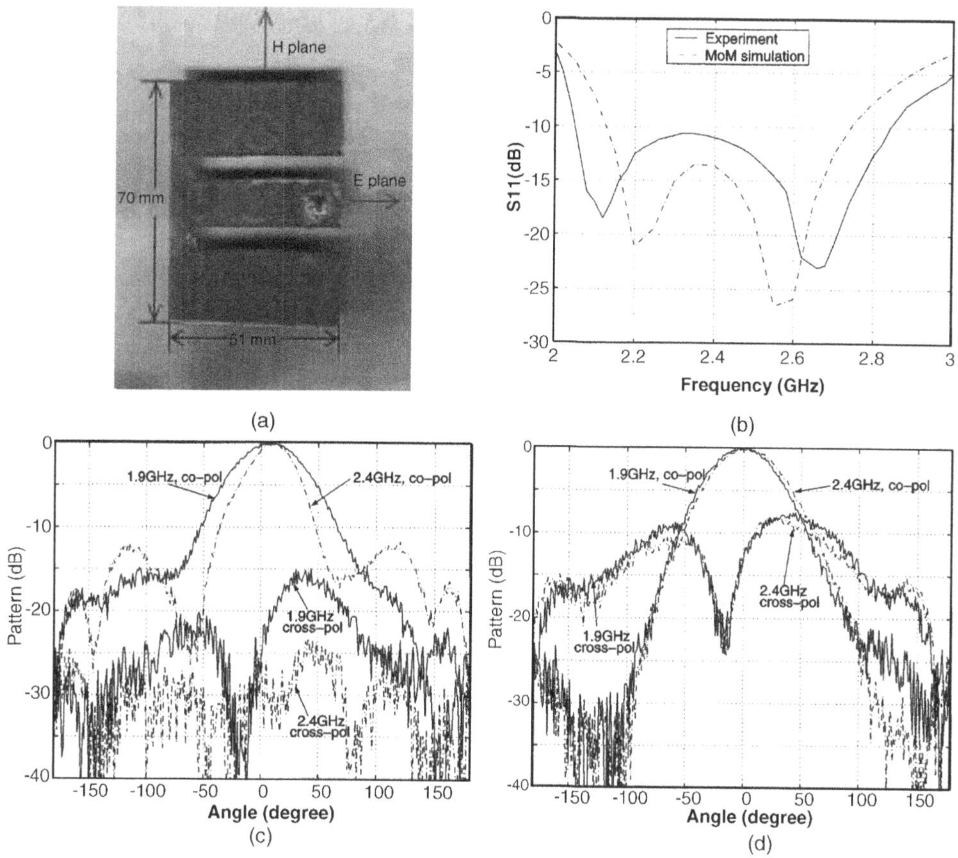

Figure 8.1 (a–b) E-shaped patch antenna on foam substrate resonating at 2.12 and 2.66 GHz, and (c–d) Measured E-plane and H-plane co-polarization and cross-polarization radiation patterns at two resonant frequencies of 1.9 GHz and 2.4 GHz for the E-shape patch antenna designed for a wireless communication application. Reproduced by permission of ©2001 IEEE [7]

strip patch antenna on a conventional microwave substrate is reported in [11]. However, this provided first pass design only, i.e. after further simulations the desired antenna performance could be achieved. This was supported through design example. Therefore, when designing a U-slot loaded patch antenna on a conventional microwave substrate material, one can follow the procedure outlined in [11].

Next, a coaxial probe fed novel Ψ-shape patch antenna is reported in [12], where an impedance bandwidth of the order of 54% was obtained with acceptable directional radiation patterns, but with the presence of higher cross-polarization towards the end of the frequency band. The antenna uses a thin conventional microwave substrate on which the antenna is etched. This thin substrate with patch is placed on top of a 6.4 mm foam substrate backed by a 75 mm × 75 mm ground plane. The antenna geometry and its impedance match are shown in Figure 8.3(a) and (b), respectively. The measured gain radiation patterns for the same at 4 GHz, 5 GHz, 6 GHz, and 7 GHz are shown in Figure 8.4(a)–(d), respectively. The presence of high

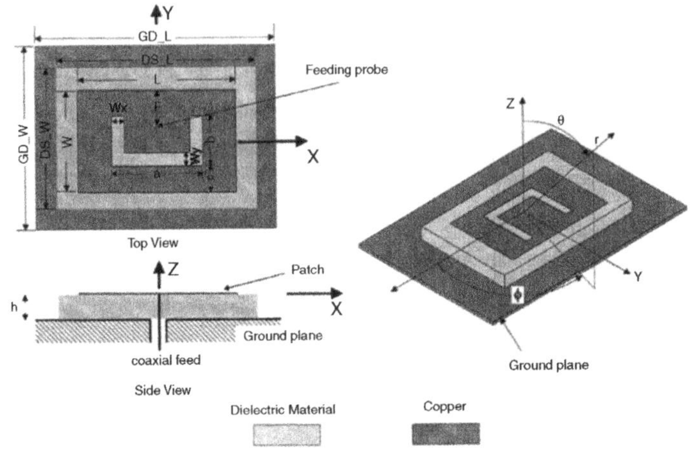

(a)

	GD_L	GD_W	DS_L	DS_W	ε_r	L	W
Antenna A	120.0	100.0	90.0	80.0	2.33	36.0	26.0
Antenna B	120.0	100.0	90.0	80.0	2.33	36.0	26.0
Antenna C	120.0	100.0	90.0	80.0	2.33	36.0	26.0
	a	b	c	h	F	w_x	w_y
Antenna A	14.0	18.0	4.0	5.0	13.0	2.0	2.0
Antenna B	14.0	18.0	4.0	6.4	13.0	2.0	2.0
Antenna C	14.0	18.0	4.0	8.0	13.0	2.0	2.0

(b)

Figure 8.2 (a) Antenna geometry with design parameters and (b) selected design parameters of the antennas with three designs namely Antennas A, B, and C. Reproduced by permission of ©2000 IEEE [10]

cross-polarization towards the higher frequency is attributed to the longer patch length and thus higher order mode generation. The antenna showed front-to-back (F/B) ratio of better than 20 dB throughout the frequency range.

8.2.2 Feed Mechanism Modification for Broadband Patch Antennas

In this section, the effect of feed mechanism modification on improving the impedance bandwidth of a patch antenna is discussed. In [13], the authors presented a broadband patch antenna with differential feed scheme implemented with the help of a folded plate pair as shown

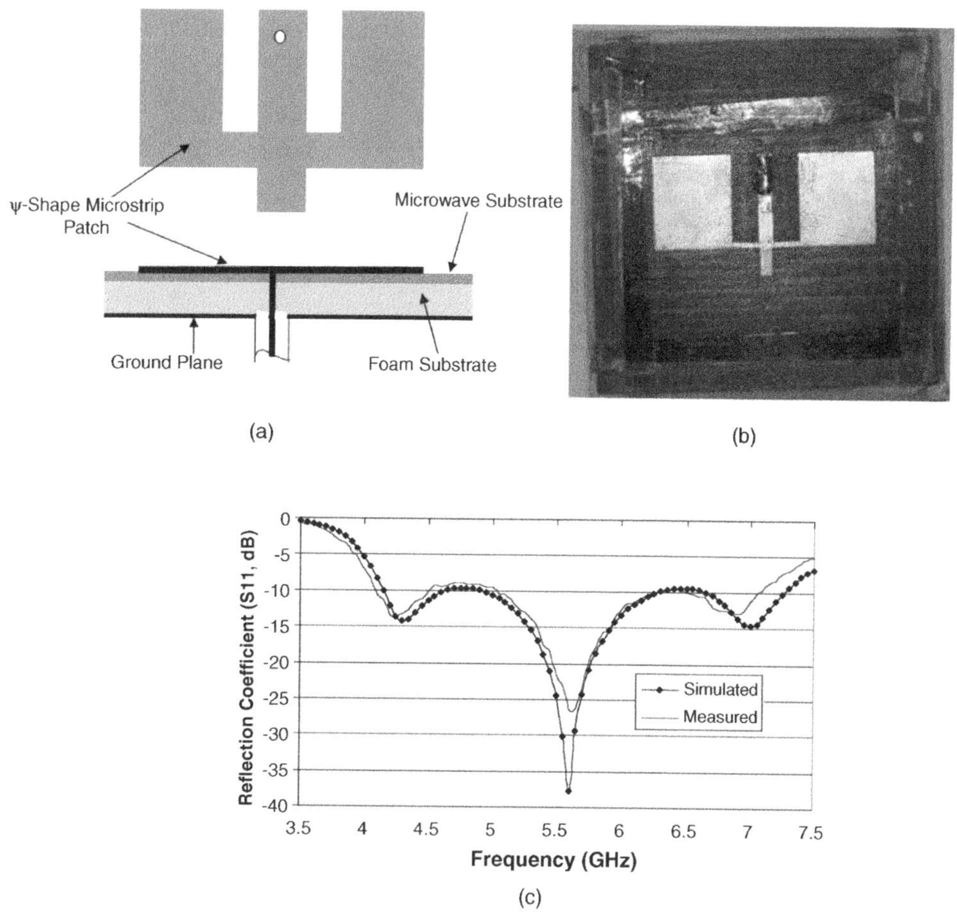

Figure 8.3 (a–b) The Ψ-patch antenna geometry and its photograph, and (c) measured and simulated impedance match performances. Reproduced by permission of ©2009 IEEE [12]

in Figure 8.5(a). By this feeding approach, long probes to the patch are avoided which consequently leads to the inductances. Further, the horizontal parts of the folded plates, incorporated with the patch and the ground plane, introduces the capacitance, which also compensates for the probe inductance. Additionally, the folded plate pair provides a good impedance matching circuit for an additional higher order mode resonance, in addition to the dominant mode resonance of the patch. This approach works as a wide impedance matching technique leading to a large bandwidth of 74%, as shown in Figure 8.5(b), which is much more than those of a conventional patch antenna and broadband patch antenna. Further, it provided stable symmetric radiation patterns with low cross-polarization levels and a gain of around 8.5 dBi, over the impedance bandwidth. The measured radiation patterns showing low cross-polarization, better than −20 dB, over the bandwidth are shown in Figure 8.6(a). For comparison, its simulated counterpart is also shown in Figure 8.6(b) illustrating cross-polarization levels below the scale of the plot.

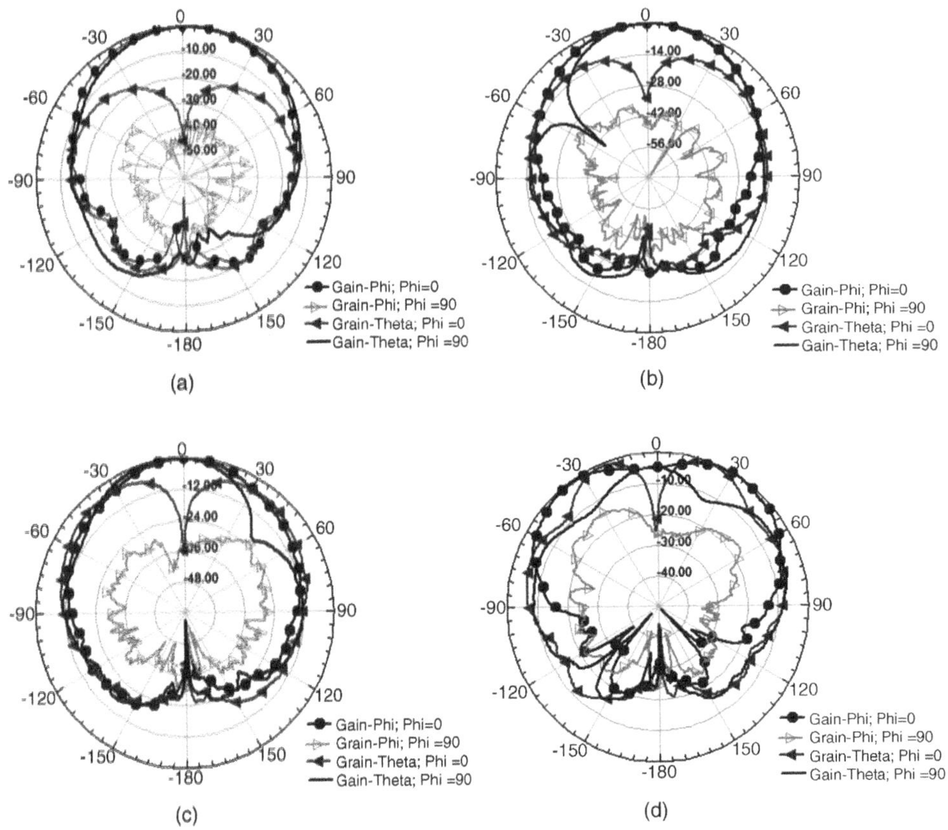

Figure 8.4 Measured normalized radiation patterns within the impedance bandwidth at (a) 4 GHz; (b) 5 GHz; (c) 6 GH; and (d) 7 GHz. Reproduced by permission of ©2009 IEEE [12]

Another feeding scheme that provided a wide impedance bandwidth with a patch antenna is reported in [14], which employed a meandered strip structure and is suitable for both planar and curved grounding structures. The achievable impedance bandwidths were 24% and 32%, respectively, for both planar and curved ground plane surfaces. This also provided low-cross-polarization and stable radiation patterns over the bandwidths. A similar paper in [15] provided a measured impedance bandwidth of 30% with a rectangular patch antenna with the help of a meandered feed structure. The side view of the antenna is shown in Figure 8.7(a) and its impedance match and gain performances are shown in Figure 8.7(b). The antenna shows measured bandwidth from 1.56 to 2.12 GHz of about 30.5%. For SWR 1.5 : 1, the impedance bandwidth of the antenna is 24% (from 1.6 to 2.04 GHz). The mechanism for low cross-polarization due to this feed is as follows. Basically, the meandering probe operates in the same way to the differential two-probe design, to suppress cross-polarization and the unwanted probe radiation. The two vertical portions of the meandering probe excite constructively the co-polarization pattern but destructively to cross-polarization pattern. The horizontal portions of the meandering probe, incorporated with the radiating patch and the ground plane, introduce

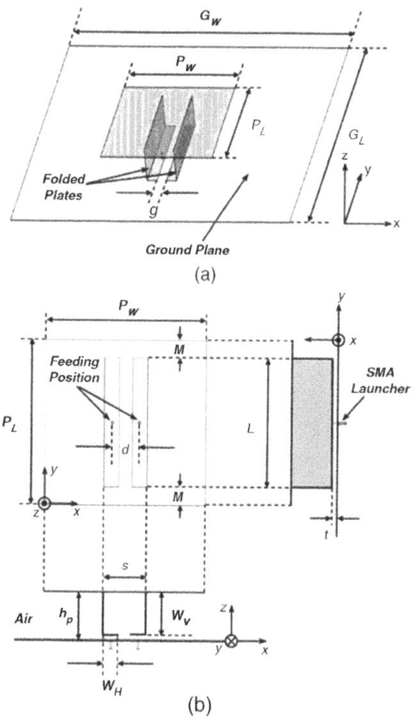

Parameters	P_L	P_W	h_p	G_L=G_W	L	M
Value (mm)	68 (0.49λ_o)	60 (0.43λ_o)	16 (0.12λ_o)	350 (2.53λ_o)	43 (0.31λ_o)	12.5 (0.18λ_o)

Parameters	W_V	W_H	s	g	d	t
Value (mm)	13.5 (0.098λ_o)	7.5 (0.054λ_o)	26 (0.19λ_o)	11 (0.08λ_o)	20 (0.14λ_o)	2.5 (0.018λ_o)

(c)

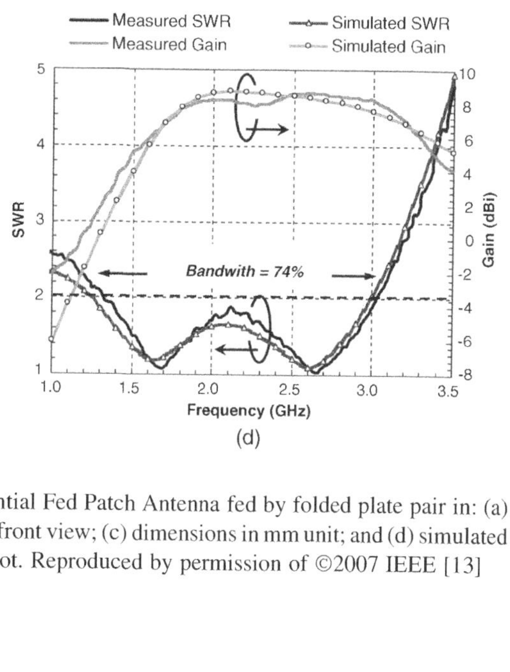

Figure 8.5 Geometry of the proposed Differential Fed Patch Antenna fed by folded plate pair in: (a) perspective view and (b) top view, side view, and front view; (c) dimensions in mm unit; and (d) simulated and measured VSWR and gain vs. frequency plot. Reproduced by permission of ©2007 IEEE [13]

capacitances. These capacitances can suppress some of the inductance contributed by the vertical portions of the probe. The wide-band characteristic is also due to the large spacing between the radiating patch and the ground plane, and is due to the use of a low permittivity substrate. The average measured antenna gain is about 9 dBi.

8.2.3 Artificial Magnetic Ground Planes for Broadband Patch Antennas

In conventional microstrip patch antennas, the ground plane is a good electric conductor, such as copper, and in simulation studies is normally assumed to be a perfect electric conductor

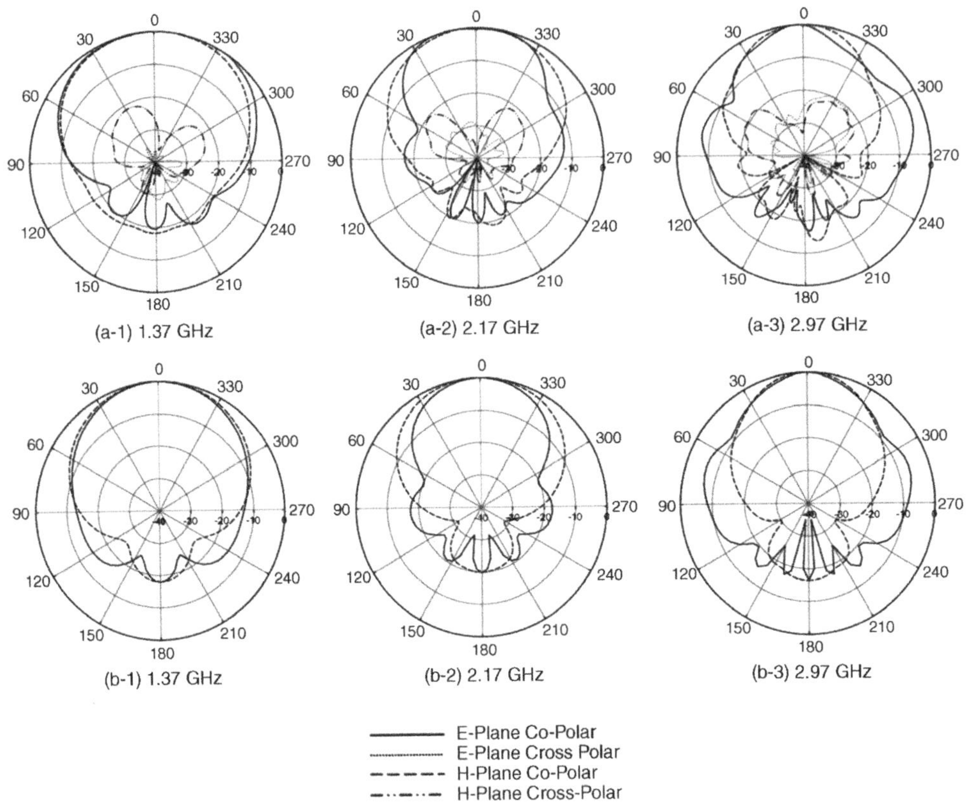

Figure 8.6 (a) Measured and (b) simulated radiation patterns of the proposed differential fed patch antenna. The simulated cross-polarized levels are vanishingly small under ideal conditions that cannot be seen. Reproduced by permission of ©2007 IEEE [13]

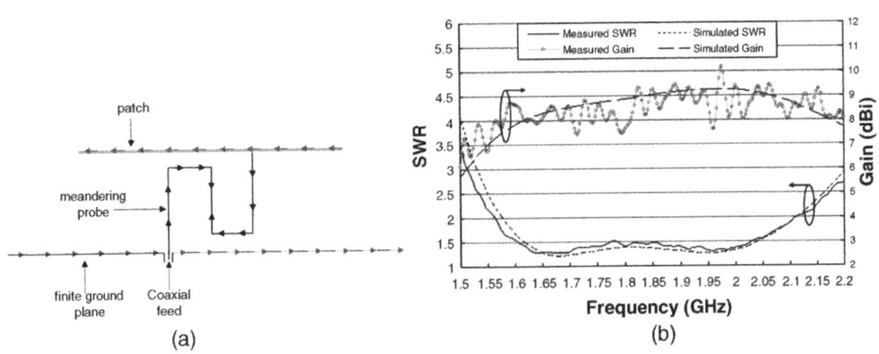

Figure 8.7 (a) Meanderline coaxial probe feeding the patch antenna; (b) simulated and measured SWR and gain vs. frequency plot of the meandering probe fed patch antenna. Reproduced by permission of ©2006 IEEE [15]

(PEC). The image effect on such a ground plane reduces the radiation efficiency of the patch and thus its bandwidth. The situation is opposite with a magnetic ground plane, where the image currents are in phase with those of the patch and the radiation efficiency and bandwidth are enhanced considerably. Since there are no natural magnetic conductors, their effects must be simulated artificially, which are accomplished by using resonant circuits. The resulting surface is therefore known as an artificial magnetic conductor, or AMC. On a planar surface, this can be done by etching tightly coupled arrays of conducting traces, such as rectangular patches, rings, or other shapes. Balancing the inductance and capacitance of the elements can result in parallel resonances and high impedance surfaces, simulating an artificial magnetic conductor. The planar geometry is also compatible with that of the patch, which can easily be fabricated with the same technology. Here we provide two examples of the patch antenna bandwidth broadening using such high impedance surfaces.

Figure 8.8 shows the geometry of the first example, which is a rectangular microstrip patch antenna over an AMC ground plane [16]. The AMC surface elements are small square conducting patches, shorted individually to the ground. These small AMC patches are not resonant at the operating frequency of the radiating patch, but interact with each other to cause a parallel resonance. The shorting pins enhance the AMC patch inductance and can reduce their size. However, without the shorting pins the surface resonance is still possible, but occurs with larger AMC patch elements. Table 8.1 shows the performance of the rectangular patch antenna over the AMC ground plane, for different locations of the feed point. Placing the same radiating patch over a conventional 80 mm × 80 mm PEC ground provided an impedance bandwidth of 3.85%, and a bore-sight gain of 3.92 dBi. However, as

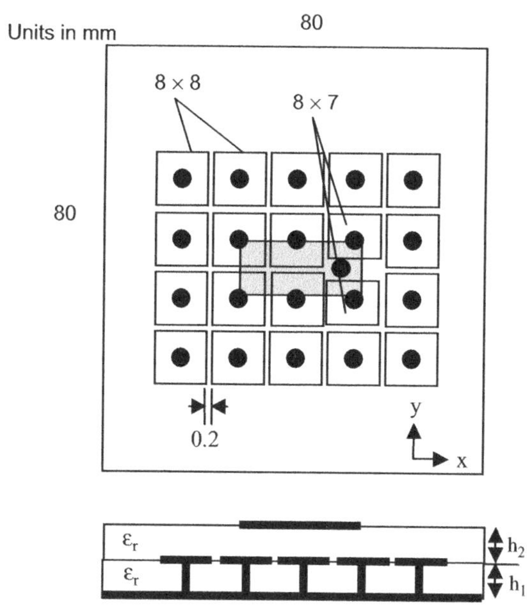

Figure 8.8 Geometry of a microstrip patch with an AMC substrate, patch length L = 14 mm, width W = 6 mm, $\varepsilon_r = 2.5$, $h_1 = h_2 = 1.59$ mm, feed distance from patch centre $L_f = 5.2$ mm. Adapted from ©2006 IEEE [16]

Table 8.1 Minimum S_{11}, bandwidth and gain for the antenna in Figure 8/8, vs. the feed location L_f

Feed location L_f (mm)	Minimum S_{11} (dB)	Impedance bandwidth	Gain at $\theta = 0°$ (dBi)
4.6	−22.44	19.91% (6.24–7.62 GHz)	10.31
5.2	−44.77	20.20% (6.23–7.63 GHz)	10.32
5.8	−35.75	16.38% (6.50–7.66 GHz)	10.32

Source: Adapted from IET *Microwaves, Antennas & Propagation,* Vol. 153, No. 6, 2006 [16].

Table 8.1 shows, over the AMC ground, the patch bandwidth has increased from 3.85% to 20.2%, and its gain from 3.92 dBi to 10.32 dBi. The results clearly show the beneficial effects of the AMC ground. With such a simple modification of the patch antenna ground plane the level of its performance enhancements, in impedance bandwidth and gain, are remarkable.

The second example is illustrated in Figure 8.9, which shows two identical patch antennas over significantly reduced sized AMC ground planes. In this example, the objective was the

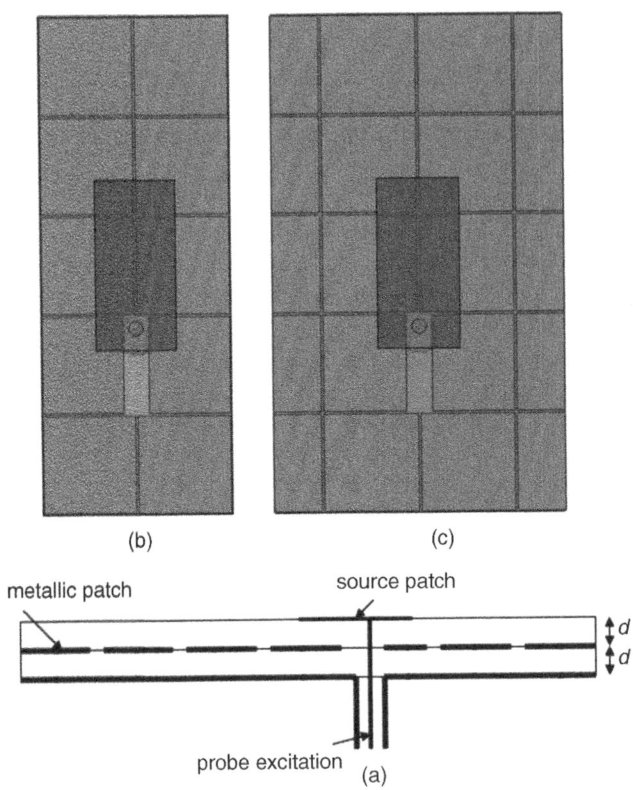

Figure 8.9 Geometry of a rectangular microstrip patch antenna over two small AMC ground planes, d = 1.59 mm, AMC patch sizes = 8 mm × 8 mm, spacing between patches = 8.2 mm, $\varepsilon_r = 2.5$, *tan* $\delta = 0.0022$, copper conductors. Reproduced by permission of ©2009 IEEE [17]

Table 8.2 Impedance bandwidth and directive gain performance of the microstrip patch antenna over PEC and AMC grounds, Antenna #1 and Antenna #2, over PEC grounds; Antenna # 3 and Antenna #4 over AMC grounds

	Ground plane type and size	Input impedance bandwidth (%)	Maximum gain (dBi)
Antenna1	Conductor 41 × 16.4 (mm^2)	2.82	5.40
Antenna2	Conductor 41 × 24.6 (mm^2)	2.82	6.19
Antenna3	AMC (5 × 2) 41 × 16.4 (mm^2)	7.30	6.71
Antenna4	AMC (5 × 3) 41 × 24.6 (mm^2)	8.64	7.03

Source: Reproduced by permission of ©2009 IEEE [17].

investigation of the AMC ground plane size and its effect on the patch antenna performance. Specifically, the study was intended to explore the effectiveness of the AMC surface, in improving the patch performance, when the AMC ground plane size becomes smaller than the wavelength [17]. The results are shown in Table 8.2, where Antenna 1 and Antenna 2 are the same radiating patch antennas over small PEC ground planes of different sizes. They can be assumed as the reference antennas. Antenna 3 is the same radiating patch antenna, but now over the AMC ground having the same ground plane dimensions as the Antenna 1. Similarly, Antenna 4 has the same ground plane size as the antenna 2. Comparing the performance results, presented in Table 8.2, it is clear that, even with such small ground plane sizes, the bandwidth enhancements due to AMC ground planes are still very significant, and at about a factor of three. The gain enhancements are of the order of 1 dB, and are limited by the back radiations. These two examples clearly demonstrate the beneficial effects of using AMC ground plane surfaces in planar antenna design. However, there are differences between conventional bandwidth broadening methods and those based on AMC surfaces. The conventional methods based on coupled resonators are reasonably well developed and yield significantly broader bandwidths. However, they do not have as many beneficial effects on the gain. On the other hand, the AMC-based methods enhance the gain significantly more, but are not as well developed yet, especially for enhancing the patch antenna bandwidth. There is still room for further bandwidth enhancements by the AMC surfaces.

8.3 Patch Antennas for Multiband Wireless Communications

Today's portable and hand-held wireless communication devices are not only compact but also multifunctional in nature. They can provide multiple types of wireless communications involving audio, video, data transfer, radio, and TV, etc. all in one device. While these features are being introduced in these devices, the challenge is always put to the antenna engineers to design antennas that are compact, low cost, efficient, capable of transferring high data rate, safe for human health, and able to handle simultaneous communications in multiple bands. Some of the different frequency bands that are important for these applications are GSM1800 (1710–1880 MHz) and UMTS (1900–2170 MHz), and WLAN and WPAN standards IEEE 802.11b/Bluetooth (2400–2485 MHz) and IEEE 802.11a (5150–5350 MHz 5725–5875 MHz). In this section, a few examples of the antennas for these applications are presented.

In [18], a compact uni-planar printed antenna is presented which is suitable for WLAN bands (2.4 GHz/5.2 GHz/5.8 GHz) and WiMAX bands (2.5 GHz/3.5 GHz/5.5 GHz). The antenna element is a coupled-fed PIFA which has a height of 9 mm and a width of 13 mm, when placed at the top edge of the display panel in a laptop computer. The desired multiband operation was achieved by using a coupling feed and by embedding a band-notching slit in the radiating arm of the PIFA. Figures 8.10(a) and (b) show the geometry of this antenna including the ground plane and antenna placement location. The simulated and measured impedance match performance results of this antenna are shown in Figure 8.10(c). Three resonant modes are excited at about 2.5 GHz, 3.5 GHz and 5.5 GHz. The resonant mode at 2.5 GHz covers 2.4 GHz WLAN and 2.5 GHz WiMAX bands. The second resonant mode covers 3.5 GHz WiMAX band, and finally the third mode covers 5.2 GHz/5.8 GHz WLAN and 5.5 GHz WiMAX bands. This antenna also showed high radiation efficiency and acceptable gain over the desired frequency bands, as shown in Figure 8.11. Figure 8.11 (a) reveals that, for both 2.4 GHz WLAN and 2.5 GHz WiMAX bands, the antenna gain is varied from about 3.5 to 4.7 dBi and the radiation efficiency is about 82–92%. For the 3.5 GHz band, the antenna gain is about 4.2 to 5.3 dBi and the radiation efficiency is about 90–95%. Similarly, from the results shown in Figure 8.11(b), one can observe that, the 5.5 GHz band shows an antenna gain of about 5.4 to 6.6 dBi and the radiation efficiency of about 92%.

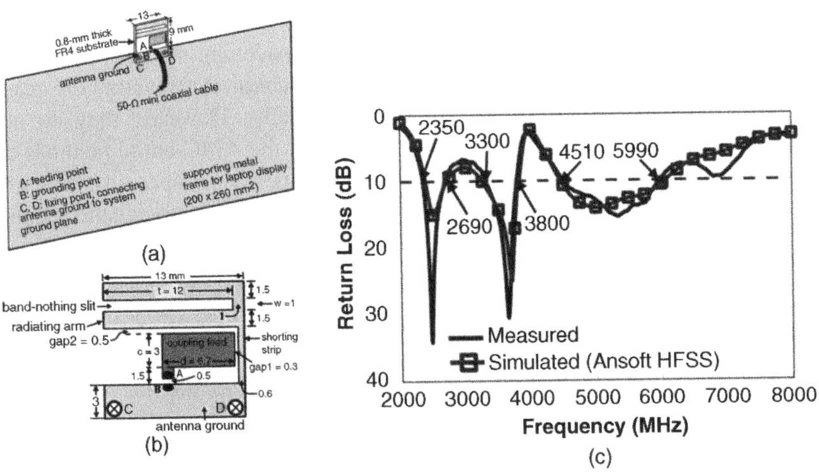

Figure 8.10 (a) Geometry of the uniplanar printed coupled-fed PIFA with a band notching slit for WLAN/WiMAX operation in the laptop computer; (b) dimensions of the metal pattern of the printed PIFA; (c) the simulated and measured return losses of the proposed PIFA. Reproduced by permission of ©2009 IEEE [18]

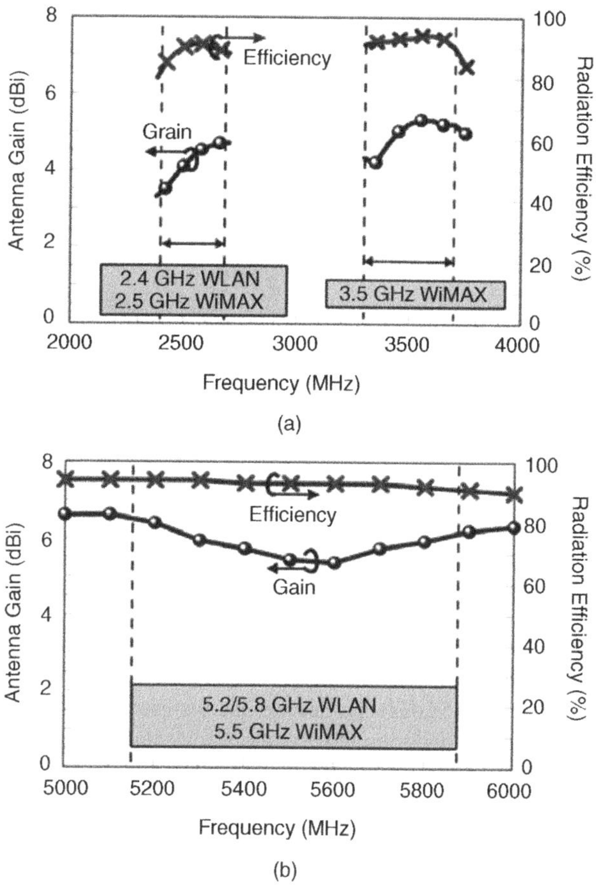

Figure 8.11 Measured gain and radiation efficiency for the proposed PIFA: (a) the 2.4 GHz WLAN and 2.5 GHz/3.5 GHz WiMAX bands; (b) the 5.2 GHz/5.8 GHz WLAN and 5.5 GHz WiMAX bands. Reproduced by permission of ©2009 IEEE [18]

A multiband antenna with seven different printed radiating elements was reported in [19], which was integrated into a small handset. Two printed inverted F-antennas (PIFAs) were designed to be used in the frequency bands namely, GSM1800, UMTS, IEEE 802.11b, and Bluetooth standards. Similarly, four IEEE 802.11a inverted F-antennas (IFAs) were integrated along the upper edges of the handset. An additional IFA for the global positioning system (GPS) with circular polarization was also integrated along one handset corner. In this design, polarization was not specified except for the GPS application, where right-hand circular polarization (RHCP) is preferred. The antenna geometry with dimensions in mm unit and the photograph of the printed antenna are shown in Figure 8.12. The antenna system occupies only about 23% of the volume of a typical handset.

Additionally, an E-shape patch [20] is reported for the multiband wireless communications. Thin E-shaped microstrip patch antennas (ESPAs) are reported to operate in the 5–6 GHz

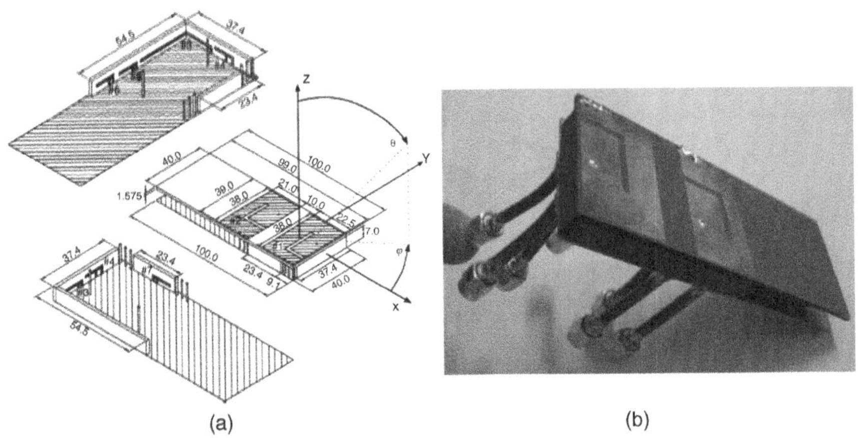

Figure 8.12 (a) Geometry and overall dimensions (in millimeters) of the multiantenna system (with antenna reference numbering); (b) photograph of the antenna. Reproduced by permission of ©2008 IEEE [19]

frequency range. They were suitable for wireless local area network (WLAN) adapter PCMCIA cards of standard 5 mm thickness for two IEEE 802.11a WLAN bands (5.15–5.35 GHz and 5.725–5.825 GHz). These antennas were investigated for same orientation for transmit and receive communications and orthogonal orientation for the spatial polarization diversity reception. Figure 8.13(a) shows the coaxial probe-fed diversity reception antennas which employ two orthogonal E-shape patches. Figure 8.13(b) shows the measured scattering parameter results, representing the reflection coefficient at the antenna ports and isolation between the antennas. It can be seen that both antennas are well matched and their isolation is better than −20 dB. The measured diversity radiation patterns for the antennas I and II

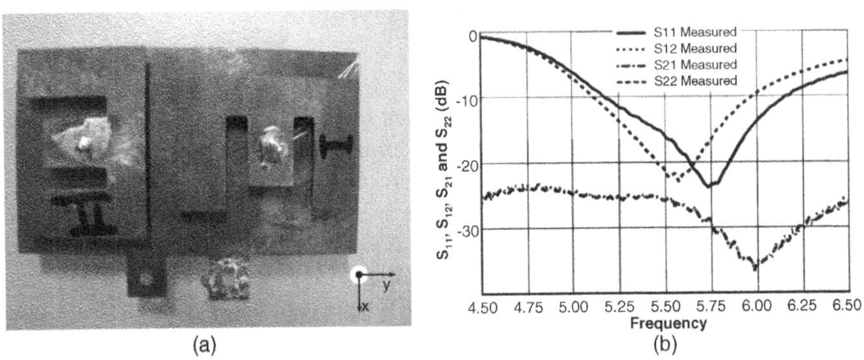

Figure 8.13 (a) Spatial polarization diversity arrangement of antennas composed of two orthogonal E-shaped patches; (b) scattering parameters of the antennas. Reproduced by permission of ©2004 IEEE [20]

at 5.25 GHz are also shown in Figure 8.14. It can be seen that the co-polarization components for the antennas I and II are reversed, i.e. co-polarization component for the antenna I become as the cross-polarization component for the antenna II and vice versa.

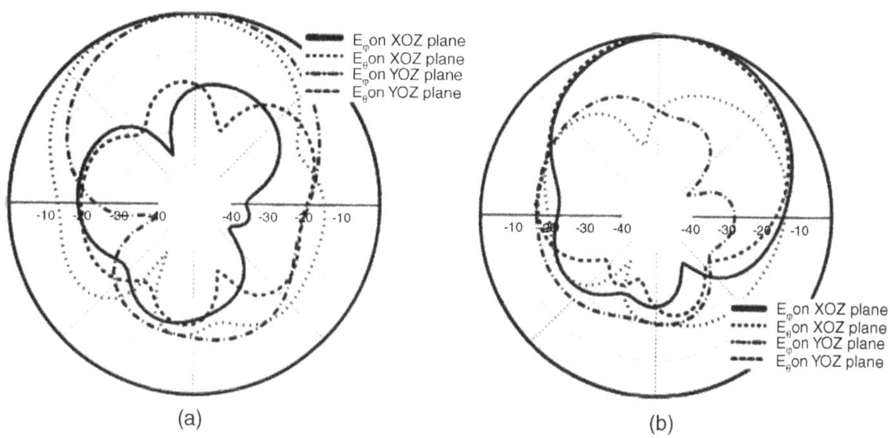

Figure 8.14 Radiation patterns (in dBi) of the diversity antenna, (a) when antenna I is excited at 5.25 GHz; (b) when antenna II is excited at 5.25 GHz. Reproduced by permission of ©2004 IEEE [20]

8.4 Enhanced Gain Patch Antennas

A novel method for increasing the gain of an aperture-coupled microstrip patch antenna was presented in [21, 22] by employing a dielectric cover that is spaced at some height above the antenna. The antenna is shown in Figure 8.15(a), which shows a grounded co-planar waveguide-fed rectangular microstrip patch antenna with a dielectric sheet as cover placed at some spacing from the patch surface. The antenna provides impedance matching and 3dB gain bandwidths large enough to be useful in some high speed wireless communications. The antenna was fabricated including a 0.254 cm thick TMM10 ($\varepsilon_r = 9.2$) cover for experimental verifications. The uncovered, $0.525\lambda_0$, and $1.05\lambda_0$ cover height cases were measured, and their results agreed well with the simulations. The maximum measured gain was 13.9 dBi, compared with 6.67 dBi for the uncovered antenna. The measured and simulated radiation patterns of the antenna with the dielectric cover of spacing $1.05\lambda_o$ are shown in Figure 8.15(b). The measured impedance matching bandwidths for the uncovered, $1/2\lambda_0$-spaced, and λ_0-spaced cases are 28.0%, 20.9%, and 18.5%, respectively. These results compare well with the 27.4%, 23.0%, and 21.0% found in the simulation. In this case, the design trade-off is between surface area and volume, which requires spacing height than surface area in the case of an array. In many applications, the dielectric cover can be realized as part of the antenna's housing. The matching and radiation properties are also compared in Table 8.3 for different cover spacings and thicknesses. While the impedance matching bandwidth increases with increase in the spacing height – approaching that of the uncovered baseline antenna – the 3 dB gain bandwidth decreases nearly monotonically. However, near the spacing of 1/2 λ_0, the impedance and 3 dB gain bandwidths

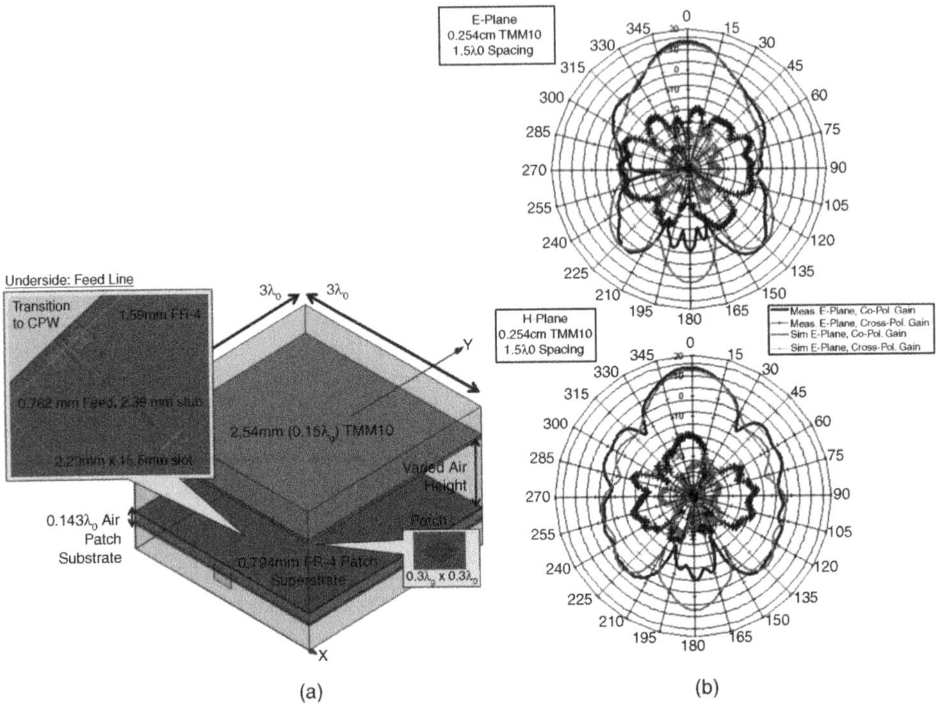

Figure 8.15 (a) Geometry of patch antenna with spaced-off dielectric cover; (b) comparison of the simulated and measured E-plane and H-plane gain radiation patterns with TMM10 ($\varepsilon_r = 9.2$) cover sheet of 0.254 cm thickness at a height of $1.05\lambda_o$ from the patch surface. Adapted from ANTEM 2009 [21, 22]

are similar around 18%. For the shorter spacing heights, below 1/4 λ_0, the dielectric cover degrades the gain since the antenna is essentially de-tuned. At the larger spacings of 1/2 λ_0, λ_0, and $2\lambda_0$, the gain increases by more than 6 dB and in the case of $2\lambda_0$ spacing, there are multiple frequencies at which the broadside gain is maximum. This is caused by the resonance structure created between the original antenna and the spaced-off cover. Only at certain angles do the radiated fields add constructively. This new resonance structure also causes variation in resonance frequency of the antenna.

The gain enhancement is also shown by employing a frequency selective surface (FSS) type structure, which has conducting patch arrays etched on a substrate layer and spaced at a height from the radiating element [23]. Figure 8.16(a) shows such an antenna structure where the size of the FSS structure is 15 cm × 15 cm. Figure 8.16(b) shows the reflection coefficient and gain plots versus frequency. It can be observed that the antenna shows enhanced gain over the matching bandwidth from 8.2 GHz to 8.6 GHz with a peak value of 19.8 dBi. With this structure the gain enhancement depends on the reflectivity of the superstrate. With a dielectric sheet cover the reflectivity depends on its relative permittivity. With an FSS surface, on the other hand, the reflectivity can be controlled in its design, which means the differential gain enhancement is no longer a limited quantity, but can be made small or large to match the design requirement of the antenna gain. However, one should note that in this antenna the gain enhancement is due to the aperture size enlargement, caused by the superstrate reflection of the

Table 8.3 Matching and radiation results from varying spacing height

Spacing Height	Matching BW	Half-Power Gain BW	Half-Power Beamwidth	Peak Gain (dBi)	Peak Gain Freq. (GHz)
<no cover>	27.8%	39.6%	90°	5.79	5.3
$0\lambda_0$ (touching)	8.5%	21.0%	27°	−4.68	4.2
$1/16\ \lambda_0$	16.7%	18.8%	18°	2.68	5.7
$1/8\ \lambda_0$	13.6%	23.7%	46°	5.76	6.3
$^*1/4\ \lambda_0$	8.2%	28.1%	34°	7.97	6.0
$(1/2 - 1/16)\ \lambda_0$	16.9%	19.4%	27°	12.6	5.3
$(1/2 - 1/32)\ \lambda_0$	19.8%	18.0%	29°	12.32	5.1
$1/2\ \lambda_0$	21.8%	17.4%	28°	12.14	5.0
$(1/2 + 1/32)\ \lambda_0$	23.9%	16.8%	29°	11.59	4.8
$(1/2 + 1/16)\ \lambda_0$	26.3%	17.2%	28°	10.86	4.7
$(1 - 1/16)\ \lambda_0$	15.9%	11.8%	20°	13.6	5.6
$(1 - 1/32)\ \lambda_0$	27.2%	12.3%	22°	13.4	5.5
λ_0	27.6%	13.1%	23°	13.2	5.3
$(1 + 1/32)\ \lambda_0$	27.6%	13.3%	24°	13.0	5.2
$(1 + 1/16)\ \lambda_0$	28.7%	13.6%	24°	13.0	5.1
$2\ \lambda_0$	26.7%	7.7%	20°	13.0	5.6

*Slightly above VSWR <2 at 5.2 GHz caused significantly lower matching bandwidth.
Source: Adapted from ANTEM 2009 [21]

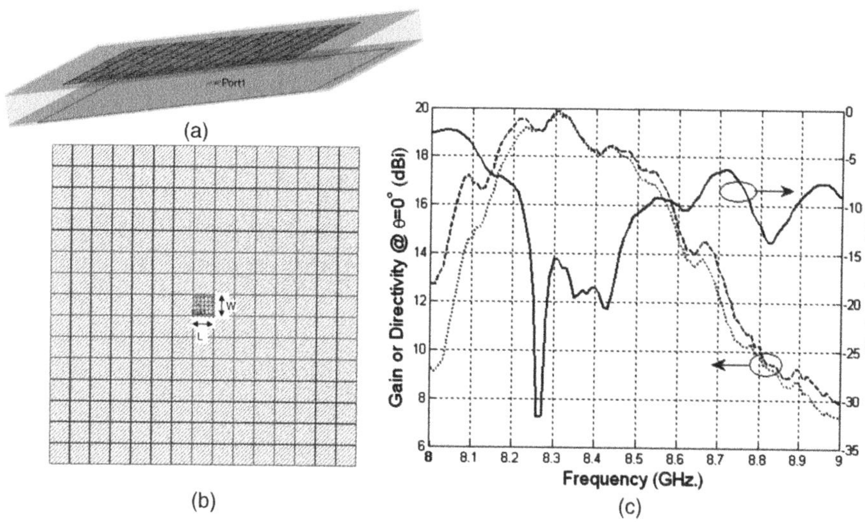

Figure 8.16 Cavity resonance antenna with truncated FSS superstrate (15 by 15 unit cells) and microstrip patch antenna: (a) 3-D view; (b) top-view. L = 10 mm and W = 8.5 mm; (c) measured antenna gain at bore-sight and return loss versus frequency. Adapted from ANTEM 2009 [23]

patch radiation. Consequently, the aperture phase error increases by the superstrate reflectivity, which in turn reduces the gain bandwidth of the antennas. Thus, as the antenna gain increases, its gain bandwidth decreases. Therefore, with this class of antennas the bandwidth limitations is no longer decided by its impedance bandwidth, but rather by its gain bandwidth.

8.5 Wideband Compact Patch Antennas

In this section, compact wideband patch antennas are discussed, which are important in mobile, especially in hand-held or portable, device applications. Several papers have dealt with the compact antennas [24, 25], and developed techniques for the size reduction. However, most designs result in bandwidth degradation, which is not the objective of this section. Maintaining the original wideband character of the antenna, even after reducing its size, is a challenging problem, which needs further research attention. One promising technique is the use of symmetry, or natural boundary conditions of the antenna, to reduce antenna size without affecting its original field distributions. Here, we discuss one such method that has recently been used successfully on both reduced size E-shape patch and U-shape slot loaded patch antennas [25]. It was shown that a half U-spaded slot loaded patch antenna also maintains wide-bandwidth behavior reaching close to the full size U-slot loaded patch element bandwidth of 30%. The antenna half-structure used shorting wall or shorting pins near the symmetry plane of the antenna to simulate full size antenna, as shown in Figure 8.17.

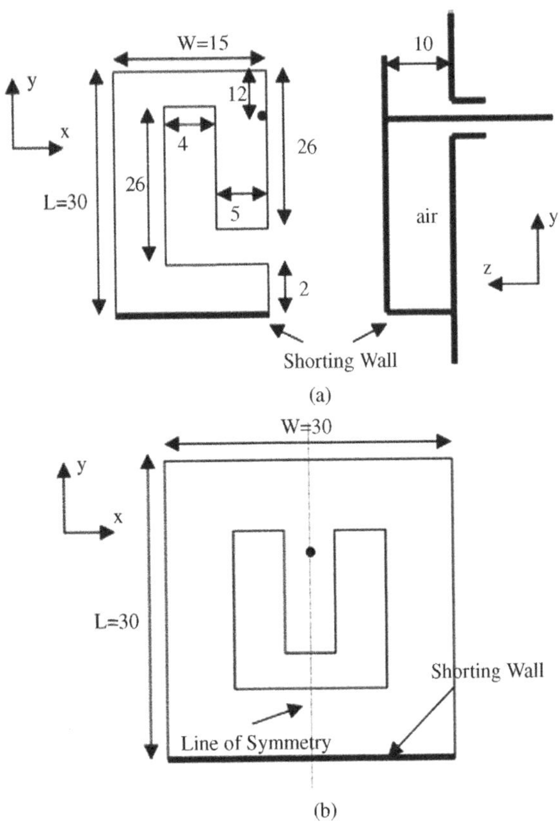

Figure 8.17 Geometries of the U-slot patch antennas with shorting pin (units in mm and not to scale): (a) half U-slot; (b) full U-slot, similarly, geometries of the E-shaped patch antennas; (c) half E-shaped patch; (d) full E-shaped patch. Reproduced by permission of ©2005 IEEE [25]

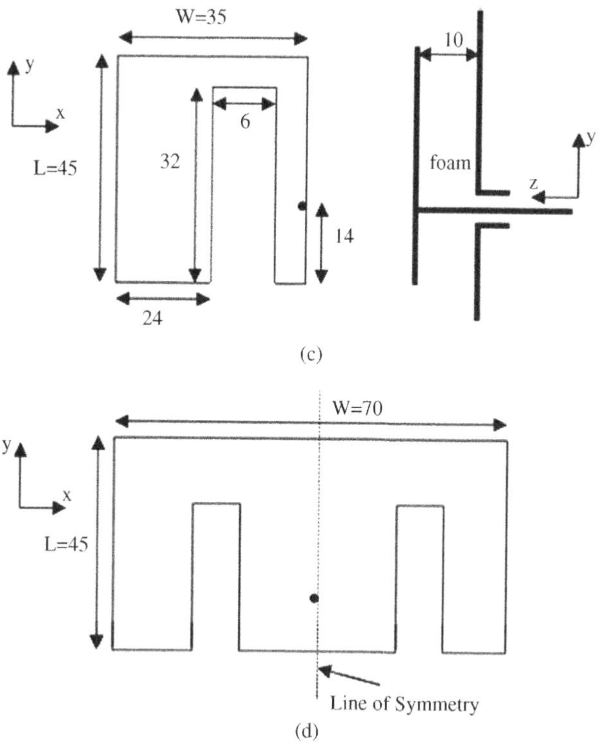

Figure 8.17 (*Continued*)

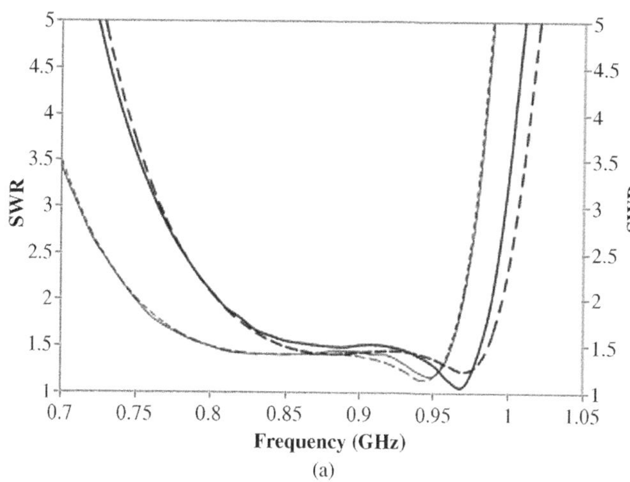

Figure 8.18 (a) Measured and simulated SWR of the full and half U-slot patch antennas (Figures 8.17 (a) and (b) with shorting pin; (b) measured and simulated SWR of the full and half E-shaped patch antennas (Figures 8.17(c) and (d)). Reproduced by permission of ©2005 IEEE [25]

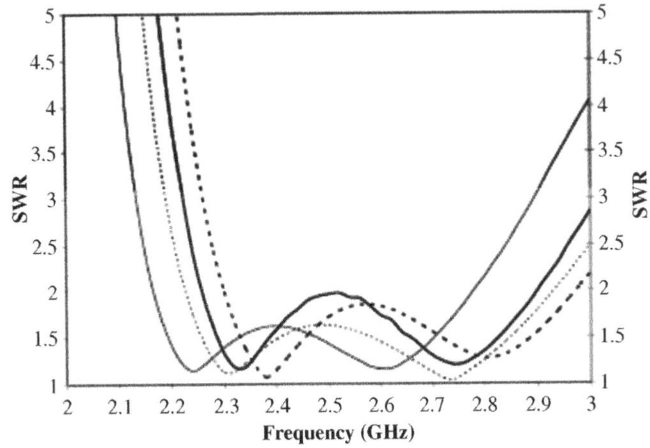

(b)

Figure 8.18 (*Continued*)

Figure 8.18 shows the simulated and measured VSWR of both antennas, i.e. the full and half size designs. It can be seen that, half antennas show similar, slightly less bandwidth than the full size patches. For example, an impedance bandwidth of 28.6% with radiation efficiency exceeding 90% is reported for the shorting pin loaded half U-slot loaded antenna, which is close to that of the full antenna structure. A similar response is also shown for the half size E-shape patch antenna.

8.6 Microstrip Slot Antennas

8.6.1 Principle of Radiation and Limitations

Microstrip slot antennas are widely used for military and commercial wireless applications due to their promising features such as their light weight, small size, low cost, and ease of conformity with planar and non-planar surfaces. Microstrip slot antenna comprises a slot cut in the ground plane of the microstrip line so that the slot is perpendicular to the strip conductor of the microstrip line [5]. The fields of the microstrip lines excite the slot. For efficient excitation of the slot, the strip conductor is short-circuited through the dielectric substrate on the edge of the slot as shown in Figure 8.19(a). The strip conductor is terminated in an open circuited stub beyond the edge of the slot as shown in Figure 8.19(b). The length L_m of the open circuited microstrip stub is approximately a quarter-wavelength long, so that an effective short circuit is realized at the outer edge of the slot. A center-fed slot antenna has a high radiation resistance, hence a matching network may be needed to match the antenna input impedance to the characteristic impedance of the microstrip line.

The microstrip-fed slot antenna has the advantage of low cross-polarization, as compared to the microstrip patch antennas. Its drawback is the inherent bidirectional radiation, which can be corrected by using a metallic reflector on one side. In comparison to patch elements, the

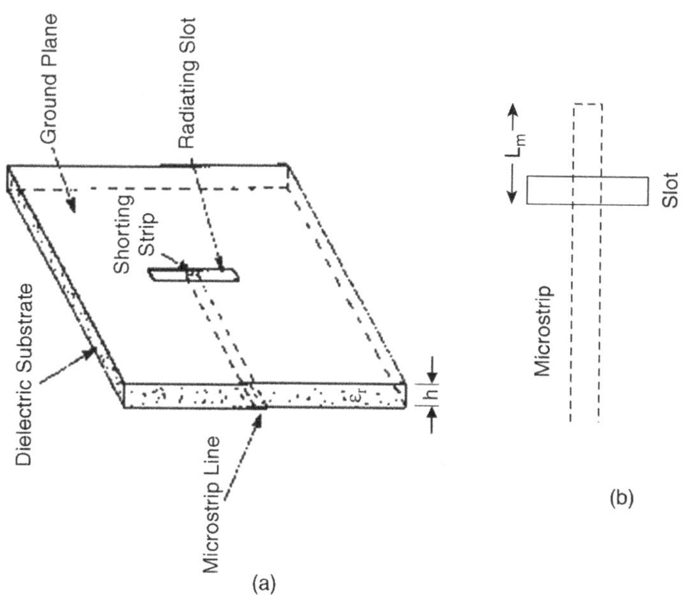

Figure 8.19 (a) A rectangular slot antenna; (b) cross-section of a microstrip slot. Reproduced by permission of ©2001 Artech House [5]

antennas with slot configurations demonstrate enhanced characteristics, including wider bandwidth, less conductor loss and better isolation. In addition, since slots can be made very narrow, multiple slots can be incorporated on the same ground plane to enable operation in several frequencies, and thus for multiband applications. Another important feature of slot antennas is their economic use of the design space. Since slot antennas are cut on the ground plane, they actually do not add to the size or weight of the structure, rather they reduce them, which are important features for mobile and hand-held devices. Next, the microstrip slot antennas with wide bandwidth and multiband characteristics are discussed.

8.6.2 Wideband Microstrip Slot Antennas and Size Reduction

As discussed earlier, the microstrip slot antennas provide bi-directional radiation patterns and have the potential to provide wide bandwidth performance. Therefore, these antennas are good candidates for wireless communication applications, also leaving enough space for the RF circuitry. Further, these antennas have the potential to achieve compactness or size reduction by controlling shape of the slots. Several wideband microstrip slot antennas were reported in [26, 27] with various slot shapes and placements such as the straight slot, the L-slot, the inverted T-slot, and dual L-slot shapes, which helped in achieving an impedance bandwidth from 40–87%. The bent shapes of the L- and T-slots reduce their height and provide more space on the ground plane.

The straight slot antenna geometry with the coordinate system is shown in Figure 8.20(a) [26]. A monopole slot of length 30 mm is cut at the top end of the finite ground plane of

80 mm × 50 mm near the edge of a microstrip structure on FR-4 substrate ($\varepsilon_r = 4.50$, tan $\delta = 0.02$). Here the slot is electromagnetically fed by a 50 Ω microstrip transmission line, excited by a 50 Ω SMA probe, and occupies around 37.5% of the ground plane length. The achieved impedance bandwidth of the antenna, when designed on an FR-4 material is 47%

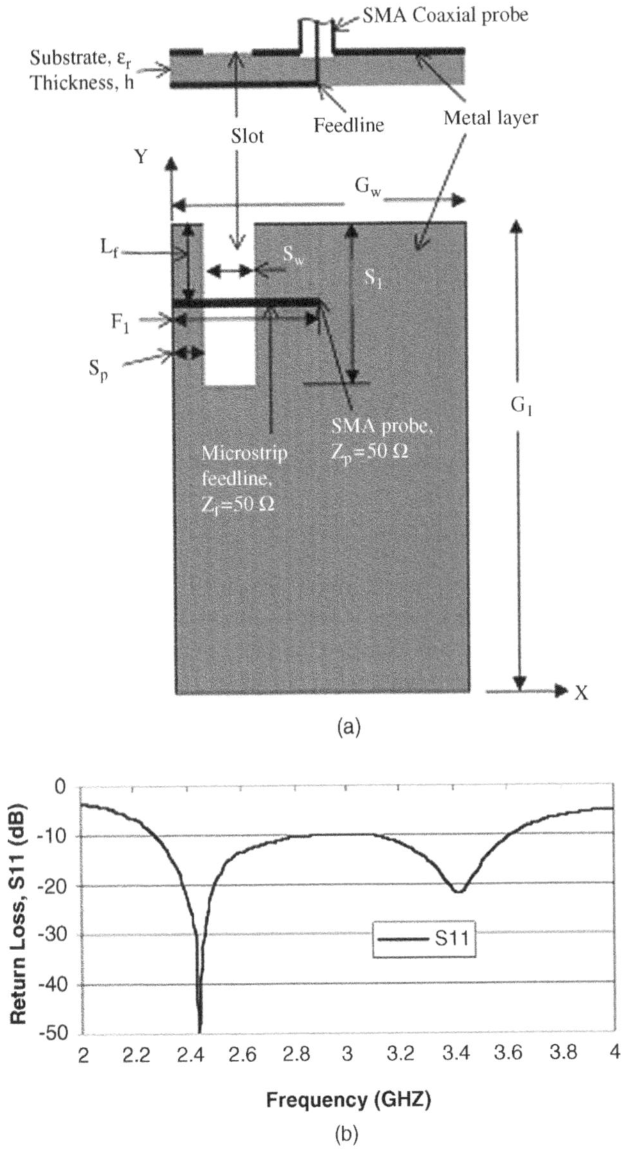

Figure 8.20 (a) Geometry and different parameters of a single slot antenna on one corner of a finite ground plane, fed using a microstrip line from bottom excited by a SMA probe; (b) measured return loss for the fabricated slot antenna on one corner of a finite ground plane. Reproduced by permission of ©2004 IEEE [26]

(2.26 to 3.36 GHz), which is evident from the measured reflection coefficient plot in Figure 8.20 (b). This antenna when designed on the low loss ($\varepsilon_r = 2.5$, $\tan\delta = 0.001$, h = 1.22 mm) dielectric material showed an improved 60% (2.30–4.25 GHz) impedance bandwidth. The measured and simulated radiation patterns of the antenna at 3 GHz are shown in Figures 8.21(a) and (b), respectively, which show bi-directional patterns. The measured and simulated data agree well. This antenna was also tried for spatial polarization diversity reception in complementary configuration, as shown in Figure 8.22(a). This design helps in receiving improved communication signal, in multipath propagation environments. The surface current distributions of the antenna (Figure 8.22(b)) show the presence of two orthogonal components that radiate from the two orthogonal slots and therefore help in improving the reception of the communication signal in faded environments.

The straight slot antennas were further studied with the ground plane of size 80 mm × 50 mm, by modifying the shape, placement location on the ground plane, and the feeding

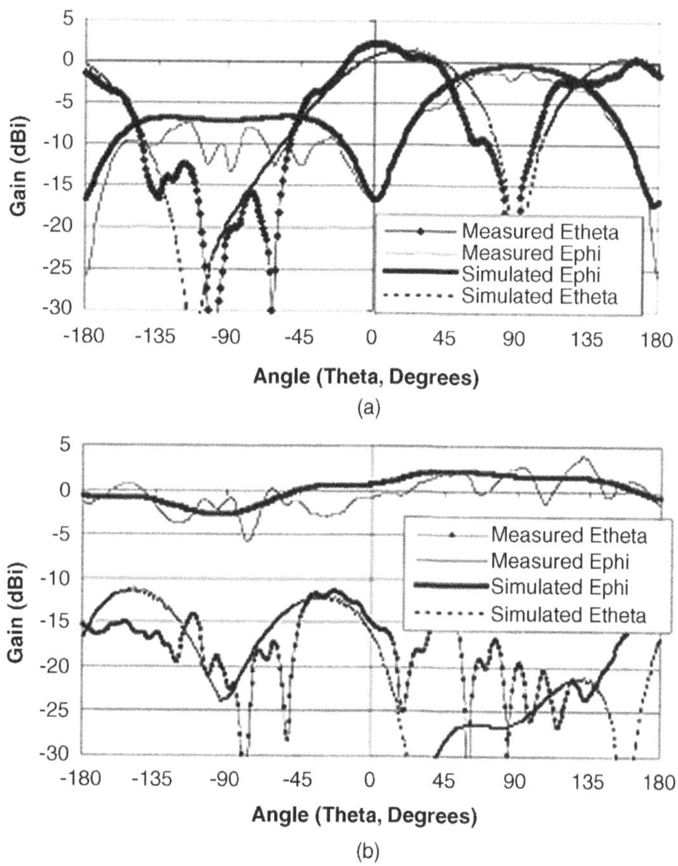

Figure 8.21 Measured and simulated gain patterns at 3.00 GHz for the antenna shown in Figure 8.20(a). Reproduced by permission of ©2004 IEEE [26]

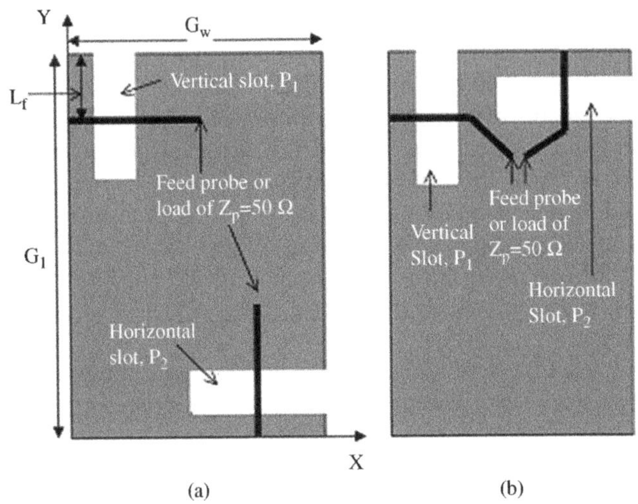

Figure 8.22 Geometries of the complementary dual orthogonal slot antennas for spatial polarization diversity on corners of a finite ground plane fed using microstrip line from bottom: (a) slots in diagonal corners; (b) slots in adjacent corners; (c) current distribution on the antenna with slots in diagonal placement. Reproduced by permission of ©2004 IEEE [26]

style [27]. The geometry of the straight slot, L-shape slot, and inverted T-shape slot antennas are shown in Figure 8.23(a)–(c), respectively, at the center of the ground plane. The L-slot is the bent version of the straight slot and comprises a vertical slot and a horizontal slot and occupies 23% of the ground plane length. Similarly, the geometry of the inverted T-slot antenna also comprises a vertical slot and a horizontal slot, which occupies only 17.5% of the ground plane length. The antennas were fabricated on a low loss substrate material ($\varepsilon_r = 2.5$, $\tan\delta = 0.001$, $h = 0.79$ mm), and experimentally tested for verification of the simulated results. The measured impedance matching for straight, L-shape slot and inverted T-shape slot are shown in Figures 8.24(a)–(c), respectively. It can be seen that the straight slot, L-shape slot and inverted T-shape slot antennas provide an impedance bandwidth of 60% (2.51–4.66 GHz), 83.7% (2.41–5.88 GHz), and 80.5% (2.74–6.43 GHz), respectively. The antennas provide bi-directional radiation patterns. These antennas were also tried for spatial polarization diversity reception in complementary configurations with almost 87% impedance bandwidth.

8.6.3 Ultra-wide Bandwidth (UWB) Slot Antennas

An UWB antenna is an antenna capable of producing a similar radiation (patterns and gain) over a very wide frequency range. According to the Federal Communications Commission (FCC), a UWB antenna should provide a gain and impedance bandwidth of 110% (3.1 GHz–10.6 GHz). In [28] an ultra-wideband microstrip slot antenna is reported for wireless communication applications on a finite ground plane of size 80 mm × 50 mm. The antenna consists of a

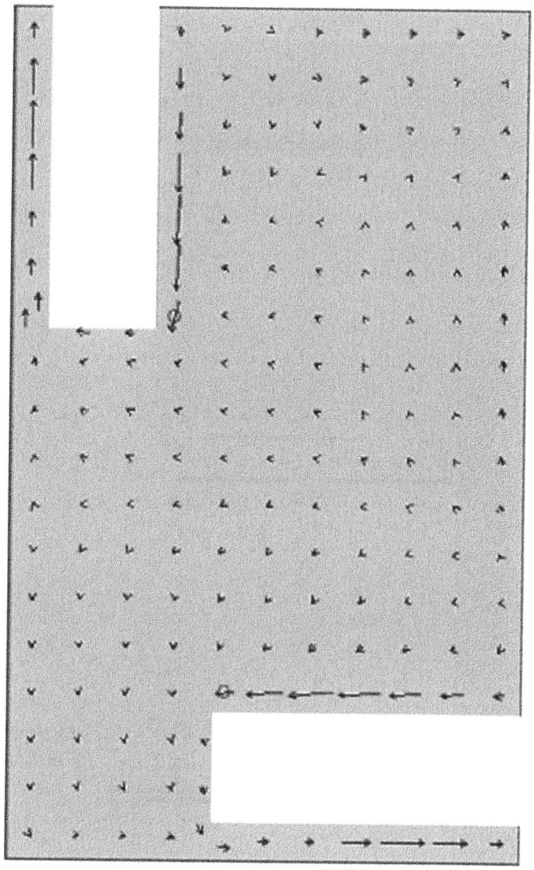

(c)

Figure 8.22 (*Continued*)

pentagon-shaped microstrip slot fed in using a microstrip line and a pentagon stub. The antenna geometry is shown in Figure 8.25(a) and photograph of the fabricated antenna is shown in Figure 8.25(b). While simulations considered low loss substrate material, antenna was re-simulated for experimental verification on FR-4 substrate material ($\varepsilon_r = 4.5$, tan $\delta = 0.02$, h = 1.40 mm). The antenna uses only 25% of the ground plane length hence leaves enough space for the RF circuitry. The feedline is rotated by 15% from horizontal. It was found that a suitable combination of slot, feed line and stub can provide a wide impedance bandwidth. The radiation pattern of the antenna is shown in Figure 8.26, which shows that as frequency increases, the antenna radiation pattern deteriorates because the same antenna size becomes electrically much larger than the monopole antenna. The simulated and measured impedance match results and measured group delay data are shown in Figures 8.27(a) and (b), respectively. The measured impedance bandwidth is 117% (2.6–10.0 GHz) about the same as the simulated results of 115% (2.5–9.3 GHz). The low loss substrate material ($\varepsilon_r = 2.2$, tan $\delta = 0.004$,

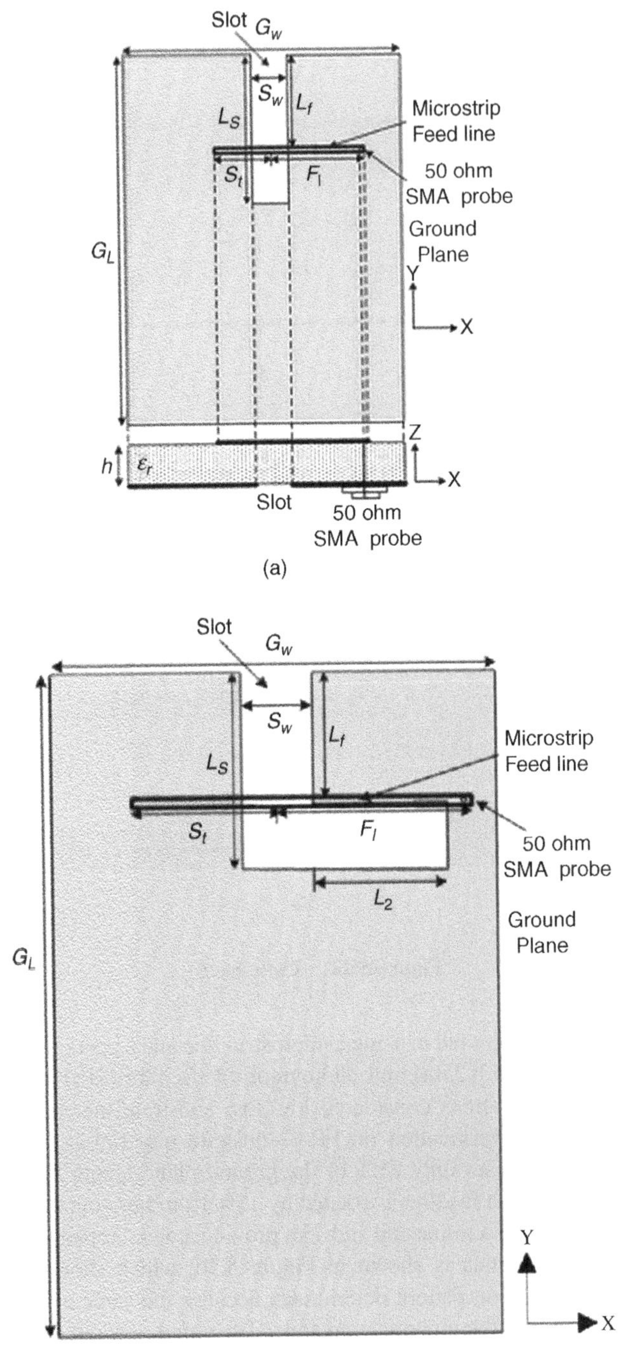

Figure 8.23 Microstrip monopole slot antennas fed-in using straight microstrip transmission lines on a finite ground plane of 80 mm × 50 mm: (a) straight slot; (b) L-shape slot; (c) inverted T-shape slot. Reproduced by permission of ©2005 IEEE [27]

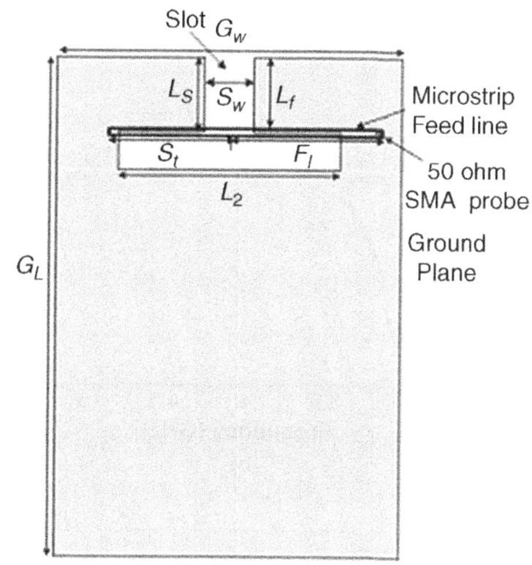

(c)

Figure 8.23 (*Continued*)

h = 1.58 mm) antenna showed 124% (2.65–11.3 GHz) of impedance bandwidth. The antenna was verified for the group delay, using the two antenna arrangement, where the transmit and receive antennas were facing each other, while connected to the two ports of a Network Analyzer. The ripple present on the group delay results is attributed to the scattering effect from the network cables. It can be observed that the group delay between the antennas is around 0.7 ns and varies by 0.125 ns within the bandwidth. Thus the antenna shows a fairly constant group delay, meeting the requirement of a UWB antenna.

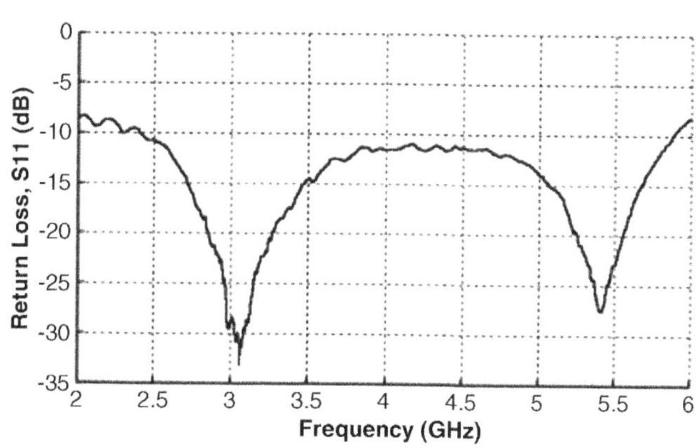

(a)

Figure 8.24 Measured reflection coefficient results of the microstrip slot antennas shown in Figure 8.23, namely (a) straight slot; (b) L-shape slot; (c) inverted T-shape slot. Reproduced by permission of ©2005 IEEE [27]

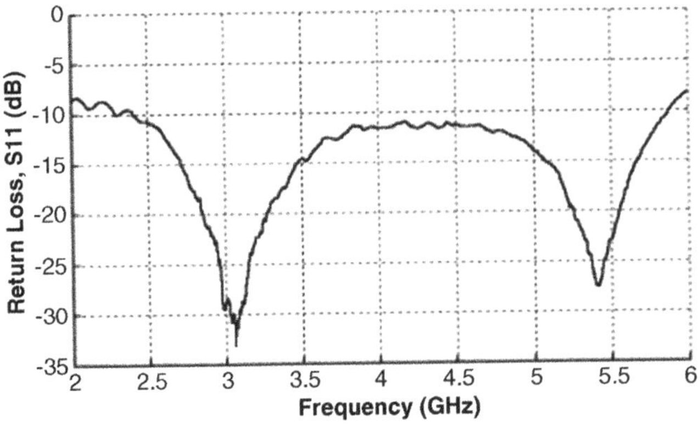

(b)

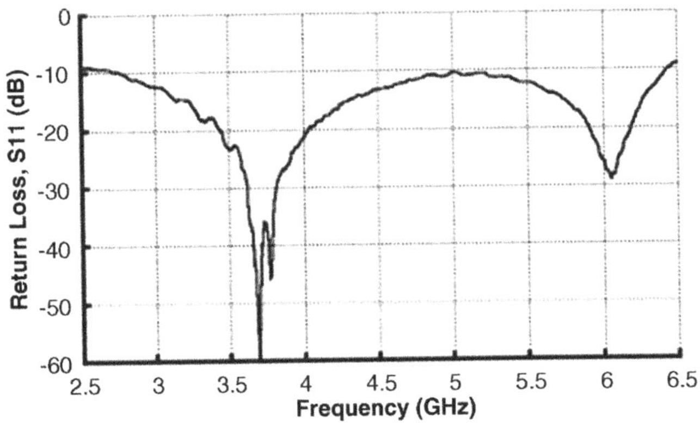

(c)

Figure 8.24 (*Continued*)

8.6.4 Multiband Slot Antennas

As was indicated earlier, narrow band resonant slots can be made using long narrow slots, cut on the ground plane conductors. The space requirements can therefore be very small, which provides an opportunity to design multiband antennas on a single substrate. Such slot antennas can be excited using microstrip lines etched on the surface of the substrate dielectric resulting in a very compact structure. Here, a systematic design approach is presented that generalizes the slot antenna design, and enables the design of a single or any multiband antenna configuration. Another important feature of a narrow-width slot antenna is its flexible shape. The slot can be bent to any shape to accommodate it in the available space, or achieve a desired polarization. Since, in a narrow slot, the polarization of the radiated field is normal to the slot length, the slot shape can be used to alter, or fix, the polarization. For instance, to achieve 45° polarization directions, the slot may be bent orthogonally in its middle point, which is illustrated in the examples shown here [29]. Also, the selected examples are designed for operation in WLAN and WiMax bands, the desired current wireless communication bands.

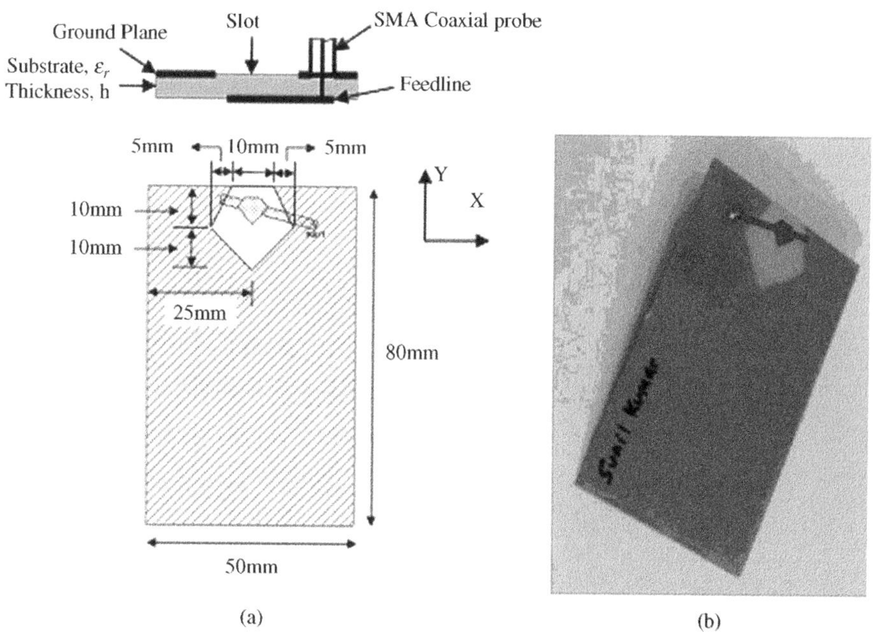

Figure 8.25 UWB pentagon shape microstrip slot antenna fed in using tilted microstrip transmission line with a pentagon shape stub. Reproduced by permission of ©2009 IEEE [28]

Figure 8.28 shows the geometry of a narrow width antennas with three slots etched on its ground plane, which are referred to as Antenna #1, Antenna #2 and Antenna #3. On the lower face of the substrate a microstrip line is etched that excites the slots. To control the polarization, and/or reduce the space requirement, the slots are bent along their length, thus providing orthogonal radiated field components. The resulting radiated field polarization is the vectorial sum of the two orthogonal components. Since the slot impedance varies along its length, its cross-over location with the microstrip line can be selected appropriately, to match its impedance to that of the microstrip line. A single slot antenna, i.e. a single band antenna, is shown in Figure 8.29. Figure 8.29(a) shows the antenna geometry and Figure 8.29(b) gives its return loss plot, which at resonance has its $S_{11} = -37$ dB. This is Antenna #1.The microstrip line is designed with $W_m = 5$ mm, on a substrate with $\varepsilon_r = 2.2$, and thickness h = 1.575 mm. The antenna design parameters are shown in the caption of Figure 8.29(b), which for $S_{11} < -10$ dB has a bandwidth between 2.4 GHz and 2.5 GHz. This provides a bandwidth of about 4%.

Similarly, Figures 8.30(a) and 8.30(b) show the dual-slot (dual-band) antenna geometry (Antenna #2) and its simulated return loss plot, clearly indicating the dual band performance. The design frequencies are 2.45 GHz and 3.5 GHz, and the achieved return losses are $S_{11} = -36$ dB and -31 dB, respectively. The impedance bandwidths ranges are between 2.4 to 2.5 GHz, and 3.4 to 3.6 GHZ. The antenna design parameters are shown in the caption of Figure 8.30(b).

The corresponding antenna geometries and return loss performances, for a tri-band (Antenna #3) and quad-band (Antenna #4) antennas, are shown in Figures 8.31 and 8.32. Again,

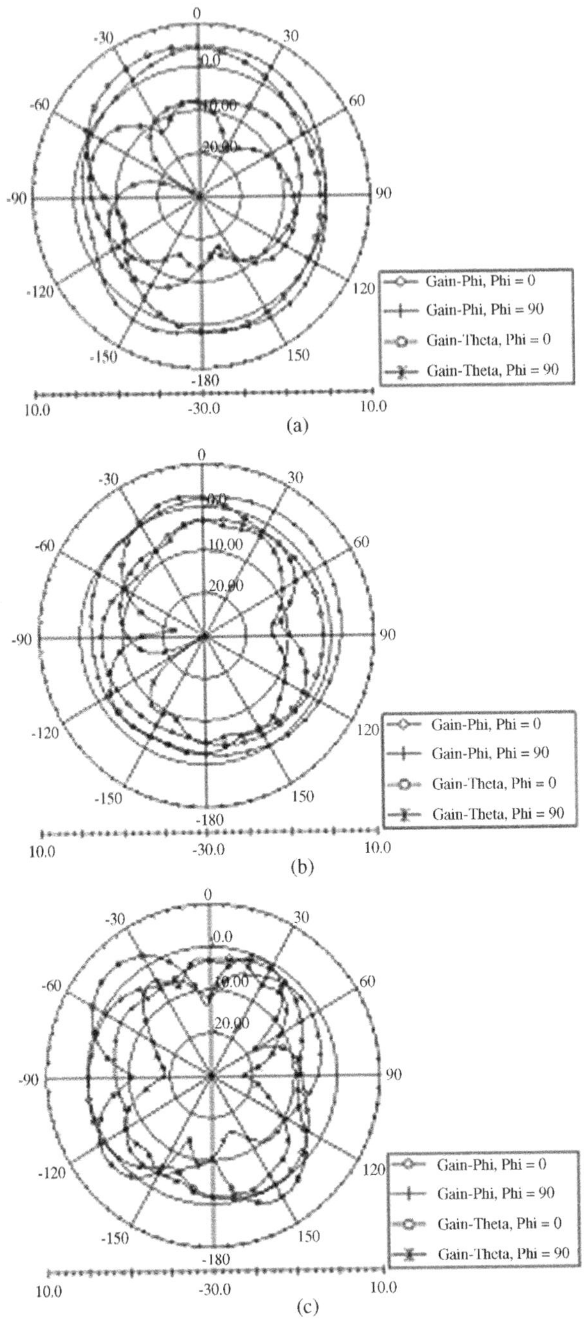

Figure 8.26 Gain radiation patterns of the antenna at (a) 4 GHz; (b) 7 GHz; (c) 10 GHz. Reproduced by permission of ©2009 IEEE [28]

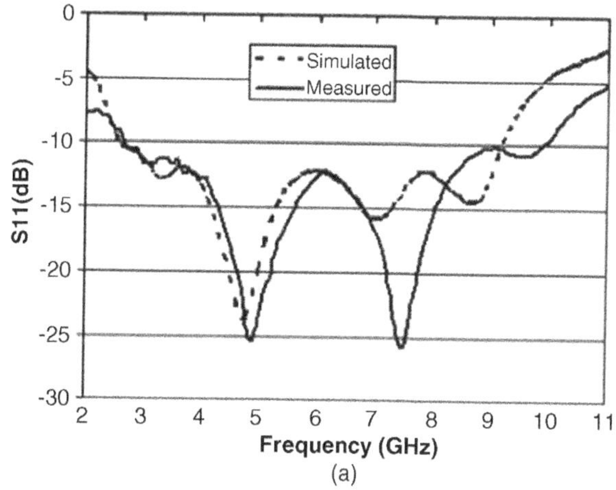

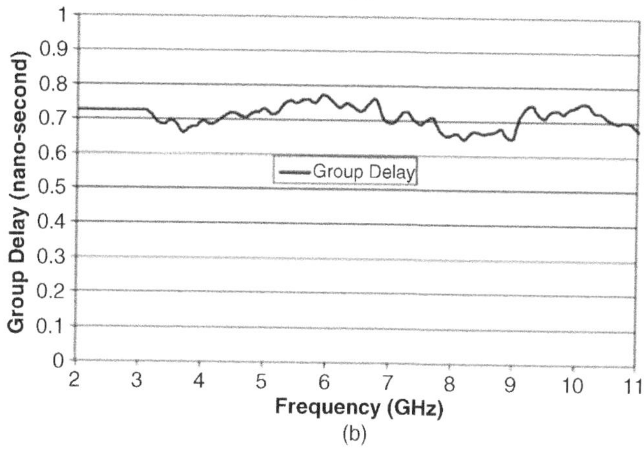

Figure 8.27 (a) Comparison of the simulated and measured reflection coefficients and (b) Measured group delay of the antenna. Reproduced by permission of ©2009 IEEE [28]

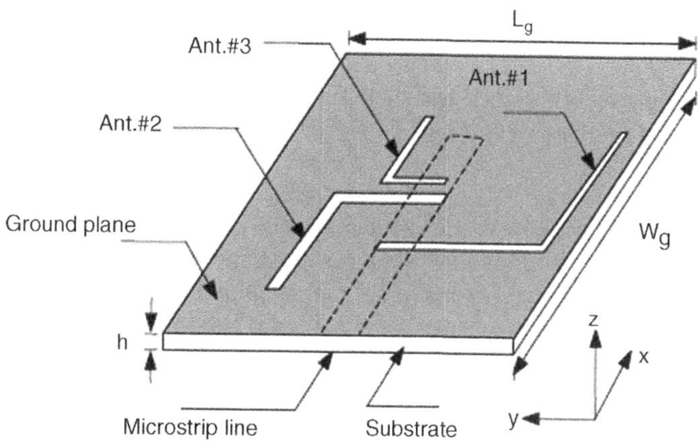

Figure 8.28 Structure of a Tri-band microstrip slot antenna. Reproduced by permission of ©2009 ETRI [29]

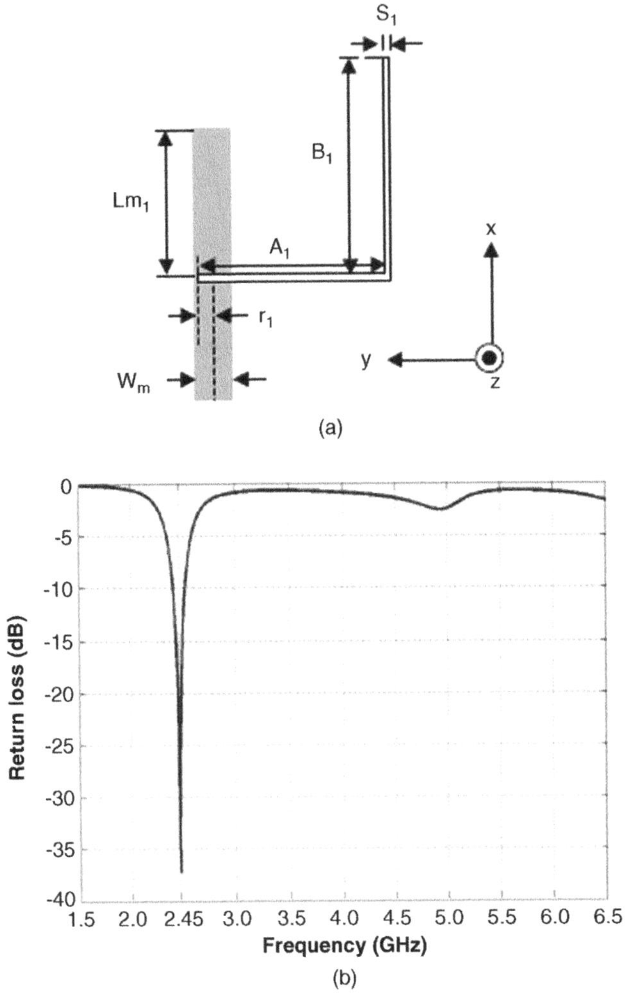

Figure 8.29 (a) Geometry of the single band right angle slot antenna; (b) simulated return loss of the antenna. Reproduced by permission of ©2009 ETRI [29]

Figures 8.31(a) and 8.32(a) show the design geometries, and Figures 8.31(b) and 8.32(b) their corresponding return loss plots. For the tri-band antenna, the design frequencies are 2.45 GHz, 3.5 GHz and 5.8 GHz, providing return losses corresponding to $S_{11} = -38$ dB, -36 dB and -23 dB. Their corresponding impedance bandwidths are 100 MHz (2.4 – 2.5 GHz), 200 MHz (3.4 – 3.6 GHz), and 400 MHz (5.6 – 6.0 GHz), respectively. The obtained impedance bandwidths can cover the 2.4 GHz WLAN band, and the 3.5 GHz WiMax band, and both 5.8 GHz WLAN and WiMax bands. These impedance matching conditions are achieved without modifying the microstrip feed line. In practice, further improved impedance matching can be achieved by using appropriate tuning stubs along the feed line. This final design step is shown below for the quad-band antenna.

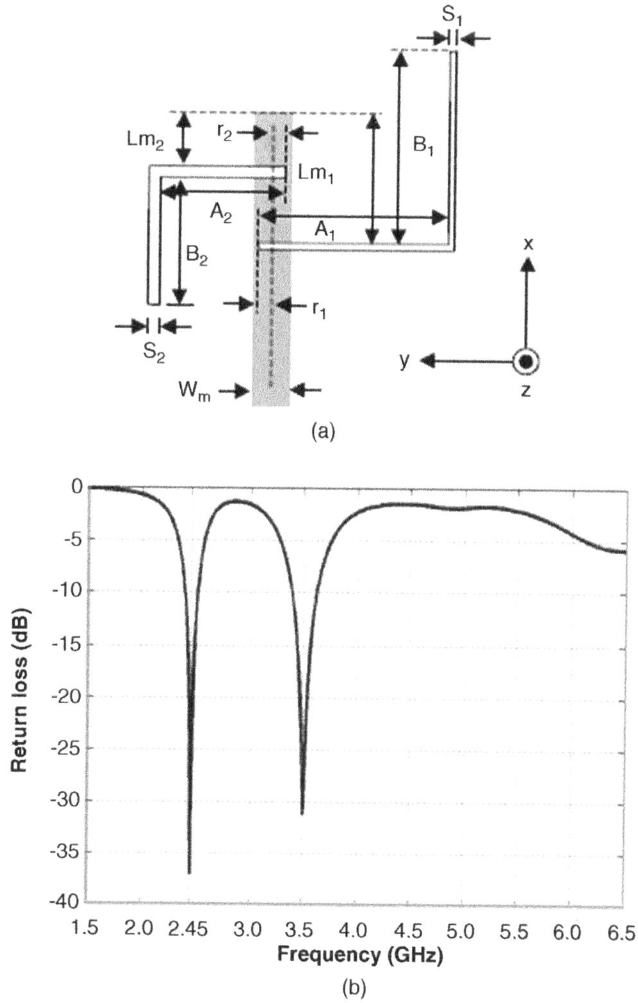

Figure 8.30 (a) Geometry of the dual band right angle slot antenna; (b) simulated return loss of the antenna. Reproduced by permission of ©2009 ETRI [29]

For the quad-band antenna (Antenna #4) the design frequencies are 2.45 GHz, 3.5 GHz, 5.09 GHz, and 5.7 GHz. The achieved return losses are shown in Figure 8.32(b), providing $S_{11} = -39$ dB, -28 dB, -23 dB, and -25 dB, respectively, with bandwidths of 100 MHz, 200 MHz, 500 MHz, and 310 MHz. The antenna design parameters are shown in the caption of Figure 8.32(b). Again, these results are obtained without the use of tuning stubs. An example of a design with final impedance tuning, using two matching stubs on the feed line of the quad-band antenna, is shown in Figure 8.33(a), which also shows its design parameters. Its return loss performance is shown in Figure 8.33(b). The design frequencies are 2.45 GHz, 3.5 GHz, 5.07 GHz, and 5.88 GHz, and the achieved return losses are -46.1 dB, -39.1 dB, -32.4 dB, and -37.4 dB, respectively. Their corresponding bandwidths are 100 MHz,

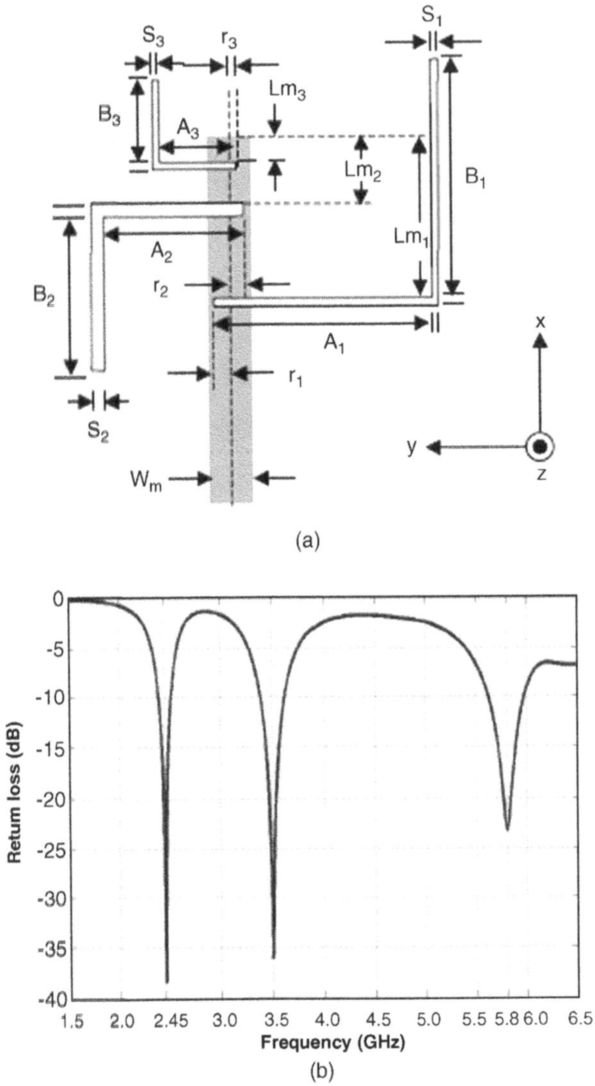

Figure 8.31 (a) Geometry of the tri-band right angle slot antenna with tuning stub; (b) simulated return loss of the antenna. Reproduced by permission of ©2009 ETRI [29]

200 MHz, 630 MHz, and 440 MHz. Figure 8.33(b) also compares the measured S_{11} performances with the simulated data. The agreement is satisfactory, which improves at the first two and the last frequencies, and slights shifts in the third frequency band. The measured radiation patterns of this antenna, in all four bands, are shown in Figure 8.34. The gain patterns are all satisfactory, providing nearly omni-directional patterns, with low cross polarizations. The peak gains of this antenna are 2.8 dBi, 3.4 dBi, 5.3 dBi and 3.9 dBi at 2.26 GHz, 3.5 GHz, 4.9 GHz, and 5.88 GHz.

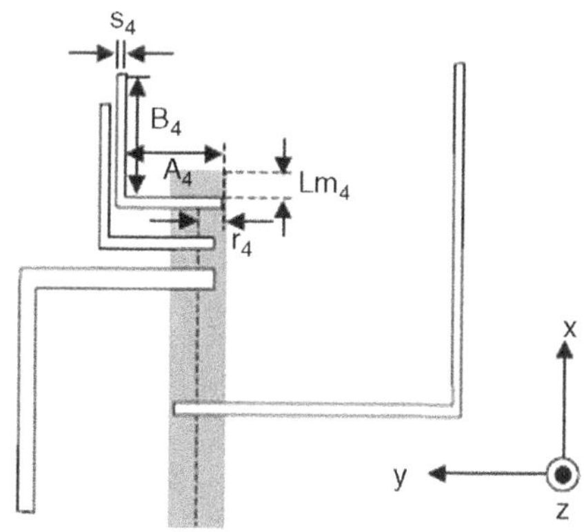

(a)

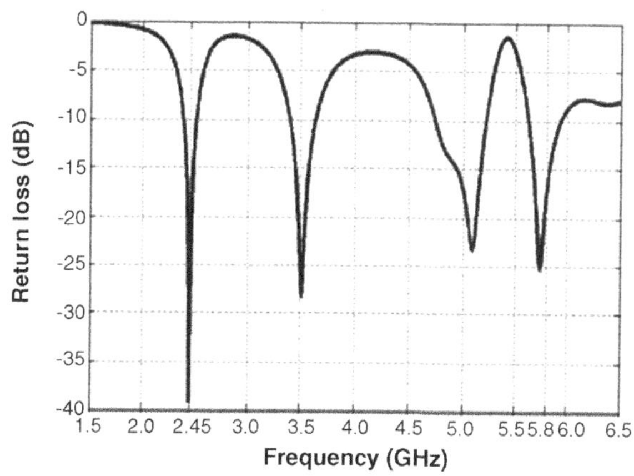

(b)

Figure 8.32 (a) Geometry of the quad-band right angle slot antenna: (b) simulated return loss of the antenna. Reproduced by permission of ©2009 ETRI [29]

8.6.5 Differential Dual-Frequency Slot Antennas

In recent years, increasing competition in the wireless communications market has generated the need for fully integrated radio frequency front-end solutions for which differential signals are preferable. Conventional antennas are single port devices, hence are difficult for differential signal implementations. It is preferable to consider such antennas as a two-port device in the design stage. Here, an example is provided, which operates in the differential mode, and the two wireless bands at 2.51 GHz and 5.38 GHz [30]. The antenna configuration is shown in Figure 8.35, which is a dual layer structure. It consists of a microstrip patch, split in two to enhance its gain, and etched on the upper surface of the upper substrate. The lower layer has a microstrip line on the upper side and a ground plane on the lower. The microstrip line supports two inputs, at its two ends, and a dog-bone slot is etched in the ground plane, orthogonal to the microstrip line. The substrate parameters are: $\varepsilon_{r1} = \varepsilon_{r2} = 4.4$, $h_1 = 1.6$ mm,

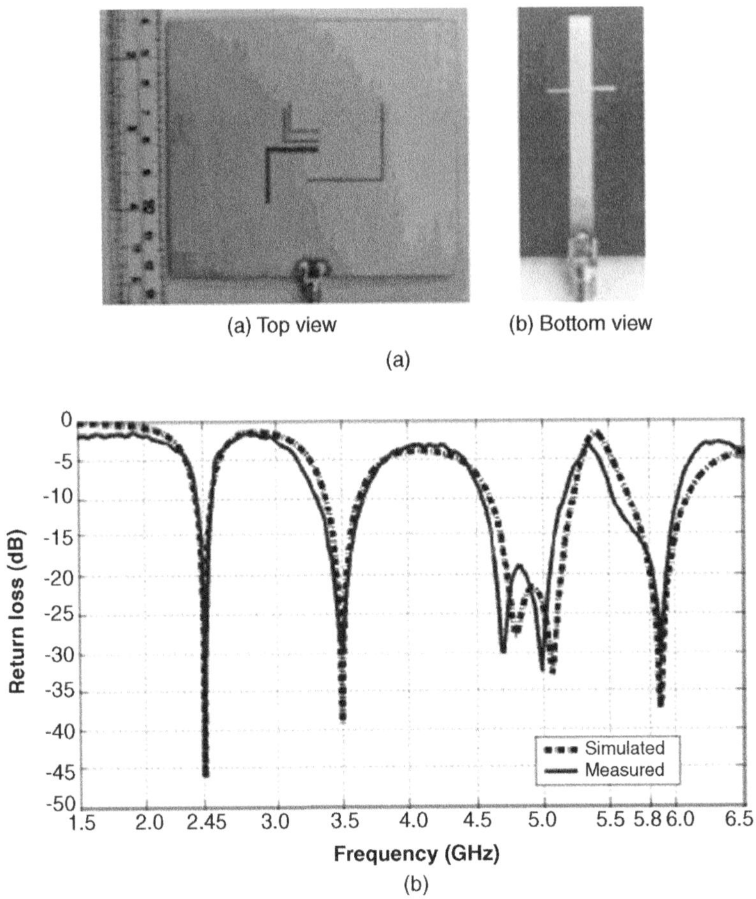

Figure 8.33 (a) Photograph of the quad-band right angle slot antenna with double tuning screw; (b) simulated and measured return loss of the antenna. Reproduced by permission of ©2009 ETRI [29]

$h_2 = 0.8$ mm, $w = 30$ mm, $l = 26$ mm, $f_l = 34$ mm, $f_w = 1.4$ mm, $S_l = 2$ mm, $s_w = 18$ mm, and $s_p = 1.4$ mm, respectively. The upper and lower photos of the assembled antenna are shown in Figure 8.36.

For the normal differential mode antenna, the patch is usually operated as a half wavelength antenna. The introduction of the near resonant slot generates a second resonant frequency, and also causes the back radiation needed to generate the omni-directional radiation patterns. For the selected dimensions, the influence of the slot and patch spacing on the resonant frequencies is shown in Figures 8.37 and 8.38. Increasing the slot length decreases both resonant frequencies. However, the patch spacing does not affect the lower resonant frequency, but increases the upper one. The measured performance of the antenna is compared with the simulation in Figure 8.39. The measured bandwidths are 4.4% at 2.51 GHz, and 3.1% at 5.38 GHz. Also, there is about 2.4% and 2.5% frequency shifts at the lower and upper frequencies, which can be attributed to the tolerances, and the gap

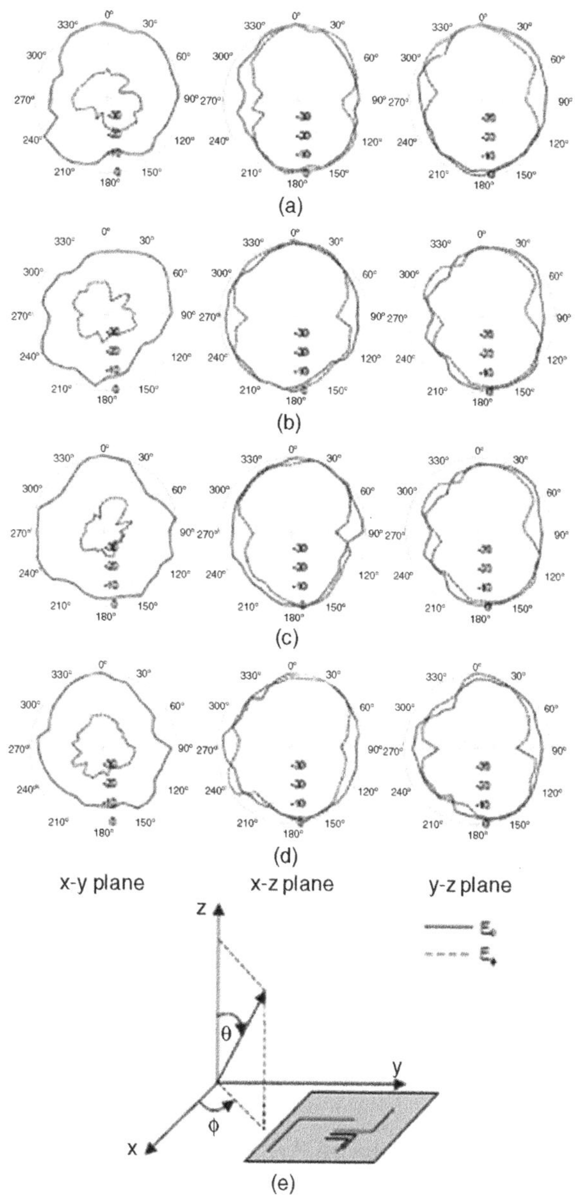

Figure 8.34 Measured radiation patterns for the quad-band antenna: (a) $f_1 = 2.46$ GHz; (b) $f_2 = 3.5$ GHz; (c) $f_3 = 4.9$ GHz; (d) $f_4 = 5.88$ GHz; (e) polar coordinates. Reproduced by permission of ©2009 ETRI [29]

between the two substrates, in the experimental model. The measured radiation patterns are shown in Figure 8.40, and the measured gain in Figure 8.41. The radiation patterns are nearly omni-directional, and the peak gains are 2.06 dBi and 3.42 dBi, respectively, at the lower and upper end frequencies.

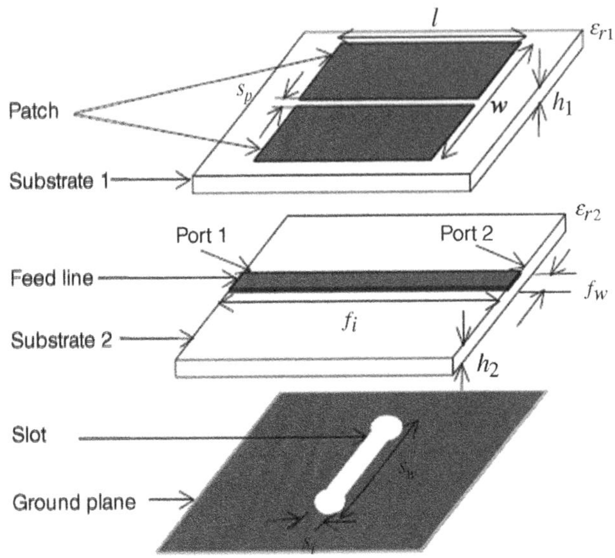

Figure 8.35 Antenna configuration of the differential dual frequency antenna. Reproduced by permission of ©2009 ETRI [30]

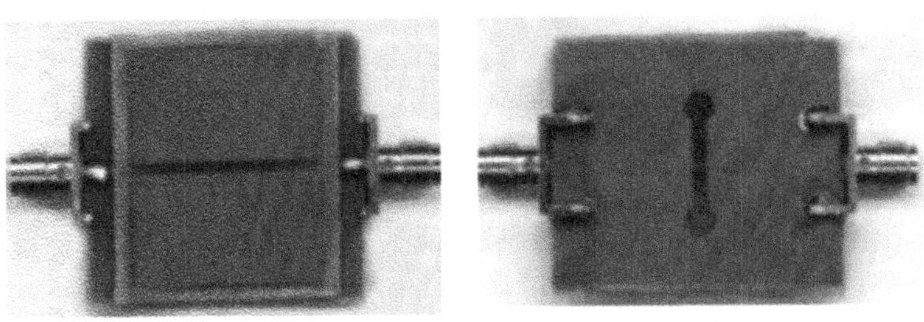

(a) (b)

Figure 8.36 Photos of the differential dual frequency antenna: (a) top view: (b) bottom view. Reproduced by permission of ©2009 ETRI [30]

8.7 Microstrip Planar Monopole Antenna

In this section, planar monopole antennas are discussed, which can also be fabricated using the printed circuit technology. Both multiband and ultra-wide bandwidth performance antennas have been designed using this type of antennas, and are suitable for use in the wireless communication devices. Printed UWB antennas are dealt in this book more elaborately in Chapter 10. In [31], the authors presented a compact multiband planar monopole antenna design, which provides coverage for 2.4 GHz and 5 GHz WLAN communication bands. The antenna radiates omni-directional radiation patterns. The antenna geometry is shown in Figure 8.42(a), which used a microstrip transmission line to excite the antenna on a FR-4 substrate material. Figure 8.42(b) shows the simulated and measured impedance match results as the VSWR plot. It

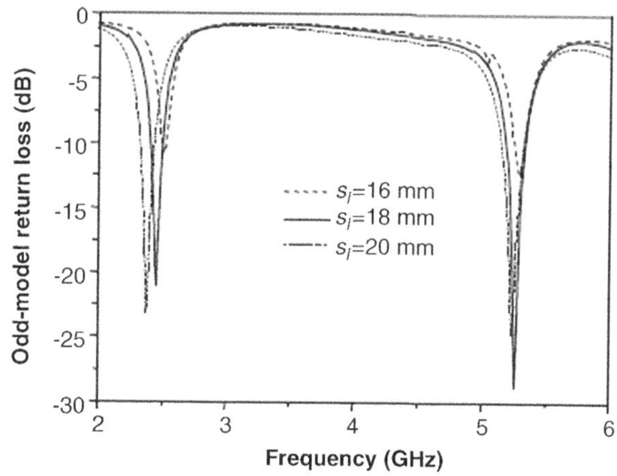

Figure 8.37 Simulated effect of slot length on odd-mode return loss. Reproduced by permission of ©2009 ETRI [30]

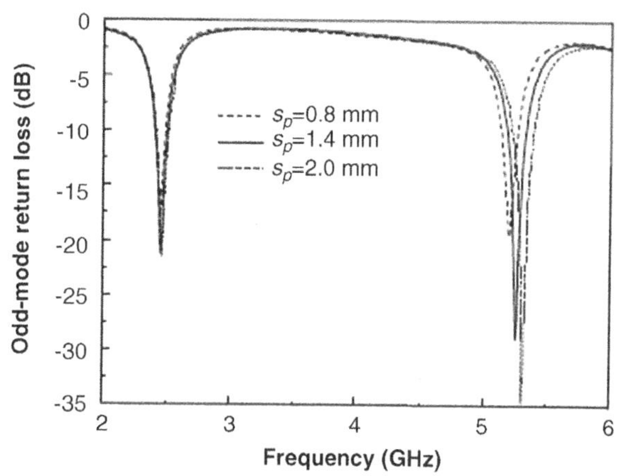

Figure 8.38 Simulated effect of patch spacing on odd-mode return loss. Reproduced by permission of ©2009 ETRI [30]

is evident that the antenna gives frequency coverage for the DCS (1710–1880 MHz), PCS (1850–1990 MHz), UMTS (1920–2170 MHz), 2.4-GHz WLAN (IEEE 802.11b in the US: 2400–2484 MHz), and 5-GHz WLAN (HIPERLAN/2 in Europe: 5150–5350/5470–5725 MHz and IEEE 802.11a in the US: 5150–5350/5725–5825 MHz) wireless applications. Figure 8.43 shows the measured radiation patterns of the antenna at 1.9 GHz, 2.4 GHz, 5.2 GHz and 5.8 GHz and confirms acceptable radiation patterns for these applications.

Next, planar monopole antennas are presented which have elliptical and circular shapes as shown in Figure 8.44(a), and provide the ultra-wide bandwidth (UWB) performance [32]. The four types (circular and elliptical of E and EC type) of the UWB printed monopole antennas are manufactured using the Rogers RT6010LM substrate with thickness of 0.64 mm. The measured

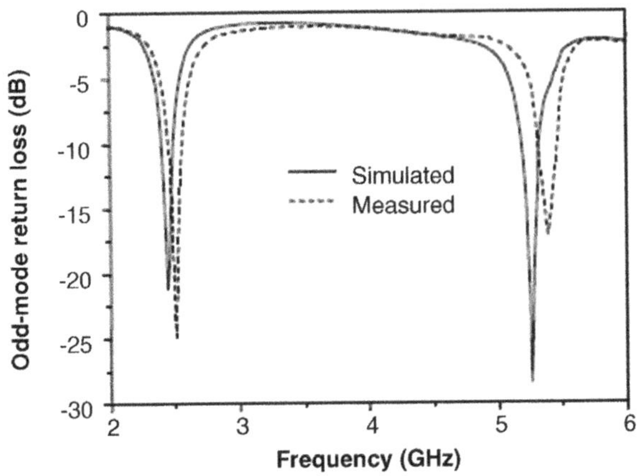

Figure 8.39 Simulated and measured odd-mode return loss. Reproduced by permission of ©2009 ETRI [30]

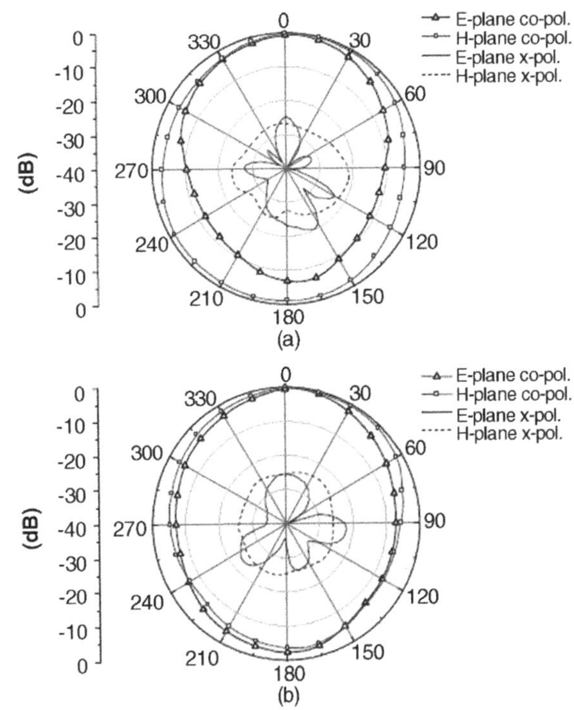

Figure 8.40 Measured radiation pattern of the antenna: (a) at lower band; (b) at upper band. Reproduced by permission of ©2009 ETRI [30]

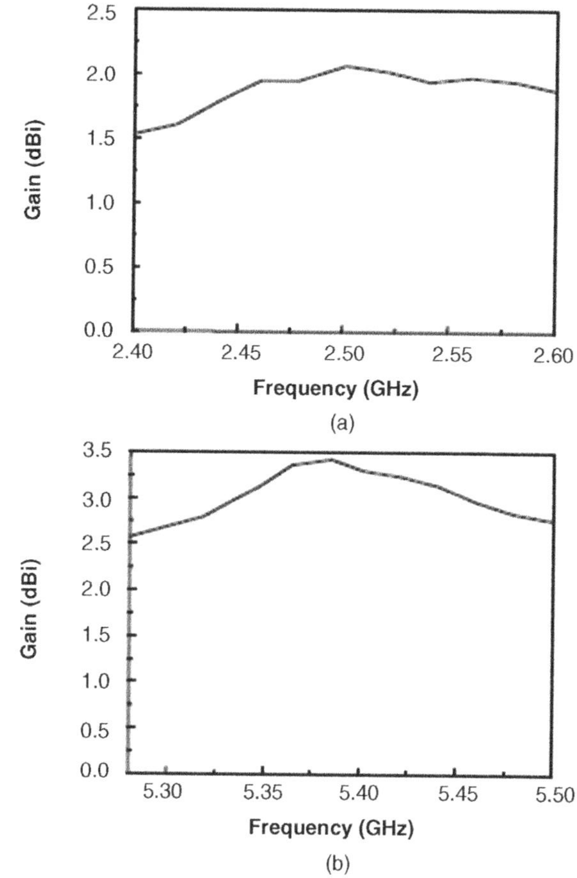

Figure 8.41 Measured gain of the antenna: (a) at lower band; (b) at upper band. Reproduced by permission of ©2009 ETRI [30]

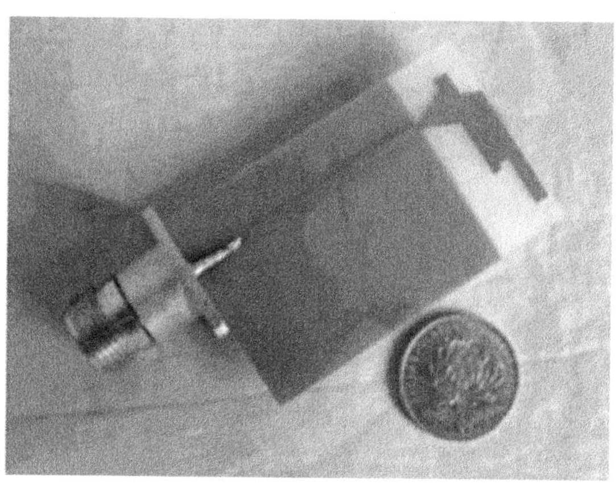

(a)

Figure 8.42 (a) Prototype of the proposed multiband antenna; (b) simulated and measured VSWRs for the multiband antenna. Standard bandwidth requirements are shown in gray. Reproduced by permission of ©2006 IEEE [31]

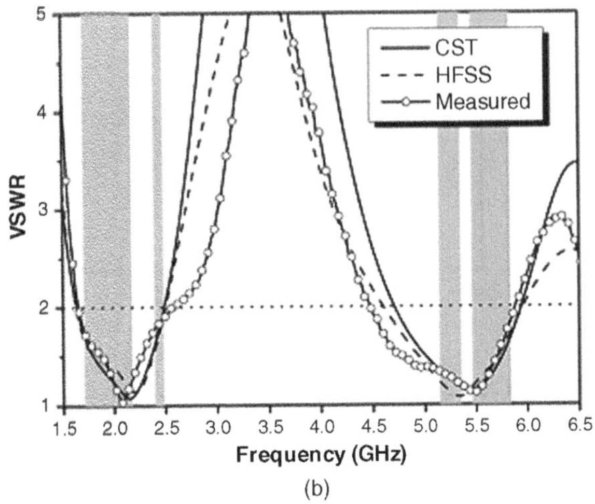

Figure 8.42 *(Continued)*

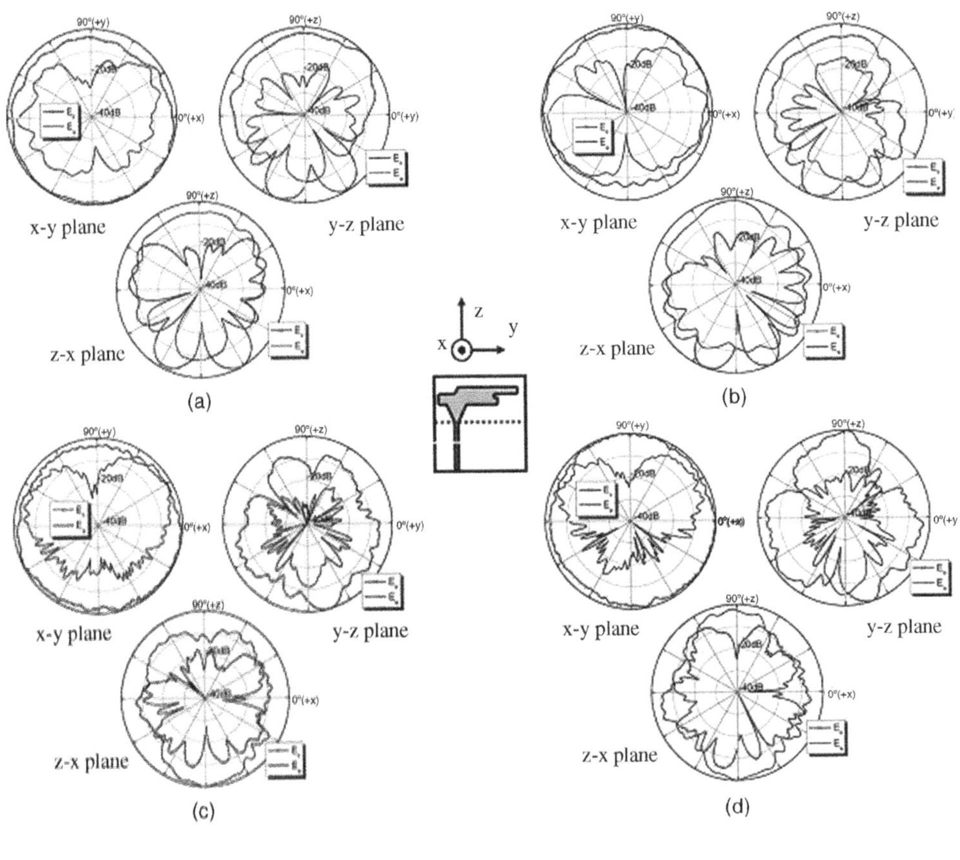

Figure 8.43 Measured radiation patterns for the multiband antenna at (a) 1.9 GHz; (b) 2.4 GHz; (c) 5.2 GHz; (d) 5.8 GHz. Reproduced by permission of ©2006 IEEE [31]

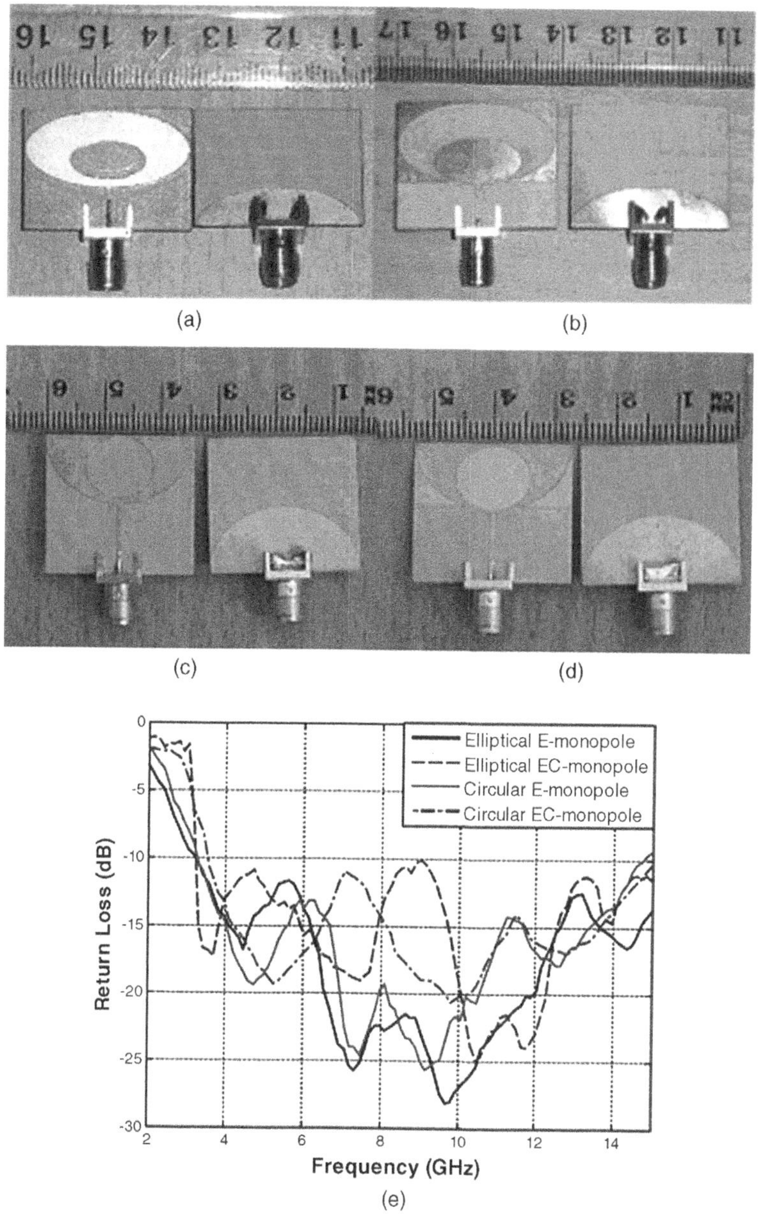

Figure 8.44 (a) Photographs of the manufactured elliptical E-monopole; (b) elliptical EC-monopole; (c) circular E-monopole; (d) the circular EC-monopole; (e) variation of measured return losses with frequency for the designed antennas. Reproduced by permission of ©2008 IEEE [32]

impedance match results are presented in Figure 8.44(b). The results indicate that the four types of antennas feature UWB behavior with a bandwidth from 3.1 GHz to more than 15 GHz. The radiation patterns are shown in Figure 8.45, which show omni-directional radiation patterns over the bandwidth at 3 GHz, 6 GHz, and 9 GHz, respectively.

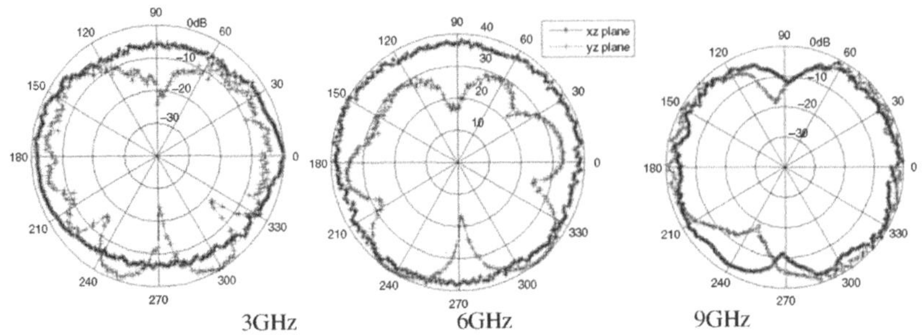

Figure 8.45 Measured radiation patterns for the E-monopole antenna at different frequencies. Reproduced by permission of ©2008 IEEE [32]

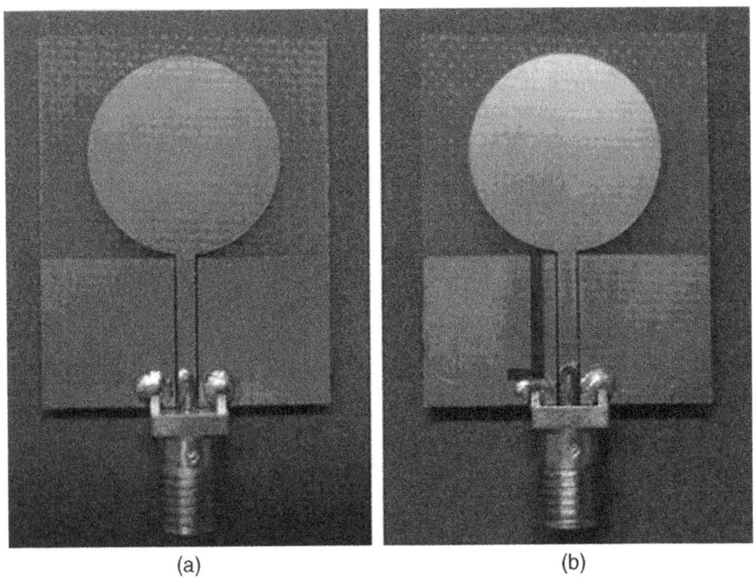

Figure 8.46 Photographs of (a) the reference circular disc monopole antenna; (b) the proposed defected-ground-plane monopole antenna; (c) simulated surface current distribution on the conductors of the defected ground-plane monopole antenna at 2.7 GHz from Ansoft HFSS. Reproduced by permission of ©2008 IEEE [33]

Another planar monopole antenna is presented next which shows UWB performance but with a notch band and is aimed at filtering some of the high power signals in the UWB communication range [33]. The compact multiband antenna requiring size of 24 mm $\times$ 28.3 mm consists of a printed circular disc monopole antenna with an L-shaped slot cut in the ground plane, forming a defected ground plane structure. The antenna geometry with planar monopole alone and planar monopole with a slit are shown in Figures 8.46(a) and (b). The current distribution on the antenna, as shown in Figure 8.46(c), reveals that at low frequencies the addition of the slot creates two orthogonal current paths, which are responsible for two

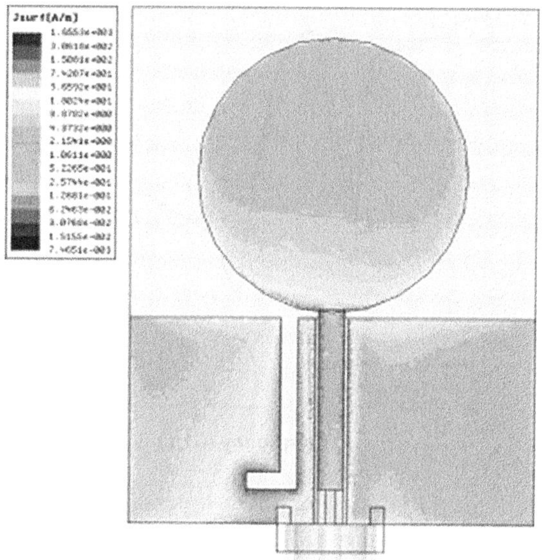

(c)

Figure 8.46 *(Continued)*

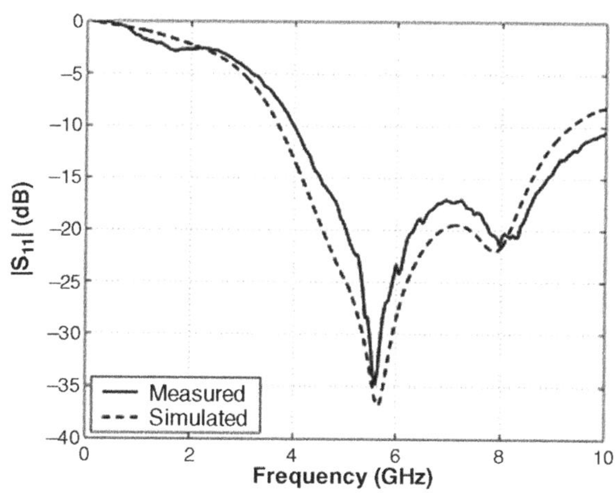

(a)

Figure 8.47 (a) Measured and simulated return losses for the reference circular disc monopole antenna shown in Figure 8.46(a); (b) measured and simulated return losses for the defected-ground-plane monopole antenna shown in Figure 8.46(b). Reproduced by permission of ©2008 IEEE [33]

additional resonances. These resonances in the antenna exhibit orthogonal pattern diversity, while enabling the adjacent resonances to be merged, forming a wideband low-frequency response and maintaining the inherent wideband high-frequency response of the monopole. The antenna exhibits a measured impedance bandwidth of 600 MHz from 2.68 GHz to

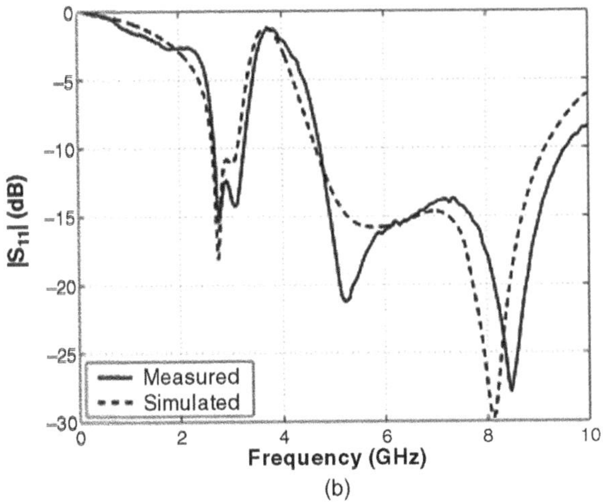

Figure 8.47 (*Continued*)

3.28 GHz, and a bandwidth of 4.84 GHz from 4.74 GHz to 9.58 GHz, as shown in Figure 8.47, for both with and without the defected structure designs, respectively.

References

1. C. A. Balanis, *Modern Antenna Handbook*, John Wiley & Sons, Inc., New York, 2008.
2. C. A. Balanis, *Antenna Theory and Design*, 3rd edition, John Wiley & Sons, Inc., New York, 2004.
3. J. L. Volakis, *Antenna Engineering Handbook*, 4th edition, McGraw-Hill Company, Maidenhead, 2007.
4. T. A. Milligan, *Modern Antenna Design*, 2nd edition, John Wiley & Sons, Inc., New York, 2005.
5. R. Garg, P. Bhartia, I. Bahl and I. Ittipiboon, *Microstrip Antenna Design Handbook*, Artech House, Boston, 2001.
6. G. Kumar and K. P. Ray, *Broadband Microstrip Antennas*, Artech House, Boston, 2003.
7. Fan Yang, Xue-Xia Zhang, Xiaoning Ye, and Yahya Rahmat-Samii, "Wide-band E-shaped patch antennas for wireless communications," *IEEE Transactions on Antennas and Propagation*, vol. 49, no. 7, July 2001.
8. N. Jin and Y. Rahmat-Samii, "Parallel particle swamp optimization and finite-difference time-domain (PSO/FDTD) algorithm for multiband and wide-band patch antenna designs," *Trans. IEEE Antennas Propagation*, vol. 53, no. 11, pp. 3459–3468, Nov. 2005.
9. K. F. Lee, et al., "Experimental and simulation studies of coaxially-fed U-slot rectangular patch antenna," *IEE Proc. Microwave, Antennas Propagation*, vol. 144, pp. 354–358, 1997.
10. Kin-Fai Tong, Kwai-Man Luk, Kai-Fong Lee and Richard Q. Lee, "A broad-band U-slot rectangular patch antenna on a microwave substrate," *IEEE Transactions on Antennas and Propagation*, vol. 48, no. 6, June 2000.
11. S. Weigand, G. H. Huff, K. H. Pan and J. T. Bernhard, "Analysis and design of broad-band single-layer rectangular U-slot microstrip patch antennas," *IEEE Transactions on Antennas and Propagation*, vol. 51, no. 3, March 2003.
12. S. K. Sharma and L. Shafai, "Performance of a novel Ψ-shape microstrip patch antenna with widebandwidth," *IEEE Antennas and Wireless Propagation Letters*, 2009.
13. Ching-Hong K. Chin, Quan Xue and Hang Wong, "Broadband patch antenna with a folded plate pair as a differential feeding scheme," *IEEE Transactions on Antennas and Propagation*, vol. 55, no. 9, September 2007.
14. P. Li, H. W. Lai, K. M. Luk and K. L. Lau, "A wideband patch antenna with cross-polarization suppression," *IEEE Antennas and Wireless Propagation Letters*, vol. 3, 2004.
15. Hau-Wah Lai and Kwai-Man Luk, "Design and study of wide-band patch antenna fed by meandering probe," *IEEE Transactions on Antennas and Propagation*, vol. 54, no. 2, February 2006.

16. D. Qu, L. Shafai and A. Foroozesh, "Improving microstrip patch antenna performance using EBG substrates," *IET Microwaves, Antennas and Propagation*, vol. 153, no. 6, pp. 558–563, December 2006.
17. A. Foroozesh and L. Shafai, "Performance enhancement of the compact microstrip antennas using AMC ground planes," *IEEE International Symposium on Antennas and Propagation and the USNC/URSI National Radio Science Meeting, APS/URSI*, Charleston, SC, June 1–5, 2009.
18. Cheng-Tse Lee and Kin-Lu Wong, "Uniplanar printed coupled-fed PIFA with a band-notching slit for WLAN/ WiMAX operation in the laptop computer," *IEEE Transactions on Antennas and Propagation*, vol. 57, no. 4, April 2009.
19. Rafał Głogowski and Custódio Peixeiro, "Multiple printed antennas for integration into small multistandard handsets," *IEEE Antennas and Wireless Propagation Letters*, vol. 7, 2008.
20. Y. Ge, K. P. Esselle and T. S. Bird, "E-shaped patch antennas for high-speed Wireless networks," *IEEE Transactions on Antennas and Propagation*, vol. 52, no. 12, December 2004.
21. C. Meagher and S. K. Sharma, "Investigations on an aperture-coupled microstrip patch antenna with a spaced dielectric cover for enhanced gain performance," *13th International Symposium on Antenna Technology and Applied Electromagnetics and the Canadian Radio Sciences Meeting 2009*, Banff, Canada, Feb. 2009.
22. C. Meagher and S. K. Sharma,"A wideband aperture-coupled microstrip patch antenna employing spaced dielectric cover for enhanced gain performance," *IEEE Trans Antennas and Propagation*, vol. 58, no. 9, September 2010.
23. A. Foroozesh and L. Shafai, "Modeling, simulation, design and fabrication of a high-gain cavity resonance antenna having a highly-reflective patch-type FSS superstrate," *13th International Symposium on Antenna Technology and Applied Electromagnetics and the Canadian Radio Sciences Meeting 2009*, Banff, Canada, Feb. 2009.
24. A. K. Shackelford, Kai-Fong Lee, and K. M. Luk, "Design of small-size wide-bandwidth microstrip patch antennas," *IEEE Antennas and Propagation Magazine*, vol. 45, no. 1, February 2003.
25. R. Chair, Chi-Lun Mak, Kai-Fong Lee, Kwai-Man Luk, and A. A. Kishk, "Miniature wide-band half U-slot and half E-shaped patch antennas," *IEEE Transactions on Antennas and Propagation*, vol. 53, no. 8, August 2005.
26. S. K. Sharma, L. Shafai and N. Jacob, "Investigation of wide-band microstrip slot antenna," *IEEE Transactions on Antennas and Propagation*, vol. 52, no. 3, March 2004.
27. S. I. Latif, L. Shafai and S. K. Sharma, "Bandwidth enhancement and size reduction of microstrip slot antennas," *IEEE Transactions on Antennas and Propagation*, vol. 53, no. 3, March 2005.
28. S. K. Rajgopal and S. K. Sharma, "Investigations on ultrawideband pentagon shape microstrip slot antenna for wireless communications," *IEEE Transactions on Antennas and Propagation*, vol. 57, no. 5, May 2009.
29. P. Rakluea, N. Anantrasirichai, K. Janchitrapongvei and T. Wakabayashi, "Multiband microstrip-fed right angle slot antenna design for wireless communication systems," *ETRI Journal*, vol. 31, no. 3, pp. 271–281, June 2009.
30. L. Han, W. Zhang, G. Han, R. Ma and L. Li, "Differential dual-frequency antenna for wireless communication," *ETRI Journal*, vol. 30, no. 6, pp. 877–879, June 2008.
31. Sheng-Bing Chen, Yong-Chang Jiao, W. Wang and Fu-Shun Zhang, "Modified T-shaped planar monopole antennas for multiband operation," *IEEE Transactions on Microwave Theory and Techniques*, vol. 54, no. 8, August 2006.
32. A. M. Abbosh and M. E. Bialkowski, "Design of ultrawideband planar monopole antennas of circular and elliptical shape," *IEEE Transactions on Antennas and Propagation*, vol. 56, no. 1, January 2008.
33. M. A. Antoniades and G. V. Eleftheriades, "A compact multiband monopole antenna with a defected ground plane," *IEEE Antennas and Wireless Propagation Letters*, vol. 7, 2008.

9

UHF Passive RFID Tag Antennas

Daniel Deavours[1] and Daniel Dobkin[2]
[1]*University of Kansas, USA*
[2]*Enigmatics, USA*

9.1 Introduction

Radio Frequency Identification (RFID) is the use of wireless communications to identify a physical object, generally without direct human intervention. RFID is widely used in a variety of fields. An RFID system is generally held to consist of an interrogator or reader, and one or more transponders or tags, which are generally physically attached to the object or objects to be identified. The tag sends a radio signal which is received by the reader, allowing it to extract a (generally unique) identifying number from the tag and thus identify the object to which the tag is attached.

RFID systems can be categorized by the frequency of operation, power source, and communications protocol in use. Most RFID systems operate in either the low frequency region (30–300 kHz), the high frequency band (3 MHz–30 MHz), or the ultra-high-frequency band (300 MHz–3 GHz). These bands are often referred to by their acronyms, LF, HF, and UHF respectively. LF and HF systems are generally inductively coupled and substantially non-radiating; the distance between reader and tag at which a tag can be read (the read range) is roughly comparable to the size of the reader antenna. UHF systems usually employ conventional radiative coupling, with read range determined by the conventional inverse square law in the case where propagation is unobstructed.

RFID tags can be active, passive, or something in between: the something in between is variously known as semi-passive, semi-active, or battery-assisted. An active tag is a conventional radio having its own power supply (typically a battery) and a transmitter. A receiver may also be included if the tag is to accept commands. In some cases, the receiver may operate at one frequency (e.g. in the LF band) while the transmitter operates at a different frequency (e.g. in the UHF band). The range of an active tag may extend to hundreds of meters with sufficient tag transmit power, although battery life, transmit power, and frequency of transmission must be traded off.

Microstrip and Printed Antennas: New Trends, Techniques and Applications. Edited by Debatosh Guha and Yahia M.M. Antar

A passive tag has no radio transmitter and no local power source; both the transmission of information to the reader and power required by the tag's circuitry (if any) are supplied by the incident radio signal from the reader. By varying the amount of power reflected back to the transmitter (backscatter modulation), a signal is sent from the tag to the reader. Passive tags using conventional integrated circuits (ICs) require incident RF power from 10–20 microwatts to a few milliwatts. Since this power level is greatly in excess of the power levels required for a conventional radio receiver to operate, passive tags are very short range compared to other radio receivers for comparable transmitting power. A passive tag operating in the UHF band typically can be read when it is 5–15 meters from a reader transmitting one watt; the comparable range for an IEEE 802.11 link operating at the same power levels and much higher data rates is hundreds of meters in unobstructed operation. Passive tags are of interest for many applications because of their low cost (potentially as little as $0.10 per tag or less), but their performance is limited by the need for large incident power.

Semi-passive tags normally include a battery used to power the IC circuitry, but do not include a radio transmitter. Backscatter modulation is used to transmit data from the tag to the reader. Because they do not need to receive the large amount of RF power needed to operate an integrated circuit, semi-passive tags have substantially longer read range than passive tags, typically varying from several tens to over a hundred meters in unobstructed operation.

Because development of RFID systems has occurred over many decades, undertaken by differing organizations focused on different applications, a variety of communications protocols exist for specifying the transport of data between tags and readers. Most of these protocols are mutually incompatible at all relevant protocol layers. However, in recent years, UHF passive tags using the EPCglobal Class 1 Generation 2 protocol, substantially identical to ISO 18000-6C, have been widely deployed.

UHF passive tags place unusual requirements on the associated antenna structures, and thus their design and fabrication involve specialized considerations not often encountered in conventional printed antennas. These considerations are the subject of the remainder of this chapter.

9.2 Application Requirements

9.2.1 Electrically Small

RFID nearly always operates in unlicensed bands in the United States, or specific allocated bands in other jurisdictions (Figure 9.1). The 2.4–2.48 GHz band, or substantial subsets thereof, is available for unlicensed use in most areas of the world, but is subject to interference from – and provides interference to – wireless local area networks, microwave ovens, cordless phones, and other sources. Most long-range RFID systems operate in the 860–960 MHz region. The United States Federal Communications Commission allows unlicensed operation in the 902–928 MHz band; European Union nations usually permit operation in the band 865–868 MHz. RFID operation in Japan is allowed at 952–954 MHz. Most other nations use subsets of FCC-like or European-like bands. Thus, most UHF tag designs must operate in the 900 MHz region, where the wavelength is about 33 cm, and a resonant dipole is about 15–16 cm long. In addition, to operate worldwide, an operating bandwidth of about 100 MHz or 11% is required.

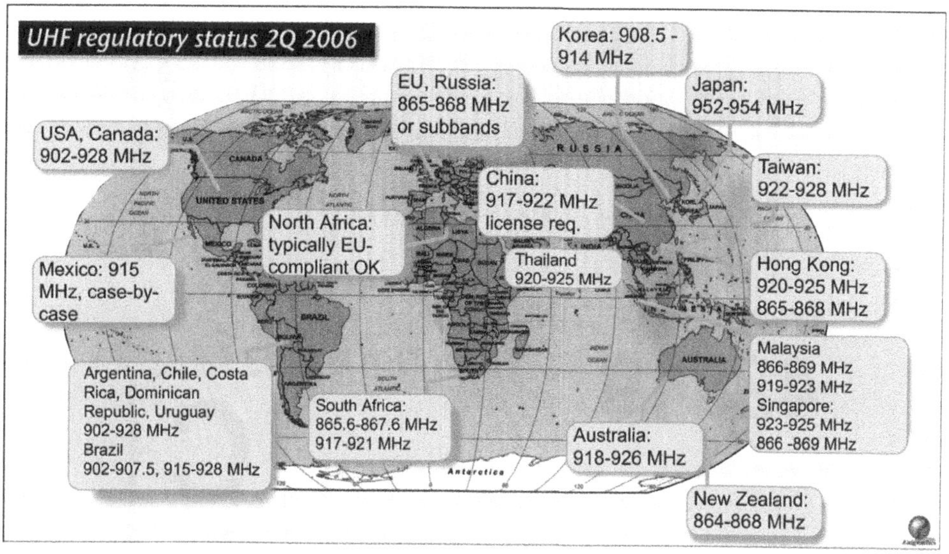

Figure 9.1 Regulatory allocation for RFID operation in the UHF bands. Map courtesy of the University of Texas at Austin Libraries. Data sourced from Yee 2004, Accenture 2004 and EPC 2006

Very few volume applications can use half-wave resonant antennas. Adhesive labels for palletized cardboard shipping boxes are generally constrained to a width of 100 mm (4 inches), roughly 1/3 of a wavelength (Figure 9.2). Still smaller labels are used to mark individual item containers. Thus, it is necessary to construct antennas with reasonable efficiency and bandwidth that are significantly smaller than the resonant size.

9.2.2 *Variable and Uncontrollable Dielectric Environment*

Many tags are embedded in adhesive labels intended to be attached to cardboard containers. Cardboard is mostly air and has little effect on antenna behavior. However, the contents of the box may not be so innocuous. Products containing aqueous solutions packed in plastic or glass containers form a substantial subset of consumer products. Pure water has a dielectric constant of about 80, decreasing modestly as less-polar constituents are added. Pure water is also moderately lossy at microwave frequencies, with loss increasing rapidly with the addition of salts. Thus, tag antennas are expected to function in close proximity to very permittive and lossy dielectrics.

Conductive packaging is also very common in consumer product transport. In addition to metallic cans and containers, metallized plastic is widely used to extend the life of perishable commodities by controlling oxygen ingress, and acts as a conductive ground plane at 900 MHz. Metal near an RFID tag is particularly troublesome since it affects the antenna performance in rather profound ways, which are discussed in more detail in various sections below.

Because of their use as labels for objects that are to be packaged and shipped, it is often the case that RFID tags must be very thin, typically < 1 mm at their greatest thickness. A thin tag will not be torn off or damaged during handling. Unfortunately, this means that the antenna

Figure 9.2 An RFID tag embedded in an adhesive label

structures are necessarily in very close proximity to disturbing materials within a case or package. Conventional microstrip (patch) antennas, which are one common approach for constructing an antenna over a ground plane, are constrained to use very thin substrates, impacting achievable bandwidth.

9.2.3 Variable Orientation

In many applications, the RFID tag may be at an arbitrary orientation with respect to the reader antenna. A typical example is the use of passive RFID to track aircraft baggage, in which the bags are of irregular size and shape and are placed on conveyors at whatever orientation is convenient at the time of loading. In this case, guaranteeing copolarization of the tag and reader antennas is difficult if both are linearly polarized. While one possible solution is to employ circularly polarized reader antennas, read range is degraded if linearly polarized tag antennas are employed, since only one half of the transmitted power will be harvested. Because linearly polarized tag antennas are less expensive, they are more often used in practice. Polarization-diverse tag antennas improve range and relax requirements on the tag orientation, but increase complexity of the IC and cost of the antenna structure.

9.2.4 Tag IC Requirements

Passive UHF tag integrated circuits typically require between 10 and 100 microwatts of DC power for normal operation. (Higher power may be required for non-volatile memory writes.) Since no battery or other power source is available, this power must be collected by the tag antenna. The DC supply voltage is derived by rectifying the received RF signal using a multistage diode charge pump; the efficiency of the charge pump in converting RF to DC power is typically 20–30% at low incident power. Thus, we need to collect about 30 to 300 microwatts of RF power to operate the tag.

The maximum power that can be collected by a resonant dipole is the product of the incident power density and the effective cross-sectional area of the antenna:

$$P_{RX} = U_{inc} A_e \tag{9.1}$$

where the incident power density is proportional to the square of the incident field:

$$U_{inc} = \frac{E_{inc}^2}{2Z_0}; \quad Z_0 = 377\,\Omega \tag{9.2}$$

The incident power a distance d away from a transmitter transmitting with power P_{TX} and gain G_{TX} is as follows:

$$U_{inc} = \frac{P_{TX} G_{TX}}{4\pi d^2} \tag{9.3}$$

and the effective cross-section for a matched resonant antenna is proportional to the square of the wavelength:

$$A_e = G_{dipole}\left(\frac{\lambda^2}{4\pi}\right) \approx 150\,cm^2 @915\,MHz \tag{9.4}$$

In US operation, the transmitting reader is allowed to provide 36 dBm effective isotropic radiated power, so at 10 meters, the antenna could collect roughly

$$P_{RX,max} \approx 4\frac{0.015}{4\pi(100)} \approx 50\,\mu W \tag{9.5}$$

Thus, an efficient antenna driving a low-power IC may be able to achieve read ranges exceeding 10 meters.

Passive tag antennas are rarely matched to a real 50-ohm load. Instead, they are connected directly to the input pads of the tag integrated circuit. The integrated circuit front end is a multistage charge pump consisting of diodes (or diode-connected transistors) driving storage capacitors. The actual instantaneous load is thus highly non-linear, but it can be roughly approximated as a lossy capacitance, if the input voltage is sufficient to turn the rectifying elements on. Typical values for commercial ICs are on the order of a picofarad of input capacitance – that is, an equivalent negative reactance of a hundred ohms – in series with 5–50 ohms input resistance. The power delivered to the load is reduced relative to that predicted for a matched load:

$$\frac{P_{RX,actual}}{P_{RX,matched}} \equiv \tau = \frac{4R_{ant}R_{load}}{|Z_{ant} + Z_{load}|^2} \tag{9.6}$$

where R_{ant} and R_{load} are the real parts of the antenna and load impedances Z_{ant} and Z_{load}. As noted above, it is usually not possible to make the antenna long enough to be resonant, so the antenna impedance alone is likely to be capacitive, and the real resistance may be significantly less than the 75 ohms or so expected for a resonant dipole. Unlike the classic case of a resonant dipole with a real impedance close to that of a standard transmission line, the tag antenna will suffer significant mismatch loss if measures are not taken to transform the antenna impedance to a more appropriate match to the IC load.

Putting it all together, we can predict the maximum read distance d of a tag according to [12]:

$$d = \frac{\lambda}{4\pi}\sqrt{\frac{P_{TX}G_{TX}G_{dipole}\tau}{P_{min}}} \tag{9.7}$$

where P_{TX} and G_{TX} are the transmit power and gain of the transmitting antenna respectively, and λ is the free-space wavelength, about 328 mm at 915 MHz. Note that the gain of the transmit and receive antennas are direction-dependent. Equation (9.7) also ignores any polarization mismatch. Commonly, a circularly polarized reader antenna and a linearly polarized dipole antenna create a polarization loss of 0.5, which would further decrease the read distance by a factor of 0.707. Finally, this equation assumes *free space*, which is almost never exists in the real world. The environment can scale the distance to nearly zero, or in some cases almost double the read distance.

So what are reasonable values for a maximum read distance? Let's consider a typical case. The transmit power in North America is limited to 30 dBm (1 watt) and a 6 dBi gain antenna (a linear gain of 2). The gain of a dipole is approximately 2.1 dBi, or about 1.5 linear, and the power transfer efficiency can be made arbitrarily close to 1. The minimum power for a modern IC is about −17 dBm (20 microwatts). Assuming a 3 dB (0.5 linear) polarization loss, we would expect a 7.14 meter (23.4 feet) read distance. A polarization-matched reader antenna and a friendly ground reflection might boost the read distance to 14.3 meters (47 feet).

9.2.5 Cost Pressures

Finally and most importantly, passive tag antennas are subject to demanding cost requirements in most applications. An RFID tag is used to identify a physical object. If the tag cost represents a significant part of the cost of the object, its presence can severely impact operating margins for the users, unless compelling reasons (such as regulatory requirements) for applying the tag exist. Passive UHF RFID tags are sold for US$0.10 in volumes of hundreds of thousands to millions of units. To meet such a price, each component of the tag must cost at most a few pennies. Thus the antenna substrate, antenna metallization, patterning, attachment of the IC, and passivation or protection, must all be accomplished at the minimum possible cost, while still ensuring tolerance of harsh environmental conditions and providing good reliability.

For example, when additive patterning techniques (screen printing or electroplating) are used to form the metallized pattern, cost pressures may drive the use of antennas with minimal metallized area, even when a larger antenna would provide superior performance. Antennas intended for use in close proximity to metallic structures are often several mm thick, and require backside metallization to provide a reproducible electrical environment and readability when not attached to the target object. Both these requirements add substantial cost to the tag fabrication. Minimizing cost is a key requirement for applications where goods in transit are to be tracked, since the RFID tag may be used for only one complete product cycle and then discarded.

Cost pressures can be ameliorated when a tag can be used repeatedly, as they are then amortized over multiple cycles. Tracking of reusable plastic bins, digital storage media, or aircraft beverage carts are examples of such applications. UHF passive tags have also been used to track large, expensive assets, such as truck trailers; in these applications, the cost of the tag is a negligible component of total system cost, and higher tag costs can be tolerated.

In summary, a passive UHF tag antenna must satisfy a number of constraints, including some unusual requirements not seen in more conventional antenna designs:

1. It must be electrically small, and in many cases < 1 mm thick.
2. It must tolerate wide variations in the local environment.
3. It must tolerate variations in polarization.
4. It must drive a substantially capacitive load.
5. It must be fabricated at extremely low cost, yet operate for years in the field.

9.3 Approaches

There are a number of common approaches for satisfying these constraints:

- meandered dipoles;
- tip-loaded dipoles;
- slot antennas;
- patch antennas;
- near-field loop antennas (not discussed here).

These approaches have been used singly and in combination to produce a wide variety of commercial antenna designs (Figure 9.3).

Common to all tag antennas is the challenge of maintaining adequate bandwidth within a minimal form factor. Recall that the quality factor Q can be generally defined as follows:

$$Q = \omega \frac{\text{Energy Stored}}{\text{Power Dissipated}} \approx \frac{1}{Bandwidth} \tag{9.8}$$

The bandwidth of an antenna is directly related to the reciprocal of the antenna's quality factor. So increasing the bandwidth of an electrically small antenna is equivalent to increasing the power dissipated (preferably by radiation rather than ohmic loss) and/or decreasing the energy stored.

Antennas can be said to store energy in the magnetic field associated with the flowing current, and the electric fields associated with the stored charge. At resonance, an equal amount of energy is stored in each of these modes, so we can examine either to understand the likely magnitude of the stored energy. For example, the inductance of a cylindrical wire of length ℓ and radius r is of the form:

$$L \approx \frac{\mu_0}{2\pi} \ell \ln[\ell/r] \tag{9.9}$$

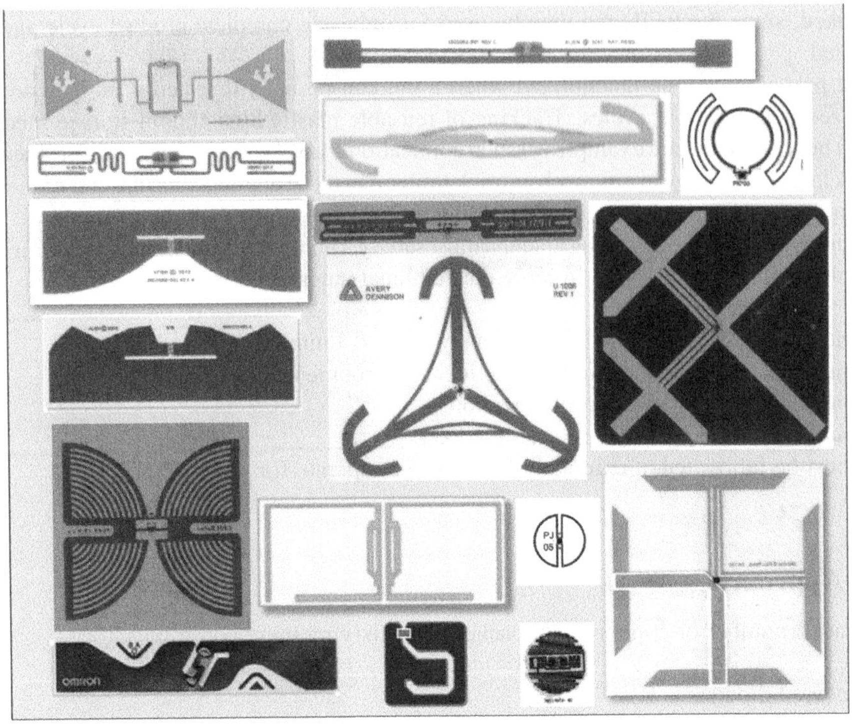

Figure 9.3 Some representative commercial tag antenna designs

Since the current along an antenna must go to zero at the ends, the effective inductance is reduced by about a factor of two (the exact value of the effect being dependent on the current distribution).

The radiation resistance for short wire antennas can be written as:

$$R_A = 80\alpha^2\pi^2\left(\frac{\ell}{\lambda}\right)^2, \tag{9.10}$$

where α is a constant that depends on the current distribution along the antenna; it is roughly half for lengths much less than a wavelength, increasing to about 0.6 for a resonant antenna, resulting in a radiating resistance of approximately 73 Ohms. The resulting quality factor of the antenna can be approximated as the ratio of the inductive reactance to the resistance; after a bit of algebra, we obtain:

$$Q \approx \frac{0.23}{\alpha^2}\frac{\lambda}{\ell}\ln\frac{l}{r} \tag{9.11}$$

That is, the quality factor increases, and bandwidth decreases, as the antenna grows shorter. A resonant dipole will have a Q of around 5–7 for typical conductor sizes. As noted previously, the resonant length for a straight dipole is about 15–16 cm for the UHF frequencies most often employed in RFID; this length is excessive for most applications. Various approaches, to be discussed below, are used to achieve acceptable impedance behavior in smaller antennas; in all cases, we must deal with the tendency of the bandwidth of the antenna to be degraded as the size is reduced.

Wire antennas are generally too expensive to employ in RFID; printed antennas are used instead. Most printed antennas use conductors whose thickness is much less than their width. A ribbon of width W is roughly equivalent to a wire of diameter W/2 in the formulas above.

9.3.1 Meander Dipole

In order to compress a resonant wire-like dipole into a space that is short compared to a half-wavelength, one can bend the wire. When multiple bends are employed, the antenna takes on the appearance of a river meandering over flat terrain: a meandered dipole (Figure 9.4).

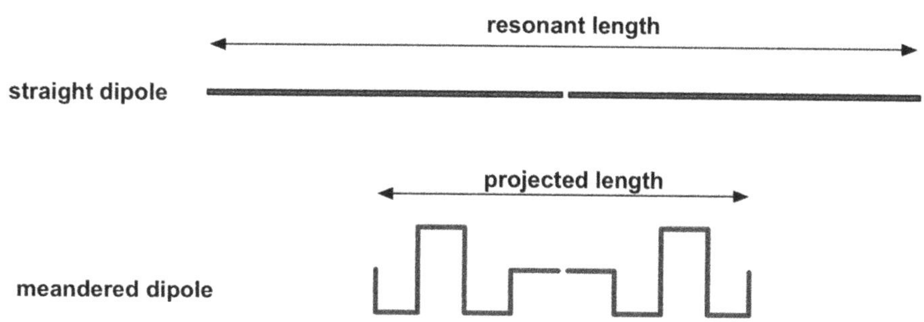

Figure 9.4 A meandered antenna has a smaller projected length

A resonant meandered dipole cannot be constructed merely by bending a half-wave dipole. Since the wire is no longer collinear, the equivalent inductance of each segment is reduced relative to what it would have been in a straight wire, while the capacitance is little affected until the meander structure becomes quite dense. Therefore, the total length of the wire required for resonance is increased over that needed for a straight wire. However, the projected length is reduced, and this is the critical dimension in applications such as smart labeling (Figure 9.2).

A consequence of reduced projected length is reduced radiated fields. In essence, the oppositely-directed currents on the bent portions nearly cancel each other out at long distances, so that the radiation from the antenna predominantly comes from those segments oriented along the length of the structure: that is, the radiation resistance scales with projected rather than actual length in Equation (9.10), and is reduced relative to a conventional dipole. Corresponding to the discussion above, reduced dissipation implies a higher ratio of stored to dissipated energy and thus a narrower bandwidth. Let us imagine that we scale the projected length to 9 cm (to fit on a 100-mm label). If the stored energy is about the same, and the dissipated energy is reduced by $(9/15)^2 \approx 0.36$, the bandwidth of the antenna is only about 1/3 of that of a conventional straight dipole, and the quality factor is increased by the same factor. Using Equation (9.11) with plausible values for the ratio of length to equivalent radius, a straight resonant dipole has a Q of about 6, and an equivalent meandered dipole would thus have a Q of about 18. The bandwidth of the compact antenna would be on the order of 50 MHz – sufficient for US operation (902–928 MHz), but inadequate for global use (860–960 MHz), and very sensitive to the near-antenna environment. It will be helpful to reduce the stored energy as much as practical to obtain a more robust antenna.

The most convenient approach is to increase the width of the conductive lines used to fabricate the antenna. For example, in Figure 9.5, we show a number of meandered antenna structures that fit within a 90 mm by 10 mm box whose linewidths vary from 0.25 mm to 1.75 mm, with a straight 152 mm long by 1 mm wide dipole for reference. In each case, the total length of wire is adjusted to ensure resonance at 915 MHz. To estimate the quality factor, we used the following expression:

$$Q_A = \frac{1}{2R_A}\left(\omega\frac{dX_A}{d\omega} + |X_A|\right) \tag{9.12}$$

where X_A is the reactance of the antenna at the frequency of interest. Note that at resonance, $|X_A| = 0$, but in case $|X_A|$ is not exactly zero, we include it in our estimate.

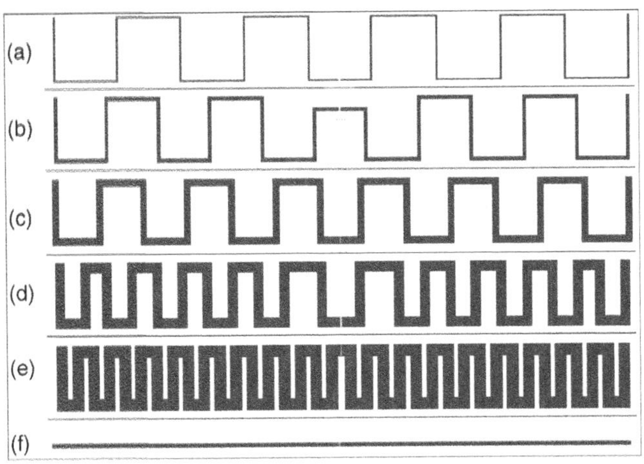

Figure 9.5 A set of meandering dipoles with (a) 0.25 mm line width, 90 mm projected length; (b) 0.5 mm line width, 90 mm projected length; (c) 1mm line width, 90 mm projected length, (d) 1.5 mm line width, 90 mm projected length; (e) 1.75 mm line width, 90 mm projected length; (f) 1.0 mm wide straight line 152 mm long

The behavior of these structures was analyzed using a conventional method-of-moments simulation technique. Model results are summarized in Table 9.1 and Figure 9.6. The quality factors are in rough agreement with the values expected from the simple model of Equation (9.11). The radiation resistance of the meandered antennas is slightly higher than one would expect from naive quadratic scaling: $69.5(90/150)^2 \approx 25$ ohms, vs. the observed 29 ohms, probably indicative of small additional radiation from the counter-directed currents, due to the finite size of the structure. Increased line width produces a substantial reduction in the antenna Q and a corresponding increase in expected bandwidth for the antenna. However, the best achievable quality factor still exceeds that obtained in a straight dipole by more than a factor of 2. A Q of 13.6 yields a 3-dB bandwidth of 67 MHz, just enough to cover 865 MHz (Europe) to 928 MHz

Table 9.1 Meander dipole quality factor and radiation resistance vs. line width

Linewidth (mm)	Estimated Q_A at 915 MHz	R_A at 915 MHz (ohms)
0.25	19.0	30.0
0.5	17.4	29.4
1	15.7	27.5
1.5	14.1	29.3
1.75	13.6	28.4
1 × 152 ribbon dipole	6.7	69.5

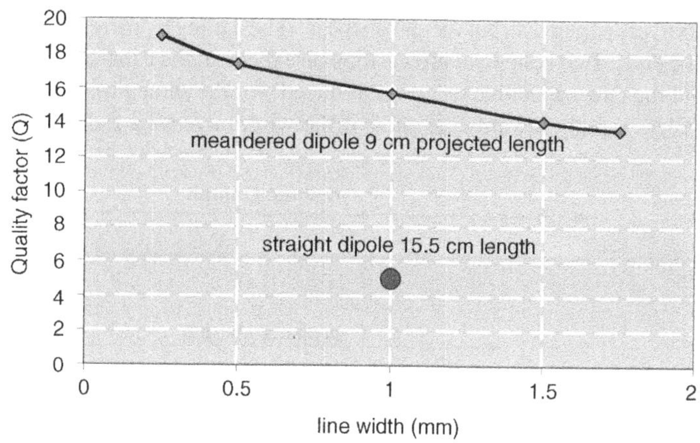

Figure 9.6 Quality factor versus linewidth for meandered dipole with 90 mm projected length, resonant at 915 MHz

(North America), but with limited margin for environmental variations. Meandered dipoles must trade bandwidth for antenna size.

An example of a practical tag antenna design for UHF operation is depicted in Figure 9.7. The structure shown is tuned for operation at 830 MHz, but can be adapted to higher frequencies by trimming the ends of the wire. The wire perimeters of the meandered sections are shown; note that while the projected length is less than 100 mm, the actual wire length of the

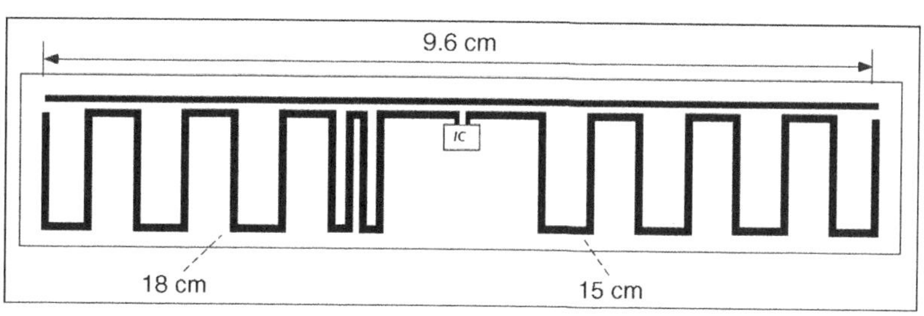

Figure 9.7 A practical meandered tag design. Reproduced by permission of ©2007 Elsevier

meandered section is about 330 mm. The antenna presents an inductive load to match to the capacitive tag IC. (We will discuss matching in more detail below.)

9.3.2 Tip Loading

In a simple straight wire dipole, most of the current flows near the center of the antenna, and most of the charge is stored near the ends. Another method of reducing the projected length of an antenna is to make more effective use of the available space to store charge (Figure 9.8). Tip loading is a very old technique for lowering the resonant frequency of the antenna; for wire antennas, a sphere is appended to the ends of the wire. For RFID tag antennas, such a structure would be prohibitively expensive, but we can do the next best thing by increasing the width of the structure near the ends. Two beneficial effects may be expected: the capacitance of the structure is increased (and thus the capacitive reactance reduced) for the same projected length, and the current falls off less rapidly towards the ends of the wire, increasing radiation resistance.

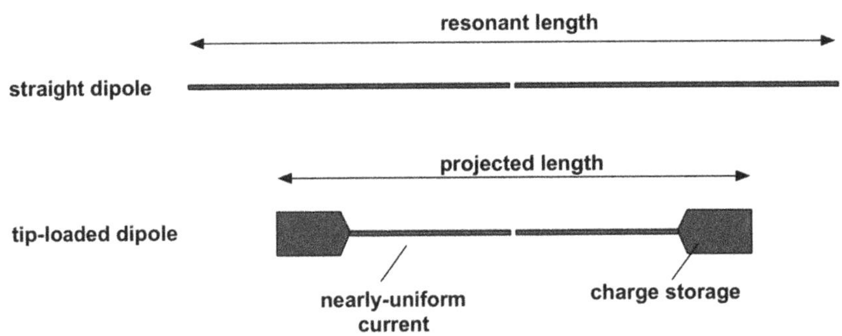

Figure 9.8 A tip-loaded structure provides additional charge storage at the ends of the wire

For flat structures, the capacitance of the tip structure is roughly proportional to its perimeter. By increasing the size of the tip load, the capacitive reactance of the structure is reduced, and thus the resonant frequency falls for the same projected length. Some measured data is depicted in Figure 9.9.

A commercial tip-loaded dipole design is shown in Figure 9.10. In addition to the large flared ends, this design also employs a bent structure. The central ring acts to transform the capacitive impedance of the main part of the antenna to match the IC load impedance; this will be discussed in more detail below.

9.3.3 Combined Meander and Load

As is apparent in, for example, Figure 9.10, actual structures employ more than one technique to reduce projected length and maximize performance. Figure 9.11 shows some representative alternatives for achieving a small projected length. The pure meander structure (a) has been discussed above, and forms our baseline for examining other approaches. In structure (b), the central wire (which contributes most of the radiation) has been straightened, with denser

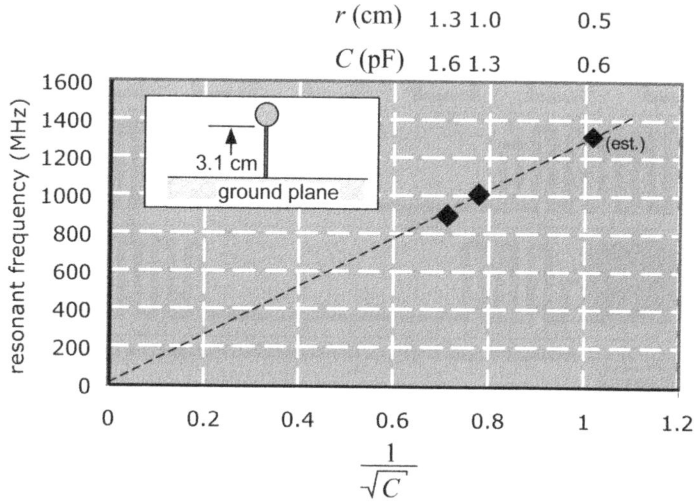

Figure 9.9 Measured resonant frequency for short wire monopole with a disk of varying diameter placed at the tip of the wire

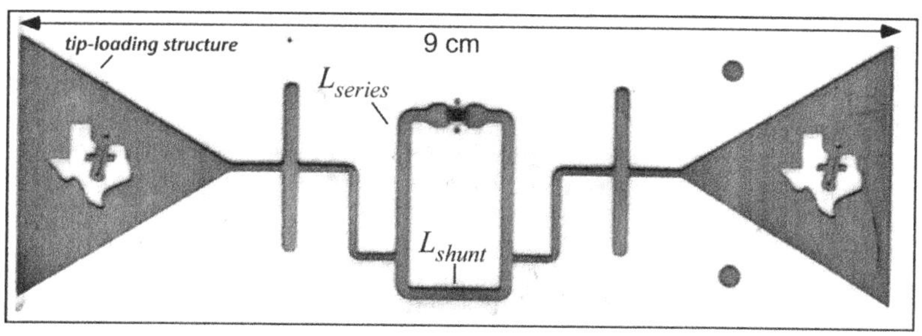

Figure 9.10 A commercial tag antenna design using tip loading to minimize projected length. The matching structure is also shown; see Section 9.3.7. Reproduced by permission of ©2007 Elsevier

meandered structures at the end to provide charge storage and inductance in the smallest projected length. Structure (c) combines a meandered wire (to increase inductance) with a tip load (to decrease capacitive reactance). Finally, structure (d) uses a spiral inductor at the end; spirals provide substantially more inductance per unit length than meandered structures do.

One way to increase α is to rearrange the meanderings. Towards the middle of the dipole, where the currents are largest, we could have fewer meanderings, and have more meanderings towards the ends of the dipole. That way, the current distribution looks nearly linear for much of the dipole, then rapidly goes to zero at the ends. Another way is to add a capacitor to the end of a short dipole through tip loading (discussed above). Also, instead of using a meandering line to create additional inductance, one can use a spiral inductor. A spiral inductor creates a much larger inductance for its size than meandering lines. In each structure, the wire lengths have been adjusted to achieve resonance at the frequency modeled.

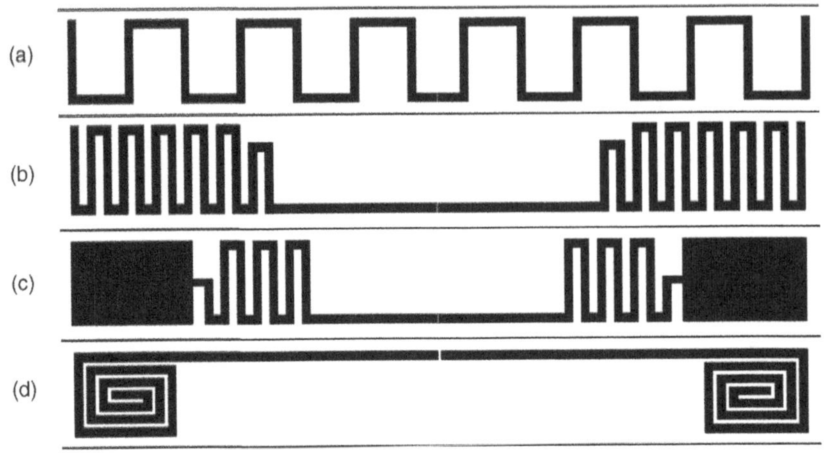

Figure 9.11 (a) A regular meander dipole; (b) inductively tip-loaded dipole using meandering lines; (c) capacitively tip-loaded dipole and inductively loaded using meandering lines; (d) spiral inductor tip-loaded dipole

We modeled each structure with a standard MoM code, and report R_A and Q_A obtained from Equation (9.12). The results are summarized in Table 9.2.

Table 9.2 Radiation resistance and quality factor for the structures in Figure 9.11

Antenna	Estimated Q_A at 915 MHz	R_A at 915 MHz (ohms)
(a)	15.7	27.5
(b)	13.8	35.4
(c)	13.8	32.0
(d)	20.4	39.2

The displaced meander structure (b) and tip + meander structure (c) both give about a 12% reduction in Q. The spiral inductor (d) provides a substantial increase in radiation resistance, but also stores substantial energy in the spiral inductance, so that the Q actually increases despite increased dissipation. If one is printing the antenna using (expensive) silver ink, then the clear winner is (b), which produces the best Q_A with the least amount of ink.

9.3.4 *Fat Dipole*

As the linewidth of the antenna is increased, the equivalent inductance is reduced, and the capacitance is increased, for a fixed length. Both these changes lead to reduced reactance – that is, reduced stored energy. Thus, all other things being equal, we expect that fatter lines will produce an antenna with lower Q and higher bandwidth.

As an example, we modeled two printed dipoles, one that is 1 mm wide and 152 mm long, and the other 10 mm wide and 148 mm long (Figure 9.12). Both are resonant at about 915 MHz.

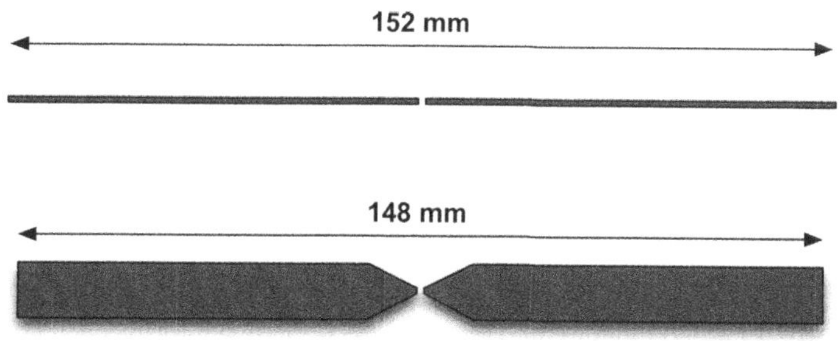

Figure 9.12 Resonant dipoles with differing linewidths

The 1 mm dipole has a Q of 6.7, and the 10 mm dipole has a Q of 3.4; these values are in good agreement with the estimates obtained from Equation (9.11).

Practical structures of this type will still resonate at frequencies above the UHF band when the length is reduced to 100 mm or less, but the magnitude of the capacitive reactance at the desired operating frequency is greatly reduced relative to the real part of the impedance, so that matching the antenna to the IC load is greatly simplified, and the resulting structure will have better bandwidth. The significant disadvantage of these structures, particularly when produced by printing, is the large area of metal employed, and consequent increased cost of manufacturing.

It is important to note that the structure of the feed region plays an important role in the behavior of these broad ribbon-type antennas. For example, if we replace the tapered feed shown in Figure 9.12 with a pair of rectangular ribbons, presuming the test port to be connected at the edges, we observe a substantial decrease in the input resistance of the resulting antenna (from around 20 ohms to 10 ohms), as shown in Figure 9.13. Thinking in terms of current flow in the ribbon, we may view this change as resulting from the capture of a substantial amount of charge at the neighboring edges, reducing the radiating current flowing in the remainder of the antenna. In lumped-element terms, this is equivalent to saying that we

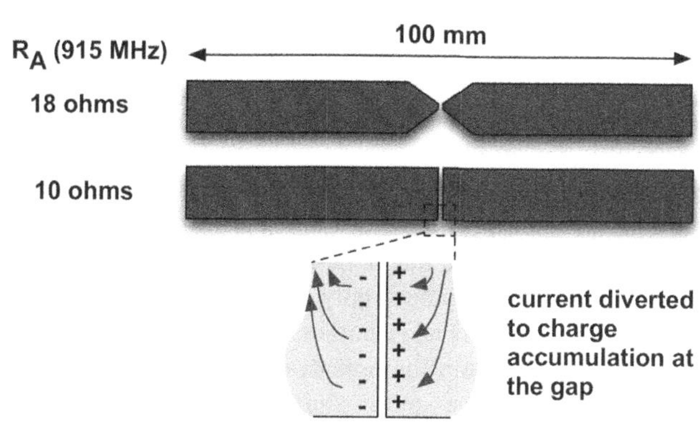

Figure 9.13 Effect of flared versus narrow feed gap

have added a shunt capacitance resulting from the neighboring edges, reducing the real impedance of the antenna as roughly:

$$R_{A,eff} = \frac{R_A}{1 + \omega^2 (R_A C_{shunt})^2} \tag{9.13}$$

9.3.5 Slot Antennas for Tags

In many T-matched designs (discussed below), there is a fairly narrow gap or slot between the element that forms the shunt inductor and the parallel element that forms the feed line to the IC. In conventional T-match designs, this slot is substantially parallel to the long axis of the antenna. A more subtle approach employs a slot with a portion perpendicular to the current flow from the IC; an example of this type of design is depicted in Figure 9.14.

Figure 9.14 Example of slot tag antenna design

Let's break up the currents around a slot into two different modes (Figure 9.15). We can think of the currents as being common-mode and differential, but for reasons that will become apparent in a moment, we will use the terms *dipole mode* and *slot mode*.

In the dipole mode, all the currents are moving in the same direction, so the radiated electric field will be aligned with the currents. Radiation from these currents will be maximized in the plane perpendicular to the slot, in classic dipole fashion. In Figure 9.15, the dipole mode radiates up, down, and out of the plane of the paper, but not left or right.

The slot mode currents are oppositely directed (and thus must circulate around the slot). The circulating current can be regarded as a magnetic-monopole-like structure. Equivalently, the radiation from the opposed currents can be simply added together. When viewed from above, the currents exactly cancel and there is no radiation, but when viewed in the plane of the slot (e.g. within the paper, looking up from the bottom of the page, or down from the top), the currents are at slightly differing distances and are thus out of phase. The resulting radiated fields do not exactly cancel. Radiation is maximized in the plane of the slot (the plane of the paper here), though the exact amount of radiation in each direction depends on the shape of the slot.

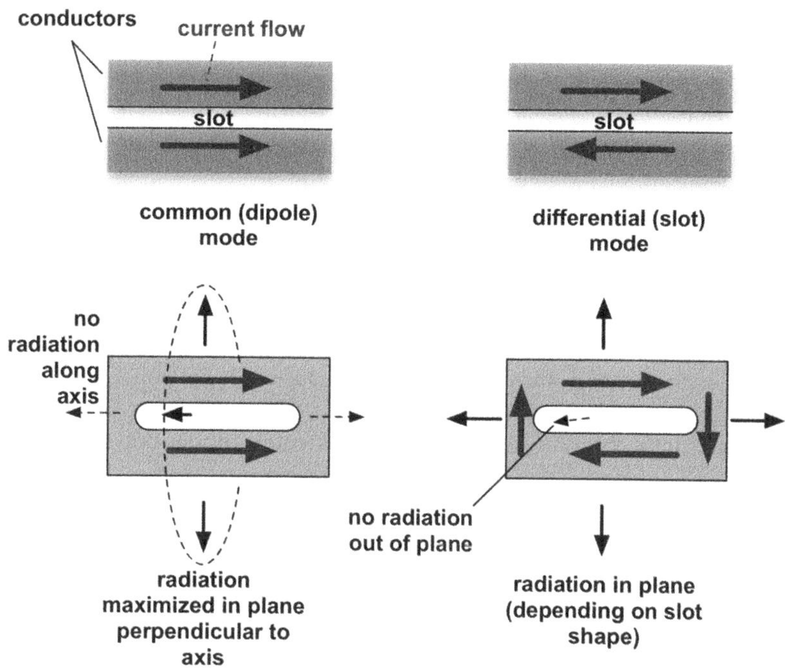

Figure 9.15 Current flow in conductors near a slot can be regarded as the sum of dipole and slot mode currents

As a consequence, looking down on the slot, dipole radiation is maximized and slot mode radiation is zero, whereas in the plane of the slot, slot mode radiation is maximized and dipole radiation may be zero. In a sense, the two modes produce orthogonal maxima! Thus, by mixing dipole mode and slot mode current flows, the antenna radiation pattern can be adjusted more flexibly than is readily achieved with a dipole alone. Thus, in the structure of Figure 9.14, making the paths on the two sides of the slot asymmetric will cause the net current flow to be a mixture of slot and dipole modes. The antenna will be closer to omnidirectional than a simple dipole, and will also have some polarization diversity. This design approach makes very efficient use of limited space to maximize radiation resistance and minimize reactance.

Unfortunately, there are some practical limits to this approach. The length of the slot is determined by the reactance of the chip, in order to ensure that the resulting inductance resonates with the chip capacitance. (Even if it weren't, wider antennas cost more money!) That length governs the radiating resistance from the slot antenna. So a slot antenna that's 15 or 20 mm long is not likely to radiate very much. However, the very small dipole antenna shown in Figure 9.14 also is pretty short, so the radiating resistance of both the dipole and half slot is similar.

9.3.6 Dual Dipole Antennas

As noted previously, it is not always possible to control the orientation of a tag antenna relative to the reader antenna. One possible response is to employ a circularly-polarized reader antenna, and this is commonly practiced. Another approach is to use a polarization-diverse tag antenna, typically implemented as a dual-dipole or tripolar structure. (See, for example, the cross-like and tripod-like structures in Figure 9.3 above.)

It is easy to see that a linearly-polarized incident wave will excite currents along the dipole structure most closely aligned to its direction of polarization. When the incident polarization is intermediate between that of the constituent dipoles, both dipole structures will have current flow. When a circularly polarized wave is incident, the wave will excite both dipoles, one delayed in phase by 90 degrees with respect to the other.

In order for a dual dipole structure to be of use, the integrated circuit must have two distinct rectifying inputs, so that the IC in effect adds the received *power* from the antennas. If the currents are summed prior to rectification, we have merely created a linear combination of the incident waves, equivalent to a rotation; the resulting structure is not polarization-diverse, but merely has a new preferred axis of polarization.

Dual-dipole antennas fact the same constraints as single-dipole designs: they must be compact, inexpensive, and reasonably broadband. The design approaches described for single-dipole structures can be used for dual-dipole structures.

9.3.7 Matching: A Case Study

In the discussion so far, we have examined resonant antenna structures (structures that present a real load). In practice, passive tag antennas must be matched to an integrated circuit. Modern commercial integrated circuits present a reactive load corresponding to approximately a picofarad in parallel with a resistance of hundreds to thousands of ohms. While some designs are constructed to present an inductive load to the IC and thus resonate out the reactance (Figure 9.7), most antennas use a variant of the traditional T-match, in which a tapped loop is employed to match the IC to the antenna (e.g. see Figure 9.10).

In this section, we will examine the modified T-match in more detail. We will present a simple circuit model that provides a reasonable description of the antenna behavior over a relevant bandwidth. First, we will examine a "naïve" implementation of a dipole antenna with a modified T-match and examine how the tag performs over a range of frequencies. Then, armed with some information about filter theory, we will reexamine the circuit model, perform some manipulations, and see how we can significantly increase the bandwidth of the antenna.

A T-match, as implemented in most RFID tag designs, consists of a loop structure in which the IC is embedded, with the main antenna either tapping into the loop, or in some cases wrapping around it (thus using mutual inductive coupling between the tuning loop and antenna). The actual workings of the T-match are rather complicated because of the mutual inductance between the various conductive lines. Fortunately, we can ignore much of this complexity. A simple lumped-element model turns out to be sufficiently good to allow accurate predictions of antenna performance. The simplified model also permits us to perform an elegant analytic transformation that will yield insight into how to optimize the bandwidth of the tag.

To begin with, let's treat the short conductive elements as lumped inductors, assuming that we can ignore any mutual inductance, or absorb it into the model values. We've superimposed little inductors over the traces in Figure 9.16 to illustrate the point. The tuning structure looks like a small shunt inductor and larger series inductors leading to the input of the IC.

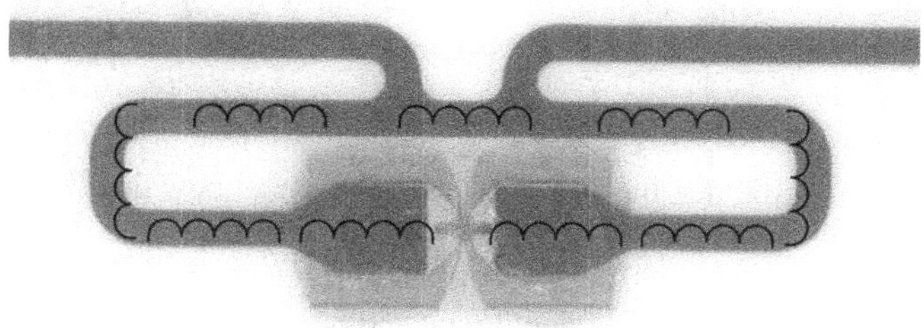

Figure 9.16 Modified T-match with an inductor-based view

Recall that we can model the dipole antenna as a series RLC circuit (at least for a relatively narrow range of frequencies). Using our simplified view of the matching structure, we obtain an equivalent circuit like that shown in Figure 9.17. Here, we use the symbol L_h to denote the shunt inductor and L_e to be the series inductor. The IC is modeled as a parallel RC circuit, which is reasonably accurate over the range of frequencies that the IC will operate. For simplicity, we depict a single-ended circuit, although the actual structure is differential. Note that, because coupling between elements is being neglected, the best-fit values of the inductors may not be identical to the inductance of a similar length of isolated conductive trace.

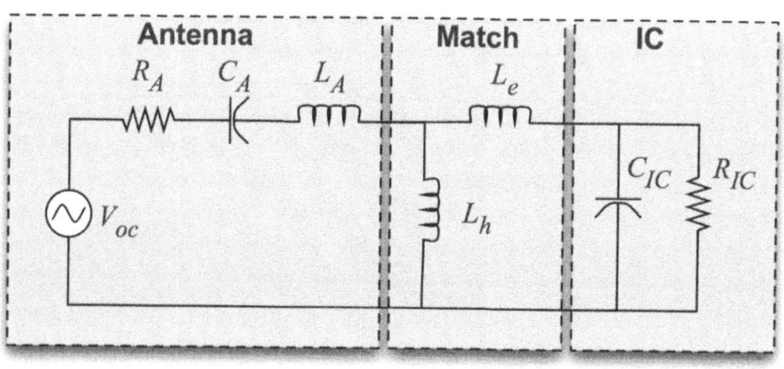

Figure 9.17 Circuit model of the modified T-match

The circuit of Figure 9.17 acts to perform an impedance transformation. That's how one can use the T-match to couple between a 20 Ohm antenna and a 2000 Ohm IC resistance. Because of this transformation, the mutual inductance effects can be folded into the values of the equivalent inductors without affecting the accuracy of the model. The impedance scaling is controlled by the ratio of L_h to the total matching inductance. Let us define a ratio parameter β:

$$\beta = \frac{L_h}{L_e + L_h} \tag{9.14}$$

We can show that the chip resistance is scaled down, or the antenna resistance is scaled up, by a factor of β^2. To cancel the chip reactance, the parallel impedance of the antenna and L_h, in series with L_e, must cancel the chip reactance. If the antenna is resonant, we should choose the matching elements resonate with the IC capacitance:

$$(L_e + L_h)C_{IC} = 1/(2\pi f_0)^2 \tag{9.15}$$

However, there is no reason that the antenna must be resonant now that we have a matching circuit available. Compact antennas are usually electrically short and present a capacitive reactance. L_e and L_h can be adjusted to offset the antenna reactance (to a point). This is one of the many flexible options that the T-match offers, and one reason it is used with such a large variety of antennas.

As an example, let's consider the antenna shown in Figure 9.18, whose modeled impedance behavior is shown in Figure 9.19. We can see that the series R-L-C circuit assumption is good for the reactance except at the extremes, but the constant resistance assumption breaks down if one considers more than 10% of bandwidth. Assume the target frequency is 915 MHz. We see that the impedance presented at 915 MHz is 28–j135 Ohms, which is considerably capacitive.

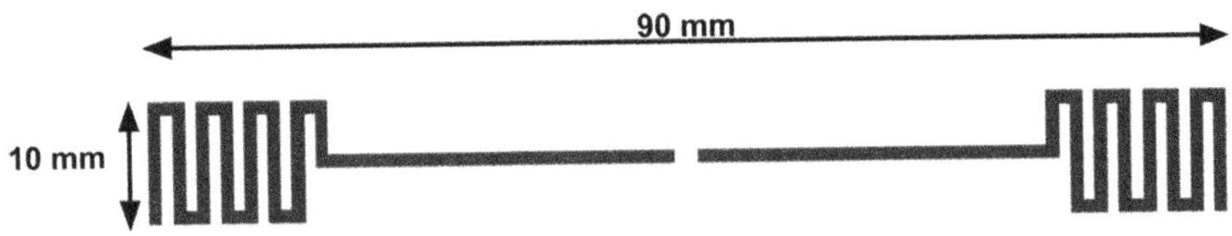

Figure 9.18 Example short dipole antenna

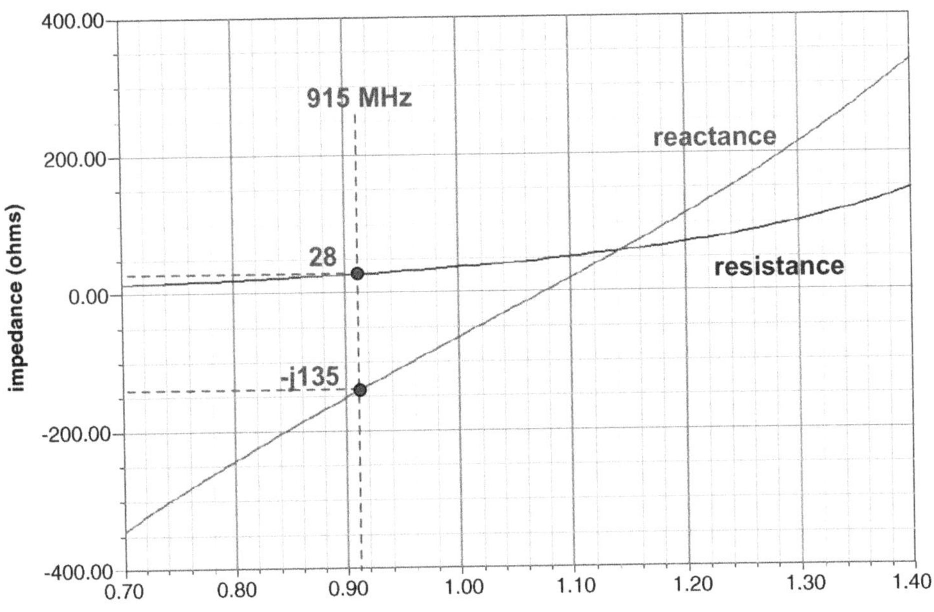

Figure 9.19 Impedance of dipole antenna shown in Figure 9.18

Let's assume that the IC impedance that we want to match to is $10 - j150$ Ohms at 915 MHz, so the object of the T-match is to transform the antenna impedance into $10 + j150$ Ohms. The T-match that gives the desired impedance is shown in Figure 9.20. The resulting simulated impedance of the antenna with the T-match was found as $9.8 + j149$ Ohms, which is close enough for our purposes.

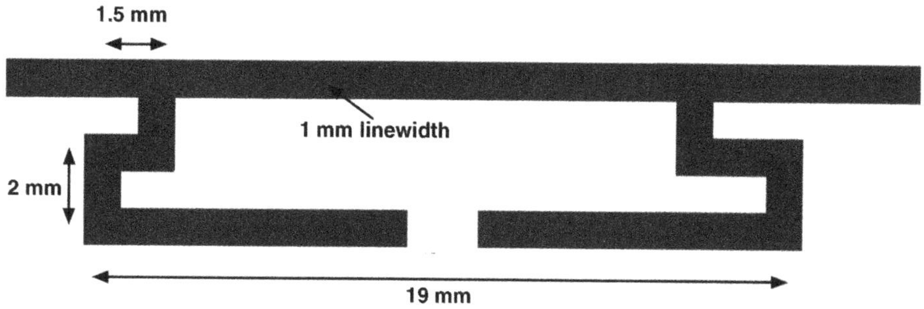

Figure 9.20 T-match used to impedance match dipole of Figure 9.18. Measurements are from center to center of the traces

An equivalent circuit of the form given in Figure 9.17, with values appropriate to this structure, is shown in Figure 9.21. The simulated antenna (including matching network) impedance from a full method of moments calculation R_a^s, X_a^s, and the values obtained from the equivalent circuit model R_a^c, X_a^c, are compared in Figure 9.22. The residual inductive reactance (≈150 Ohms) is chosen to resonate out the IC capacitance. The simulated and circuit values show excellent agreement over a large range of frequencies. The modest divergence between the two at high frequencies is mainly assignable to the limitations of the R-L-C model for the antenna. It is really quite surprising that the circuit model is as accurate as it is.

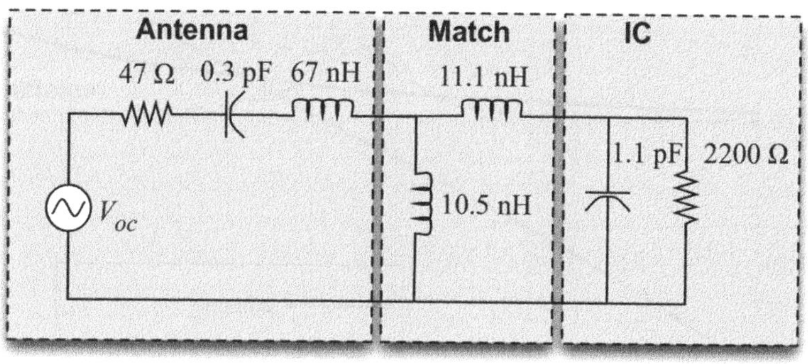

Figure 9.21 Circuit equivalent to the RFID antenna with matching network

In Figure 9.23, we compare the impedance of the antenna itself, and the combination of the antenna and the T-match. We can see that the matching circuit looks like a pure inductor at low and high frequencies, with something like a parallel resonance in between. It is also apparent

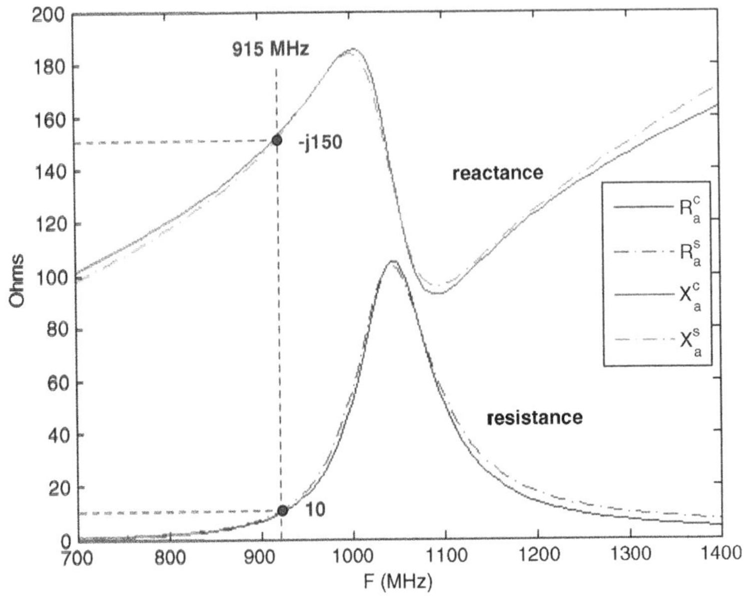

Figure 9.22 Impedance of the dipole with T-match, as seen from the IC input, derived from simulations (s) and equivalent circuit model (c)

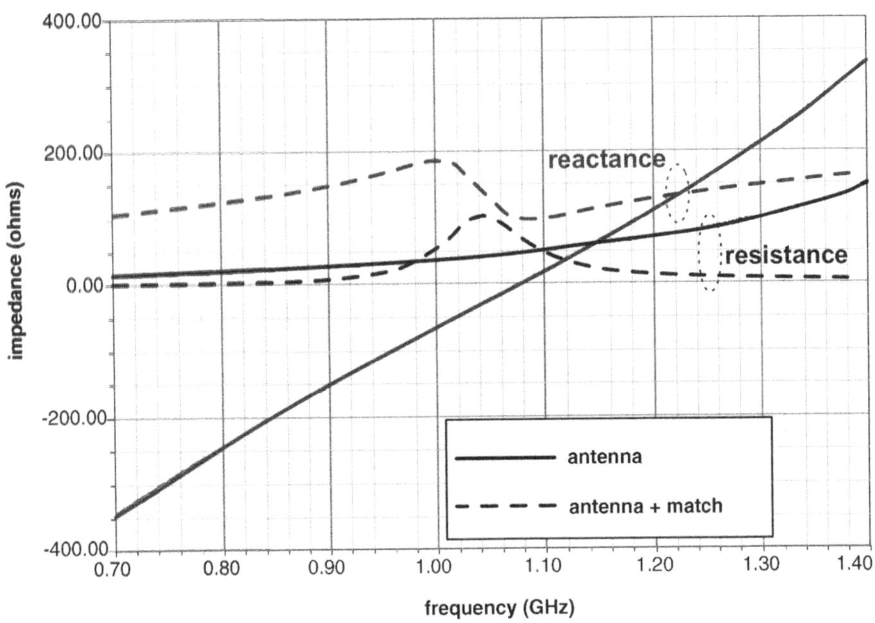

Figure 9.23 Comparison of the unmodified antenna impedance (solid lines) and the impedance of the combined antenna and T-match (dashed lines)

that the frequency at which the impedance is matched to the IC load is located in a region in which the impedance of the structure is changing rapidly with frequency: this is not a very broadband match. How can we do better?

In order to clarify what's going on, let us consider the circuit transformation shown in Figure 9.24. The transformation is controlled by β, where

$$\beta = \frac{L_h}{L_e + L_h}; \quad L_{se} = \beta L_e; \quad L_{sh} = \beta^2 (L_e + L_h) \tag{9.16}$$

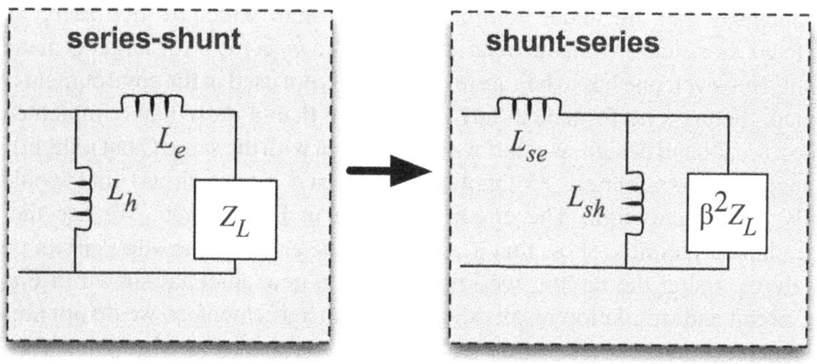

Figure 9.24 Simple circuit transformation

If we apply this transformation to the circuit in Figure 9.17, we arrive at the circuit shown in Figure 9.25, which the reader may recognize as a series-parallel bandpass filter. The series LC circuit has a reactance with a positive slope at resonance (it is moving from a series capacitance to a series inductance with increasing frequency), while the parallel LC circuit has a negative

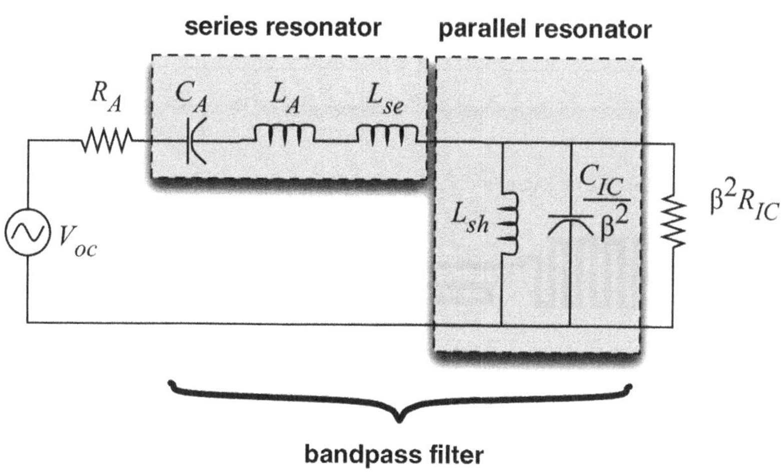

Figure 9.25 Transformed circuit equivalent of the RFID antenna with T-match

reactance slope at resonance (it is moving from a shunt inductor to a shunt capacitor with increasing frequency). If the two circuits (parallel and series) resonate at the same frequency and have nearly the same slope, then the reactance of the two circuits will cancel over a range of frequencies, providing broadband response.

Good bandwidth is obtained when the two resonators have about equal Q, and the resistive loads are related as shown in Equation (9.16):

$$\beta^2 R_{IC} = 1.4 R_A \tag{9.17}$$

Fortunately, many ICs have Q values that are in the range of 10 to 15, which is similar to the Q values of antennas that are about 90mm long and 10mm wide. In fact, there may be an advantage to get *more* bandwidth out of an antenna with a *larger* Q if the larger Q matches to the Q of the chip! However, one has to be careful. If the tag is not used in the environment in which it was designed, the tag's performance may suffer more than a more naïve implementation.

To create a broadband design, we start with an antenna with the same Q but a slightly reduced R_A (estimated at approximately 25 Ohms), and devised a new circuit that would give us significantly more bandwidth. The circuit is shown in Figure 9.26. We use the required impedance transformation to show that $\beta = 0.15$, and determined the other circuit parameters as shown above. Using the circuit, we implemented a new antenna, shown in Figure 9.27. Again, the circuit and simulation results show excellent agreement, so we do not present them here.

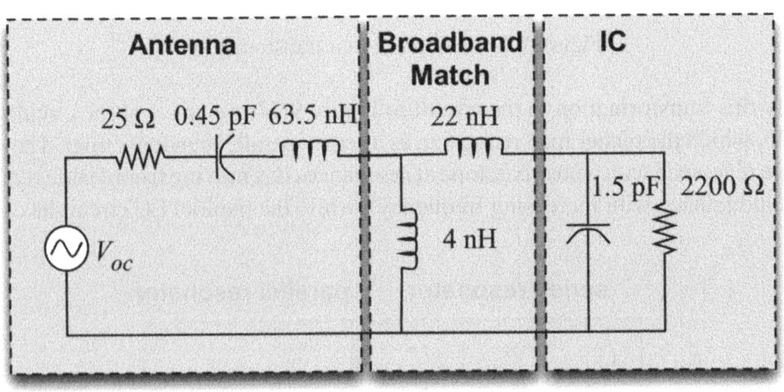

Figure 9.26 Circuit of antenna and T-match designed for broadband performance

Figure 9.27 Wideband antenna design

The power transfer efficiency as a function of frequency of both the naïve antenna (Figure 9.20) and the antenna developed from the bandpass filter (Figure 9.27) are compared in Figure 9.28. We can see that the range of frequencies over which the antenna transfers at least 90% of the incident power to the load is extended by a factor of about 3. Note that this enhanced

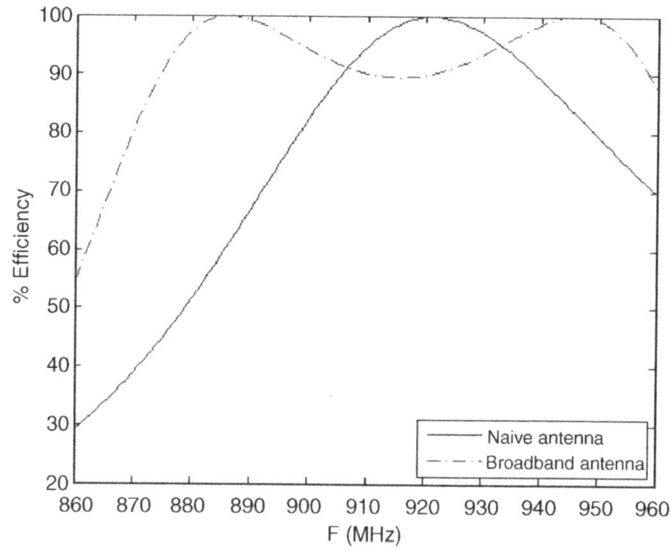

Figure 9.28 Efficiency of naïve and broadband antenna versus frequency

bandwidth is "free" in the sense that we only had to be more clever about how we designed the antenna and matching network to get improved bandwidth; the antenna's projected length is about the same, and the cost of fabrication hasn't changed.

Finally, we plot the radiation pattern of the dipole antenna (dB scale) in Figure 9.29, in which the dipole length is oriented along the X axis. As expected, the radiation is strong in the Y-Z plane, and minimal in the positive and negative X direction.

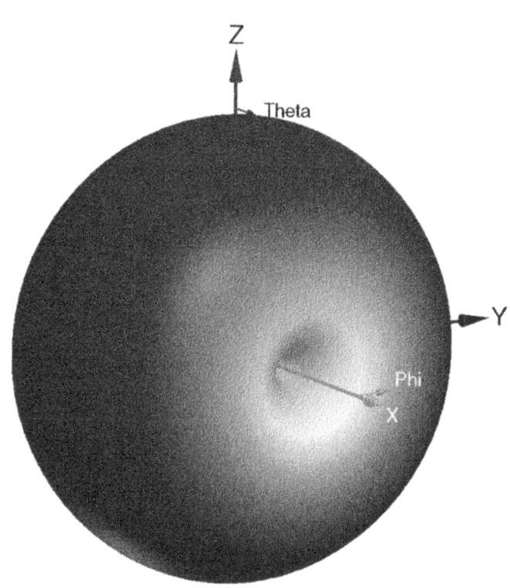

Figure 9.29 Radiation pattern of the dipole

9.3.8 Microstrip Patch Antennas

There have been numerous, ingenious ways in which microstrip antennas have been modified for use with RFID. In this section, we'll briefly describe what a microstrip antenna is and why it's useful. Then, we'll explain a simple way in which the familiar structures used in dipole antennas can be readily adapted to microstrip antennas. To simplify this section, we will not take the "classical" approach to describing microstrip antennas, but don't worry – you won't miss anything important.

First, let's consider a "normal" dipole antenna that parallel and close to a large, flat conducting surface. The conducting surface acts as a mirror image, and, unfortunately, the radiation from mirror image antenna cancels the radiation from the real antenna. This cancellation causes a drastic reduction in the radiating resistance. To be precise, let h be the distance from the antenna and ground plane, and $k = 2\pi/\lambda$ be the free space wave number (about 19.2 radians per meter). Then the radiating resistance is reduced by a factor of $sin^2(kh) \approx (kh)^2$ for small h. A 10mm separation from the metal ground plane will reduce a 25 Ohm radiating resistance to about 1 Ohm. Further reducing the spacing to 3 mm reduces the resistance by an additional factor of 10. Clearly, metal dramatically changes the behavior of dipole antennas. Below, we investigate what can be done about it.

A *microstrip* antenna is a planar antenna that is separated from a ground plane (a large, flat, metal surface) by a dielectric material (possibly air) and is designed to operate above the ground plane. Obviously, we need some way of getting energy into or out of the antenna, so there is some "feed" structure. A common way to feed a microstrip antenna is from the bottom using a coaxial probe in which the ground of the coaxial cable is connected to the ground plane, and the center conductor of the coaxial cable extends through the dielectric and connects to the antenna. Another way to feed the antenna is from the side through the use of a microstrip transmission line, which is a strip of metal separated from the ground plane by a dielectric. Figure 9.30 shows a microstrip antenna that is fed along the radiating edge by a microstrip transmission line. The electric fields along the edge of the antenna are illustrated.

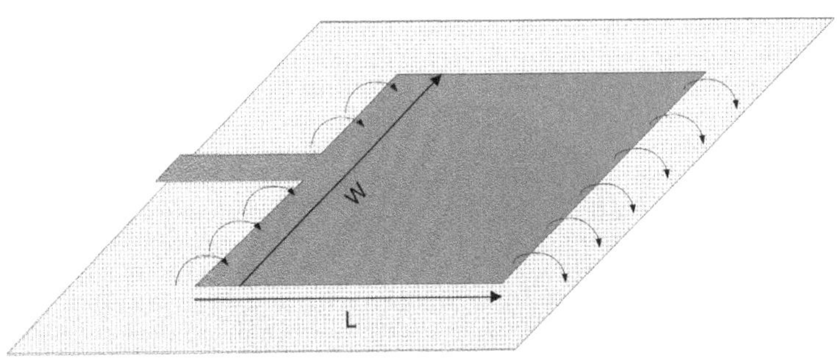

Figure 9.30 Edge-fed microstrip antenna

Notice the curvature to the field lines, or "fringing" of the lines. It is this fringing where radiation escapes from the antenna. This is different from a dipole antenna that essentially radiates everywhere along the antenna. One thing that you'll note about Figure 9.30 is that the antenna looks like a rectangle, and that it may be wider than it is long. This is common for microstrip antennas because in this manner the radiating area is increased, and thus the quality factor is reduced; otherwise the Q of the antenna is so large that its bandwidth is impractically small. This kind of antenna is often called a *patch antenna* because it consists of a large rectangular patch of metal (although more complex shapes are also used).

Why are microstrip antennas interesting for RFID? If the tag is intended to be attached to a flat metal object such as a tool container, it makes sense to use an antenna that's been designed with a conductive ground plane in mind. Unshielded tag antennas are also strongly affected by the proximity of water (with a relative dielectric constant of about 80) and other aqueous liquids. A microstrip antenna's ground plane shields the antenna from the underlying material, approximating a universal tag that can be used on any object.

The problem with a conventional patch antenna is that it is single-ended: well-suited to a coaxial cable connection, but less appropriate for attachment to an integrated circuit, which prefers a differential signal. To solve this problem, we are led to consider a microstrip *dipole* antenna. A microstrip dipole antenna is nothing more than a dipole antenna that is parallel and close to a ground plane, separated from the ground plane by some dielectric (possibly air). A microstrip dipole is different from a "normal" dipole antenna in three important ways:

1. The resonant frequency of the dipole antenna decreases slightly, or more depending on a number of factors including the permittivity of the material, thickness, and geometry of the antenna. A conventional 900 MHz resonant dipole may resonate at as low as 780 MHz near a ground plane.
2. The radiating resistance becomes *very* small, as little as one Ohm for a typical structure. We can regard this as a consequence of contacting a patch in the middle instead of at the edge. A conventional patch antenna can be viewed as a quarter-wave transmission line, transforming the very small central load to a much larger impedance at the edge. When we place the load at the center, we observe this very small impedance.
3. The Q of the antenna becomes *very* large: we have observed values of several hundred.

As a consequence, microstrip dipoles are difficult to match to the IC load, and have a very narrowband response. To make a useful design, we must surmount both these problems.

Achieving broadband performance is a fundamentally difficult challenge for microstrip antennas, which has been the topic of copious research for over 50 years. Thousands of research papers and dozens of books have been written on the topic. Three primary factors control the bandwidth of such structures:

- the volume formed between the antenna and ground plane;
- the dielectric constant of the dielectric spacer;
- the efficiency of the antenna.

Bandwidth is roughly proportional to the volume of the antenna. Thus, a wider antenna will have more bandwidth than a narrower antenna, and a thicker spacer will give more bandwidth than a thinner spacer. The dielectric constant of the dielectric also impacts bandwidth. Recall that Q is the ratio of energy stored to power dissipated. A material with a large dielectric constant will store more energy in the electric field than a material with a low dielectric constant, and thus will have a larger Q. Air is the ideal, but foamed polyethylene or polystyrene (which is mostly air with a small amount of low dielectric constant plastic) is popular. Finally, efficiency can be traded for bandwidth. Recall again the definition of Q: the ratio of energy stored to power dissipated. The definition doesn't indicate *how* power is dissipated. Ideally the dissipated power ought to be in the form of radiation, but power dissipated as ohmic loss still reduces antenna Q.

We have seen that achieving sufficient bandwidth is challenging even for a conventional short dipole. With a Q an order of magnitude or two or larger, microstrip dipoles present a fundamentally different set of challenges. Moderately-sized microstrip RFID antennas are either inefficient or narrowband (or both!). Fortunately, microstrip antennas tend to be much more *directive* than a standard dipole: most of the radiated energy is in the direction perpendicular to the ground plane, with very little power radiated along or behind the ground plane. Figure 9.31 shows the typical radiation pattern (dB scale) of a microstrip antenna placed on a 30 cm^2 metal plate. The radiation intensity in the positive Z direction (away from the metal plate) is about four times that of a dipole, while very little (but yet some) is emitted in the negative Z direction (the direction of the metal plate).

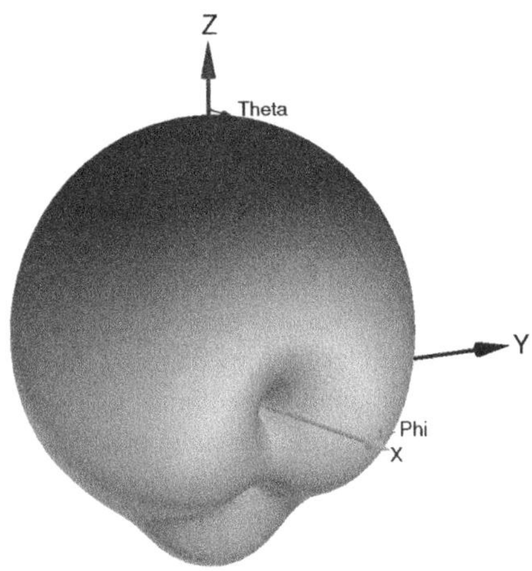

Figure 9.31 Radiation pattern of a typical microstrip antenna

Matching to the small real impedance of the patch dipole seems challenging, but fortunately the versatile T-match is up to the task. Remember that the IC impedance is multiplied by β^2. All that is necessary is to use a sufficiently small β to compensate for the small radiating resistance.

To illustrate the point, we constructed a hand-cut antenna shown in Figure 9.32. The antenna was 138 mm long, 110 mm wide, and separated from the ground plane by a 15 mm polystyrene substrate ($\varepsilon_r = 1.04$). You can see that in this case the T-match uses the patch as the shunt inductor (and thus the shunt inductance is very small!). We can roughly treat the inductance of the segments as proportional to the product of length and the logarithm of the aspect ratio (Equation (9.9)). The series inductance legs are around 16 mm long and 4.5 mm wide; the shunt "inductor" can be approximated as a hemispherical current flow of radius 8mm, with "inductance" proportional to half the radius. We then divide by two to obtain the single-ended equivalent shunt inductor. The quantity β is thus about:

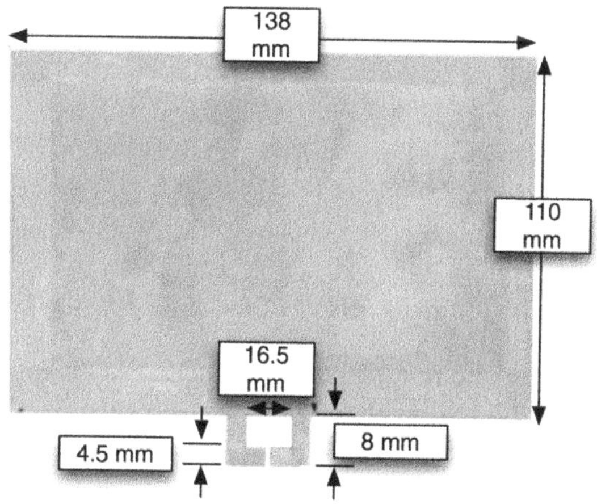

Figure 9.32 Hand-cut wideband patch antenna using a T-match for impedance matching

$$\beta = \frac{L_h}{L_e + L_h} \approx \frac{2}{2 + 16\ln(64/4.5)} \approx 0.045 \tag{9.18}$$

The match is capable of transforming the 2200 ohm IC resistance to match a (single-ended) patch resistance of around 4 ohms.

By using the optimal T-match designs similar to those described in the previous section, we were able to get 90 MHz of bandwidth at 900 MHz from this antenna with better than 90% efficiency.

Unfortunately, this antenna is too large for many passive RFID applications. Practical tags must typically be less than 10 cm in length, and 2 to 4 cm in width, to minimize cost and maximize utility. A common scenario is an inexpensive antenna attached to a foam backing – a Foam Attached (FAT) tag. Tag antennas are laminated to the flexible foam backing with a high-speed reel-to-reel process, resulting in an inexpensive tag ideally suitable for attachment to metallic objects. The application demands that the tag be readable both in air (for testing, validation, and other benign tagging applications) and on metal. We need an antenna that can act as both a conventional and a patch dipole.

An example antenna designed for this application uses a flexible 3.2 mm thick HDPE foam ($\varepsilon_r = 1.095$). We specially designed this kind of FAT antenna so that it would operate in free space as a normal, impedance-matched dipole, but once it was placed on metal, it would operate as an impedance-matched microstrip dipole. Figure 9.33 shows one. This antenna is shorter, but much narrower and thinner than the previously shown microstrip antenna. Its Q when over a ground plane is *huge*, estimated to be more than 200. In order to get a practical bandwidth, we need to introduce loss.

Let's briefly examine how this kind of antenna works. First, we constructed a circuit model for the tag, including antenna, matching network, and IC. First, let's look at how the antenna behaves in air. From the design and modeling process, we deduce the equivalent circuit shown in Figure 9.34. The broad ribbon structure has relatively low reactances to compensate for the small value of radiation resistance, so Q is about 6, giving good broadband behavior. The antenna alone resonates at about 1.2 GHz, so it presents a capacitive load at 915 MHz.

Figure 9.33 A free space/microstrip "FAT" tag

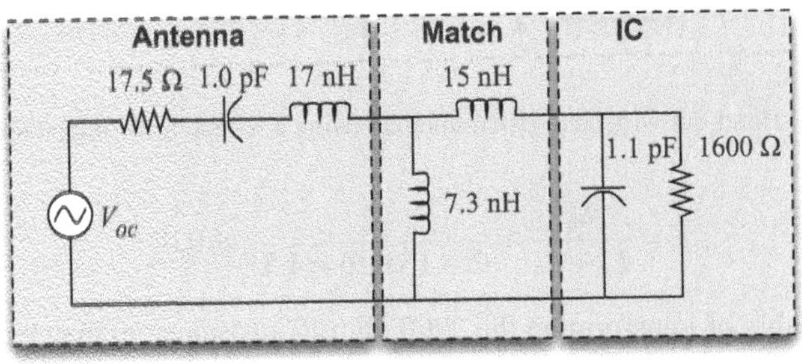

Figure 9.34 Equivalent circuit of free space/microstrip FAT tag

After the T-match, the reactance is transformed to about j150 to j160 ohms, providing a good conjugate match with the IC (with an input reactance of about −j150 ohms). The real impedance is about ohms at 900 MHz, increasing to 12 ohms at 930 MHz, vs. the IC input of about 16 ohms. We thus expect power transfer from the antenna to the IC to be good. This expectation is confirmed in Figure 9.35. This tag as not as broadband as some of the other antennas we've looked at. However, the antenna only gives up a maximum of 1 dB of inefficiency from 900 to 930 MHz, so with all the imperfections, its performance is not noticeably different from an "optimal" tag. Commercial tags for specific applications often give up more than 1 dB of performance in free space to work better in special applications.

Let us place the *same* tag antenna over a metal ground plane and see what happens. The new equivalent circuit model for this condition is shown in Figure 9.36. The antenna resistance has fallen from about 17 Ohms to 0.3 Ohms – a 59-fold reduction in radiating resistance! (Note that with such a small radiating resistance, a significant amount of the antenna resistance is due to the resistivity of the conductor.) Note also that the inductances have fallen modestly, and the antenna capacitance has increased, so that the resonant frequency of the antenna hasn't changed much. The result of falling radiation resistance and decreasing reactance is a huge increase in Q: to about 200 for the antenna alone! We shall need to make some trade-offs to deal with this extreme value.

The T-match transforms the very tiny radiation resistance of the antenna to a modestly higher value of around 1 ohm at 900 MHz, increasing to 2 ohms at 930 MHz. The reactance is about j135 ohms at 900 MHz, increasing to j175 ohms at 930 MHz; a conjugate match with the chip

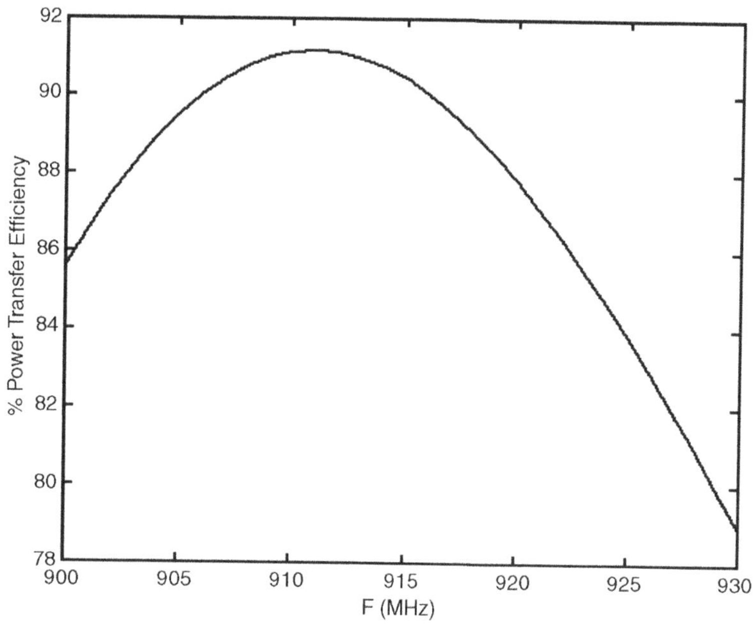

Figure 9.35 Power transfer efficiency of the FAT tag from Figure 9.33 suspended in air

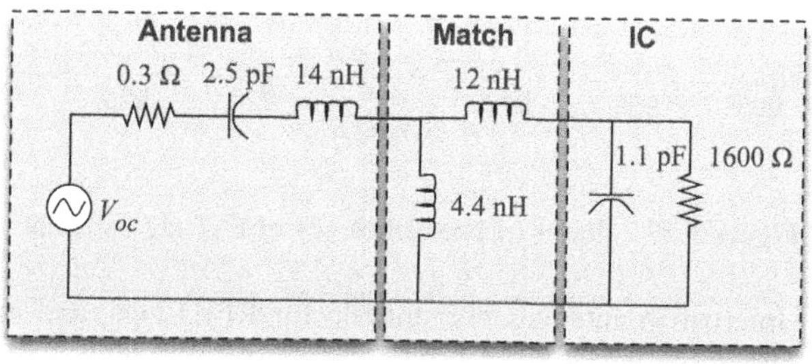

Figure 9.36 Equivalent circuit model of FAT tag on large flat metal surface

occurs at 915 MHz. From Equation (9.16) at the center of the frequency band we obtain $\tau \approx 4(2)(16)/(18)^2 = 40\%$. So the T-match is very helpful in improving power transfer efficiency. The computed power transfer efficiency is shown in Figure 9.37; it reaches about 40% at resonance, as expected, but drops off very rapidly towards the edges of the band, due to the substantial net reactance contribution to the impedance (or equivalently, due to the large value of Q).

Note that when radiation resistance is this small, the resistive losses in the antenna structure can be significant, reducing the radiative efficiency. These losses depend on the materials and means of construction used; radiative efficiency can be as low as 25–50% for this antenna, which uses wide copper traces; it will be much worse with narrow silver-ink traces. On the other hand, when the antenna is over a conductive ground plane, it becomes much more directive. As long as the reader is in the beam of the tag antenna, these effects nearly compensate. The net result is that at band center, the tag performance is only degraded by about 2 to 4 dB relative to free space. Since range goes as the square root of received power, this corresponds to only a 10–40% decrease in read range; we have observed ranges up to 10 meters for FAT tags on metal under good conditions.

At the band edges, read range is much shorter, but in those cases where sufficient time is available, we can rely upon the reader to hop to a frequency where the tag can be read. The increased directivity also means poorer performance at high angles to the normal direction. Thus, these tags may be quite acceptable for applications such as asset tracking, but perform poorly with a high speed conveyor reader.

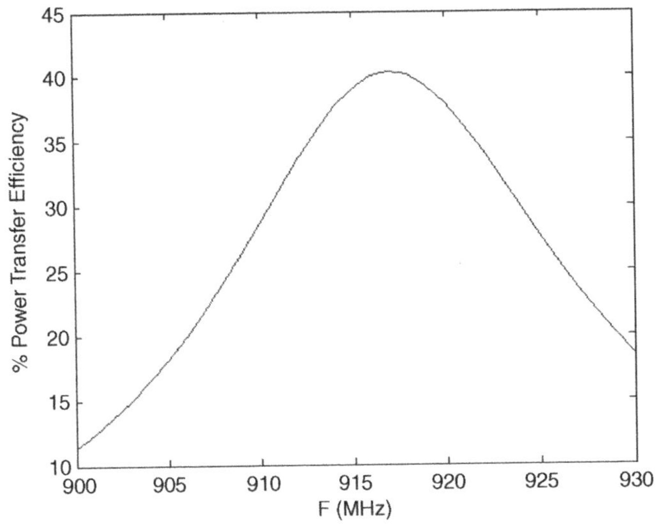

Figure 9.37 Power transfer efficiency of FAT tag on metal

We can see that microstrip antennas are suitable for RFID tags, and moreover that it is possible to design structures that can act both as microstrip antennas and as dipoles in air. Although some performance trade-offs occur, the increase in directivity that results from the presence of a metal ground plane partially compensates the reduced efficiency resulting from the decrease in radiation resistance.

9.4 Fabrication

The driving force in determining fabrication approaches for passive RFID tags is, unsurprisingly, minimizing cost. Cost reduction drives us to consider additive patterning techniques (which reduce the use of conductive materials), and antennas with reduced conductive areas (to further limit costly conductors). Fabrication cost is also reduced by minimizing the number of steps required to form the completed device.

The completed tag consists of an integrated circuit, sometimes mounted on a conductive strap, attached to an antenna structure placed on a substrate, typically a thin plastic. This is known as an inlay. An inlay may be employed in that form, or embedded within an adhesive label, which is then referred to as a *smart label* (Figure 9.2). Less commonly, the inlay is placed within a plastic package for outdoor use.

9.4.1 Antenna

9.4.1.1 Plating and Etching

Conventional photolithography and etching of electroplated copper have been used for fabrication of RFID tag antennas. While this process can yield excellent results and has been a standard in the industry for decades, some problems arise in RFID applications. The plating solutions and etchants are awkward to dispose of in an environmentally-sound fashion. In addition, the process removes most of the copper that was deposited, and thus the most expensive input is used rather wastefully. Most RFID tags produced today do not use a plate-and-etch approach, but it can be very useful for short production runs of specialty tags, or tags that require high conductivity. The thin microstrip antenna shown in Figure 9.38 is an example where maximum conductivity will have a noticeable impact on performance.

Figure 9.38 An etched-copper antenna

9.4.1.2 Silver Ink Screen Printing

Silver inks are an intriguing alternative to photolithographic methods. Silver inks are basically small silver particles, either flakes or spheres, which are suspended in a solution. The inks can then be applied by any number of very simple processes. Silk screen printing is possible and common, for example. Other processes such as gravure are possible. Ink jet printing is possible, but can be challenging because the silver particles might clog very small nozzles. Some inks use very small "nano" silver particles that can be used in an ink jet printer, but sacrifice some cost and performance.

After printing, the ink must be cured, usually in an oven at an elevated temperature. During the curing, the silver particles migrate towards each other, and in some cases, the surface of the silver particles start to become physically attached to the next particle, forming a dense, conductive web. Curing times and processes vary, and is the only difficult part of the process.

The obvious benefits of silver inks are that they are easy to use, and it doesn't require the purchase of a large set of expensive equipment and chemicals to get started. A silk screen printer and an oven are sufficient. Unfortunately, there are some downsides. First, silver ink is made from silver, and silver is expensive. Second, though the cured silver inks are conductive, studies show that printed conductive ink is only about 1/10th as conductive as copper, and 1/6th as conductive as aluminum of comparable thickness. To get similar conductivities, one needs relatively wide, thick traces, which drives up costs. Third, there is some concern about the potential environmental effects of silver ink, though proponents of the technology dismiss the fear as unfounded. Finally, silver corrodes upon exposure to air; an example of at least cosmetic degradation in a printed antenna after extended air exposure is depicted in Figure 9.39. (The antenna appears to continue to still perform well despite the corrosion.) Copper also corrodes, though more slowly, and while aluminum oxidizes rather rapidly, the layer of aluminum oxide forms a protective layer that usually prevents further corrosion.

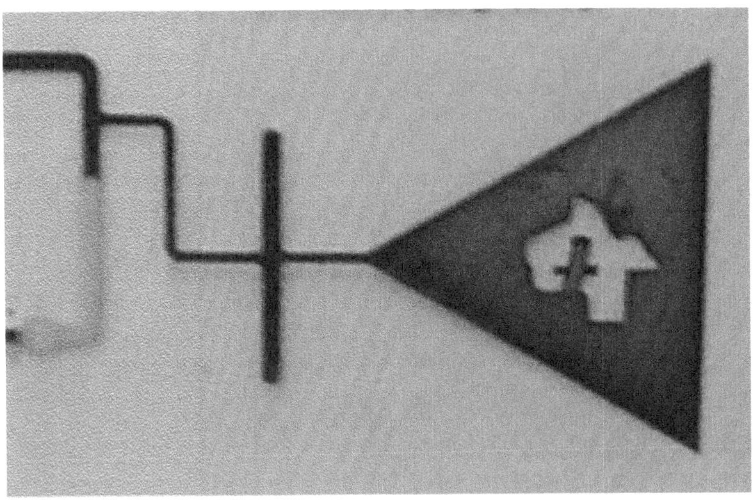

Figure 9.39 A five-year-old silver-printed tag showing signs of oxidization

9.4.1.3 Vapor Deposition

Metal deposition via evaporation or sputtering is well known and widely available. Evaporation is typically performed at very low pressures ($< 10^{-5}$ Torr), where the mean free path is large compared to the size of the vacuum chamber, and metal atoms can be regarded as traveling in straight lines. An antenna can be formed by evaporating aluminum through a shadow mask. This approach is simple and potentially inexpensive, in that large batches can be prepared in a single deposition step. However, the metal will accumulate on the shadow mask, necessitating periodic chamber cleaning. Sputtering can be used to deposit copper and other metals, but as it is performed at a higher pressure (a few Torr) in argon, the metal vapor scatters readily and cannot be accurately patterned during deposition.

9.4.1.4 Deposition and Laser Ablation

In a laser ablation process, one again starts with a thin copper or aluminum layer on a polyester substrate. The metal layer is usually very thin, so it may be formed on the polyester through evaporation or sputtering. Then, a high powered laser literally vaporizes the metal off of the surface of the polyester. Obviously, this requires special infrastructure and a high quality, uniform metal layer. The benefit of this approach is that it is highly automated, and lends itself to custom fabrication; it is in principle possible to make every antenna unique, and small production runs of hundreds or thousands of antennas are perfectly practical. However, the method is relatively slow and inappropriate for high-volume, low-cost fabrication.

9.4.1.5 Printing and Plating

Selective electroplating can be used to form RFID tag antennas. The electroplating process starts with deposition of a thin conductive layer of some printable material, such as a silver ink or a carbon ink, on to a polyester film. Then, the film is dipped into a chemical bath containing copper sulfate, and a negative voltage is applied to the conductive layer, causing the copper to be selectively deposited on the conductive region to form an antenna.

Selective electroplating of a continuous roll of RFID tags requires specialized equipment and safety precautions, and has not been widely adopted. The authors are not aware of any commercial use of electroplated RFID tag antennas.

9.4.1.6 Electroless (Chemical) Deposition

Electroless plating, in which the oxidation and reduction reactions take place locally within the solution, can be used to form metal layers without the need for an external current source (Figure 9.40). With electroless plating, one again starts by printing a pattern using a special catalyst ink, followed by a curing step. Then, the pattern is placed in a chemical bath containing an electroless plating solution. Once the process starts, the process continues as long as the antenna is immersed in the solution. The copper thickness is thus proportional to the time spent in the solution. The deposited material is not pure copper, though it retains at least 90% of its conductivity.

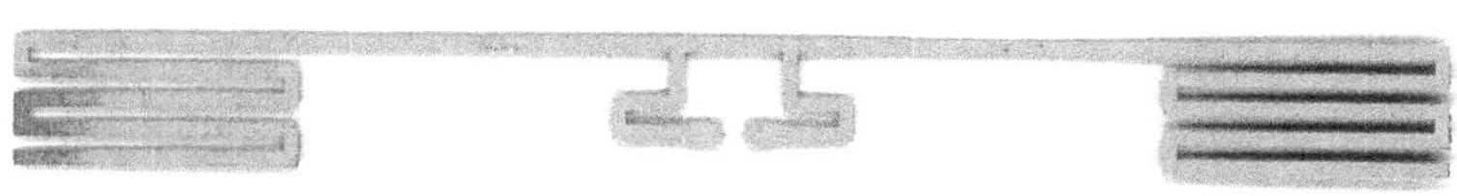

Figure 9.40 An electroless-copper antenna

The benefit of this approach is that it is much less expensive to set up than a selective electroplating process. Unfortunately, the metal deposition rate is low, so the antenna may need to be in the solution for a long time, reducing throughput and increasing cost.

9.4.1.7 Die Cut

Many people who have experimented with antenna design find that it's easy to build a "quick and dirty" antenna using copper foil, a design printed on a label, a sharp knife, and a straight edge. The tolerance of the antenna isn't particularly good, but you can do a lot of exploratory work without any expensive equipment, and it only takes 10–15 minutes and simple tools to go from concept to implementation. Figure 9.41 shows one of the early patch antennas we designed using such a technique.

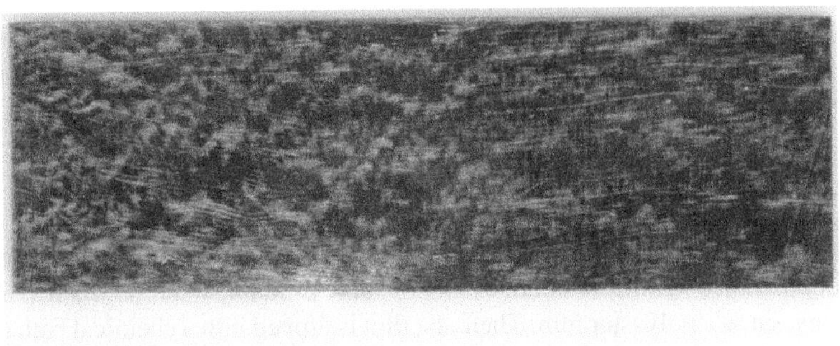

Figure 9.41 An example of an antenna made by hand by cutting copper foil

Not surprisingly, a more precise cutting edge can produce remarkably precise antennas. One problem with thin metal foils is that they stretch easily, and that's not a property that you want from an antenna. So it's common to use copper or aluminum that's been bonded to polyester. Polyester is inexpensive, resists stretching, but is flexible, which is why it is so popular with RFID. So one typically starts with a polyester film and forms a thin layer of copper or aluminum by vapor deposition or adhesion. Then, the polyester side is laminated with a thin layer of pressure-sensitive adhesive and a release liner. Next, the antenna pattern can be cut using a rotary die cutter, laser, or some other means, and the unwanted metalized polyester is stripped away, leaving the antenna. An optional step may include cutting the same pattern out of another polyester film of identical thickness and removing the antenna pattern, leaving only the negative of the antenna pattern. The two webs may be joined in order to provide a film of uniform thickness independent of the antenna pattern. Figure 9.42 shows an example of a die cut antenna.

Figure 9.42 A die-cut antenna

The benefit of this approach is that it can be performed fairly inexpensively and at very high speeds using roll-to-roll processes; the appropriate equipment is widely available. However, resolution is limited, and the IC must be mounted on a strap. Finally, the antenna design cannot have a void that is completely surrounded by metal, since there is no easy way to remove the void once it has been cut.

9.4.2 *Assembly*

9.4.2.1 Bondwires

Wire bonding has been widely used in electronic assembly for decades. A thin (50 to 100 micron diameter) wire of aluminum or gold is pressed onto a conductive pad, with the aid of ultrasonic scraping and/or heating to break through the native oxide and other surface contaminants, achieving intimate electrical contact. To bond an IC to an antenna, the IC must be placed on the antenna face-up, and held in place by an adhesive, in close proximity to a pair of conductive pads whose surfaces are exposed.

Wire bonding is reliable, mechanically robust, and provides a low-inductance contact if proper design techniques are used. However, the chip must be precisely placed, and each bond is created individually. Even when highly automated bonding machines are employed, the cost is substantial compared to the target cost of a few pennies for a completed RFID inlay. Wire bonding is appropriate for small-volume applications.

9.4.2.2 Flip-Chip and Anisotropic Conductive Adhesives

Flip-chip assembly has been widely explored since being introduced by IBM in the late 1960s. In this approach, after IC fabrication is complete, metal bumps are formed on the surface of the contact pads to the chip. The bumps are several tens of microns high, and may be formed of solder, or of a solderable metal such as tin-plated copper. The chip is then inverted over appropriate conductive pads, possibly also covered with solder, and heated. The solder melts, and surface tension causes the chip to self-center over the appropriate pad regions. Alternatively, the bumps may be placed directly on to an antenna and held in place by a small epoxy bond, which is cured rapidly by heat while pressure is applied to the IC. The resulting electrical connection may be less reliable, but cheaper to produce.

Flip chip assembly based on soldering is reliable and provides low-inductance connections. However, creation of the appropriate solder pads on typical RFID substrates such as polyimide is awkward and relatively expensive. A less costly approach, widely employed in the fabrication of flat-panel displays, is to use anisotropic conductive adhesives for the attachment step. These adhesives contain a low-concentration dispersion of conductive particles. They are substantially insulating under normal circumstances, but when the material is compressed under the metal pad of an IC, the conductive particles are pressed into each other and the metal surfaces, creating a contact between the chip and the substrate. Anistropic conductive adhesives are inexpensive to place, and have the additional benefit that the contact region is self-aligned to the mating of the chip and substrate conductive pads.

9.4.2.3 Straps

Straps are essentially a simple, low-cost package for an IC. The strap itself is much larger than the IC; an IC might be 0.5mm on a side, while a strap can be 4 by 8mm or larger. In the simplest form, shown in Figure 9.43, the strap contains an IC that has been attached using a flip-chip assembly. A small antenna-like pattern is printed or etched used one of the techniques described above (those in Figure 9.43 are etched aluminum) on a narrow, repeating pattern, then the chip is placed on the web using a flip-chip assembly method. In the resulting package, the IC is

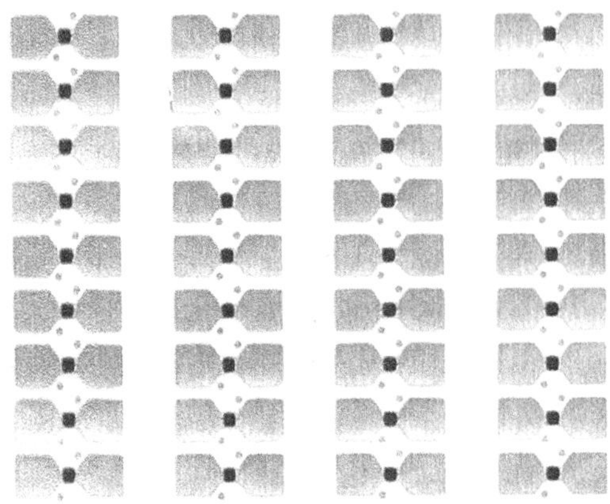

Figure 9.43 Example of straps with bare chips

exposed, as is a relatively large conducting surface. These straps are then slit into a narrow web, rolled, and placed on a spool (similar to an old-style projector spool).

In a laboratory environment, straps are very convenient because they can be placed onto antennas by hand using conductive adhesives. Several of the prototype antennas shown in this chapter are designed to use straps that are manually attached. (The clear tape on top of the strap to physically secure the strap, as shown in Figure 9.41 for example, is a sure sign of a hand-attached strap.) The antenna and strap can be misaligned by a large fraction of a millimeter with only modest effect on performance, so it's a process that can easily be done by hand. In contrast, misaligning a chip by a hundredth as much – a few microns – is the difference between a good connection and a tag that doesn't work.

Straps are also used in high volume manufacturing. The concept is that the IC-to-strap process can be specialized and performed at high speeds because of the density of straps (number of straps per square cm). Then, straps can be attached to antennas using more traditional roll-to-roll processing similar to the way a label can be applied to a piece of paper. High speed processes are important to reducing costs because the same number of machines and people can produce more tags per hour. High volume manufacturers, including Avery Dennison and Alien, use straps almost exclusively in their high volume product lines. Alien and Avery Dennison represent a large fraction of the total UHF RFID tags sold globally. Both companies have committed significant resources to using straps to make tags.

9.4.2.4 Fluidic Self-assembly

Fluidic self-assembly is a method developed by Alien Technology for very high production throughput of straps. The "traditional" strap is formed using any number of chip-bonding techniques, especially flip-chip assembly. This is a sequential process in which a robot arm picks up the IC, flips it over, then applies the IC to the strap substrate, and proceeds to the next. While a number of robot arms can work in parallel, this is still a fairly sequential process. The concept with FSA is that this process can be done in a massively parallel way using natural processes.

A cut-away view of the FSA strap is shown in Figure 9.44. The IC is not cut square, but rather cut with a slight taper. This taper can be put into the silicon by etching; the etchant will not etch the silicon straight down, but because of the very careful controlled crystalline structure of silicon, the etchant will etch away the silicon at a very predictable angle. The inverse of that IC angle can be removed from a plastic film, which will serve as the substrate. A large number of ICs are suspended in a fluid and circulated within a system. When a sheet or roll of the substrate is placed in the system, the IC-filled fluid is moved across the surface of the substrate. Occasionally, an IC will find the opening in the substrate and slide in. Once in the substrate, the IC will remain in the recess. Because large numbers of chips can be passed over the substrate in a relatively short amount of time, it doesn't take very long for most of the substrate to be filled with chips.

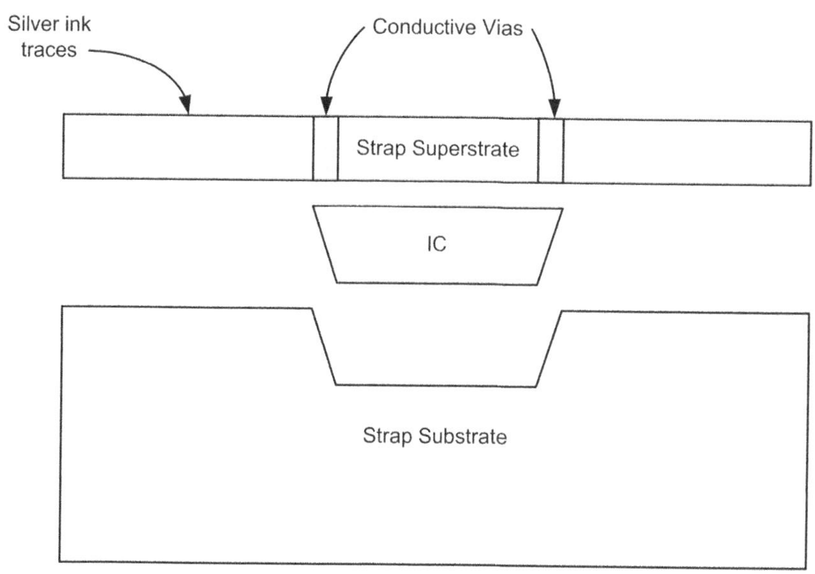

Figure 9.44 Cut-away view of a FSA-assembled strap

Once the great majority of holes are filled with chips, the substrate is removed and processed. Next, the superstrate layer is laminated to the substrate, fully encapsulating the IC. Holes are made and filled with conductive material, and silver ink traces are printed on the top side of the strap to give a large conductive area to attach to the antenna. The resulting sheet of straps is shown in Figure 9.45.

Obviously, the FSA process allows large amounts of ICs to be placed into straps rapidly, and straps can be placed onto antennas rapidly, so this FSA process is amenable to a very high volume production process. Alien has used this kind of FSA since its initial entry into the RFID market. Unfortunately, as of May 2009, it is not clear whether the world-wide market is sufficiently large to support this process.

Another added benefit of this process is that the strap presents a robust package to the IC. The IC is completely encapsulated by plastic, and therefore less vulnerable to the environment. Silicon is very brittle and will break relatively easily under impact, and the flip-chip bond of a tiny IC is vulnerable and weak because of the small bonding area. The strap fully protects the

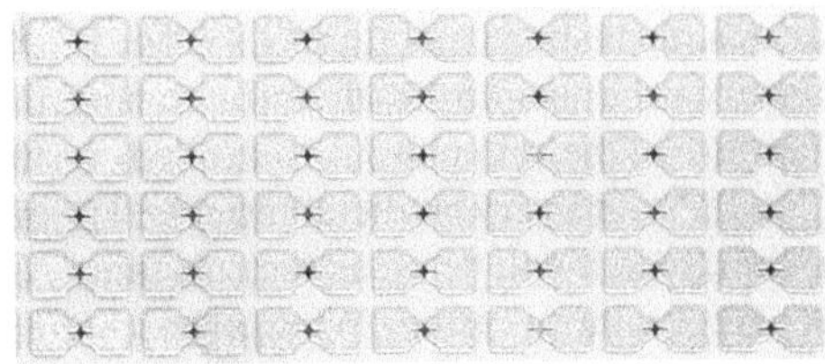

Figure 9.45 A sample of straps produced by FSA. Note that some substrate holes are not populated with chips

silicon from a significant amount of impact, and the surface of the strap presents a very large area for robust bonding. As a result, tags made from this method tend to be more rugged than their flip-chip counterparts.

9.5 Conclusion

In this chapter, we have examined printed antennas applied specifically to RFID. One of the largest drivers in RFID is price, and printed antennas allow RFID to be price-competitive. We described the theoretical factors that affect printed dipole antennas, and ways to mitigate the negative effects of electrically short dipole antennas. We showed how to construct an antenna using short conductive traces, and presented the circuit theory that can be used to obtain broadband impedance matching. We cover some of the basics of impedance matching for microstrip antennas, showing that the same techniques for dipole antennas are applicable. We present an example of a particular antenna that behaves both as a dipole and microstrip dipole, depending on the environment as an example of the kinds of trade-offs that are required for electrically small, narrowband microstrip antennas. Finally, we cover some of the manufacturing issues with RFID, including fabricating the antenna and methods of attaching the IC to the antenna.

It is not possible to cover all of the issues related to RFID tags in one chapter, and space constraints cause us to neglect a number of interesting and contemporary issues, such as near-field UHF antennas. We hope that the reader comes away with an appreciation of the fundamental limits of RFID antennas, and techniques that are used to overcome some of the challenges, as well as the pragmatic requirements around keeping tag costs as small as possible. After reading this chapter, readers should be able to apply their new knowledge and be able to make useful and informed observations about the tags presented in Figure 9.3.

Acknowledgments

Some images in this chapter are taken from *The RF in RFID* [18], and are used by the kind permission of the publisher.

References

1. P. Raumonen, L. Sydanheimo, L. Ukkonen, M. Keskilammi, and M. Kivikoski, "Folded dipole antenna near metal plate," *IEEE Antennas and Propagation Society International Symposium*, vol. 1, pp. 848–851, 2003.

2. L. Ukkonen, D. Engels, L. Sydanheimo, and M. Kivikoski, "Planar wire-type Inverted-F RFID tag antenna mountable on metallic objects," *IEEE Antennas & Propagation Symposium Monterey*, Tampere University/MIT, 2004.
3. J. Siden, P. Jonsson, T. Olsson, and G. Wang, "Performance degradation of RFID system due to the distortion in RFID tag antenna," *11th International Conference on Microwave and Telecommunications Technology (CriMiCo 2001)*, p. 371, Sevastopol, Crimea, Ukraine, September. 2001.
4. S. R. Aroor and D. D. Deavours, "Evaluation of the state of passive UHF RFID: an experimental approach," *IEEE Systems Journal*, vol. 1, no. 2, pp. 168–176, 2007.
5. K. Rao, D. Duan, and H. Heinrich, "On the read zone analysis of radio frequency identification systems with transponders oriented in arbitrary directions," *APAC Microwave Conference*, Intermec/IBM, p. 758, 1999.
6. S. Basat, K. Lim, I. Kim, M. Tentzeris, and J. Laskar, "Design and development of a miniaturized embedded UHF RFID Tag for automotive tire applications," *Electronic Components and Technology Conference*, vol. 1, p. 867, 2005.
7. P. Nikitin, K. Rao, S. Lam, V. Pillai, R. Martinez, and H. Heinrich, "Power reflection coefficient analysis for complex impedances in RFID tag design," *IEEE Transactions on Microwave Theory and Techniques*, vol. 53, no. 9, p. 2721, 2005.
8. P. Nikitin, S. Lam, and K. Rao, "Low cost silver ink RFID tag antennas," *IEEE Antennas and Propagation Society International Symposium*, vol. 2B, p. 353, 2005.
9. C. Chihyun, C. Hosung, and I. Park, "Design of UHF small passive tag antennas," *IEEE Antennas and Propagation Society International Symposium*, vol. 2B, p. 349, 2005.
10. L. Ukkonen, D. Engels, A. Sydanheimo, and M. Kivikoski, "Reliability of passive RFID of multiple objects using folded microstrip patch-type tag antenna," *IEEE Antennas and Propagation Society International Symposium*, vol. 2B, p. 341, 2005.
11. N. A. Mohammed, M. Sivakumar, and D. D. Deavours, "An RFID tag capable of free-space and on-metal operation," *IEEE Radio and Wireless Symposium*, San Diego, CA, January 18–22, 2009.
12. K. Rao, P. Nikitin, and S. Lam, "Antenna design for UHF RFID tags: a review and a practical application," *IEEE Transactions on Antennas and Propagation*, vol. 53, no. 12, pp. 3870–3876, 2005.
13. K. Rao, P. Nikitin, and S. Lam, "Impedance matching concepts in RFID transponder design," *Fourth IEEE Workshop on Automatic Identification Advanced Technologies*, p. 39, Oct. 2005.
14. H. Kwon and B. Lee, "Meander line RFID tag at UHF band evaluated with radar cross-sections", *Proceedings of the 2005 Asia Pacific Microwave Conference*, 2005.
15. J. Joe and M. Chia, "Voltage, efficiency calculation and measurement of low power rectenna rectifying circuit," *IEEE Antennas and Propagation Society International Symposium*, vol. 4, p. 18554, 1998.
16. M. Sivakumar and D. Deavours, "A dual-resonant microstrip antenna for paperboard in the cold chain," *IEEE Sarnoff Symposium*, 2008.
17. D. D. Deavours, "Analysis and design of wideband passive UHF RFID tags using a circuit model," *IEEE RFID*, Orlando, FL, April 27–28, 2009.
18. D. M. Dobkin, *The RF in RFID*, Chapter 7, Elsevier-Newnes, Oxford, 2007.

10

Printed UWB Antennas

Zhi Ning Chen, Xianming Qing and Shie Ping See
Institute for Infocomm Research, Singapore

10.1 Introduction

This chapter covers the basic issues related to ultra-wideband (UWB) technology, such as the emission masks for UWB systems and promising applications of UWB technology. The design challenges in UWB antenna design from an engineering point of view are also highlighted. A short review of the state-of-the-art UWB antennas, especially for printed planar UWB antenna designs is given. Following this Introduction, an update on the latest technologies developed for printed UWB antennas is provided. The innovative technology that has been developed to reduce the ground plane effect on the performance of small printed UWB antennas is introduced. Based on this technology, a slim printed UWB antenna for wireless universal serial bus (USB) dongle and a compact diversity antenna are illustrated. Last, a printed wide slot/aperture UWB antenna and its variations are discussed in detail. All the antennas in the case studies have been numerically optimized and experimentally verified.

The term ultra-wideband (UWB) system has been synonymously associated with terms such as impulse, carrier-free, baseband, time-domain, non-sinusoidal, orthogonal function, and large-relative-bandwidth radio/radar systems. The basic feature of the UWB system is the occupancy of extremely wide operating bandwidth due to the use of impulse signals as compared to conventional radios. Since the late 1960s, the technologies related to UWB such as transmitters, receivers, radio frequency signals, antennas and systems have been presented and developed mainly for radio and radar systems, in particular, for military applications [1].

For the UWB technology to benefit the public and at the same time protect against the possible interference to other existing electronic systems, the Federal Communications Commission (FCC), USA released the First Report and Order on February 14, 2002 [2]. The order is for the unlicensed use of UWB systems below the 900 MHz band and 3.1–10.6 GHz band. The unlicensed commercial applications include imaging systems, vehicular radar

Microstrip and Printed Antennas: New Trends, Techniques and Applications. Edited by Debatosh Guha and Yahia M.M. Antar

systems as well as communication and measurement systems. In order to regulate the emission from UWB systems, the effective isotropic radiated power (EIRP) emission for indoor and outdoor UWB based communication and measurement systems should be less than -41.3 dBm/MHz as shown in Figure 10.1. The realized 10-dB fractional bandwidth for communication systems is about 109%.

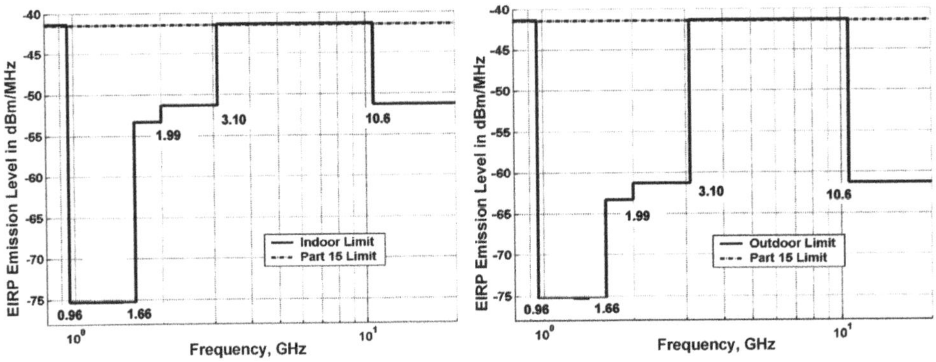

Figure 10.1 Emission masks for commercial UWB communications and measurement systems by FCC. Reproduced by permission of ©2010 I^{2}R

$$\text{Bandwidth} = 2 \times \frac{\text{frequency}_{\text{upper edge}} - \text{frequency}_{\text{lower edge}}}{\text{frequency}_{\text{upper edge}} + \text{frequency}_{\text{lower edge}}} \times 100\% \approx 109\% \tag{10.1}$$

It should be noted that to avoid the possible interference between UWB devices and other electronic systems, particularly the wireless communications systems operating within the UWB band of 3.1–10.6 GHz, several countries have avoided the usage of the spectrum of 4.8–6.2 GHz. This is due to the possible interference between the UWB systems and other systems such as the USA Public Safety system at 4.9 GHz and wireless fidelity (WiFi) system at 5.150–5.825 GHz.

Due to the unique features of UWB technology, there have been many promising UWB-based applications in communications. For example, in wireless communications, the extremely wide operating bandwidth has the potential for high data-rate connections, e.g. up to 480 Mbps. However, the very low emission level has limited the wireless connection range to a few meters. As a result, the UWB radio technology is more likely to be applied in consumer electronics such as handset/laptop-centric and home networks which require short-range but high data-rate wireless solutions. The possible UWB-enabled devices include the wireless USB dongles or/and next-generation Bluetooth modules which can be embedded in handsets, laptops, and other consumer electronic devices as shown in Figure 10.2.

From the promising applications of UWB technology in wireless communication systems, it is expected that UWB-enabled devices will be deployed in portable/wearable consumer electronics. Such applications add to the challenges in UWB antenna design. These challenges

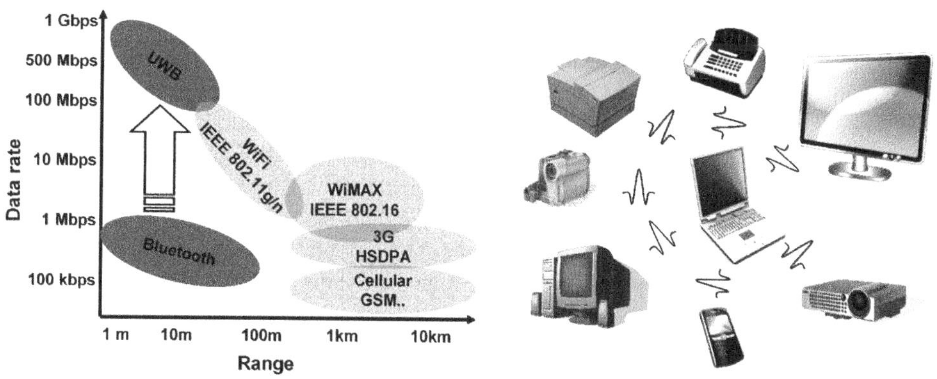

Figure 10.2 UWB-based wireless communication technology and their promising applications in consumer electronics. Reproduced by permission of ©2010 I^2R

differentiate the UWB antenna design from conventional narrowband and even broadband antenna design as summarized in Table 10.1. The high data-rate requires an operating bandwidth that is as wide as possible not only in terms of the impedance and radiation-related performance, but also the phase response in impulse-based systems [3]. The antennas for portable or wearable devices have severe size constraints due to limited space in the devices. Therefore, small and compact UWB antennas are highly desirable. The miniaturization of UWB antenna can be conducted by designing antennas that have a physically small form factor which usually leads to an electrically small design, thus making the design much more challenging than conventional narrowband antennas. Another method to miniaturize the antenna is to have functionally small designs where multiple antennas are used to incorporate more functions in the device without any additional increase in size. The low-cost solution is one of the critical factors for the successful adoption of UWB devices in consumer electronic products. The antenna engineers have to take the factors listed in Table 10.1 into consideration during the early stages of their design so as to make the antennas acceptable for mass production.

Table 10.1 Challenges and important considerations in UWB antenna design [3]

Challenges	Key Considerations
Ultra-wideband	• Frequency range: 3.1–10.6 GHz, 3.1–4.8 GHz, or 6.2–9.8 GHz • Gain: Consistent over the designed frequency band • Impedance matching: VSWR < 2 for 50 ohms input impedance • Beamwidth: Consistent • Polarization: Unchanged • Phase: Linear
Small size	• Physically small: Embeddable, easy integration with other circuits. Low profile • Electrically small: Overall size smaller than $\lambda/4$ (e.g. 25 mm @ 3 GHz) • Functionally small: Diversity, filtering, band-notch
Low cost	• Materials: Inexpensive • Process/Tooling: Simple mechanical structure • Installation: Easy to install with high reliability and robustness • Tuning: Minimal tuning required to shorten the time-to-market

Source: Reproduced by permission of ©2010 I^2R.

Over the years, there have been many broadband and also frequency-independent antenna designs, such as the log-periodic antennas, spiral antennas, Vivaldi antennas, bi-conical antennas, disk-cone antenna, TEM horns, and self-similar spiral antennas, etc. [4–8]. However, these antennas are unsuitable for UWB-based wireless communication systems since they suffer from either the frequency-dependent phase center or bulky configurations. Planar monopole antennas have been investigated for broadband applications for many years [9–19]. These antennas feature broad bandwidth, small size, and low cost. The broadband planar monopoles are usually installed above the ground plane or circuit board, resulting in a high profile which limits the application of such antennas in portable/wearable consumer electronics.

For UWB-based wireless communication systems, a printed UWB antenna is preferred. The small and low profile printed antenna is easy to integrate with the UWB radio, which lowers the assembly cost and makes the system more robust. Therefore, much effort has been devoted to the design of printed UWB antenna [20–24].

Figure 10.3 shows three typical printed UWB antenna designs which have been widely studied and used in UWB systems. The design in Figure 10.3(a) is a planar elliptic monopole printed onto a piece of printed circuit board (PCB) slab and fed by a microstrip line. A piece of conductor that is etched on the bottom end at the back of the PCB slab is grounded to form the "ground plane" of the antenna. The upper radiator can be of any shape. The antenna radiates in a similar way to a dipole antenna. The overall size of the antenna including the ground plane controls the lower edge frequency of the operating bandwidth. The impedance matching of the antenna is sensitive to the gap between the bottom edge of the upper radiator and top edge of the ground plane. The design in Figure 10.3(b) is a co-planar waveguide (CPW) fed wide slot antenna which features bi-directional radiation. The T-shaped feeding structure excites the

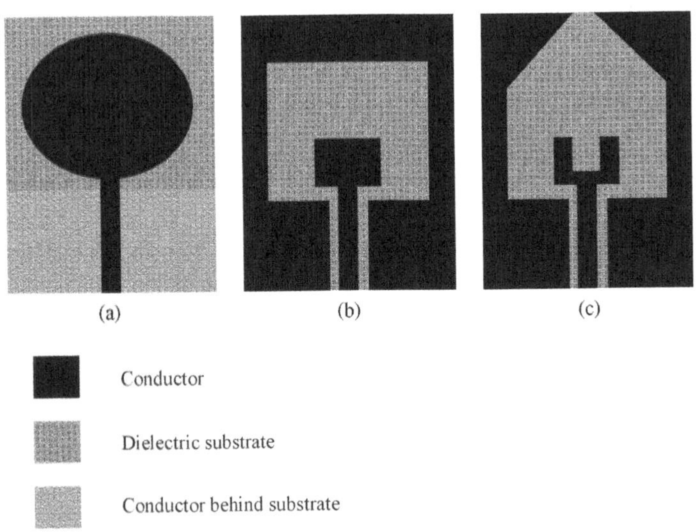

Figure 10.3 Three typical printed UWB antennas: (a) a microstrip-fed monopole antenna; (b) a CPW-fed wide slot antenna; (c) a CPW-fed monopole-like slot antenna. Reproduced by permission of ©2010 I^2R

slot. The feeding structure can be optimized using arbitrary shapes to achieve a wideband impedance matching. Similarly, the wide slot antenna can also be excited by a microstrip feeding structure similar to that shown in Figure 10.3(a). The design in Figure 10.3(c), also known as a monopole-like slot antenna, is an improvised variation of Figure 10.3(b). The slot is open at the top end and the antenna is smaller than the original version.

With broadband feeding structures, many printed UWB antennas are capable of providing the desired performance across the entire UWB band of 3.1–10.6 GHz, lower (3.1–4.8 GHz) or upper (6.2–9.8 GHz) UWB band as listed in Table 10.1. Some of the latest progress in UWB antenna design will be addressed in this chapter. First, the technology to greatly reduce the dependence of small UWB antenna performance on the ground plane, which is one of most challenging issues related to small antenna design, will be highlighted. Next, the printed UWB antenna with a slim profile will be proposed and applied to a compact diversity antenna design. Finally, two designs, namely the wide slot antenna and monopole-like slot antenna will be introduced and elaborated. All the antennas have been studied by simulation and validated by measurement. The important information derived from the parametric study will also be provided in detail.

10.2 "Swan" Antenna with Reduced Ground Plane Effect

As mentioned above, when the antenna is designed with a small physical size, usually the overall size of the antenna will also be electrically small. The antennas used in portable devices do not have a large ground plane but a small floating system ground plane which is used as a zero-potential reference for systems. The system ground plane actually acts as the other radiating element of the antenna, therefore the overall size should include both the antenna and system ground plane [25]. Figure 10.4(a) shows an example of a printed UWB antenna on a finite-sized system ground plane. The antenna has achieved good impedance matching across the 3.1–10.6 GHz band when the length of the ground plane is 9 mm as shown in Figure 10.4(b). When the length l is varied from 4 mm to 29 mm, the impedance matching significantly changes. In particular, the first resonance shifts down when the length l increases. This implies

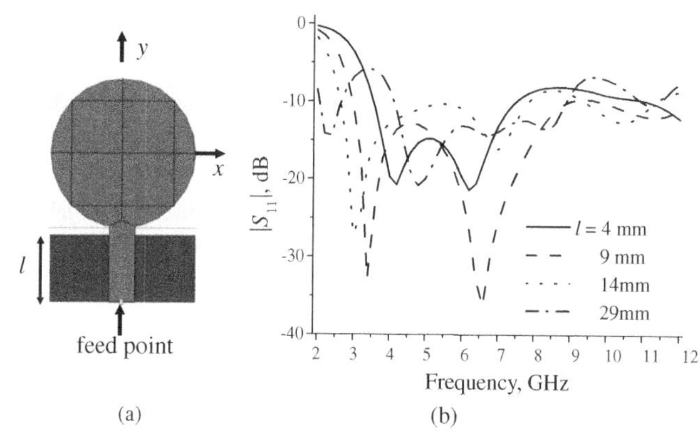

Figure 10.4 (a) A printed microstrip-fed monopole UWB antenna; (b) $|S_{11}|$ response against frequency with varying overall (ground plane) lengths. Reproduced by permission of ©2010 I^{2}R

that the antenna operates as an asymmetrical unbalanced-fed dipole as shown in Figure 10.5. Therefore, any changes made to the ground plane in terms of the shape, size, and position of the upper radiator will affect the performance of printed antennas, especially when the overall size of the antennas is electrically small.

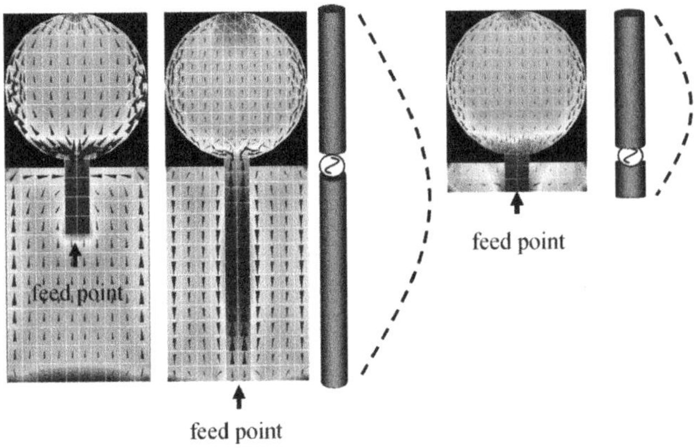

Figure 10.5 Printed UWB monopole antennas acting as asymmetrical unbalanced dipoles. Reproduced by permission of ©2010 I^2R

The small printed UWB monopole or wide slot antennas also face such design challenges, where the ground plane has severe effects on antenna performance in terms of impedance matching, resonant frequency, gain, and radiation patterns during measurement [23]. Such effects result in the difficulty of maintaining the desired antenna performance, especially when they are installed or embedded in devices where there is a change in the ground plane conditions. Another challenge lies in measuring the antenna performance with acceptable accuracy in the presence of the RF connector and cable when the ground plane is electrically small in size.

Differential antennas, such as the dipole-like and loop-like designs have been proposed to reduce the effect of the ground plane on antenna performance [26, 27]. The size of such antennas is larger than original monopole versions. The effect of the ground plane is partially alleviated because the ground plane is still close to the antennas and acts as a reflector or parasitic element. During the measurement, the currents flowing on the RF connectors and cables can be suppressed by using an RF choke [28, 29] or absorbed by a ferrite core [23]. However, these methods can only alleviate the dependence of antenna performance on ground plane during the measurement of small antennas; the designs themselves are still inherently dependent on the ground plane.

Section 10.2.1 will present a method to reduce the ground-plane effect on the performance of a small printed UWB antenna by notching the upper radiator. The radiator and ground plane of the antenna are etched onto a piece of printed circuit board (PCB) with an overall size of 25 mm × 25 mm × 1.5 mm. A notch is cut from the radiator while a strip is asymmetrically attached to the radiator. In particular, the ground plane effect on the impedance performance is greatly reduced because the electric currents on the ground plane are significantly suppressed, especially at the lower edge operating frequency.

10.2.1 Antenna Design

Figure 10.6 shows the printed UWB antenna with a rectangular notch on the radiator [30]. The antenna is etched onto a piece of PCB (RO4003, $\varepsilon_r = 3.38$, $\tan\delta = 0.002$ and 1.52 mm in thickness). The notch ($w_s \times l_s$) is cut onto the upper radiator close to the attached strip ($w_{rs} \times l_{rs}$) at a distance, d_s. The impedance matching can be achieved by changing the feed gap, g, or/and the position of feed point, d. Two bevels are cut to further improve the impedance matching, particularly at the higher frequencies. The length of the ground plane, l_g was designed to keep the overall size of the antenna less than 25 mm × 25 mm. The optimal parameters of the design are: $w_s \times l_s = 4\,\text{mm} \times 12\,\text{mm}$, $w_{rs} \times l_{rs} = 2\,\text{mm} \times 6\,\text{mm}$, $d = 6\,\text{mm}$, $d_s = 4\,\text{mm}$, $g = 1\,\text{mm}$, and $l_g = 9\,\text{mm}$. The feeding strip is 3.5 mm in width as shown in Figure 10.6(a). A Cartesian coordinate system (x, y, z) is oriented such that the bottom surface of the PCB lies in the x-y plane. With the optimal parameters, the design was fabricated as shown in Figure 10.6(b).

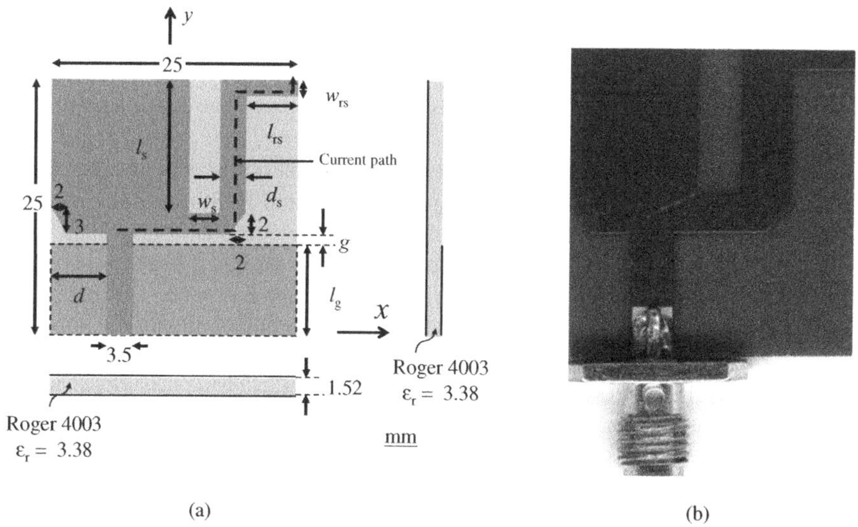

Figure 10.6 A printed notched UWB antenna: (a) geometry; (b) prototype. Reproduced by permission of ©2010 I^{2}R

Figure 10.7 compares the simulated and measured $|S_{11}|$ response against frequency of the antenna, which are in good agreement. It is seen that the design achieves good impedance matching above 2.95 GHz. Well-matched multiple resonances are excited simultaneously for an ultra-wideband operation.

It is of interest to note that in the simulation, the antenna is fed by a delta-type source at the bottom end of the feeding strip with a 50-Ω internal resistance, and without any RF connector and feeding cable. However, in the measurement, a radio frequency (RF) cable through a 50-Ω SubMiniature version A (SMA) connector was connected to the bottom end of the feeding strip and grounded to the edge of the ground plane of the antenna on the reverse side of the PCB. As discussed previously, during such small antenna measurement, the RF cable usually affects the performance of the antenna greatly, particularly at the lower resonant frequency.

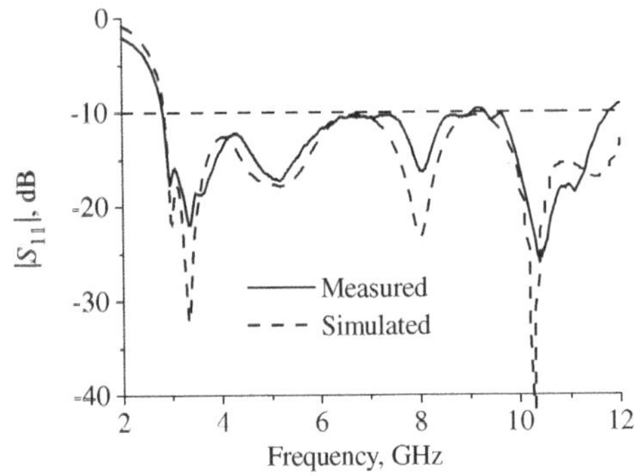

Figure 10.7 Comparison of simulated and measured $|S_{11}|$. Reproduced by permission of ©2010 I^{2}R

However, from Figure 10.7, the presence of the RF cable hardly affects the lower edge of the operating band at around 3 GHz. This strongly suggests that without any additional devices or increase in the size of the antenna, the design has achieved ground plane dependence in terms of the impedance matching.

Figure 10.8 illustrates the simulated current distributions on the antennas with/without the notch. The current at Region A in both Figures 10.8(a) and 10.8(b) is much weaker than that on Region B. In Figure 10.8(a), it is evident that the electric current around the right portion of the radiator (Region B) where the notch was cut is stronger than the current on the feeding strip, while the current on the left portion of the radiator as well as the ground plane (Region A) are much weaker in comparison. The notch has significantly affected the current distribution at the lower operating frequency. For the "Swan" antenna, the radiation from the portion around the notch is dominant, while the radiation is mainly from the region around the feed gap of the antenna without notch as shown in Figure 10.8(b). As a result, the

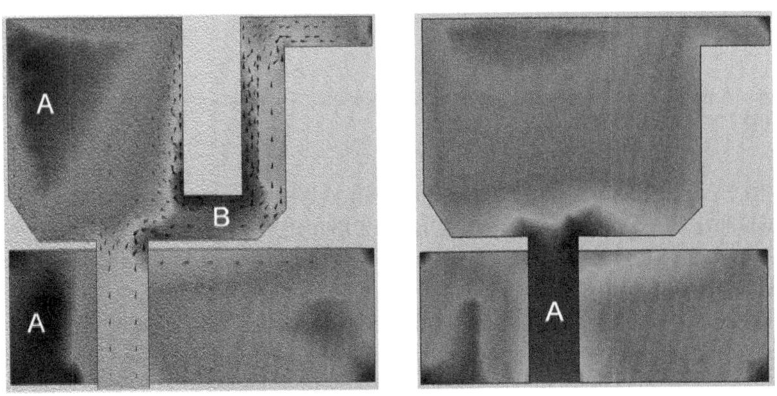

(a) (b)

Figure 10.8 Distribution of electric currents on antennas at 3 GHz: (a) with notch; (b) without notch. Reproduced by permission of ©2010 I^{2}R impedance matching of the "Swan" antenna is more sensitive to the notch than the ground plane because of the weaker electric currents on the ground plane. Therefore, the effects of the ground plane and RF connector/cable on the antenna performance can be alleviated significantly.

From the three-dimensional radiation measurement, the antenna efficiency of 79% to 95% across the operating bandwidth of 3.1–10.6 GHz was obtained. For such omni-directional small antenna design, the antenna efficiency is of greater interest than the peak gain.

10.2.2 Parametric Study

The parametric study shows that the performance of the antenna is mainly affected by geometrical and electrical parameters, such as the dimensions related to the notch, top strip, feeding strip, feed gap, ground plane as well as the dielectric constant of the substrate. The notch-related geometrical parameters including the dimensions of the notch (w_s, l_s) and location of the notch d_s have significant effects on the impedance matching, especially at the lower operating frequencies. As shown in Figure 10.9, increasing l_s shifts down the lower edge frequency by extending the current path. However, the width and location of the notch (w_s, d_s) hardly affect the lower edge frequency. The top strip has the size of $w_{rs} \times l_{rs}$. Figure 10.10 shows that a longer top strip l_{rs} shifts down the lower edge frequency by increasing the overall size of the antenna. However, the width of the top strip w_{rs} hardly affects the impedance matching. Finally, the impedance response is also affected by the dielectric constant ε_r as shown in Figure 10.11. It can be seen that a change in the dielectric constant ε_r leads to a shift in the characteristic impedance of the feeding strip from 50-Ω.

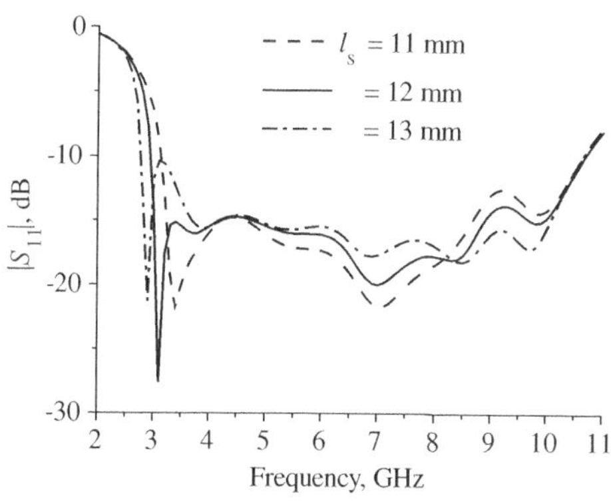

Figure 10.9 Effects of varying notch length l_s on impedance matching. Reproduced by permission of ©2010 I^2R

The impedance matching is very sensitive to the feed gap, g, especially at higher frequencies. Also, the length of the ground plane l_g affects the impedance matching more significantly at higher frequencies than at lower frequencies.

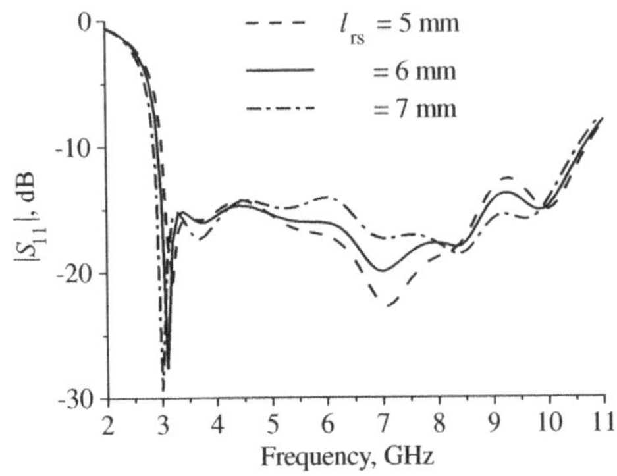

Figure 10.10 Effects of varying top strip length l_{rs} on impedance matching. Reproduced by permission of ©2010 I^2R

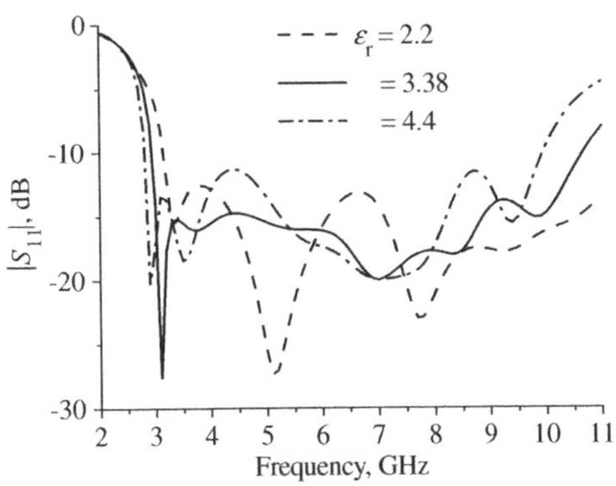

Figure 10.11 Effects of varying ε_r on impedance matching. Reproduced by permission of ©2010 I^2R

The key feature of this design is the notch on upper radiator. The characteristics of the two designs, namely with and without the notch are compared. In order to examine the effect of the notch on the impedance matching, the notch on the "Swan" antenna was removed while all the other dimensions were kept unchanged. The antenna without the notch has been optimized while maintaining the overall size of 25 mm × 25 mm. As shown in Figure 10.12, it can be seen that the antenna without the notch is only able to achieve a lower edge frequency of 3.7 GHz. Therefore, the notch not only reduces the effect of ground plane on the antenna performance but also miniaturizes the antenna size by increasing the length of the effective current path.

In conclusion, the printed UWB antenna can be designed with the reduced ground plane effect by notching the upper radiator. This is because the notch is able to enhance the radiation from the upper radiator, in particular, at the lower operating frequency. Compared

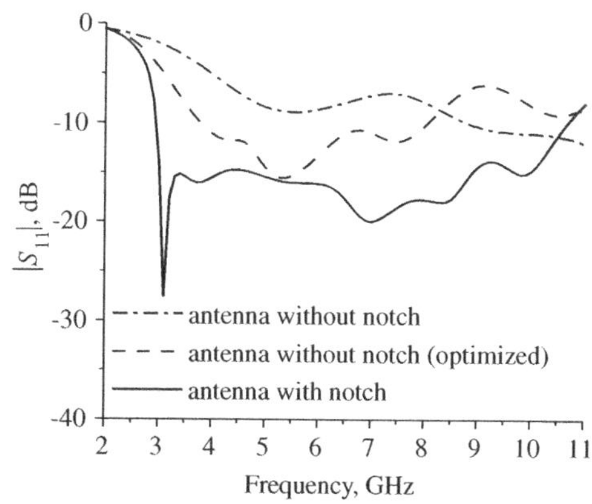

Figure 10.12 Effect of the notch on impedance matching. Reproduced by permission of ©2010 I^2R

to the existing technologies, this design also features the advantages of miniaturized overall size and simple antenna configuration.

10.3 Slim UWB Antenna

One of the greatest challenges in UWB antenna design is to miniaturize the antennas but maintain broad impedance bandwidth and high radiation efficiency. In this section, a slim UWB antenna is presented for wireless USB applications. The UWB antenna design for wireless USB has been a challenging task due to the size constraint [31].

10.3.1 Antenna Design

Figure 10.13(a) shows the geometry of the UWB antenna and a Cartesian coordinate system. The antenna is printed onto an 11 mm × 45 mm dielectric substrate. The dielectric substrate (Rogers 4003C) has a thickness of 0.8 mm and a relative dielectric constant of 3.38. The antenna consists of a triangular radiating element and an L-strip of 1 mm width extended from the end of the radiator for impedance matching. The length of the horizontal and vertical sections of the L-strip is 11 mm and 3 mm, respectively. The feed gap between the radiator and the system ground plane is 1 mm. The radiator is excited by a 50-Ω coaxial probe of radius 0.6 mm at the end of the microstrip line of a 1.85 mm width. In practical terms, the feed point of the antenna can be connected to the RF circuit on the reverse side of the PCB, or to other layers in a multilayer configuration with the use of a via. A 2 mm-wide slot is cut on the radiator in order to reduce the antenna size and effect of the cable on the impedance matching, especially at the lower operating frequency. The slot is able to increase the length of the current path so that the lower edge frequency can be reduced.

On the other side of the PCB, the ground plane has a vertical strip as well as a horizontal strip extended perpendicularly from the vertical strip which can be used to control the impedance matching of the antenna. The antenna area is approximately 20.5 mm × 11 mm, which leaves

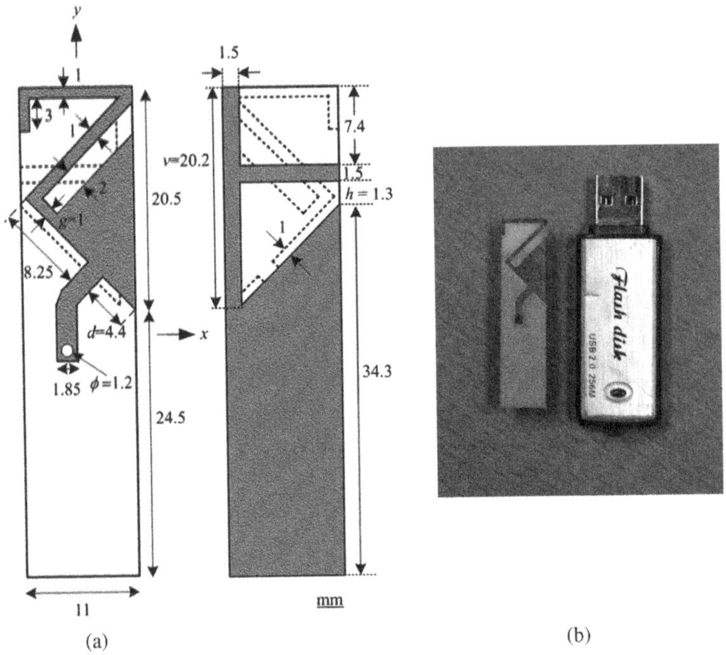

Figure 10.13 Slim UWB antenna: (a) antenna geometry; (b) photo of antenna and wireless USB dongle. Reproduced by permission of ©2010 I^{2}R

an area of 24.5 mm × 11 mm for other circuits or modules. Figure 10.13(b) shows the antenna prototype placed beside a typical USB dongle.

From Figure 10.14, it can be seen that the antenna has achieved an impedance bandwidth of 3.1–5 GHz for $|S_{11}| < -10$ dB. Besides, it is noted that the lower edge frequency corresponds closely to the simulations which implies that the effect of the cable has been minimized.

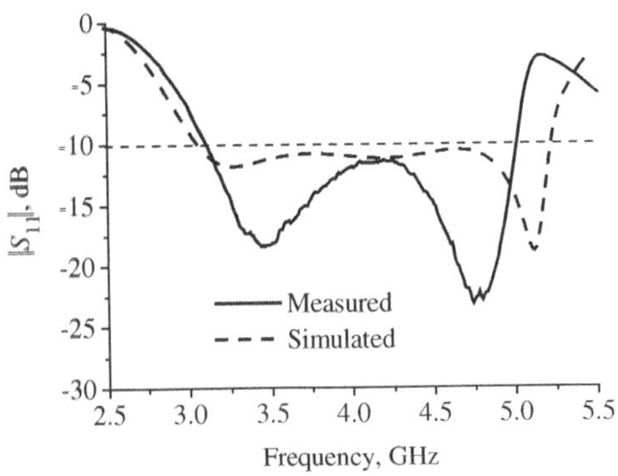

Figure 10.14 Measured and simulated $|S_{11}|$. Reproduced by permission of ©2010 I^{2}R

The radiation characteristics of the antenna were investigated across the impedance bandwidth of 3.1–5 GHz. Figure 10.15 shows the measured radiation patterns at 3.1 GHz, 4 GHz, and 5 GHz in the three principal planes, namely the *x-z*, *y-z* and *x-y* planes.

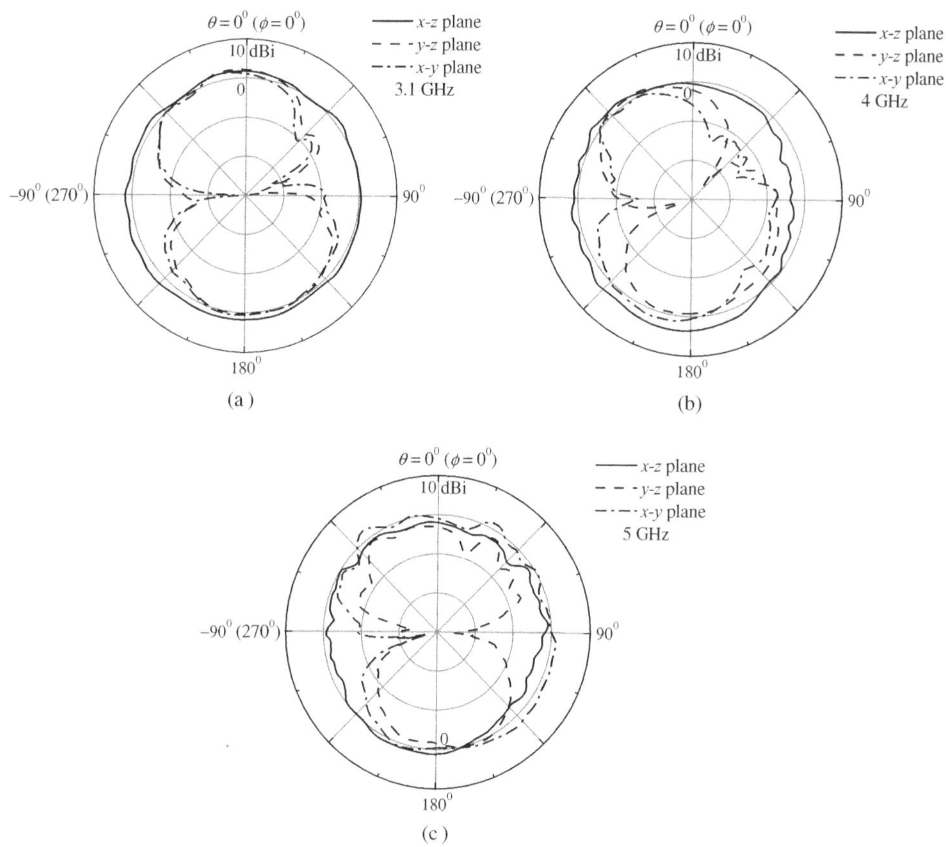

Figure 10.15 Measured radiation patterns in the *x-z*, *y-z*, and *x-y* planes at: (a) 3.1 GHz; (b) 4 GHz; (c) 5 GHz. Reproduced by permission of ©2010 I^{2}R

The measured antenna efficiency over 3–5 GHz is at least 60% across the bandwidth. On average, the antenna has an efficiency of about 75%. The peak efficiency is 88% at 3.25 GHz.

The measured maximum and average total gain in the *x-y*, *x-z*, and *y-z* planes across the impedance bandwidth are shown in Figures 10.16(a) and (b), respectively. The total gain is defined as the sum of the E_θ and E_ϕ components. From the measured results, it is seen that the maximum total gain in the *x-z* plane is less than 3.7 dBi, with a variation of 2.4 dBi across the bandwidth. In the *x-y* and *y-z* planes, the maximum total gain is lower than 3.6 dBi and 2.5 dBi, respectively. As the antenna is to be used in portable devices or integrated into the wireless USB dongle in an indoor environment, it will be useful to know the average gain as well. In order to calculate the average gain, the gain for both the E_θ and E_ϕ components at a certain point along a specific cut is measured, and the gain for the total field is then calculated. With this process

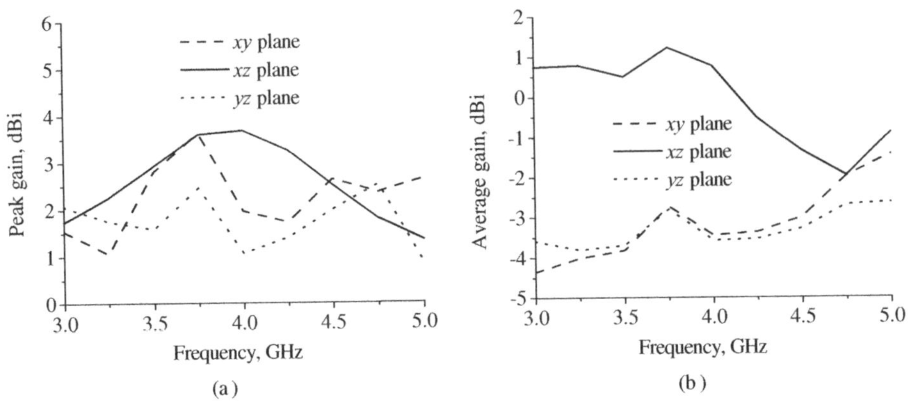

Figure 10.16 Measured maximum and average gain: (a) maximum gain; (b) average gain. Reproduced by permission of ©2010 I^{2}R

repeated N times, the average gain for the total field can be computed by taking the geometric mean of the gain along a specific plane as shown in Equation (10.2):

$$G_{average}(f, \theta, \phi) = \frac{\sqrt[N]{\prod_{n=1}^{N} G_n(\theta_n, \phi_n, f)}}{N} \tag{10.2}$$

where, G_{average} stands for the average gain of the total field, namely the sum of the E_θ and E_ϕ components along a specific plane; $G_n(\theta_n, \phi_n, f)$ is the gain measured at a particular orientation (θ_n, ϕ_n); and N is the total number of measured G_n.

It can be observed that the variation of the average gain in the x-z plane is from -2 dBi to 1.5 dBi. The average gain in the y-z and x-y planes is lower, which varies from -4.5 dBi to -1.5 dBi. The lower average gain is due to the radiation nulls along the y-axis.

10.3.2 Parametric Study

A parametric study was carried out to investigate the effects of the important parameters on the impedance matching. The effects of varying the width of the vertical strip (v) and the position of the horizontal strip (h) on the ground plane, the length of the notch (g) on the radiator as well as the position of the feed point (d) are studied. In the simulations, the values of the parameters are the same as that shown in Figure 10.13(a), except for the parameter under investigation.

The introduction of the notch on the radiator has a significant influence on the impedance matching and isolation at the lower edge frequency as shown in Figure 10.17(a) [30]. With the notch, the size of the antenna can be reduced due to the majority of the current predominantly concentrated near the notch region at the lower operating frequency of 3 GHz. Generally, an increase in the notch length, which corresponds to a decrease in g, is able to decrease the lower edge frequency.

The position of the feed point (d) can be used to control the impedance matching of the pass-band from 3.5 GHz to 4.5 GHz as observed from Figure 10.17(b). With a decrease in the f, the

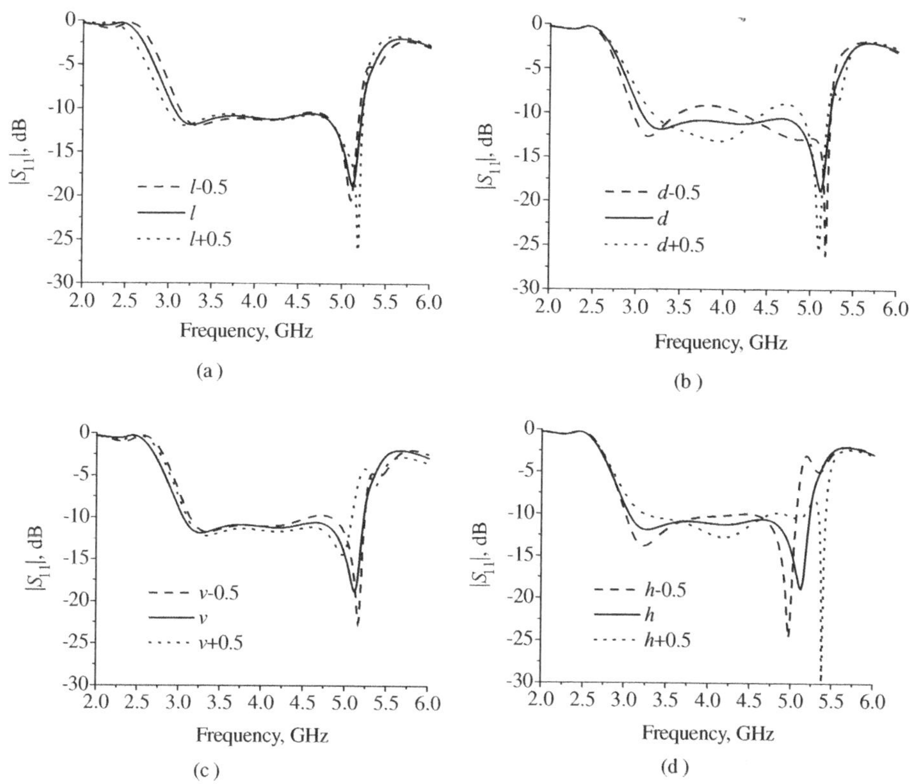

Figure 10.17 Effects of varying the (a) *g*; (b) *d*; (c) *v*; (d) *h* on the impedance matching. Reproduced by permission of ©2010 I^{2}R

lower edge frequency can be reduced but at the expense of degrading the matching at the center frequency.

Also, as the two resonances are spaced further apart, the matching at the center frequency will be deteriorated. The shape of the ground plane can be varied in terms of the width of the vertical strip (v) as well as the position of the horizontal strip (h). As shown in Figure 10.17(c), the lower edge frequency decreases slightly with an increase in the width. However, increasing the width further will lead to an increase in the lower edge frequency and a decrease in the upper edge frequency, resulting in a reduction in the impedance bandwidth. The two resonances are also sensitive to the position of the horizontal strip. From Figure 10.17(d), a decrease in h will improve the matching at the lower resonance but the upper edge frequency is reduced significantly.

10.4 Diversity Antenna

As in the case of conventional radio systems, the multi-band OFDM-based UWB systems are vulnerable to multipaths. As such, the signals may combine destructively at a receiver, causing fading to occur. However, the reliability of the system can be improved through the use of diversity techniques, which is achieved by using the information from the different branches available to the receiver so as to increase the signal-to-noise (S/N) ratio at the decoding stage. In an indoor environment, pattern (spatial) diversity is used due to the random polarization and direction of the incoming signals [32–34]. In pattern diversity, the antennas with different beam patterns on each branch are used. The multipath components are weighted differently at each channel, creating different interference patterns of the signal at each branch. Therefore, each channel will receive the transmitted signal with different strengths depending on the branch pattern and the propagation characteristics at that moment in time.

The diversity antenna designed for portable devices in consumer electronics, such as PCs and mobile devices will be a challenging task due to the space constraint. It will be essential to maintain low mutual coupling when the antenna elements are placed in close proximity to each other, as well as stable radiation performance across a broad operating bandwidth. The close proximity of the antennas not only makes the antenna system more compact, but also enhances the pattern directivity which is desirable for pattern diversity. However, due to the increased mutual coupling, the correlation between the channels will be increased and the system efficiency will be reduced. On the other hand, the diversity antennas have been applied in PCMCIA cards in laptop computers but with large size in order for the antenna elements to be spaced sufficiently far apart so as to achieve good isolation [35–37].

In this section, a compact UWB diversity antenna for portable devices is presented. As an example, the antenna is designed to cover the lower UWB band of 3.1–5 GHz and has high average gain. A method to calculate the transfer function of the antenna will also be discussed, from which the radiated pulses are obtained and analyzed.

10.4.1 Antenna Design

Figure 10.18 shows the geometry of the UWB diversity antenna and the Cartesian coordinate system. The antenna was printed onto a 37 mm × 45 mm × 0.8 mm PCB slab. The PCB (Rogers 4003C) has a relative dielectric constant of 3.38. The two identical radiators are

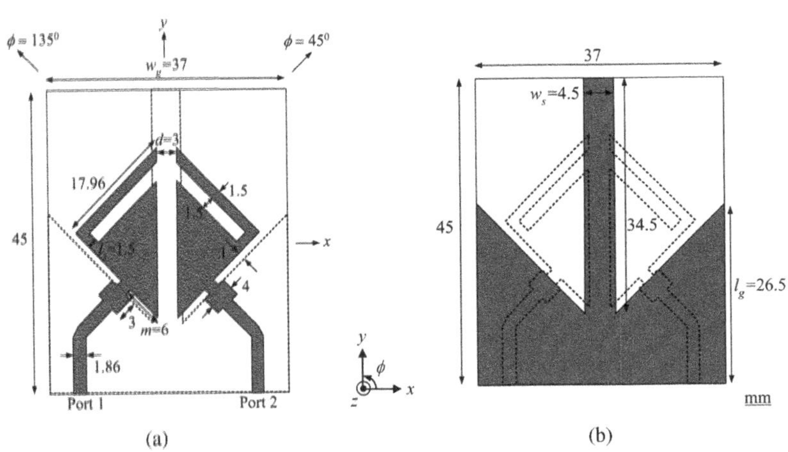

Figure 10.18 Antenna geometry: (a) front side; (b) reverse side. Reproduced by permission of ©2010 I^{2}R

separated 3 mm from each other and are symmetrically positioned with respect to the y-axis. The radiators are excited via 50-Ω microstrip lines of 1.86-mm width in a $\phi = 45°$ and 135° configuration. Each radiator has a 1.5-mm-wide notch etched in a direction parallel to the feeding strip. The purpose of cutting the notches is to concentrate most of the current on the radiators instead of the system ground plane, especially at the lower operating frequency, so that the effect of the ground plane and coaxial probe on the impedance matching and radiation can be greatly minimized [30, 38]. A stub measuring 4 mm × 3 mm on the feeding strip near to the radiator is used to enhance the impedance matching. There is a gap of 1 mm between the ground plane on the reverse side of the PCB and the bottom side of the radiators. In addition, the mutual coupling between the two radiating elements can be reduced by having a central strip that extends vertically from the ground plane [35].

The simulated and measured input reflection coefficient $|S_{11}|$ and isolation $|S_{21}|$ in Figure 10.19 show that the isolation within the impedance bandwidth is greater than 20 dB over the achieved bandwidth of 3.1–5 GHz. Furthermore, the measured lower edge frequencies for both the impedance matching and isolation correspond closely to the simulated ones, which imply that the effect of the cable has been minimized.

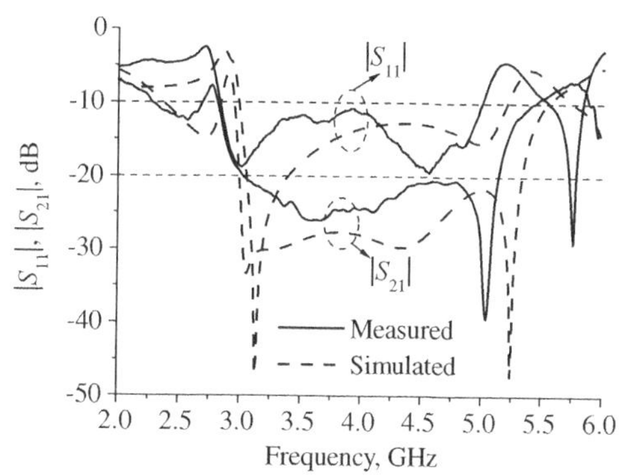

Figure 10.19 Measured and simulated input reflection coefficient and isolation. Reproduced by permission of ©2010 I^{2}R

Figure 10.20 shows the measured radiation patterns at 4 GHz in the three principal planes, namely the $\phi = 45°/225°$, 135°/315°, and $\theta = 90°$ (x-y) planes. In the measurements, Port 2 of the antenna was terminated with a 50-Ω load when Port 1 was excited, and vice versa. Comparing the patterns when Port 1 and Port 2 were separately excited, it can be observed that they cover complementary spatial regions symmetrically with respect to the y-z plane. Also, the radiation patterns are relatively stable across the impedance bandwidth.

Figure 10.21 shows the average gain and peak gain at the principal planes. In this study, only the radiation in one half-space will be of interest. For the $\phi = 45°/225°$ plane when Port 1 was excited and the $\phi = 135°/315°$ plane when Port 2 was excited, the average gain is computed for the right half-space region (i.e. $\theta = 0°–90°–180°$), whereas for the $\phi = 45°/225°$ plane when Port 2 was excited and the $\phi = 135°/315°$ plane when Port 1 was excited, the left half-space region from $\theta = 0°–90°–180°$ is considered. In the x-y plane, the average gain for the total field is calculated for the left half-space region (i.e. $\phi = 90°–180°–270°$) when Port 1 was excited

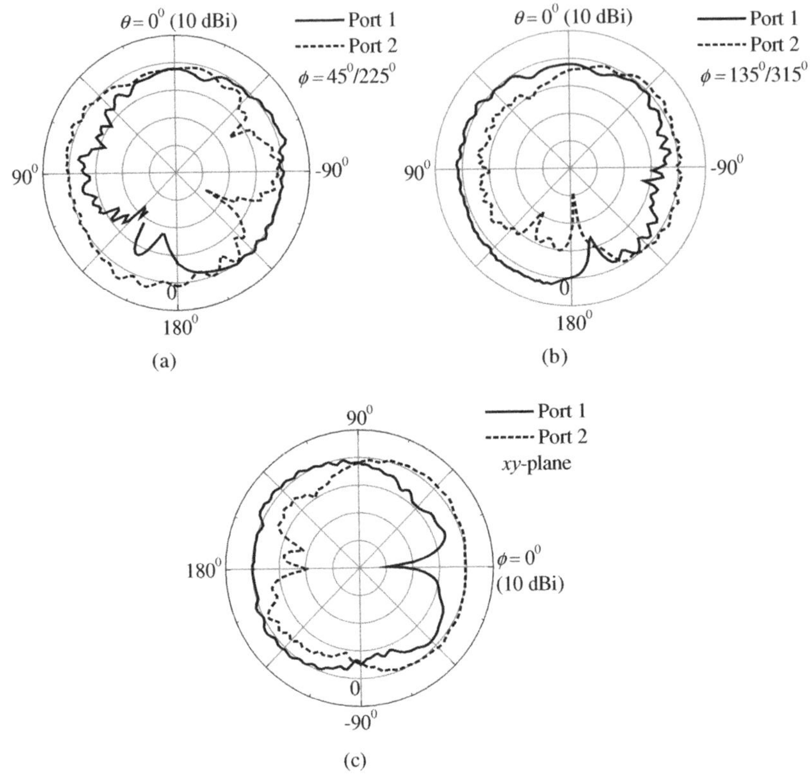

Figure 10.20 Measured radiation patterns with Port 1 and Port 2 excitation in the (a) $\phi = 45°/225°$ plane; (b) $\phi = 135°/315°$ plane; (c) $\theta = 90°$ (*x-y*) plane at 4 GHz. Reproduced by permission of ©2010 I²R

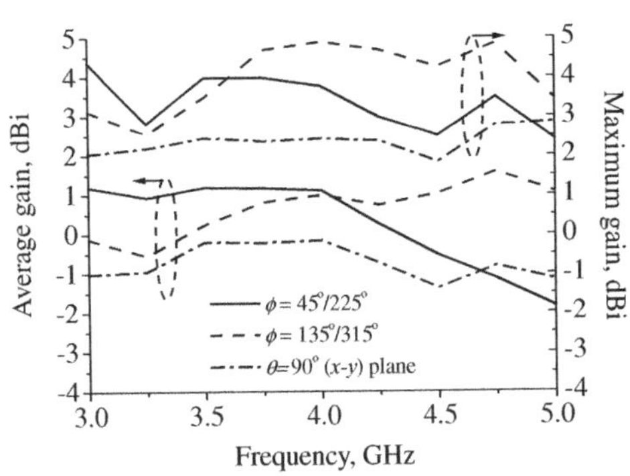

Figure 10.21 Measured average gain and maximum gain when Port 1 was excited. Reproduced by permission of ©2010 I²R

and for the right half-space region ($\phi = 90°$–$0°$–$270°$) when Port 2 was excited. The value of N used in the calculation of the average gain is 91. The measured peak gain is 2–3 dB higher than average gain over the bandwidth. This suggests that the radiation in such three planes is directional. The variation of the average gain is a good indicator of the stability of the radiation patterns within the bandwidth. From the figure, it can be observed that the average gain is greater than -2 dBi across the bandwidth. Also, it is noted that the variation in the average gain is within 3 dB at the three principal planes. Due to the symmetrical structure, the average and peak gain when Port 2 was excited will be the same as that in Port 1.

The measured efficiency is at least 70% across the bandwidth. On average, the antenna has a measured efficiency of about 80% with a peak efficiency of 93% at 4.75 GHz. The lower measured efficiency as compared to the simulation is due to the additional losses in the connectors and cables.

The size of the radiator can be reduced and yet maintain the lower edge operating frequency at 3 GHz as most of the current is concentrated around the notch and the central vertical strip instead of the ground plane, especially at the lower frequency of 3 GHz [30, 38]. In addition, since there is little current on the ground plane around the right radiator, mutual coupling between the two antenna elements is significantly reduced. For the different lengths of the ground plane, the current is mainly concentrated around the notch at 3 GHz. This suggests that good isolation can be maintained with little variation in the impedance matching at the lower operating frequency. In this way, the radiator can be optimized for a particular ground plane and used on another ground plane of a different length depending on the application requirements.

The envelope correlation coefficient is a figure-of-merit which can be used to gauge the diversity performance of the antenna. It measures the degree of similarity between the beam patterns of two antennas. When $\rho = 1$, the two patterns are identical. This implies that the signal received by Port 1 will be the same as that in Port 2, *i.e.* any fading will be experienced by both the two ports simultaneously. On the other hand, when $\rho \to 0$, there will be no overlap between the patterns of the two antennas. In this case, the incoming signal from any direction will only be received by at most only one of the two antenna elements. Although the fading might still occur depending on the direction of the incoming signal, it will be more robust than using a single omni-directional antenna. The low correlation may imply that there is little overlapping between the two beam patterns. The correlation coefficient shown in Figure 10.22 can be calculated using the S-parameters as shown in Equation (10.3) [39]. From Figure 10.22, the measured envelope correlation coefficient is lower than -25 dB across the UWB band of 3.1–5 GHz.

$$\rho = \frac{|S_{11}^* S_{12} + S_{21}^* S_{22}|^2}{(1-(|S_{11}|^2 + |S_{21}|^2))(1-(|S_{22}|^2 + |S_{12}|^2))} \tag{10.3}$$

Generally, for a given source pulse, the radiated pulses can be computed if the transfer function of the antenna-under-test (AUT) is known. It was proposed that by using a pair of identical antennas, the transfer function can be obtained by using the *ABCD* and S-parameters [40]. However, small antennas are usually less directive and have lower gain, thereby affecting the transmission performance. It will be desirable if the gain of a transmit or receive antenna is high in order to achieve a good transmission or S_{21} for a particular distance. As such, it will be useful if the transfer function of the AUT can be obtained when a pair of different antennas is used.

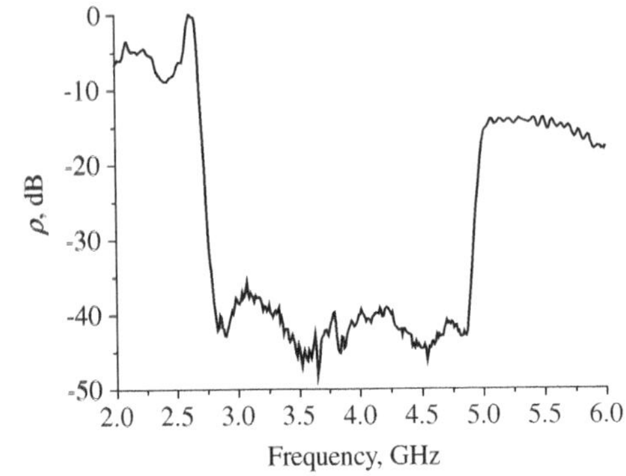

Figure 10.22 Measured envelope correlation. Reproduced by permission of ©2010 I^{2}R

In this study, in order to calculate the radiated pulses, the AUT is the transmit antenna. The receive antenna is a standard gain horn antenna (SGA). In general, the antenna system transfer function, $H(\omega)$ can be expressed as [40]:

$$H(\omega,\theta,\varphi,r)=\frac{Z_l(\omega)}{Z_l(\omega)+Z_{in2}(\omega)}H_{Rx}(\omega,\theta,\varphi).\frac{e^{-jkr}}{r}H_{Tx}(\omega,\theta,\varphi)\frac{Z_{in1}(\omega)}{Z_{in1}(\omega)+Z_s(\omega)}\hat{\rho}_{Rx}\bullet\hat{\rho}_{Tx} \tag{10.4}$$

where H_{Tx} and H_{Rx} are the transfer functions of the transmit and receive antennas. The impedances Z_s and Z_l are the reference source and load impedances of Ports 1 and 2, respectively; Z_{in1} and Z_{in2} are the input impedances of the transmit and receive antennas, respectively; r is the distance between the transmit and receive antennas (which ensures the transmit and receive antennas are in the far-field zone of each other); k is the free space wave number, and $\hat{\rho}_{Tx}$ and $\hat{\rho}_{Rx}$ are the unit vectors that indicate the polarization direction of the transmit and receive antennas, respectively.

Normalizing Z_{in1} and Z_{in2} with respect to the source and load impedances, respectively,

$$H(\omega,\theta,\varphi,r)=\frac{1}{1+\bar{Z}_{in2}(\omega)}H_{Rx}(\omega,\theta,\varphi)\cdot\frac{e^{-jkr}}{r}H_{Tx}(\omega,\theta,\varphi)\frac{\bar{Z}_{in1}(\omega)}{\bar{Z}_{in1}(\omega)+1}\hat{\rho}_{Rx}\bullet\hat{\rho}_{Tx} \tag{10.5}$$

where $\bar{Z}_{in1}$ and $\bar{Z}_{in2}$ represent the normalized input impedances of the transmit and receive antennas, respectively. The normalized input impedances can be expressed in terms of the S-parameters,

$$H(\omega,\theta,\varphi,r)=\left(\frac{1-S_{22}}{2}\right)H_{Rx}(\omega,\theta,\varphi)\cdot\frac{e^{-jkr}}{r}H_{Tx}(\omega,\theta,\varphi)\left(\frac{1+S_{11}}{2}\right)\hat{\rho}_{Rx}\bullet\hat{\rho}_{Tx} \tag{10.6}$$

Taking the ratio of the system transfer functions, where $H_{AUT\text{-}Tx,SGA\text{-}Rx}$ is the system transfer function with the AUT as the transmit antenna and the SGA as a receive antenna, and $H_{SGA\text{-}Tx,SGA\text{-}Rx}$ is the system transfer function when the SGA is used for both the transmit and receive antennas.

Since

$$H(\omega, \theta, \varphi, r) = \frac{S_{21}}{2} \tag{10.7}$$

where S_{21} is the complex transmission coefficient which takes into account the polarization matching factor, if $r_1 = r_2$,

$$\frac{S_{21(AUT-Tx,SGA-Rx)}}{S_{21(SGA-Tx,SGA-Rx)}} = \frac{H_{AUT-Tx}(\omega, \theta, \varphi)\left(\frac{1+S_{11(AUT-Tx)}}{2}\right)}{H_{SGA-Tx}(\omega, \theta, \varphi)\left(\frac{1+S_{11(SGA-Tx)}}{2}\right)} \tag{10.8}$$

The transfer function ($H_{SGA\text{-}Tx}$) of the transmit SGA can be calculated as [40]:

$$H_{SGA-Tx} = \sqrt{\frac{2S_{21(SGA-Tx,SGA-Rx)}}{(1+S_{11(SGA-Tx)})(1-S_{22(SGA-Rx)})}\frac{j\omega}{2\pi c}re^{jkr}} \tag{10.9}$$

The radiated pulse can then be obtained by multiplying the spectrum of the source pulse defined in Equation (10.10) with the transfer function of the AUT and the distance factor, $\frac{e^{-jkr}}{r}$, before taking the inverse Fourier transform. The normalized source pulse is plotted in Figure 10.23(a). Using Equation (10.9), the radiated pulses at ($\phi = 45°/225°$, $\theta = -90°$), ($\phi = 135°/315°$, $\theta = 90°$), and ($\theta = 90°$, $\phi = 180°$) were calculated and shown in Figures 10.23(b)–(d). These points were selected which correspond to the strong radiated fields when Port 1 was excited according to Figure 10.20. As an example, a Gaussian monocycle with $\sigma = 55$ ps was selected as the source pulse with the peak of its spectrum located at 4 GHz.

$$s(t) = -\frac{2t}{\sigma^2}e^{-\left(\frac{t}{\sigma}\right)^2} \tag{10.10}$$

In order to assess the ability of a coherent receiver to detect the received signal, a parameter known as fidelity was calculated using a template to correlate with the source pulse [14]. The distortion of the radiated pulses against the source pulse can be calculated by choosing the template to be the source pulse. Table 10.2 shows the fidelity calculated from the radiated pulses in Figure 10.23 when the template is the source pulse. Also, by selecting the optimal pulse parameter σ for the Gaussian monocycle, maximum fidelity can be obtained.

10.5 Printed Slot UWB Antenna and Band-Notched Solutions

A conventional printed microstrip narrow-slot antenna shows inherent narrow impedance bandwidth [41], while printed microstrip wide-slot (aperture) antennas have a much wider bandwidth. By increasing the width of the slot and positioning an open-ended feed line within the slot, the coupling between the feeding structure and the wide slot significantly enhances the bandwidth of the antenna [42]. A variety of wide-slot antennas have been reported [43–49]. Figure 10.24 illustrates some typical wide-slot antenna configurations, where the wide-slot is etched onto the ground plane of a PCB and fed by a microstrip line or CPW. The shape of the wide-slot can be square, rectangular, triangular, circular, elliptical, polygonal, etc. The typical feeding structures can be shaped as T, cross, fork, circle, triangle, square, arc, etc.

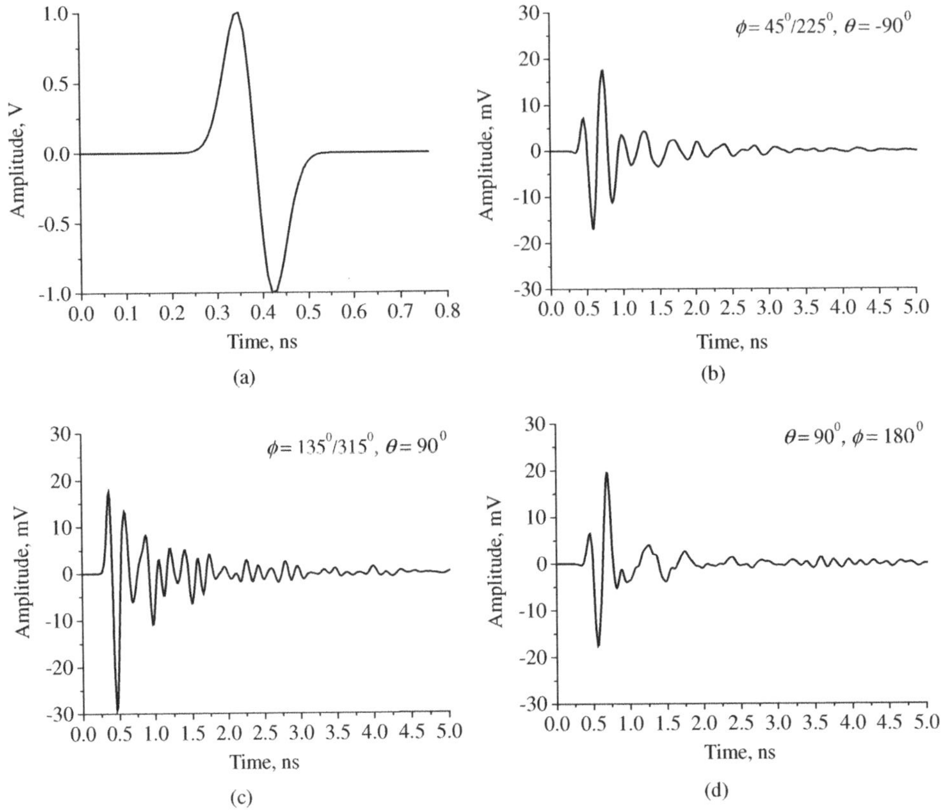

Figure 10.23 (a) Normalized source pulse. Radiated pulses when Port 1 was excited for (b) $\phi = 45°/225°$, $\theta = -90°$; (c) $\phi = 135°/315°$, $\theta = 90°$; (d) $\theta = 90°$, $\phi = 180°$. Reproduced by permission of ©2010 I^2R

Table 10.2 Fidelity of radiated pulses with varying templates

	$\phi = 45°/225°$, $\theta = -90°$	$\phi = 135°/315°$, $\theta = 90°$	$\theta = 90°$, $\phi = 180°$
Fidelity (template = source pulse)	0.693	0.700	0.775
Optimum fidelity, *F*/source pulse parameter, σ	$F = 0.803/\sigma = 80$ ps	$F = 0.709/\sigma = 62$ ps	$F = 0.869/\sigma = 76$ ps

Source: Reproduced by permission of ©2010 I^2R

10.5.1 Wide-Slot UWB Antenna

The printed wide-slot antenna is one of the most promising candidates for UWB applications due to the merits of wide bandwidth, low profile, low cost, lightweight, ease of fabrication and integration into other devices. Many wide-slot antenna designs have been reported for UWB

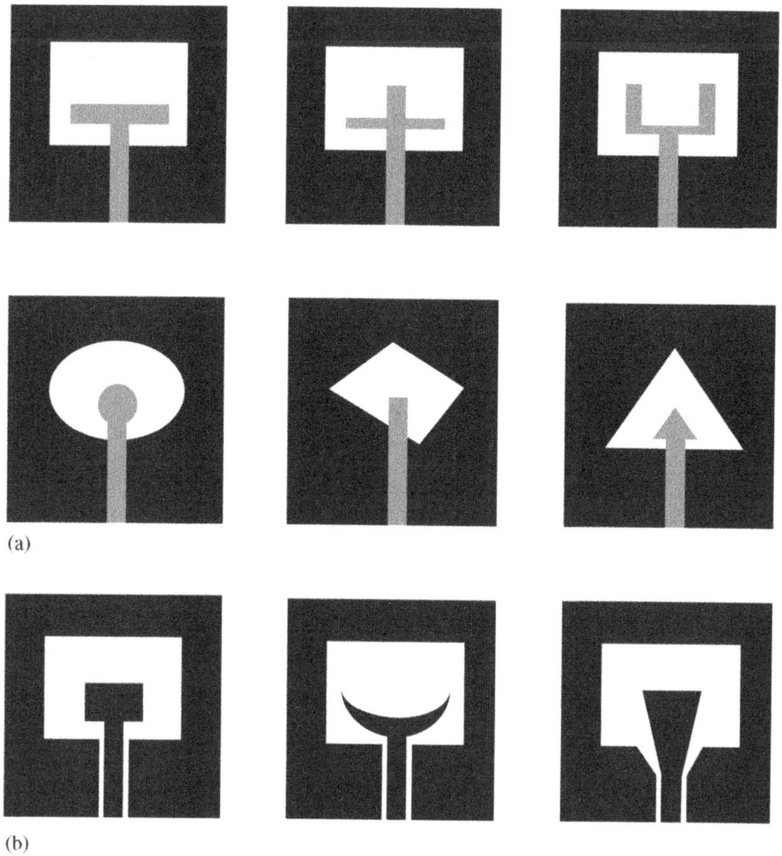

Figure 10.24 Typical wide-slot antenna configurations fed by: (a) a microstrip line; (b) co-planar waveguide. Reproduced by permission of ©2010 I^2R

applications to cover the 3.1–10.6 GHz band, where the open-ended feed structure is positioned within the slot to provide the coupling between the feeding structure and wide slot [50–57]. As an example, a microstrip fed UWB wide-slot antenna is demonstrated. Figure 10.25 shows the antenna comprising a rectangular slot and a fork-shaped microstrip feeding structure. The slot with dimensions of $l_s \times w_s$ is cut onto the upper side of a PCB. The fork-shaped feeding structure comprises one horizontal stub, two vertical stubs, and a 50-Ω microstrip line etched on the opposite side of the PCB and with the parameters of w_f, w_{f1}, w_{f2}, l_{f1}, l_{f2}. The slot and the fork-shaped feeding structure are positioned symmetrically with respect to the centerline (y-axis) of the slot. The parameters of the optimized antenna are $\varepsilon_r = 3.38$, $\tan\delta = 0.002$, $h = 0.8128$ mm, $w = 42$ mm, $l = 43.5$ mm, $w_s = 32$ mm, $l_s = 19.5$ mm, $l_g = 19$ mm, $l_{g1} = 5$ mm, $w_f = 1.86$ mm, $w_{f1} = 1.86$ mm, $w_{f2} = 1$mm, $l_{f1} = 9.5$ mm, $l_{f2} = 13$ mm, and $g = 0.8$ mm.

The antenna design was prototyped with RO4003C substrate. The simulation was carried out using CST Microwave Studio [58]. Figure 10.26 shows the simulated and measured return loss. The measured return loss is greater than 10 dB over the frequency range of 2.9–10.9 GHz or 115.9%. There are multiple nulls in the return loss response, which indicates that the antenna

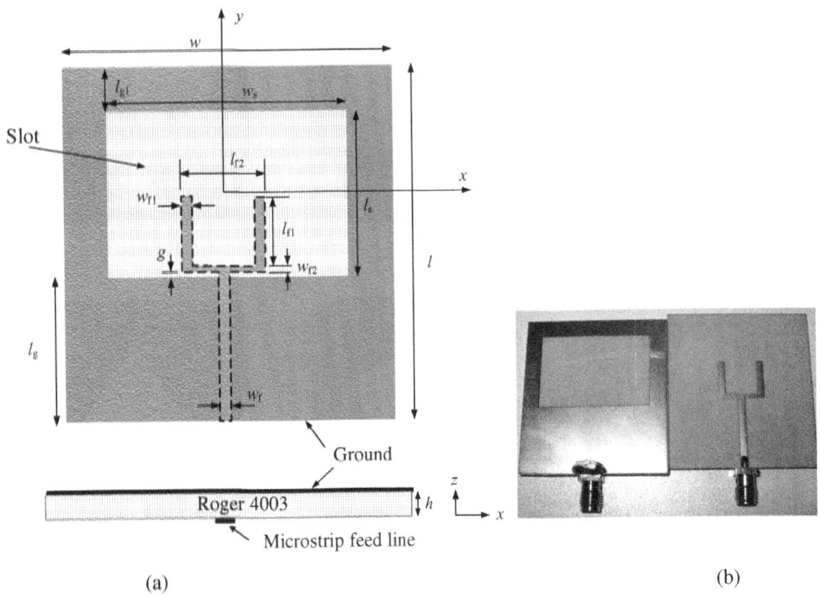

Figure 10.25 Geometry of a microstrip fed wide-slot antenna: (a) antenna configuration; (b) antenna prototype. Reproduced by permission of ©2010 I^{2}R

operates at several dominant resonances generated by the slot, the feeding stubs, and the electromagnetic coupling between them. Figure 10.27 exhibits the bore-sight gain ($\theta = 0°$, $\phi = 0°$), which ranges from 1.15 to 3.96 dBi across the frequency band of 3.1–10.6 GHz. Figure 10.28 illustrates the measured radiation patterns at 4 and 7 GHz. Quasi-omnidirectional patterns are obtained in the H-plane (x-z plane) and typical "figure-of-eight" pattern is observed in the E-plane (y-z plane), which are similar to that of conventional printed monopole antennas.

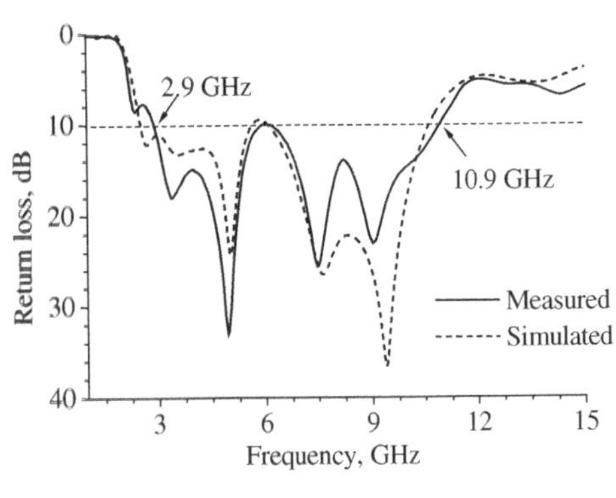

Figure 10.26 Measured and simulated return loss. Reproduced by permission of ©2010 I^{2}R

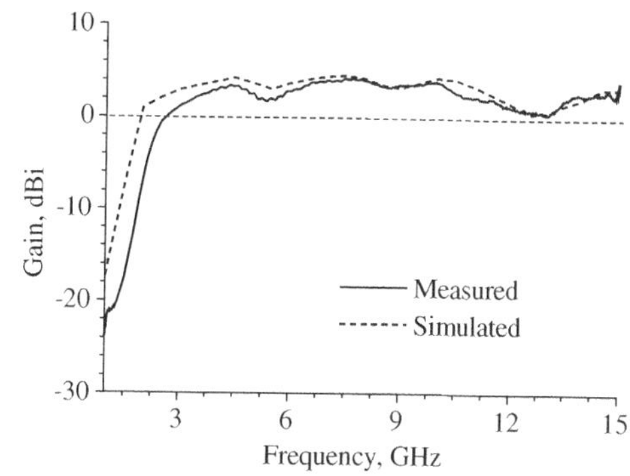

Figure 10.27 Measured and simulated gain at bore-sight ($\theta = 0°$, $\phi = 0°$). Reproduced by permission of ©2010 I²R

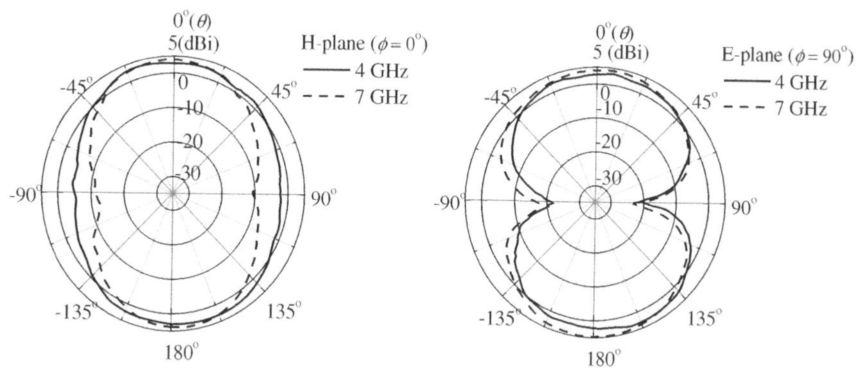

Figure 10.28 Measured radiation patterns at 4 and 7 GHz. Reproduced by permission of ©2010 I²R

In UWB communication applications, the antenna system transfer function with a flat magnitude and a linear phase response (or constant group delay) is desired [40, 59]. Figure 10.29 shows the measured system transfer function and group delay of a pair of wide-slot antennas. The antennas are positioned face-to-face with a separation of 0.7 m. The relatively flat magnitude response of around -40 dB and a constant group delay of about 15 ns at the bore-sight are observed over the entire UWB band of 3.1–10.6 GHz.

The bandwidth of the wide-slot antenna is dependent on the geometrical parameters of the slot and the feeding structure. It has been found that the width of the slot, the length and separation of the vertical stubs of the fork-shaped feeding structure greatly affect the impedance bandwidth, as shown in Figure 10.30. For larger w_s, the operating frequency range is extended. Increasing l_{f1} shifts down the operating frequency band but barely affects the bandwidth. The l_{f2} does not affect the lower edge but significantly affects the upper edge of the operating frequency band so that the bandwidth is enhanced. The other parameters only affect the bandwidth slightly but can be used to optimize the impedance matching.

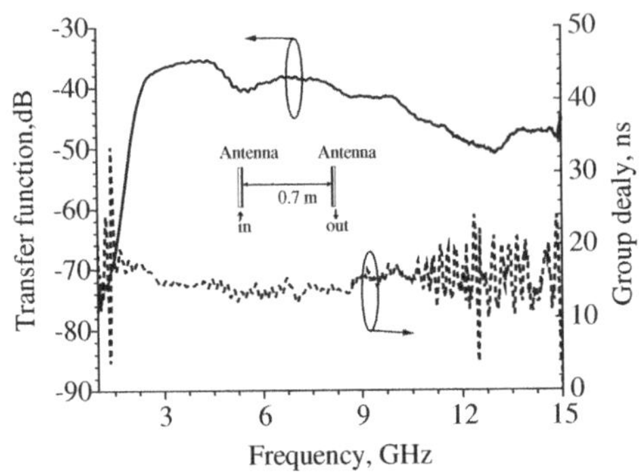

Figure 10.29 Measured system transfer function and group delay at bore-sight ($\theta = 0°$, $\phi = 0°$). Reproduced by permission of ©2010 I^{2}R

10.5.2 Monopole-Like Slot UWB Antenna

Apart from the wide-slot antennas, monopole-like slot antennas have alternatively been reported recently [60–67]. As shown in Figure 10.31, the monopole-like slot antennas consist of an open slot and a feeding structure. The ground plane surrounding the slot is open, which offers more freedom in the antenna design and the possibility to reduce the size of the antenna. Similar to the wide-slot antenna, the monopole-like slot antenna can be fed by microstrip line or CPW with the T-shaped or fork-shaped feeding structures.

As a design example, a compact CPW-fed monopole-like slot antenna is demonstrated, as shown in Figure 10.32. The slot and the feeding structure are printed onto the same side of a piece of FR4 PCB. The slot follows the shape of a complementary planar monopole antenna and is surrounded by narrow ground strips. Tapers are added at the top corners of the folded ground strips to enhance the impedance matching, especially at lower frequencies. The antenna is fed by a 50-Ω CPW line with a T-shaped feeding stub. The slot and the feeding structure are positioned symmetrically with respect to the *y*-axis. The optimized parameters of the antenna are $\varepsilon_r = 4.4$, $\tan\delta = 0.02$, $h = 0.8128$ mm, $w = 25$ mm, $l = 25$ mm, $w_s = 23$ mm, $l_s = 19$ mm, $l_1 = 11$ mm, $l_2 = 10$ mm, $l_g = 6$ mm, $w_f = 3.6$ mm, $s = 0.35$ mm, $w_t = 13$ mm, $l_t = 6$ mm, and $g = 3$ mm.

Figures 10.33 and 10.34 show the simulated and measured return loss as well as gain at bore-sight ($\theta = 0°$, $\phi = 0°$), respectively. The measured return loss is larger than 10 dB over the frequency range of 3.0 to 14.3 GHz. The measured bore-sight gain exhibits slight variation and the lower gain at the lower frequencies is caused by beam squinting. The measured radiation patterns illustrated in Figure 10.35 are similar to those of the wide-slot antenna and printed monopole antenna. The patterns are quasi-omnidirectional in the H-plane (*x*-*z* plane) and a typical "figure-of-eight" in the E-plane (*y*-*z* plane). However, the E-plane patterns exhibit beam squint, especially at the lower frequency such as 4 GHz. Since the ground plane of the antenna is small, part of the current on the ground plane flows onto the RF cable such that the radiation patterns are affected. The measured system transfer function of a pair of CPW-fed monopole-like antennas is shown in Figure 10.36. The antennas are positioned face-to-face and with a separation of 0.7 m. A flat magnitude response of around -42.5 dB and constant group delay of about 14 ns have been achieved over the entire UWB band of 3.1–10.6 GHz.

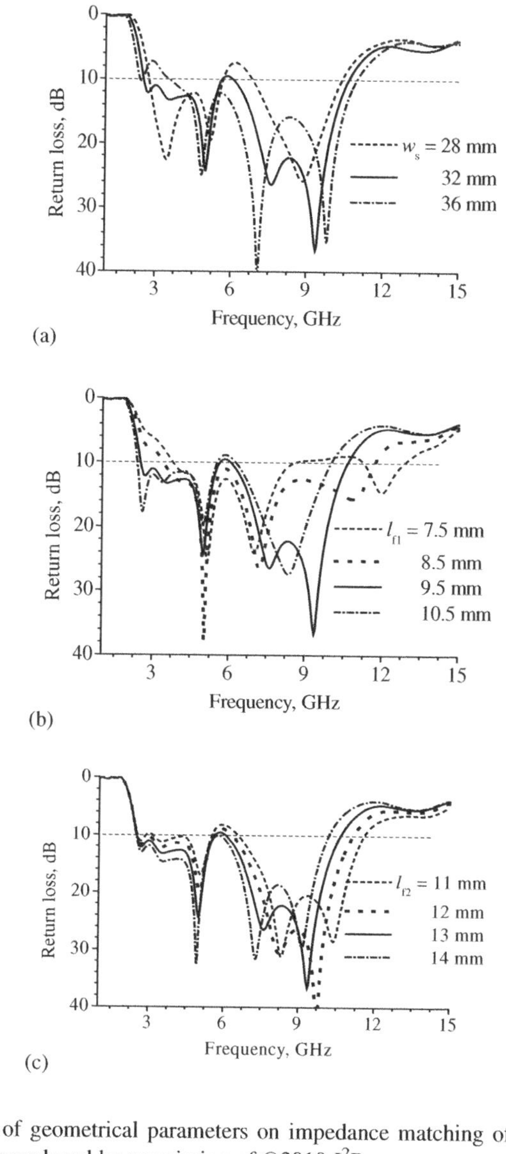

Figure 10.30 Effects of geometrical parameters on impedance matching of the wide-slot antenna: (a) w_s; (b) l_{f1}; (c) l_{f2}. Reproduced by permission of ©2010 I^2R

10.5.3 Band-Notched UWB Antennas

One of the major design considerations of a practical UWB antenna is the co-existence with existing narrowband communication systems. For instance, the 5-GHz band wireless local area network (WLAN) operates at the bands of 5.150–5.825 GHz (HIPERLAN/2: 5.150–5.350, 5.470–5.725 GHz, IEEE 802.11a: 5.150–5.350, 5.725–5.825 GHz), which will most probably interfere with UWB systems. Adding filters can suppress the interference while increasing the UWB system complexity as well as size and cost. Instead, a UWB antenna

Figure 10.31 Typical monopole-like slot antenna configurations. Reproduced by permission of ©2010 I^2R

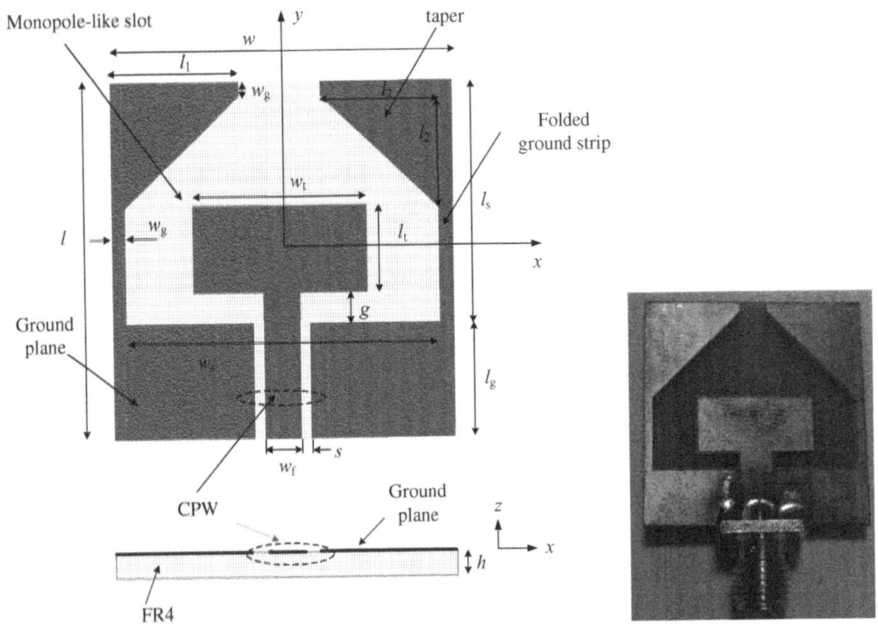

Figure 10.32 (a) Geometry of the CPW-fed monopole-like slot antenna with a T-shaped feeding structure; (b) photo of antenna prototype. Reproduced by permission of ©2010 I^2R

with band-notched characteristics is preferable to simplify the system design and reduce the size and the cost.

Generally, the band-notched characteristics are realized by introducing an additional resonant structure to the antenna body in two ways. First, the added resonant structure degrades the impedance matching of the antenna and thus less energy can be transmitted to/received from the free space. Second, the added resonant structure changes the current distributions on the antenna, which may cause the canceling of radiation in the far-field zone. Many techniques have been proposed to design band-notched UWB antennas, including etching slots or slits on the radiators, placing parasitic strips in close proximity of the radiators, using split ring resonator, and integrating filters as shown in Figure 10.37 [68–77].

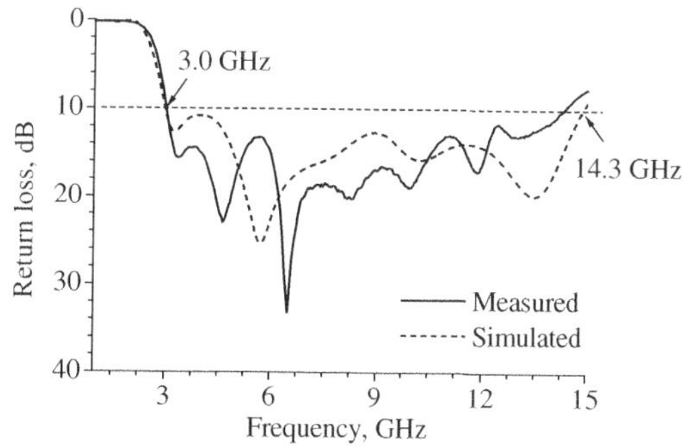

Figure 10.33 Simulated and measured return loss. Reproduced by permission of ©2010 I^2R

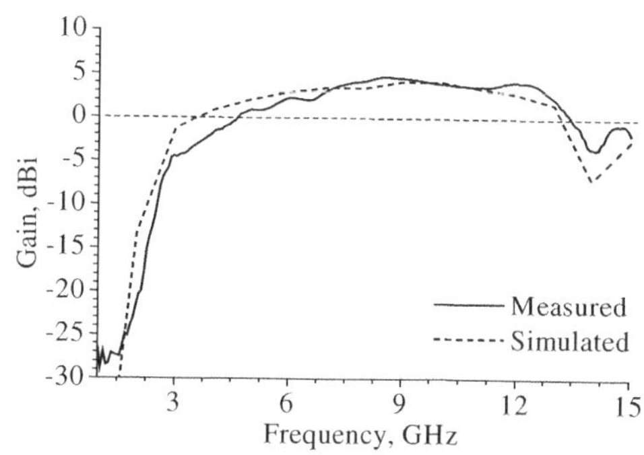

Figure 10.34 Simulated and measured gain at bore-sight ($\theta = 0°$, $\phi = 0°$). Reproduced by permission of ©2010 I^2R

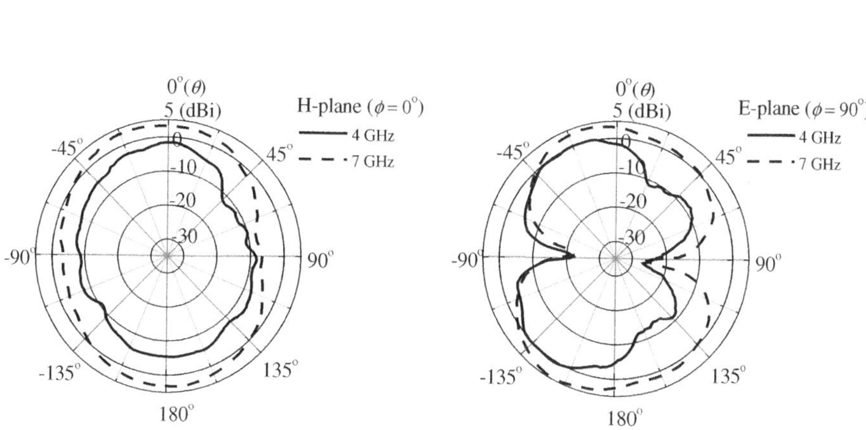

Figure 10.35 Measured radiation patterns at 4 and 7 GHz. Reproduced by permission of ©2010 I^2R

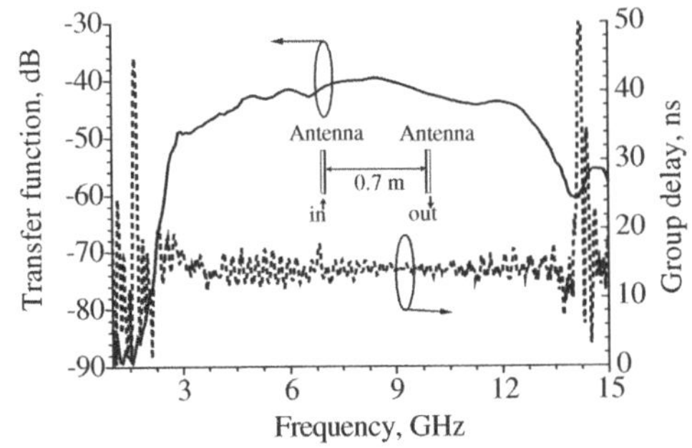

Figure 10.36 Measured system transfer function and group delay at bore-sight ($\theta = 0°$, $\phi = 0°$). Reproduced by permission of ©2010 I^{2}R

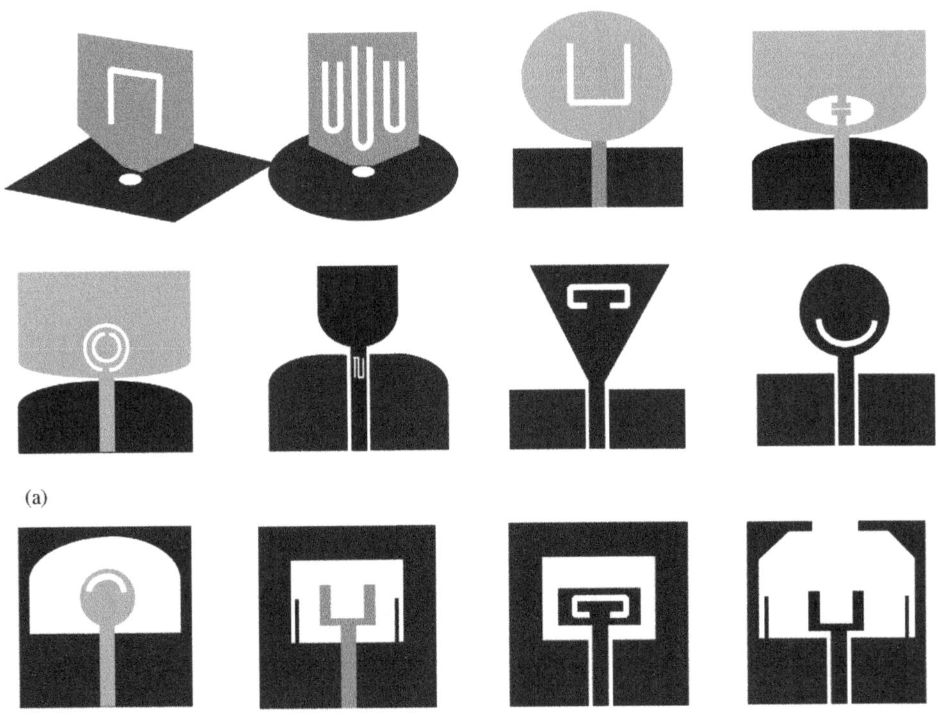

Figure 10.37 Typical band-notched UWB antenna configurations: (a) band-notched UWB monopole antennas; (b) band-notched UWB slot antennas. Reproduced by permission of ©2010 I^{2}R

A compact band-notched UWB antenna design is shown in Figure 10.38. The antenna is similar to that shown in Figure 10.32 except for the microstrip fork-shaped feeding structure. The band-notched characteristic at 5–6 GHz can be realized by adding two grounded open-circuited stubs while maintaining the desirable performance across the lower/upper UWB bands of 3.1–4.8 GHz/6.2–9.8 GHz. The length of the stubs is $\lambda/4$ at the notched frequency and symmetrically positioned at the upper portion of the bottom ground plane with respect to the y-axis. The grounded stubs produce currents which are opposite in direction to those on the folded ground strips and the fork-shaped feeding stubs, thus causing the radiation from the open-circuited stubs, folded ground strips, and fork-shaped feeding stubs to be canceled out in the far-field at the notched frequency band. The parameters of the optimized antenna are $\varepsilon_r = 4.4$, $tan\,\delta = 0.02$, $h = 0.8128$ mm, $w = 25$ mm, $l = 25$ mm, $w_s = 23$ mm, $l_s = 17$ mm, $l_g = 19$ mm, $l_1 = 11.5$ mm, $l_2 = 6$ mm, $w_g = 1$ mm $l_g = 8$ mm, $l_{f1} = 7.2$ mm, $l_{f2} = 11$ mm, $w_{f1} = 1.5$ mm, $w_{f2} = 0.8$ mm, $g = 0.8$ mm, $l_{bn} = 8$ mm, $w_{bn} = 0.3$ mm, and $d = 1$ mm.

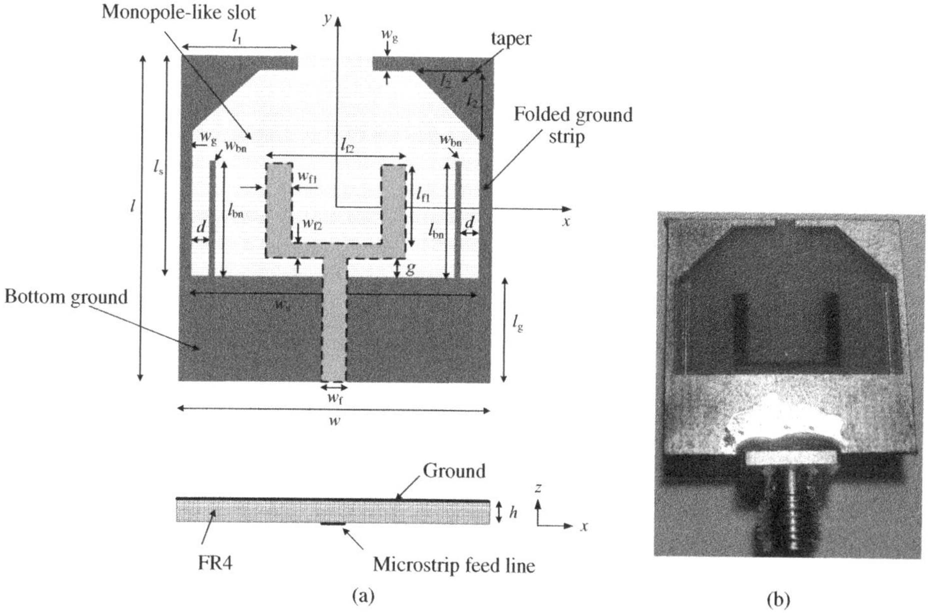

Figure 10.38 (a) Geometry of the band-notched monopole-like slot antenna using open-circuited stubs; (b) photo of antenna prototype. Reproduced by permission of ©2010 I^{2}R

Figure 10.39 shows that the antenna achieves the band-notched performance in the frequency range of 4.85–6.2 GHz, while the impedance matching degrades moderately with a maximum return loss of 7.5 dB in the frequency band of 2.7–4.85 GHz. Figure 10.40 shows that the measured antenna gain has fallen to -8 dBi over the frequency range of 5.15–5.55 GHz, which is sufficient to notch out one of the WLAN bands at 5 GHz. Figure 10.41 illustrates the measured radiation patterns at 4.0, 5.35, and 7.0 GHz. The notched characteristic at 5.35 GHz is evident.

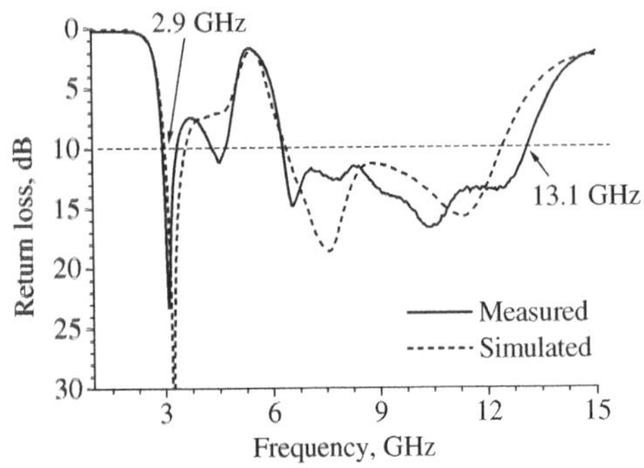

Figure 10.39 Simulated and measured return loss of the band-notched monopole-like slot antenna using open-circuited stubs. Reproduced by permission of ©2010 I^2R

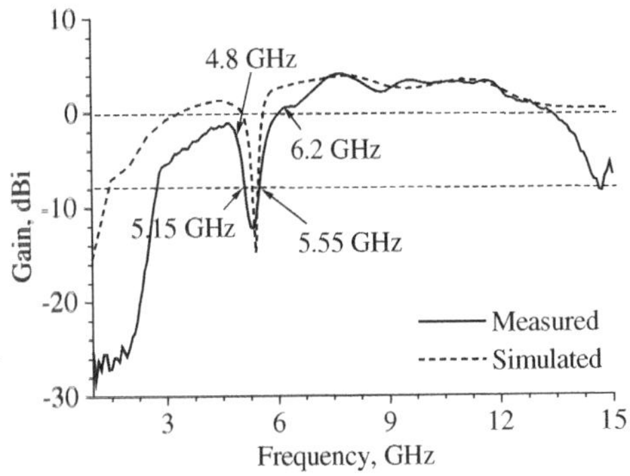

Figure 10.40 Simulated and measured gain of the band-notched monopole-like slot antenna using open-circuited stubs at bore-sight ($\theta = 0°$, $\phi = 0°$). Reproduced by permission of ©2010 I^2R

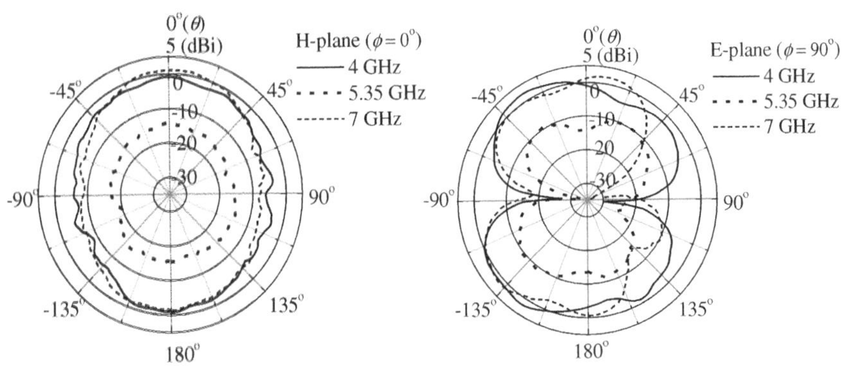

Figure 10.41 Measured radiation patterns of the band-notched monopole-like slot antenna using open-circuited stubs at 4.0, 5.35, and 7.0 GHz. Reproduced by permission of ©2010 I^2R

Figure 10.42 exhibits the transfer function and group delay of the antenna system comprising two identical band-notched antennas positioned face-to-face with a separation of 0.8 m. It can be seen that the transfer function features a relatively constant magnitude of around -48 dB over the lower UWB band of 3.1–4.8 GHz, and -44 dB over the upper UWB band of 6.2–9.8 GHz; the variation is about ± 2 dB except at the notched 5-GHz band where a drop of about 24 dB is obtained at 5.35 GHz. The group delay is also stable at both the lower and upper UWB bands at around 18 ns except for the large variation at the notched band of 4.8–6.2 GHz.

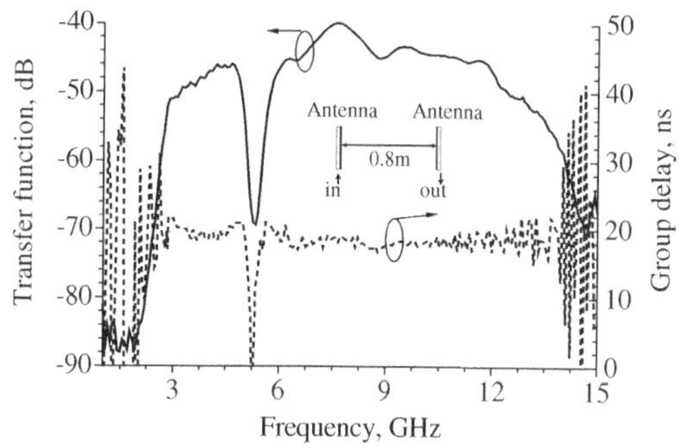

Figure 10.42 Measured antenna system transfer function of the band-notched antenna pair at bore-sight ($\theta = 0°$, $\varphi = 0°$). Reproduced by permission of ©2010 I^{2}R

The band-notched UWB antenna designs with an additional single resonant structure suffer the drawbacks of limited notched bandwidth, unacceptable skirt characteristics, and inadequate gain suppression over the notched band. In addition, the bandwidth controllability of most band-notched antenna designs is relatively limited. To enhance the notched bandwidth, band-notched antenna designs using multiple resonant structures have been reported to achieve the multiple band-notched characteristics or wideband band-notched characteristics [78–83].

A design example using multiple resonant structures to enhance the notched bandwidth of a UWB antenna is shown in Figure 10.43, which is similar to that shown in Figure 10.38. The wideband band-notched characteristic is achieved by using double open-circuited stubs and embedding two spurlines in the vertical fork-shaped feeding stubs, respectively. The notched frequency bandwidth can be controlled by adjusting the length, width, and position of the open-circuited stubs, as well as the length and width of the spurlines.

The antenna is prototyped using a piece of FR4 PCB with an overall size of 26 mm × 29 mm × 1.52 mm. The optimized parameters are $\varepsilon_r = 4.4$, $tan\,\delta = 0.02$, $h = 1.52$ mm, $w = 26$ mm, and $l = 29$ mm, $w_s = 24$ mm, $l_s = 19$ mm, $l_1 = 11.5$ mm, $l_2 = 9$ mm, $w_g = 1$ mm, $l_g = 10$ mm, $l_{f1} = 8.2$ mm, $l_{f2} = 12$ mm, $l_{s1} = 6$ mm, $w_{f1} = 2$ mm, $w_{f2} = 0.5$ mm, $w_{f3} = 0.75$ mm, $w_{s1} = 0.5$ mm, g = 0.8 mm, $w_f = 3.6$, $s = 0.35$ mmmm, $l_{bn1} = 8$ mm, $l_{bn2} = 7.8$ mm, $w_{bn1} = 0.3$ mm, $w_{bn2} = 0.3$ mm, $d_1 = 0.35$ mm, and $d_2 = 0.85$ mm.

Figure 10.44 shows that the antenna realizes good impedance matching with a return loss larger than 10 dB across the band of 2.6–13.0 GHz, except for the notched band of 4.9–5.9 GHz. The measured gain is less than -8 dBi over the entire 5-GHz WLAN frequency range of 5.15–5.85 GHz as shown in Figure 10.45. In addition, the antenna achieves the gain of

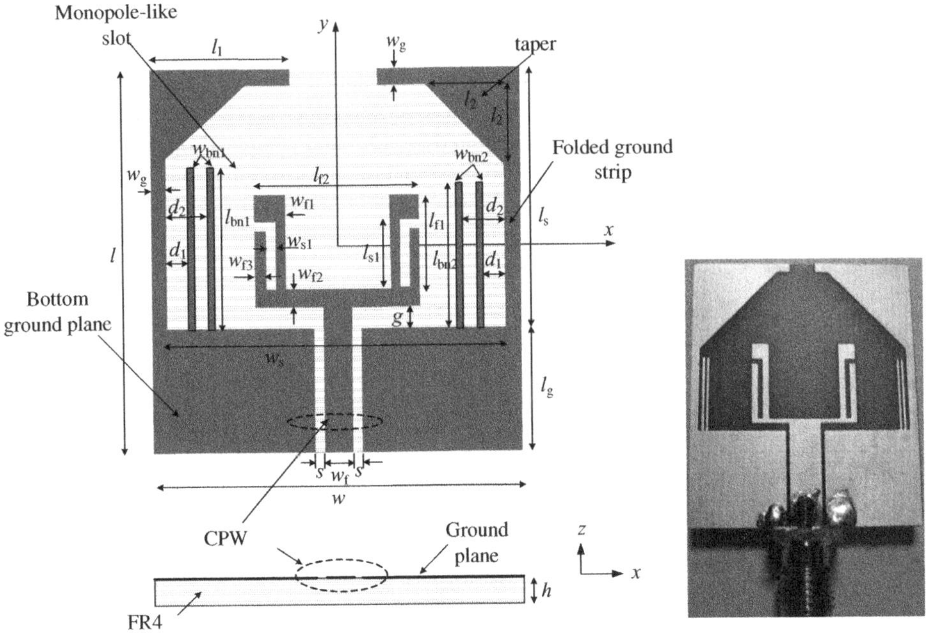

(a) (b)

Figure 10.43 (a) Geometry of the band-notched monopole-like slot antenna using double open-circuited stubs and slit fork feeding structure; (b) photo of antenna prototype. Reproduced by permission of ©2010 I^{2}R

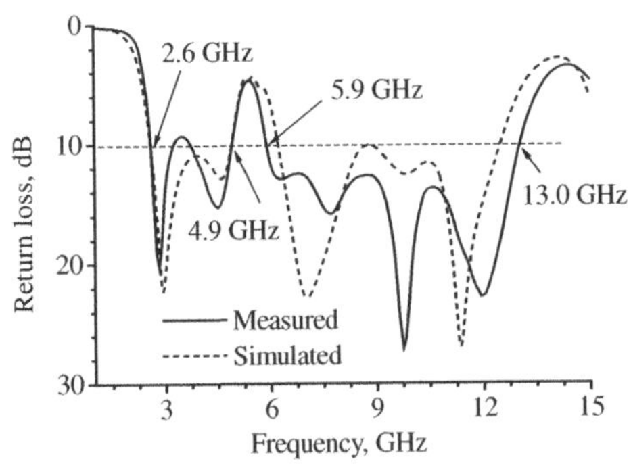

Figure 10.44 Simulated and measured return loss of the band-notched monopole-like slot antenna using double open-circuited stubs and slit fork feeding structure. Reproduced by permission of ©2010 I^{2}R

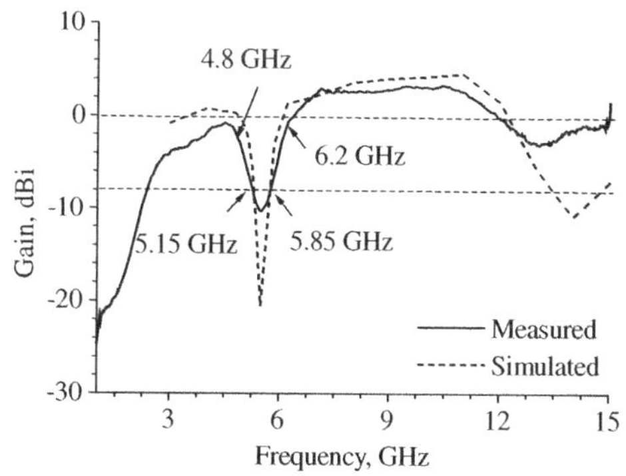

Figure 10.45 Simulated and measured gain band-notched monopole-like slot antenna using double open-circuited stubs and slit fork feeding structure at bore-sight ($\theta = 0°$, $\phi = 0°$). Reproduced by permission of ©2010 I^{2}R

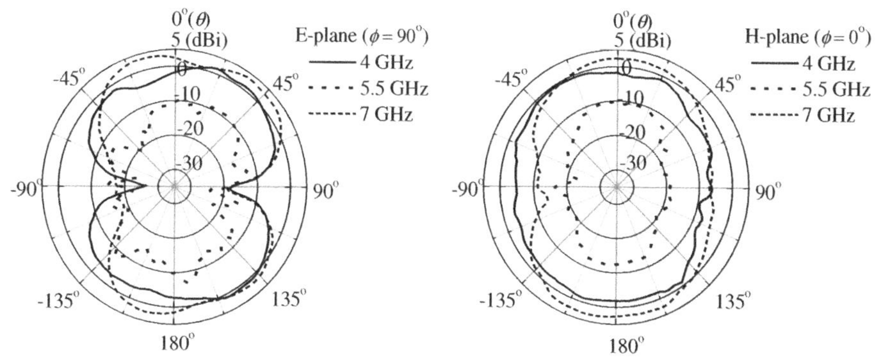

Figure 10.46 Measured radiation patterns at 4.0 GHz, 5.5 GHz, and 7.0 GHz band-notched monopole-like slot antenna using double open-circuited stubs and slit fork feeding structure. Reproduced by permission of ©2010 I^{2}R

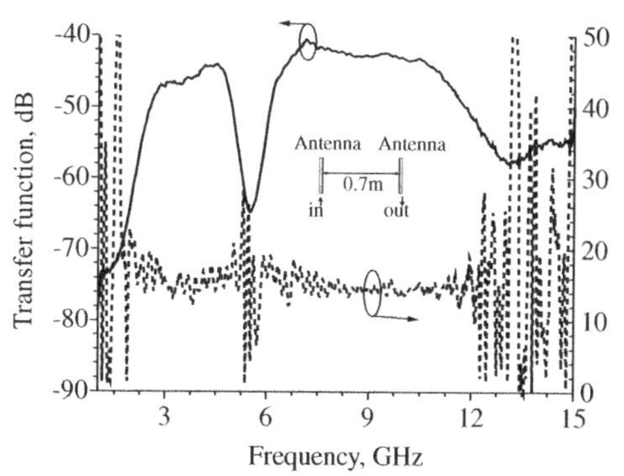

Figure 10.47 Measured antenna system transfer function of the band-notched antenna pair at bore-sight ($\theta = 0°$, $\varphi = 0°$). Reproduced by permission of ©2010 I^{2}R

around -2 dBi and 3 dBi at the lower and upper UWB bands, respectively. Figure 10.46 illustrates the measured radiation patterns at 4.0, 5.5, and 7.0 GHz, and the notched characteristic at 5.5 GHz is clearly observed.

From Figure 10.47, it can be seen that the transfer function features the constant magnitude of around -46 dB and -42 dB over the lower UWB band of 3.1–4.8 GHz and the upper UWB band of 6.2–9.8 GHz, respectively; the variation is about ± 2 dB except for the notched 5 GHz band where there is a drop at the 5–6 GHz band. The group delay is 16 ns at the lower and upper UWB bands except for the large variation at the notched band of 4.8–6.2 GHz.

References

1. T. W. Barrett, "History of ultra-wideband (UWB) radar and communications: pioneers and innovators," *Progress in Electromagnetics Symposium* (PIERS), Cambridge, MA, July 2000.
2. Federal Communications, Commission, *First Order and Report, Revision of Part 15 of the Commission's Rules Regarding UWB Transmission Systems,* FCC 02-48, April. 22, 2002.
3. Z. N. Chen, X. H. Wu, N. Yang, and M. Y. W. Chia, "Considerations for source pulses and antennas in UWB radio systems," *IEEE Trans. Antennas Propagat*, vol. 52, no. 7, pp. 1739–1748, July 2004.
4. E. A. Theodorou, M. R. Gorman, P. R. Rigg, and F. N. Kong, "Broadband pulse-optimized antenna," *Proc. Inst. Elec. Eng*, vol. 128, pt. H, pp. 124–130, June. 1981.
5. S. Evans and F. N. Kong, "TEM horn antenna: input reflection characteristics in broadband pulse-optimized antenna," *Proc. Inst. Elec. Eng*, vol. 128, pt. H, pp. 124–130, June 1981.
6. C. W. Harrison, Jr. and C. S. Williams, Jr., "Transients in wide-angle conical antennas," *IEEE Trans. Antennas Propagat*, vol. 13, pp. 236–246, Mar. 1965.
7. T. T. Wu, and R. W. P. King, "The cylindrical antenna with nonreflecting resistive loading," *IEEE Trans. Antennas Propagat*, vol. 13, pp. 369–373, May 1965.
8. D. L. Senguta and Y. P. Liu, "Analytical investigation of waveforms radiated by a resistively loaded linear antenna excited by a Gaussian pulse," *Radio Sci*, vol. 9, pp. 621–630, June 1974.
9. S. Honda, M. Ito, H. Seki, and Y. Jinbo, "A disk monopole antenna with 1:8 impedance bandwidth and omnidirectional radiation pattern," *ISAP'92*, Sapporo, Japan, pp. 1145–1148 1992.
10. M. Hammoud, P. Poey, and F. Colombel, "Matching the input impedance of a broadband disc monopole," *Electron. Lett*, vol. 29, pp. 406–407, 1993.
11. N. P. Agrawall, G. Kumar, and K. P. Ray, "Wide-band planar monopole antenna," *IEEE Trans. Antennas Propagat*, vol. 46, pp. 294–295, Feb. 1998.
12. M. J. Ammann, "Square planar monopole antenna," *Int. Conf. Antennas & Propagat*, York, England, pp. 37–40 1999.
13. Z. N. Chen, "Impedance characteristics of planar bow-tie-like monopole antennas," *Electron. Lett*, vol. 36, pp. 1100–1101, 2000.
14. Z. N. Chen, M. J. Ammann, M. Y. W. Chia, and T. S. P. See, "Annular circular planar monopole antennas," *IEE Proc. Microw. Antennas Propagat*, vol. 49, no. 4, pp. 200–203, Aug. 2002.
15. M. J. Ammann, and Z. N. Chen, "A wideband shorted planar monopole with bevel," *IEEE Trans. Antennas Propagat*, vol. 51, no. 4, pp. 901–903, Apr. 2003.
16. M. J. Ammann, and Z. N. Chen, "Wideband monopole antennas for multi-band wireless systems," *IEEE Antennas Propagat. Mag*, vol. 45, no. 2, pp. 146–150, Apr. 2003.
17. X. H. Wu, and Z. N. Chen, "Comparison of planar dipoles in UWB applications," *IEEE Trans. Antennas Propagat*, vol. 53, no. 6, pp. 1973–1983, June 2005.
18. Z. N. Chen, "Broadband roll monopole," *IEEE Trans. Antennas Propagat*, vol. 51, no. 11, pp. 3175–3177, 2003.
19. Z. N. Chen, "A new bi-arm roll antenna for UWB applications," *IEEE Trans. Antennas Propagat*, vol. 53, no. 2, pp. 672–677, Feb. 2005.
20. D. H. Kwon and Y. Kim, "CPW-fed planar ultra-wideband antenna with hexagonal radiating elements," *IEEE Int. Symp. Antennas Propagat*, vol. 3, pp. 2947–2950, June 2004.
21. J. Liang, C. C. Chiau, X. Chen, and C. G. Parini, "Printed circular ring monopole antennas," *Microw. Opt. Techn. Lett*, vol. 45, no. 5, pp. 372–375, June 5, 2005.

22. H. S. Choi, J. K. Park, S. K. Kim, and J. Y. Park, "A new ultra-wideband antenna for UWB applications," *Microw. Opt. Technol. Lett*, vol. 40, no. 5, pp. 399–401, Mar. 5, 2004.
23. K. Chung, H. Park, and J. Choi, "Wideband microstrip-fed monopole antenna with a narrow slit," *Microw. Opt. Technol. Lett*, vol. 47, no. 4, pp. 400–402, Nov. 20, 2005.
24. Y. Zhang, Z. N. Chen, and M. Y. W. Chia, "Effects of finite ground plane and dielectric substrate on planar dipoles for UWB applications," *IEEE Int. Symp. Antennas Propagat*, pp. 2512–2515, June 2004.
25. Z. N. Chen, N. Yang, Y. X. Guo, and M. Y. W. Chia, "An investigation into measurement of handset antennas," *IEEE Trans. Instrum. Meas*, vol. 54, no. 3, pp. 1100–1110, June 2005.
26. H. Morishita, Y. Kim, H. Furuuchi, K. Sugita, Z. Tanaka, and K. Fujimoto, "Small balance-fed helical dipole antenna system for handset," *Proc. IEEE 51st Veh. Technol. Conf*, vol. 2, pp. 1377–1380, Spring, Tokyo, May 15–18, 2000.
27. S. Hayashida, T. Tanaka, H. Morishita, Y. Koyanagi, and K. Fujimoto, "Built-in half size of folded dipole antenna for handsets," *IEEE Antennas Propag. Soc. Int. Symp*, vol. 3, pp. 2759–2762, June 20–25, 2004.
28. C. Ichenlin, J. Ollikainen, and P. Vainikainen, "Reducing the influence of feed cables on small antenna measurements," *Electron. Lett*, vol. 35, no. 15, pp. 1212–1214, July 1999.
29. C. Ichelin, J. Krogerus, and P. Vainikainen, "Use of balun chokes in small-antenna radiation measurements," *IEEE Trans. Instrum. Meas*, vol. 53, no. 2, pp. 489–506, Apr. 2004.
30. Z. N. Chen, T. S. P. See, and X. Qing, "Small printed ultrawideband antenna with reduced ground plane effect," *IEEE Trans. Antennas Propagat*, vol. 55, no. 2, pp. 383–388, 2007.
31. T. S. P. See, and Z. N. Chen, "A small UWB antenna for wireless USB," *Int. Conf. for Ultra-Wideband*, pp. 198–203, Sep. 2007.
32. E. Jones, "Measured angle-diversity performance of the wire-grid lens antenna," *IEEE Trans. Antennas Propagat*, pp. 484–486, May 1967.
33. E. W. Allen, "Angle diversity test using a single aperture dual beam antenna," *IEEE ICC* 88, vol. 3, pp. 1626–1632, June 1988.
34. R. G. Vaughan, "Beam spacing for angle diversity," *IEEE GLOBECOM* 98, vol. 2, pp. 928–933, Nov. 1998.
35. K. L. Wong, S. W. Su, and Y. L. Kuo, "A printed ultra-wideband diversity monopole antenna," *Microw. Opt. Technol. Lett*, vol. 38, no. 4, pp. 257–259, 2002.
36. L. Liu, H. P. Zhao, T. S. P. See, and Z. N. Chen, "A printed ultra-wideband diversity antenna," *Int. Conf. for Ultra-Wideband*, pp. 351–356, Sep. 2006.
37. H. P. Zhao, L. Liu, T. S. P. See, Z. N. Chen, and M. J. Ammann, "A printed UWB diversity antenna," *Int. Symp. Antennas and Propagat*, Nov. 2006.
38. Z. N. Chen, *Antennas for Portable Devices*, Chapter 7, John Wiley & Sons, New York, 2007.
39. S. Blanch, J. Romeu, and I. Corbella, "Exact representation of antenna system diversity performance from input parameter description," *Electron. Lett*, vol. 39, no. 9, pp. 705–707, May 2003.
40. X. M. Qing, Z. N. Chen, and M. Y. W. Chia, "Characterization of ultrawideband antennas using transfer functions," *Radio Sci*, vol. 41, RS1002, 2006.
41. Y. Yashimura, "A microstrip slot antenna," *IEEE Trans. Antennas Propag*, vol. 20, no. 11, pp. 760–762, 1972.
42. S. M. Shum, K. F. Tong, X. Zhang, and K. M. Luk, "FDTD modeling of microtrip-line-fed wide slot antenna," *Microw. Opt. Technol. Lett*, vol. 10, no. 2, pp. 118–120, 1995.
43. M. K. Kim, K. Kim, Y. H. Suh, and I. Park, "A T-shaped microstrip-line-fed wide slot antenna," *Proc. IEEE Antennas and Propag. Soc. Int. Symp*, vol. 3, pp. 1500–1503, July 2000.
44. Y. W. Jang, "Broadband cross-shaped microstrip-fed slot antenna," *Electron. Lett*, vol. 36, no. 25, pp. 2056–2057, 2000.
45. Y. H. Suh and I. Park, "A broadband eccentric annular slot antenna," *Proc. IEEE Antennas and Propag. Soc. Int. Symp*, vol. 1, pp. 94–97, July 2001.
46. J. Y. Sze and K. L. Wong, "Bandwidth enhancement of a microstripline-fed printed wide-slot antenna," *IEEE Trans. Antennas Propag*, vol. 49, no. 7, pp. 1020–1024, 2001.
47. H. D. Chen, "Broadband CPW-fed square slot antennas with a widened tuning stub," *IEEE Trans. Antennas Propag*, vol. 51, no. 8, pp. 1982–1986, 2003.
48. Y. F. Liu, K. L. Lan, Q. Xue, and C. H. Chan, "Experimental studies of printed wide-slot antenna for wide-band applications," *IEEE Antennas Wireless Propag. Lett*, vol. 3, no. 1, pp. 273–275, 2004.
49. J. Y. Jan and J. W. Su, "Bandwidth enhancement of a printed wide-slot antenna with a rotated slot," *IEEE Trans. Antennas Propag*, vol. 53, no. 6, pp. 2111–2114, 2005.

50. X. Qing, M. Y. W. Chia, and X. H. Wu, "Wide slot antenna for UWB applications," *Proc. IEEE Antennas and Propag. Soc. Int. Symp*, vol. 1, pp. 834–837, June 2003.
51. P. Li, J. Liang, and X. Chen, "Study of printed elliptical/circular slot antennas for ultrawideband applications," *IEEE Trans. Antennas Propag*, vol. 54, no. 6, pp. 1670–1675, 2006.
52. C. Marchais, G. L. Ray, and A. Sharaiha, "Stripline slot antenna for UWB communications," *IEEE Antennas Wireless Propag. Lett*, vol. 5, no. 1, pp. 319–322, 2006.
53. X. Chen, W. Zhang, R. Ma, J. Zhang, and J. Gao, "Ultra-wideband CPW-fed antenna with round corner rectangular slot and partial circular patch," *IET Microw. Antennas Propag*, vol. 1, no. 4, pp. 847–851, 2007.
54. S. Cheng, P. Hallbjörner, and A. Rydberg, "Printed slot planar inverted cone antenna for ultrawideband applications," *IEEE Antennas Wireless Propag. Lett*, vol. 7, no. 1, pp. 18–21, 2008.
55. D. Krishna, M. Gopikrishna, C. K. Aanandan, P. Mohanan, and K. Vasudevan, "Ultra-wideband slot antenna for wireless USB dongle applications," *Electron. Lett*, vol. 44, no. 18, pp. 1057–1058, 2008.
56. S. K. Rajgopal, and S. K. Sharma, "Investigations on ultrawideband pentagon shape microstrip slot antenna for wireless communications," *IEEE Trans. Antennas Propag*, vol. 57, no. 5, pp. 1353–1359, 2009.
57. M. N. Moghadasi, A. Danideh, R. Sadeghifakhr, and M. Reza-Azadi, "CPW-fed ultra wideband slot antenna with arc-shaped stub," *IET Microw. Antennas Propag*, vol. 3, no. 4, pp. 681–686, 2009.
58. CST-Microwave Studio, User's Manual, 2007.
59. X. Qing, Z. N. Chen, and M. Y. W. Chia, "Network approach to UWB antenna transfer functions characterization," *35th European Microw. Conf*, pp. 1751–1754, Paris, France, 2005.
60. S. K. Sharma, L. Shafai, and N. Jacob, "Investigation of wide band microstrip slot antenna," *IEEE Trans. Antennas Propag*, vol. 52, no. 3, pp. 865–872, 2004.
61. S. I. Latif, L. Shafai, and S. K. Sharma, "Bandwidth enhancement and size reduction of microstrip slot antenna," *IEEE Trans. Antennas Propag*, vol. 53, no. 3, pp. 994–1003, 2005.
62. W. J. Lui, C. H. Cheng, and H. B. Zhu, "Experimental investigation on novel tapered microstrip slot antenna for ultra-wideband applications," *IET Microw. Antennas Propag*, vol. 1, no. 2, pp. 480–487, 2007.
63. X. Qing, and Z. N. Chen, "Monopole-like slot UWB antenna on LTCC," *IEEE Int. Conf. on Ultra-Wideband (ICUWB)*, vol. 2, pp. 121–124, Hannover, Germany, 2008.
64. X. Qing and Z. N. Chen, "Compact ultra-wideband T-shaped CPW-fed monopole-like slot antenna," *IEEE Asia Pac. Microw. Conf*, pp. 1–4, Hong Kong, China, 2008.
65. X. Qing, and Z. N. Chen, "Compact monopole-like slot antenna and band-notched design for ultrawideband applications," *Radio Sci*, vol. 44, RS2018, doi: 10.1029/2008RS003872.
66. M. Gopikrishna, D. D. Krishna, C. K. Anandan, P. Mohanan, and K. Vasudevan, "Design of a compact semi-elliptic monopole slot antenna for UWB systems," *IEEE Trans. Antennas Propag*, vol. 57, no. 6, pp. 1834–1837, 2009.
67. X. Qing, and Z. N. Chen, "Compact coplanar waveguide-fed ultra-wideband monopole-like slot antenna," *IET Microw. Antennas Propag*, vol. 3, no. 5, pp. 889–898, 2009.
68. A. Kerkhoff, and H. Ling, "A parametric study of band-notched UWB planar monopole antennas," *Proc. IEEE Antennas and Propag. Soc. Int. Symp*, vol. 2, pp. 1768–1771, June 2004.
69. W. Choi, J. Jung, K. Chung, and J. Choi, "Compact microstrip-fed antenna with band-stop characteristic for ultra-wideband applications," *Microw. Opt. Technol. Lett*, vol. 47, no. 1, pp. 89–92, 2005.
70. J. Qiu, Z. Du, J. Lu, and K. Gong, "A planar monopole antenna design with band-notched characteristic," *IEEE Trans. Antennas Propag*, vol. 54, no. 1, pp. 288–292, 2006.
71. J. Kim, C. S. Cho, and J. W. Lee, "5.2 GHz notched ultra-wideband antennas ring slot-type SRR," *Electron. Lett*, vol. 42, no. 6, pp. 315–316, 2006.
72. H. K. Yoon, Y. Lim, W. Lee, Y. J. Yoon, S. M. Han, and Y. Kim, "UWB wide slot antenna with band-notch function," *Proc. IEEE Antennas and Propag. Soc. Int. Symp*, pp. 3059–3062, July 2006.
73. Y. C. Lin, and K. J. Hung, "Compact ultrawideband rectangular aperture antenna and band-notched designs," *IEEE Trans. Antennas Propag*, vol. 54, no. 11, pp. 3075–3081, 2006.
74. S. W. Qu, J. L. Li, and Q. Xue, "A band-notched ultrawideband printed monopole antenna," *IEEE Antennas Wireless Propag. Lett*, vol. 5, no. 1, pp. 495–498, 2006.
75. K. Chung, S. Hong, and J. Choi, "Ultrawide-band printed monopole antenna with band-notch filter," *IET Microw. Antennas Propag*, vol. 1, no. 2, pp. 518–522, 2007.
76. T. G. Ma, and S. J. Wu, "Ultrawideband band-notched folded strip monopole antenna," *IEEE Trans. Antennas Propag*, vol. 55, no. 9, pp. 2473–2479, 2007.

77. C. Hong, C. Ling, I. Tarn, and S. Chung, "Design of a planar ultrawideband antenna with a new band-notch structure," *IEEE Trans. Antennas Propag*, vol. 55, no. 12, pp. 3391–3397, 2007.
78. K. Chang, H. Kim, and Y. J. Yoon, "Ultra-wideband antenna with improved gain characteristics," *IET Microw. Antennas Propag*, vol. 2, no. 5, pp. 512–517, 2008.
79. W. S. Lee, D. Z. Kim, K. J. Kim, and J. W. Yu, "Wideband planar monopole antennas with dual band-notched characteristics," *IEEE Trans. Microw. Theory Tech*, vol. 54, no. 6, pp. 2800–2806, 2006.
80. Y. Zhang, W. Hong, C. Yu, Z. Q. Kuai, Y. D. Don, and J. Y. Zhou, "Planar ultrawideband antennas with multiple notched bands based on etched slots on the patch and/or split ring resonators on the feed line," *IEEE Trans. Antennas Propag*, vol. 56, no. 9, pp. 3063–3068, 2008.
81. Y. D. Dong, W. Hong, Z. Q. Kuai, C. Yu, Y. Zhang, J. Y. Zhou, and J. X. Chen, "Development of ultrawideband antenna with multiple band-notched characteristics using half mode substrate integrated waveguide cavity technology," *IEEE Trans. Antennas Propag*, vol. 56, no. 9, pp. 2894–2902, 2008.
82. T. G. Ma, R. C. Hua, and C. F. Chou, "Design of a multiresonator loaded band rejected ultrawideband planar monopole antenna with controllable notched bandwidth," *IEEE Trans. Antennas Propag*, vol. 56, no. 9, pp. 2875–2883, 2008.
83. M. Gopikrishna, D. D. Krishna, and C. K. Aanandan,"Band notched semi-elliptic slot antenna for UWB systems," in *38th European Microwave Conf*, pp. 889–892, Oct. 2008.

11

Metamaterial Antennas and Radiative Systems

Christophe Caloz
École Polytechnique, Montreal, Canada

11.1 Introduction[1]

Metamaterials [2–6]; are commonly defined as artificial effective electromagnetic structures with unusual properties not readily found in nature. In this definition, the term "effective" means that the size p of the structure's unit cell, or period, is much smaller than the guided wavelength λ_g of the propagating wave ($p \ll \lambda_g$), so the fields are averaged across the unit cells and really "see" the structure as a homogenous medium at a macroscopic scale [7]. The term "unusual" is relative. In its broad sense, metamaterials should include artificial dielectrics [8–10], since they constitute sub-wavelength unit cell structures. However, "unusual" is generally meant to imply material properties which have been hardly explored before the turn of the 20th century, in particular negative refractive index (NRI), first explicitly theorized by Veselago back in 1967 [11] but experimentally demonstrated, following Pendry's works on electrically negative [12, 13] and magnetically negative [14] artificial dielectrics, only in 2001 by the team of Smith et al. [15]).[2]

[1] This chapter was partly adapted from [1].
[2] The NRI structure demonstrated in [15] was in fact first reported in [16]. However, the argument of the latter paper, concluding that the composite wire-ring medium exhibited NRI because it had a pass-band in the frequency range where the wire and ring media were separately negative, was questionable. In fact, this turned out to be incidentally correct, as rigorously verified in [15], because of the symmetry of the structure, which cancelled interactions between the two types of particles.

Microstrip and Printed Antennas: New Trends, Techniques and Applications. Edited by Debatosh Guha and Yahia M.M. Antar

The metamaterials reported to date may be classified in two categories: resonant particle (RP) metamaterials and transmission line (TL) metamaterials. RP metmaterials were reported first [15, 16, 17]. They generally consist of sub-wavelength thin wires, providing a negative permeability, and split ring resonators, providing a negative permeability, within a restricted frequency range, so as to lead overall to an NRI response.[3] However, RP structures have unfavorable characteristics rendering them unsuitable for most microwave applications: narrow bandwidth and large loss due to their resonant (lorentzian) [18] nature, and bulky size due to their volumetric configuration. The discovery in June 2002 of non-resonant transmission line (TL) NRI metamaterials, independently reported by Caloz and Itoh [19, 20], Iyer and Eleftheriades [21], and Oliner [22], offered a solution to this problem, and quickly led to a large number of novel microwave applications [2–4, 6]. TL metamaterials are artificial structures constituted of sub-wavelength inductors and capacitors, where negative permittivity is provided by shunt inductors and negative permeability is provided by series capacitors. By operating far away from their cutoff frequencies, TL metamaterials do not suffer of the narrow bandwidth and absorption loss inherent to RP metamaterials. Moreover, they are ideally suited to planar structures, components and systems, although they may also be made volumetric by stacking. It is important to realize that RP and TL metamaterials are intimately related: the former are the limiting case of the latter when the particles get closely coupled to one another [23], as will be shown in Section 11.2.3.

TL metamaterial antennas, in particular based on composite right/left-handed (CRLH) structures [2], have been particularly prolific in the area of antennas [1]. This chapter provides an overview of some recent advances in this field. Section 11.2 presents the fundamental theory of metamaterials, including RP and TL metamaterials, and their relation. Sections 11.3 and 11.4 deal with leaky-wave and resonant metamaterial antennas, respectively. Very recent and somewhat exotic applications of metamaterial antenna are described in Section 11.5.

11.2 Fundamentals of Metamaterials

11.2.1 Resonant Particle (RP) Metamaterials

Figures 11.1(a) and 11.1(b) show typical RP NRI metamaterials, which are constituted of sub-wavelength thin wires [12, 13, 24] and split ring resonators [14, 25] periodically distributed in space with a period p. The thin wires and split ring resonators exhibit an electric Drude dispersion and a magnetic Lorentz dispersion [17], respectively, when the electric field (**E**) is parallel to the wires, so as to induce an electric dipole moment leading to an electric response [frequency-dependent permittivity $\varepsilon(\omega)$], and the magnetic field (**H**) is perpendicular to the plane of the rings, so as to induce a magnetic dipole moment leading to a magnetic response [frequency-dependent permeability $\mu(\omega)$]. Assuming the time harmonic dependence $e^{+j\omega t}$, the dispersive constitutive parameters read

$$\varepsilon(\omega) = \varepsilon_0\left[1-\frac{\omega_{pe}^2}{\omega^2-j\omega\Gamma_e}\right] = \varepsilon'-j\varepsilon'', \tag{11.1a}$$

[3] Several variations of these particles are possible and have been reported, including folded wires, square or circular single or double, edge-coupled or broadside-coupled rings.

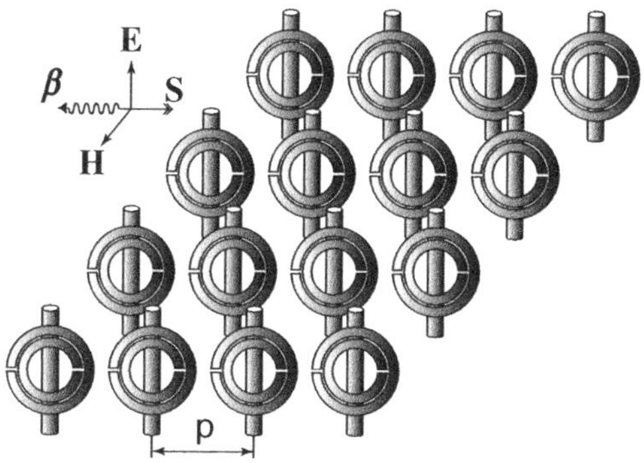

(a) Mono-dimensionally LH structure.

(b) Bi-dimensionally LH structure.

Figure 11.1 Resonant particle (RP) negative refractive index (NRI) metamaterial structures, constituted of sub-wavelength thin wires and split ring resonators. Reproduced by permission of ©2007 IEEE

$$\mu(\omega) = \mu_0\left[1-\frac{\omega_{pm}^2}{\omega^2-\omega_{rm}^2-j\omega\Gamma_m}\right] = \mu'-j\mu'', \tag{11.1b}$$

where ω_{pe} and ω_{pm} are the electric and magnetic plasma frequencies, respectively, ω_{rm} is the magnetic resonance frequency, and Γ_e and Γ_m are the electric and magnetic damping factors, respectively. These parameters are given by

$$\omega_{pe} = c\sqrt{\frac{2\pi}{\ln(p/a)}}, \quad \Gamma_e = \varepsilon_0\frac{p\omega_{pe}}{\pi a^2\sigma_e}, \tag{11.2a}$$

$$\omega_{pm} = \sqrt{\pi}\frac{r_i}{p}\omega, \quad \omega_{rm} = c\sqrt{\frac{3p}{\pi\ln(2wr_i^3/\delta)}}, \quad \Gamma_m = \frac{2pR'_m}{a\mu_0}, \tag{11.2b}$$

where c is the speed of light, a is the radius of the metal wires, σ_e is the conductivity of the wires, r_i is the inner radius of the smaller ring, w is the width of the rings, δ is the radial spacing between the rings, and R'_m is the resistance per unit length of the metal rings.[4]

The real and imaginary parts of the dispersion relations of Equations (11.1) are plotted in Figure 11.2. According to Equations (11.1), $\varepsilon' < 0$ and $\mu' < 0$ in the frequency range $\omega_{rm} < \omega < min(\omega_{pe}, \omega_{pm})$.[5] The simultaneous negativity of $\varepsilon' < 0$ and $\mu' < 0$ implies that

[4] Strictly speaking, on grounds of causality, ω_{pm} should be a constant, and not a function of ω in a Lorentz dispersion response [26]. However, the definition of effective constitutive parameters is meaningful only over a restricted frequency range and therefore Kramers-Kröing relations [26], involving frequency integration from $\omega = 0$ to $\omega = \infty$, do not apply. Nevertheless, it may be verified that the ω dependence of ω_{pm} is negligible in Equation (11.2b) for $r_i/p \ll 1$, so that the relation may be termed "quasi-Lorentzian".

[5] These bounds are exact when $\Gamma_e = \Gamma_m = 0$ and approximate otherwise.

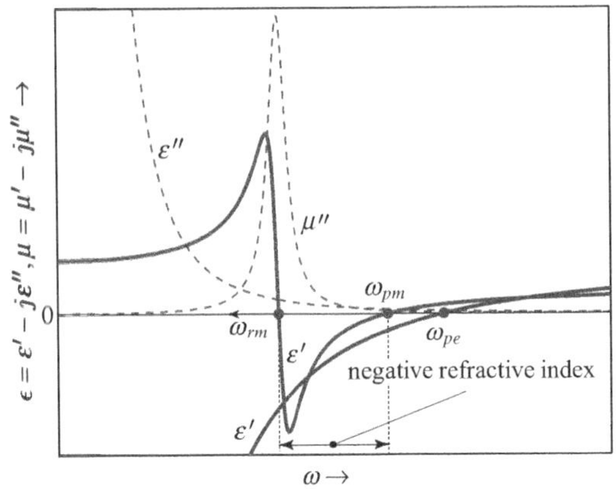

Figure 11.2 Electric-Drude and magnetic-Lorentz dispersion relations for the resonant metamaterial of Figure 11.1 computed by Equations (11.1). Reproduced by permission of ©2007 IEEE

the refractive index n is also negative, $n < 0$,[6] hence the terminology "NRI." NRI metamaterials are also called *left-handed* (LH) [11] because of the LH triad formed by the propagating waves they support, ($\mathbf{E}$, $\mathbf{H}$, $\boldsymbol{\beta}$), where $\boldsymbol{\beta} = nk_0\hat{\mathbf{k}}$ is the propagation vector, easily shown by applying Maxwell's equation to a plane wave ($e^{-j\boldsymbol{\beta}r}$), and noting that $\varepsilon' = -|\varepsilon'|$ and $\mu' = -|\mu'|$. Due to the ring resonators, RP metamaterials are resonant, and therefore exhibit a narrow frequency bandwidth and high losses (μ'') at large permeability (μ') values, as required by causality [18, 27].

11.2.2 Transmission Line (TL) Metamaterials

From basic transmission line theory, a purely LH metamaterial structure (i.e. a medium exhibiting $\varepsilon < 0$ and $\mu < 0$) could apparently be obtained by cascading a sub-wavelength unit cell constituted of a series capacitance C_L and a shunt inductance L_L [28]. However, such an idealized structure cannot be physically realized due to unavoidable parasitic series inductance L_R and shunt capacitance C_R, associated with the electric and magnetic fluxes, respectively, produced by the propagating wave. Consequently, a tentative LH material becomes in reality a composite medium including LH *and* RH contributions, leading to the richer concept of composite right/left-handed (CRLH) metamaterials, introduced in [29] and developed in [2]. In addition, losses (dielectric, ohmic, diffraction), possibly including desired leaky-wave

[6] This may be verified by writing the first-order Taylor approximation of the expression $n = \sqrt{\varepsilon\mu}/c$ for the case of low losses ($\varepsilon'' \ll \varepsilon'$ and $\mu'' \ll \mu'$) using Equations (11.1). This yields $n = n' - jn'' = \sqrt{(\varepsilon' - j\varepsilon'')(\mu' - j\mu'')}/c = \sqrt{\varepsilon'\mu'}\sqrt{(1 - j\varepsilon''/\varepsilon')(1 - j\mu''/\mu')}/c$. In the case of low loss, this expression becomes $n \approx \sqrt{\varepsilon'\mu'}[1 - \frac{j}{2}(\varepsilon''/\varepsilon' + \mu''/\mu')]/c$. To have the required physical decay in the direction of the positively growing spatial variable r from the forward wave function $e^{-j\beta r} = e^{-jnk_0 r} = e^{-j(n' - jn'')k_0 r} = e^{-jn'k_0 r}e^{-n''k_0 r}$, we must have $n'' > 0$. In the double negative range $\varepsilon', \mu' < 0$, $\sqrt{\varepsilon'\mu'}$ is real and therefore $n'' = -\sqrt{\varepsilon'\mu'}(\varepsilon''/\varepsilon' + \mu''/\mu')/(2c)$. Consequently, since $\varepsilon'', \mu'' > 0$, $n'' > 0$ requires $\sqrt{\varepsilon'\mu'} = n'c < 0$, which proves that $n' < 0$.

radiation "losses," are accounted for by the resistance R and the conductance G. The CRLH circuit model is presented in Figure 11.3(a), while Figures 11.3(b), 11.3(c) and 11.3(d) show 1D, 2D and 3D CRLH metamaterial structures, respectively.

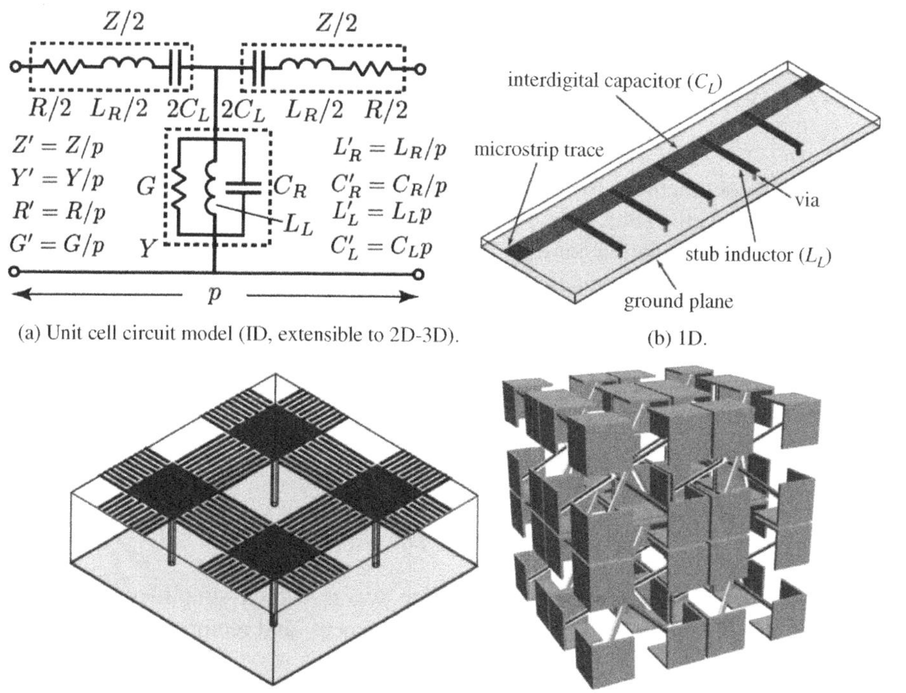

(a) Unit cell circuit model (ID, extensible to 2D-3D). (b) 1D. (c) 2D. (d) 3D.

Figure 11.3 Composite right/left-handed (CRLH) transmission line (TL) metamaterial structures, constituted of sub-wavelength inductors and capacitors forming a band-pass type structure. A CRLH metamaterial is constituted of a series resonant tank, made of the inductor L_R and capacitor C_L and related to the material permeability μ via the impedance $\mu = Z/(j\omega p)$, and of a shunt anti-resonant tank, made of the capacitor C_R and inductor L_L and related to the material permittivity ε via the admittance $\varepsilon = Y/(j\omega p)$. Reproduced by permission of ©2007, 2008 IEEE

The CRLH transmission-line unit cell [Figure 11.3(a)] is characterized by the immittances

$$Z = R + j\left(\omega L_R - \frac{1}{\omega C_L}\right) = R + j\frac{(\omega/\omega_{se})^2 - 1}{\omega C_L} \overset{p/\lambda_g \to 0}{\rightarrow} Z'p, \tag{11.3a}$$

$$Y = G + j\left(\omega C_R - \frac{1}{\omega L_L}\right) = G + j\frac{(\omega/\omega_{sh})^2 - 1}{\omega L_L} \overset{p/\lambda_g \to 0}{\rightarrow} Y'p, \tag{11.3b}$$

where the primed variables represent per-unit-length (Z', Y', L'_R, C'_R) and times-unit-length (L'_L, C'_L) quantities, $\omega_{se} = 1/\sqrt{L_R C_L} = 1/\sqrt{L'_R L'_R}$ and $\omega_{sh} = 1/\sqrt{L_L C_R} = 1/\sqrt{L'_L C'_R}$ are the series and shunt resonances, respectively, and where $p/\lambda_g = \beta p/(2\pi) \rightarrow 0$ represents the infinitesimal limit of a perfectly uniform TL or homogeneous metamaterial structure.

The general transmission matrix for a two-port network unit cell of the type shown in Figure 11.3(a) is obtained by multiplying the individual element matrixes

$$\begin{pmatrix} A & B \\ C & D \end{pmatrix} = \begin{pmatrix} 1 & Z/2 \\ 0 & 1 \end{pmatrix}\begin{pmatrix} 1 & 0 \\ Y & 1 \end{pmatrix}\begin{pmatrix} 1 & Z/2 \\ 0 & 1 \end{pmatrix} = \begin{pmatrix} 1+ZY/2 & Z(1+ZY/4) \\ Y & 1+ZY/2 \end{pmatrix}, \quad (11.4)$$

where $A = D = 1+ZY/2$ due to symmetry and $AD-BC = l$ due to reciprocity.

The Bloch-Floquet theorem is next applied to describe the periodic structure obtained by cascading the unit cell. Thus, $\psi(z+p) = e^{-\gamma_B p}\psi(z)$, where $\gamma_B = \alpha_B + j\beta_B$ is the Bloch (periodic) complex propagation constant, and ψ represents either the voltage or the current. This leads to the determinant equation

$$e^{+\gamma_B p} + (AD-BC)e^{-\gamma_B p} - (A+D) = 0, \quad (11.5)$$

which simplifies here to $\cosh(\gamma_B p) = A$ by symmetry and reciprocity. Similarly, the Bloch impedance is obtained by the ratio of the periodic voltage and current at either port of the unit cell, and reads

$$Z_B = \frac{2B}{D-A \pm \sqrt{(A+D)^2 - 4(AD-BC)}}, \quad (11.6)$$

where $\pm$ refers to positively/negatively traveling waves, which simplifies here to $Z_B(\omega) = B/\sqrt{A^2-1}$ (positively traveling wave case) by symmetry and reciprocity. Summarizing:

$$\gamma_B(\omega) = \frac{1}{p}\cosh^{-1}\left(1+\frac{ZY}{2}\right) \overset{\gamma_B p \rightarrow 0}{\rightarrow} \gamma(\omega) = \sqrt{Z'Y'}, \quad (11.7a)$$

$$Z_B(\omega) = \sqrt{\frac{Z}{Y}}\sqrt{1+ZY/4} \overset{\gamma_B p \rightarrow 0}{\rightarrow} Z_c(\omega) = \sqrt{\frac{Z'}{Y'}}, \quad (11.7b)$$

where the infinitesimal results were obtained by Taylor series approximations. The specific CRLH Bloch propagation constant and impedance are derived by substituting Equations (11.3) into Equations (11.7a) and (11.7b), respectively. In the low-loss first-order approximation, the *ZY* and *Z*/*Y* parameters for the CRLH case read

$$ZY = -\left(\frac{\omega}{\omega_R}\right)^2 - \left(\frac{\omega_L}{\omega}\right)^2 + \omega_L^2\left(\frac{1}{\omega_{se}^2} + \frac{1}{\omega_{sh}^2}\right) + \delta \overset{p/\lambda_g \rightarrow 0}{\rightarrow} Z'Y' = \frac{ZY}{p}, \quad (11.8a)$$

$$\frac{Z}{Y} = Z_L^2 \frac{(\omega/\omega_{se})^2 - 1 + \delta_{\text{num}}}{(\omega/\omega_{sh})^2 - 1 + \delta_{\text{den}}} \overset{p/\lambda_g \rightarrow 0}{\rightarrow} \frac{Z'}{Y'} = \frac{Z}{Y}, \quad (11.8b)$$

where $\omega_R = 1/\sqrt{L_R C_R}$, $\omega_L = 1/\sqrt{L_L C_L}$, $Z_L = \sqrt{L_L/C_L}$, and $\delta = \delta(R, C)$ is a loss term that is much smaller than each of the *R*,*G*-independent terms.

In the *balanced resonance* case, $\omega_{se} = \omega_{sh}$ (equal series and shunt resonances), the above relations reduce to

$$ZY = -\left(\frac{\omega}{\omega_R} - \frac{\omega_L}{\omega}\right) + \delta, \tag{11.9a}$$

$$\frac{Z}{Y} = Z_L^2 = \frac{L_R}{C_R} = Z_R^2. \tag{11.9b}$$

respectively. Equations (11.7) take then the simple form

$$\gamma_B(\omega) = \frac{1}{p}\cosh^{-1}\left\{1 - \frac{1}{2}\left[\left(\frac{\omega}{\omega_R} - \frac{\omega_L}{\omega}\right)^2 + \delta\right]\right\} \overset{\gamma_B p \to 0}{\longrightarrow} \beta(\omega) = \frac{\omega}{\omega_R} - \frac{\omega_L}{\omega}, \tag{11.10a}$$

$$Z_B(\omega) = Z_L\sqrt{1 + \frac{1}{4}\left[\left(\frac{\omega}{\omega_R} - \frac{\omega_L}{\omega}\right)^2 + \delta\right]} \overset{\gamma_B p \to 0}{\longrightarrow} Z_c(\omega) = Z_L = Z_R, \tag{11.10b}$$

where it appears that in the infinitesimal limit, the dispersion relation, $\beta(\omega)$, is purely real and therefore gap-less. It splits into a negative hyperbolic LH contribution $(-\omega_L/\omega)$, dominating at lower frequencies, and a positive linear RH contribution $(+\omega/\omega_R)$, dominating at higher frequencies. The characteristic impedance is independent of frequency $(Z_c = Z_L = Z_R)$, and therefore allows broadband matching to constant impedance (generally $50\,\Omega$) ports. The dispersion relation and Bloch impedance given by Equation (11.10) are plotted in Figure 11.4, where the scanning law $\theta(\omega) = \sin^{-1}[\beta_B(\omega)/k_0]$ is also shown for later use. Not surprisingly, the CRLH Bloch dispersion relation exhibits a band-pass response, since a CRLH structure [Figure 11.3(a)] is a periodic band-pass filter.[7] The high-pass LH cutoff ω_{cL} and low-pass RH cutoff ω_{cR} are given in the 1D case by [2]

$$\begin{aligned}\omega_{cL} &= \omega_0\sqrt{\frac{\left[\kappa + (2/\omega_L)^2\right]\omega_0^2 - \sqrt{\left[\kappa + (2/\omega_L)^2\right]^2\omega_0^4 - 4}}{2}} \\ &\overset{\omega_{se}=\omega_{sh}}{=} \omega_R\left|1 - \sqrt{1 + \frac{\omega_L}{\omega_R}}\right| \overset{\omega_L \ll \omega_R}{=} \frac{\omega_L}{2} = \frac{1}{2\sqrt{L_L C_L}},\end{aligned} \tag{11.11a}$$

[7] It is well-known in filter theory that periodic configurations, originally designed by the "image parameter method," nowadays obsolete, do not lead to optimal filters [30]. However, when the order of a standard filter designed by modern synthesis techniques (prototyping functions, transmission poles location and coupling matrix method) [31] becomes very large, corresponding to a large number of elements in the filter ladder structure, all the elements become identical. Therefore, a large-order filter is inherently a periodic structure. Thus, it makes sense to use periodic configurations in metamaterial structures, which generally include a large number of cells. A second reason to use periodic arrangements, even when the structure includes a relatively small number of cells, is the possibility to define a refractive index in terms of the fundamental space harmonic, as will be discussed later. Finally, a periodic structure, especially for dimensions larger than one, is much easier to manufacture than a non-periodic one.

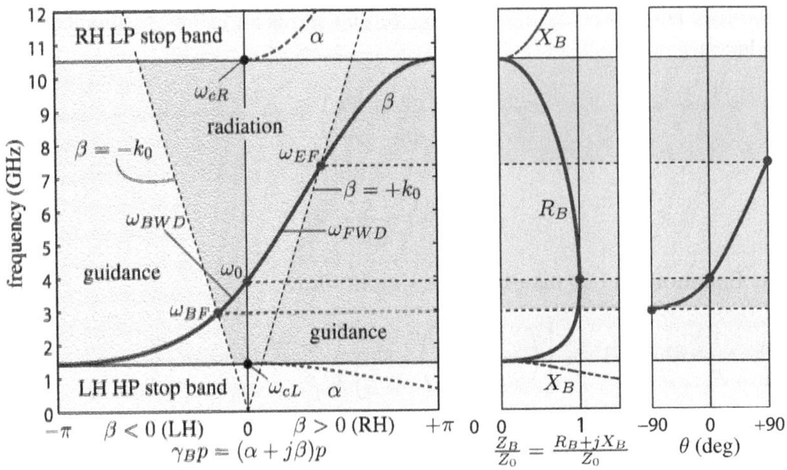

(a) Dispersion relation. (b) Impedance. (c) Scanning law.

Figure 11.4 Bloch dispersion relation $\gamma_B(\omega)$ [Equation (11.10a)], impedance $Z_B(\omega)$ [Equation (11.10b)], and scanning law $\theta(\omega) = \sin^{-1}[\beta_B(\omega)/k_0]$ of a balanced ($\omega_{se} = \omega_{sh}$) TL CRLH metamaterial. Dissipation loss is neglected. Reproduced by permission of ©2001 IEEE

$$\omega_{cR} = \omega_0 \sqrt{\frac{\left[\kappa + (2/\omega_L)^2\right]\omega_0^2 + \sqrt{\left[\kappa + (2/\omega_L)^2\right]^2 \omega_0^4 - 4}}{2}}$$
$$= \omega_R\left(1 + \sqrt{1 + \frac{\omega_L}{\omega_R}}\right) \overset{\omega_L \ll \omega_R}{=} 2\omega_R + \frac{\omega_L}{2} = \frac{2}{\sqrt{L_R C_R}} + \frac{1}{2\sqrt{L_L C_L}}, \tag{11.11b}$$

where $\kappa = L_R C_L + L_L C_R$ and $\omega_0 = 1/\sqrt[4]{L_R C_L L_L C_R} \overset{\omega_{se}=\omega_{sh}}{=} \omega_{se} = \omega_{sh}$.

By mapping telegrapher's equations to Maxwell's equations [8, 32], the equivalent material parameters are given as $\varepsilon(\omega) = Y'(\omega)/(j\omega) \approx Y/(j\omega p)$ and $\mu(\omega) = Z'(\omega)/(j\omega) \approx Z(\omega)/(j\omega p)$ ($p/\lambda_g \ll 1$), which yields, using the CRLH immittances in Equations (11.3),

$$\varepsilon(\omega) = \varepsilon_0\left[1 - \frac{\omega_{pc}^2}{\omega^2 - j\omega\Gamma_e}\right]$$
$$\text{with} \quad \varepsilon_0 = \frac{C_R}{p}, \quad \omega_{pe} = \frac{1}{\sqrt{L_L C_R}} = \omega_{sh} \quad \Gamma_e = \frac{G}{C_R}, \tag{11.12a}$$

$$\mu(\omega) = \mu_0\left[1 - \frac{\omega_{pm}^2}{\omega^2 - \omega_{rm}^2 - j\omega\Gamma_m}\right] = \mu' - j\mu'',$$
$$\text{with} \quad \mu_0 = \frac{L_R}{p}, \quad \omega_{pm} = \frac{1}{\sqrt{L_R C_L}} = \omega_{se}, \quad \Gamma_m = \frac{R}{L_R}, \quad \omega_{rm} = 0. \tag{11.12b}$$

In the case of low loss ($\Gamma_e, \Gamma_m \ll 1$), which may be achieved practically in TL metamaterials, these relations are formally identical to those of Equations (11.1), but with $\omega_{rm} = 0$. Thus, a TL metamaterial is a *double-Drude* medium. The absence of resonance in the magnetic response avoids an overall resonant behavior of the structure, and allows broadband low-loss propagation.

As already pointed out, CRLH metamaterials are band-pass filters, with a spectrum lying between their high-pass LH cutoff (ω_{cL}) and low-pass RH cutoff (ω_{cR}) frequencies. However, the Drude modeling of Equations (11.12) is accurate in the metamaterial range, which is centered at the transition frequency ω_0 between the LH and the RH ranges, where $\beta = 0$ and therefore $\lambda_g = 2\pi/\beta = \infty$ (perfectly homogeneous medium frequency).

Since the metamaterials under discussion are periodic structures, their fields are rigorously expressed as an expansion of an infinite number of *space harmonics*, which may be written in 1D as $\beta_m(\omega) = \pm[\beta_0(\omega) + 2\pi m/p]$ [33]. In the Bragg regime, where p is in the order of $\lambda_g/2$ or larger, all these space harmonics contribute significantly to the physical fields, which leads to fields distributions forming a diffraction pattern [34, 35]. In contrast, in the metamaterial regime, which occurs in the balanced CRLH case at the frequency ω_0, the contribution of the space harmonics $m > 1$ is generally negligible, so that only the fundamental space harmonic $\beta(\omega) \approx \beta_0(\omega)$ contributes significantly, and the refractive index may be defined in terms of this fundamental space harmonic as $n(\omega) = \beta_0(\omega)/k_0 = \beta_0(\omega)c/\omega$ [23].

The TL approach of CRLH metamaterials presented in this section assumes known LC parameters (L_R, C_R, L_L, C_L) and a perfectly CRLH structure [Figure 11.3(a)]. Although such parameters may be extracted by efficient tools based on microwave network theory [2], from scattering parameters obtained by full-wave analysis or measurement, and although this approach has been demonstrated very accurate in practical situations, more specific full-wave analyzes of actual CRLH structures, with their specific layout and all possible electromagnetic effects, are profitable for further confirmation and insight [36].

11.2.3 Relation between RP and TL Metamaterials

As already seen by comparing Equations (11.1) and (11.12), and as indicated by a leftward arrow in Figure 11.2, the dispersion relation of a TL metamaterial is the limit of the dispersion relation of a RP material with a magnetic resonance frequency ω_{rm} tending to zero. This similarity in the dispersion relations is not fortuitous: RP and TL metamaterials have a fundamental relation [23]. This relation is depicted in Figure 11.5.

In RP metamaterials (Figure 11.1), the wire and split ring particles are sufficiently distant from each other to experience negligible coupling. Therefore, the dispersion parameters of the overall material can be determined from the electric and magnetic dipole moments (**p** and **m**) of a unique cell, as done in Figure 11.2 and illustrated in the left-hand side of Figure 11.5, where the wire has been folded at its two ends, as often done with dipole antennas for compactness [37]. When the folded wires and split rings are brought in close proximity to one another, as shown in the center of Figure 11.5, they become strongly electrically (capacitively) and magnetically (inductively) coupled, respectively. The right-most mutation in Figure 11.5 shows that the coupled folded wires and split rings, which can provide only $\varepsilon < 0$ or $\mu < 0$ when isolated, both transform into a CRLH TL metamaterial when their mutual couplings are

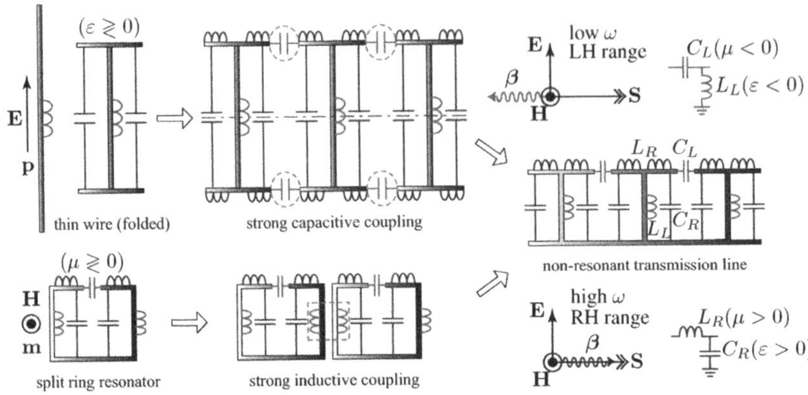

Figure 11.5 Relation between RP (thin wire and split ring) and TL (CRLH inductors and capacitors) metamaterials. The dash-dotted line in the middle of the folded wires indicate an electric symmetry plane. Reproduced by permission of ©2004 IEEE

strong enough. Thus, *TL metamaterials are the limit case of resonant metamaterials with strongly coupled particles.*

It should be noted that, although they are often most convenient in their planar form for microwave applications, 1D and 2D CRLH TL metamaterials [Figures 11.3(b) and 11.3(c)] may also be arrayed and stacked to form bulk (volumetric) one-dimensionally and two-dimensionally LH metamaterials with the same response as RP metamaterials [Figures 11.1(a) and 11.1(b)], but with a dramatically enhanced bandwidth [38].

11.3 Leaky-Wave Antennas

11.3.1 Fundamentals

Leaky-wave antennas, either in uniform or periodic configurations, have been abundantly studied for over half a century [39, 40]. They essentially provide the benefit of high directivity without requiring a complex feeding network, as antenna arrays do. However, they suffer of major limitations in their scanning capabilities, which have limited their applications to date. CRLH meta-structures have essentially suppressed these limitations, and thereby opened novel perspectives for leaky-wave antennas.

A leaky-wave antenna is a traveling-wave structure with a complex propagation constant $\gamma(\omega) = \alpha_{LW}(\omega) + j\beta(\omega)$, where $\alpha_{LW}(\omega)$ is the leakage factor [typically, $\alpha_{LW}/k_0 < 0.02$, i.e., a length of at least $10\lambda_0$ is required to radiate 90% of the power [40] and $\beta(\omega)$ is the dispersion relation. When the wave velocity is faster than the velocity of light (or $\beta(\omega) < k_0$), the main beam of the antenna radiates in the direction

$$\theta_{MB}(\omega) = \sin^{-1}\left[\frac{\beta(\omega)}{k_0}\right] = \sin^{-1}\left[\frac{c\beta(\omega)}{\omega}\right], \tag{11.13}$$

where θ_{MB} is the elevation angle from the direction normal to the structure [2, 40]. This formula shows that the main beam may be scanned with frequency, if the structure is dispersive with a dispersion of the type $\beta(\omega) = \beta_0 + \beta_1(\omega-\omega_0) + \beta_2(\omega-\omega_0)^2 + \cdots$. Conventional leaky-wave antennas are restricted to strictly positive θ_{MB}, due to strictly positive β in uniform configurations, or to a discontinuous range of negative or positive θ_{MB} excluding broadside ($\theta_{MB} = 0$), due to the standing-wave nature of the wave at $\beta = 0$ in periodic configurations (using space harmonics).

11.3.2 Leaky-Wave Properties of CRLH Metamaterials

A CRLH metamaterial exhibits a unique dispersion curve that extends across the dispersion diagram all the way from the region $\beta < -k_0$ to the region $\beta > +k_0$ [Figure 11.4(a)]. When the structure is open to free space, this dispersion gives rise to the four distinct regions, shown in Figure 11.4(a): I: LH guiding; II: LH radiating; III: RH radiating; IV: RH guiding. Moreover, in the balanced case ($\omega_{se} = \omega_{sh}$), the $\beta = 0$ transition from the LH band to the RH band is seamless. This transition is characterized by the frequency ω_0 where $\beta = 0$ with $v_g = \partial\omega/\partial\beta \neq 0$, allowing an infinite-wavelength ($\lambda_g = 2\pi/\beta$, or infinite-phase velocity $v_p = \omega/\beta = \infty$) traveling wave.

11.3.3 Fan Beam

As a consequence of the CRLH dispersion characteristic, a CRLH leaky-wave antenna scans the entire space from $\theta = -90°$ to $\theta = +90°$, including $\theta = 0°$, as frequency is varied from $\omega = -\beta c$ to $\omega = +\beta c$ (CRLH dispersion regions II and III), as shown in Figure 11.4(c). Moreover, it may be excited by an elementary and efficient (simple transmission line) feeding mechanism, due to the fact that it operates in the fundamental space harmonic. This backfire-to-endfire antenna was discovered in [41], and later extensively studied and applied [e.g. [2]].

Figure 11.6(a) shows a CRLH leaky-wave antenna and illustrates its full-space scanning capability, while Figure 11.6(b) presents typical scanning capabilities for this type of antenna. Instead of being frequency scanned, the antenna may also be electronically scanned at a fixed frequency, as required in many applications, using capacitors or inductors controlled by a bias field. In this case, it is also possible to equalize the beam by using a nonuniform biasing distribution along the structure [42]. The antenna considered is here a 1D structure. Therefore, it provides scanning only in one plane. While the beam is very directive in this plane [xz in Figure 11.6(a)], it is fat in the perpendicular direction [y in Figure 11.6(a)]: this is a *fan beam*.

This antenna represents a breakthrough in leaky-wave antennas. It is the first leaky-wave antenna capable of scanning the entire space, continuously and including broadside. It exhibits several characteristics of the antenna arrays without requiring a feeding network. Due to its exceptional flexibility, it was also successfully applied to various passive and active smart reflectors, a few of which are described in [2].

11.3.4 Conical Beam

Because it is essentially "seen" by electromagnetic waves as a uniform medium, a 1D CRLH structure may be straightforwardly extended to 2D, just like a narrow strip maybe extended to a rectangular patch. This is illustrated in Figure 11.3(c). When excited in its center, for instance

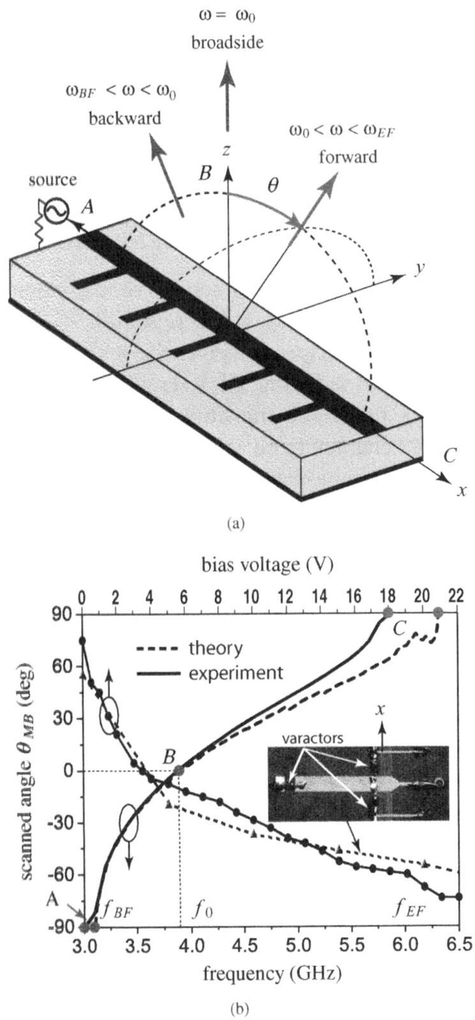

Figure 11.6 Frequency-scanned leaky-wave antennas. (a) Schematic representation of an open CRLH structure with its three radiation regions [Figure 11.4(c)]: backward ($\beta < 0$, LH range), broadside ($\beta = 0$, transition frequency f_0), and forward ($\beta > 0$, RH range). (b) Backfire-to-endfire frequency and electronic scanning relation $\theta(\omega)$ of a typical antenna. Reproduced by permission of ©2005 Dr. Shenghui Zhang and ©2008 IEEE

by a coaxial probe, such a 2D CRLH structure supports in its metamaterial regime a perfectly circular wave. When this wave has a phase velocity faster than the speed of light (CRLH dispersion regions II and III), it radiates in a leaky-wave manner. This results in a conical-beam antenna (i.e. maximum radiated power on a $\varphi = 0 \ldots 2\pi$ circle under an elevation θ around the normal axis), as illustrated in Figure 11.7(a) and demonstrated in Figure 11.7(b). As frequency varies, the opening angle of the conical pattern varies, following the dispersion relation

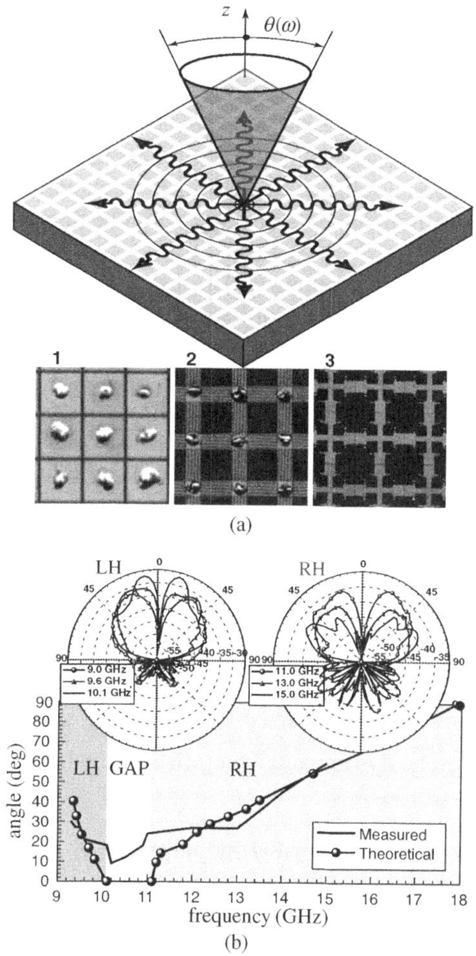

Figure 11.7 Conical-beam leaky-wave antenna. (a) Conical-beam leaky-wave radiation produced by a fast isotropic (metamaterial range) radial wave and three corresponding 2D CRLH mushroom-type [2] structures (top view): 1: simple mushroom structure, 2: interdigital mushroom structure, 3: Stepped-impedance balanced structure. (b) Radiation patterns and scanning law obtained with structure 1. The maximum LH and RH gains are 17.3 dB at 10.1 GHz and 12.4 dB at 11 GHz, respectively. Reproduced by permission of ©2004, 2008 IEEE

(11.10a), which leads to beam scanning. Due to azimuthal symmetry, the LH and RH ranges provide the same functionality, but with opposite scanning slopes ($d\theta_{MB}/d\omega$).

A fundamental difference between this antenna and conventional 2D leaky-wave antennas is that the latter use a partially reflecting sheet array [43] or a dielectric superstrate layer [44]. These are typically very sensitive to fabrication tolerances, whereas the former is an easy-to-fabricate uniplanar structure.

11.3.5 Pencil Beam

In point-to-point communications, utilized in applications such as local multi-point distribution service (LMDS), WiMAX, and satellite communications, a pencil beam [i.e., maximum radiated power in a unique direction (θ, ϕ) of space] is often required. Such a beam cannot be produced efficiently with dynamic scanning capability in a 2D structure of the type discussed in Section 11.3.4, even with edge excitation. On the other hand, the conventional phased-array option requires a complex, cumbersome, lossy, and dispersive 2D feeding network.

Several interesting pencil-beam scanning CRLH leaky-wave antennas have been reported [45–49]. These antennas consist of arrays of leaky-wave elements, using a combination of frequency tuning and phase-shift tuning to achieve pencil-beam scanning. [45] used heterodyne mixers and delay lines with filters for scanning. [46] used a Butler matrix. [49] used varactor diodes and a Rotman lens, [48] used a 2D surface, of the type discussed in Section 11.3.4, but excited at two of its edges by tuned power levels. We present here a particularly economical and flexible 2D radiating-aperture solution, involving CRLH structures playing two distinct roles.

This antenna is depicted in Figure 11.8(a), and its full-space scanning capability is demonstrated in Figure 11.8(b) [50]. It consists of a phased array of frequency-scanned (case depicted in Figure 11.8(a), with subsequent LO mixer control) or electronically-scanned (LO control replaced by varactor control) CRLH leaky-wave antennas. These antennas are fed by a series uniform ($\lambda_g = \infty$) boxed (to suppress leaky-wave radiation) [51] CRLH series power divider [52–54]. This solution does not require any corporate feeding network. In addition, it allows an arbitrary number of antenna elements, which may be arbitrarily spaced, and the number of which may even be dynamically controlled for real-time beam shaping [54].

11.3.6 Efficiency Enhancement by Power Recycling

As pointed out in Section 11.3.1, leaky-wave antennas are often designed as electrically very large ($5-15\lambda_0$) in order to provide maximal directivity, resulting from maximal effective aperture, and at the same time exhibit maximal radiation efficiency from complete power radiation. However, in many applications, circuit board space constraints prohibit such large structures and require a trade-off between size and directivity. In such cases, the antenna may be in the order of $2-3\lambda_0$ long, so a substantial fraction of the input power is wasted in the matched load, which may lead to prohibitively low radiation efficiency.

In order to mitigate this problem, the wasted power may be recycled to be reradiated [55]. The power reaching the end of the leaky-wave antenna is re-injected into adjacent leaky-wave antennas instead of being simply dumped into the load, as shown in Figure 11.9. Compared to the pencil-beam antenna of Section 11.3.5, this configuration maintains the pencil-beam radiation pattern capability while dramatically improving the radiation efficiency as well as the gain, as demonstrated in Table 11.1.

The main beam in this pencil-beam antenna may be either frequency scanned or electronically scanned in the xz plane, and yz plane scanning capability may be provided by adding phase shifters between array elements. This, this architecture can provide full-space scanning pencil-beam capability as the antenna of the type of Section 11.3.5 but with strongly enhanced radiation efficiency.

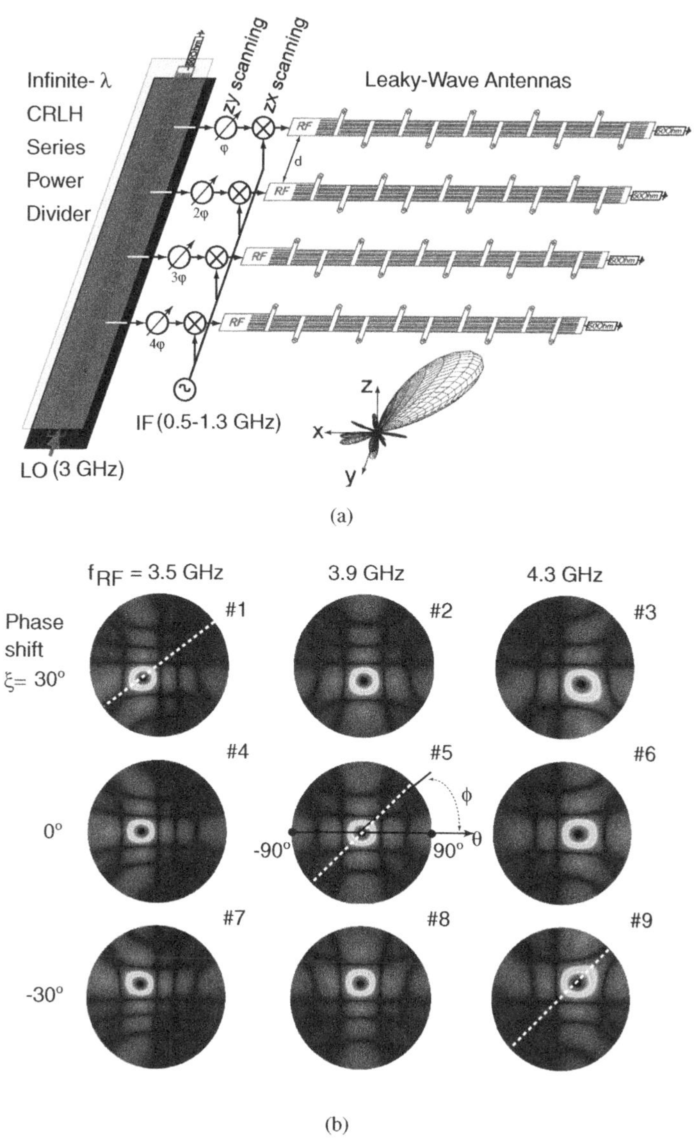

Figure 11.8 Pencil-beam leaky-wave antenna. (a) Architecture. (b) Full-wave radiation plots (radius = elevation angle θ, azimuth = azimuth angle ϕ) demonstrating the full-space scanning capability of the antenna. Reproduced by permission of ©2004, 2008 IEEE

11.3.7 Active Beam Shaping

The exceptional flexibility of CRLH metamaterials and leaky-wave antennas has inspired several ideas incorporating active elements [56]. Due to the sub-wavelength nature of the unit cell, low-gain transistors may be ideally integrated along the structure to manipulate the magnitude of the signal along it, in addition to its phase.

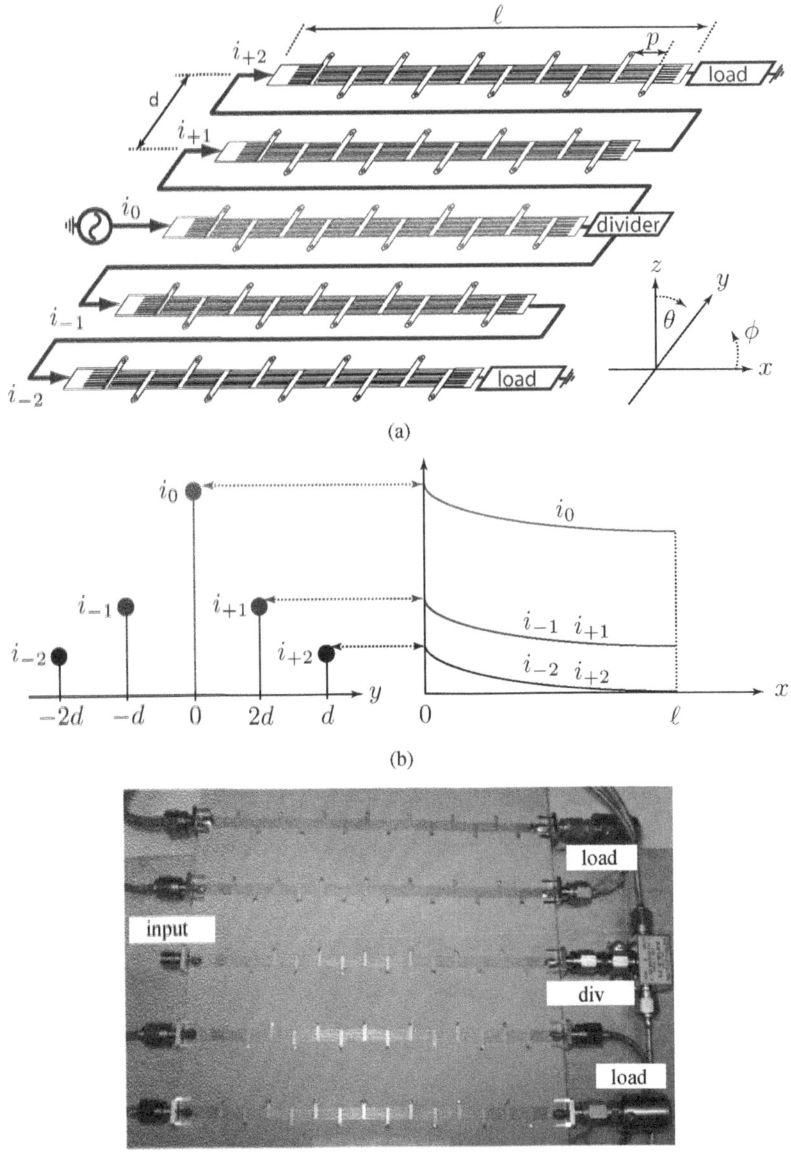

(c)

Figure 11.9 Leaky-wave antenna array with improved efficiency using a series feeding network for power recycling. (a) Perspective view. (b) Current distribution in the x- and y-directions. (c) Prototype (5 14-cell CRLH elements with period $p = 8.34$ mm, spacing $d = 28$ mm, and a 3-dB Wilkinson power divider). Reproduced by permission of ©2003, 2009 IEEE

Table 11.1 Simulated (Ansoft Designer) gain (G), directivity (D), half-power beam width (HPBW) and efficiency (η) of the CRLH array of Figure 11.9

Number of array elements	G	D	HPBW		η
			xz-plane	yz-plane	
1	6.84 dB	11.15 dB	30°	96°	37.60%
3	10.72 dB	12.53 dB	31°	62°	65.89%
5	12.45 dB	13.51 dB	31°	45°	78.29%

The beam width of a leaky-wave antenna is controlled by its leakage factor. For a given passive structure, this factor is fixed. Beyond the length for which all of the power has been radiated, no further increase of effective aperture, and therefore of directivity, may be achieved. In order to suppress this limitation, an active CRLH leaky-wave antenna, incorporating amplifiers as repeaters (or power regenerators) was proposed in [57]. This idea had been implemented in a conventional leaky-wave antenna before in [58]. The resulting antenna is virtually capable of providing an arbitrarily high directivity – since the effective aperture is unlimited – with single and simple TL excitation. Massive gain improvement compared to the case of a corresponding passive antenna is easily achieved (8.3 dB in [57]) both from increased directivity and from increased efficiency achieved by the reflection-canceling unilateral nature of the amplifiers.

A beam-shaping generalization of this concept was discussed in [57] and demonstrated in [59]. It is well known that the radiation pattern of an antenna is essentially the Fourier transform of its aperture field distribution [37]. Therefore, the shape of the radiated beam of a metamaterial antenna – again, due to its sub-wavelength structure and feeding-network-less configuration – may be easily manipulated by approximating the desired aperture distributions using gain-controlled amplifiers. This concept, taking into account the exact exponentially decaying nature of the leaky sections in the design, is called active "digitized aperture" beam forming. It is illustrated in Figure 11.10(a), verified experimentally in Figure 11.10(b) for the case of a uniform distribution, and illustrated in Figure 11.10(c) for the compared cases of a uniform distribution (maximum directivity) and a binomial distribution (minimum sidelobe level).

CRLH leaky-wave structures incorporating active elements may lead to a quasi-"universal" type of antenna. Such an antenna could provide simultaneously beam-scanning and-shaping functionality, controlled in real time by a digital signal processor. For instance, the idea of dynamic radiation pattern diversity (DRPD) multiple input multiple output (MIMO) was recently introduced in [60]. In this scheme, efficient and low-cost CRLH antennas perform a real-time scanning calibration to the scattering environment for channel optimization and data-rate maximization.

11.4 Resonant Antennas

11.4.1 Fundamentals

Leaky-wave antennas offer the advantage of high directivity, without requiring a complex feeding network, and high radiation efficiency, due to the continuity of the radiation aperture.

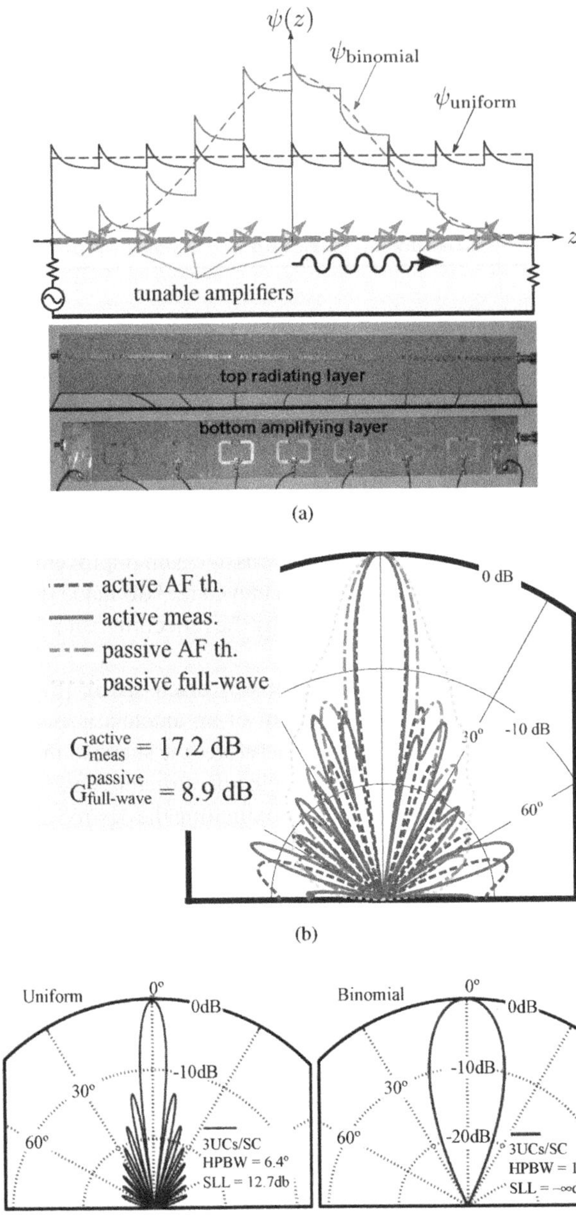

Figure 11.10 Active beam-forming leaky-wave antenna. (a) Active "digitized aperture" beam-forming concept and active CRLH antenna prototype (a microstrip ground plane with via transition slots between the top and the bottom). (b) Radiation pattern at 3.7 GHz for the active leaky-wave antenna shown in Figure 11.10(a) with a uniform aperture distribution. The "passive" case represents the same antenna without any amplifier. (c) Radiation patterns (array-factor results) corresponding to the aperture distributions shown in Figure 11.10(a) (uniform: maximally directive, binomial: minimal side-lobe level) for $N = 48$ unit cells with one amplifier every third cell. Reproduced by permission of ©2007, 2008 IEEE

CRLH implementations render leaky-wave antennas particularly flexible in terms of dynamic scanning capability. However, all leaky-wave antennas suffer of the drawback of limited bandwidth, related to beam squinting, $\partial\theta/\partial\omega$, in fixed frequency applications. Although this problem might find a solution using enhanced group velocity, possibly even super-luminal, with relatively high dispersion, and losses compensable by active elements [61], the practicality of this approach is questionable. In this case, resonant CRLH antennas exhibit complementary properties and offer other benefits, compared to their leaky-wave counterparts.

11.4.2 Resonant Properties of CRLH Metamaterials

A CRLH resonant antenna is obtained by reactively terminating a CRLH TL structure that is open to free space by a short or an open circuit. Such an antenna may then be designed in the same way as a conventional uniform-metallization antenna (e.g. a patch antenna), but with the effective wavelength and frequency response of a CRLH metamaterial.

The resonant modes of a CRLH structure of length ℓ are given by $\ell = |m|\lambda_g/2$ ($m \in \mathbb{N}$) or, equivalently, by $\beta_m\ell = m\pi$, where m can be both positive (RH band) or negative (LH band), and even zero (transition frequency), following $\beta(\omega)$. More specifically, each positive ($m > 0$) resonance mode has a twin negative ($m < 0$) resonance mode, and a zeroth order ($m = 0$) mode exists at the transition frequency, ω_0. Due to its discrete nature and subsequent finite bandwidth, a CRLH structure has a finite number of $2N$ ($2N-1$ in the balanced case) resonances. These correspond to $\beta_m p = \beta_m(\ell/N) = m\pi/N$, where N the number of unit cells of the resonator. These resonances are found by calculating the frequencies ω_m, mapping the abscissas $\beta_m = m\pi/(Np)$ of the dispersion diagram (Equation 11.10a), as

$$\frac{\omega_m^2}{\omega_R^2} + \frac{\omega_L^2}{\omega_m^2} = 2\left[1-\cos\left(\frac{m\pi}{N}\right)\right] + \frac{1}{\omega_R^2}\left(\omega_{se}^2 + \omega_{se}^2\right), \tag{11.14}$$

where $n = 0, \pm 1, \ldots, \pm(N-1)$. The discrete spectrum of a CRLH resonator is shown in Figure 11.11. While the resonance frequencies depend only on the dispersion relation, their coupling factor to external sources naturally depends on the Bloch impedance [Figure 11.4(b)], and on the excitation mechanism.

11.4.3 Multi-Band

Figure 11.12(a) shows a dual-band half-wavelength ($m = \pm 1$) open-ended resonant antenna, the dual-band property of which is an immediate consequence of the positive and negative resonance pairs [Equation (11.14)] of a CRLH structure [62]. This antenna is back-fed by a coaxial line at an off-center location for 50 Ω matching at the frequencies $f_{\pm 1}$.

In principle, all of the $2N-1$ resonances may be excited, with the exception of ω_{se} due to the absence of series currents for open-end termination [2], and matched to the source with proper excitation. The modes of each ± pair have the same guided wavelength and field distribution, as illustrated for $m = \pm 1$ in Figure 11.12(a). They therefore present very similar input impedances. This allows efficient dual-band operation from a single resonator. The return loss of the dual-band half-wavelength antenna is plotted in Figure 11.12(b) (where the weakly excited mode ω_{sh} is also visible, whereas, as expected, no ω_{se} resonance exists).

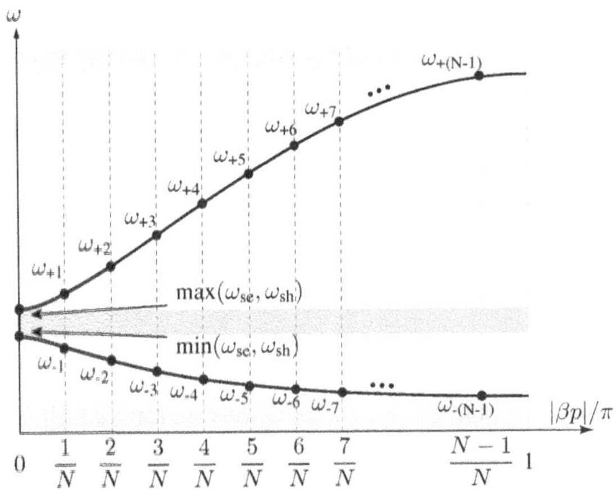

Figure 11.11 Resonance spectrum of a CRLH resonant structure comprised of N unit cells (here, the unbalanced situation, with two distinct $\beta = 0$ resonances, $(\omega_{se} \neq \omega_{sh})$. Reproduced by permission of ©2005, 2008 IEEE

This dual-band antenna has its polarization (**E** field) along the axis of the structure [x in Figure 11.12(a)]. It exhibits a half-wavelength patch type of radiation pattern, with gains of 4.2 dB and 4.5 dB for f_{-1} and f_{+1}, respectively, and cross-polarization lower than −20 dB.

Tri-band [63] and quad-band [64] (or even higher multiple-band) operation is also possible with higher-order CRLH transmission lines. However, control of the parasitic contributions becomes increasingly challenging as the number of bands increases. Since no balancing is required for resonant antennas, a simple CRLH structure may be designed as a tri-band antenna by exploiting the relaxation of this constraint [65].

11.4.4 Zeroth Order Resonance

The zeroth order CRLH resonance ($\ell/\lambda_g = 0, m = 0$) is particularly unique and interesting [66]. Figure 11.13(a) shows a zeroth order CRLH antenna, shorted at its output by via holes, and using at its input an inter-digital capacitance with a high-impedance transmission line transforming the impedance to a quasi-short of a few ohms [67]. Being short-circuited at both ends, this antenna operates in the ω_{se} mode (no ω_{sh} mode is excited, due to the absence of shunt currents). The corresponding return loss is plotted in the same figure, while Figure 11.13(b) shows the full-wave-simulated uniform current distribution and radiation pattern of this ω_{se} mode. The polarization is similar to that of the half-wavelength antenna presented in Section 11.4.3.

Figure 11.14 and Table 11.2 compare the CRLH zeroth order antenna of Figure 11.13 with a half-wavelength CRLH antenna (Section 11.4.3) and with a conventional patch antenna. Figure 11.14 shows that while the resonance frequency varies with the size of the structure for the half-wavelength and patch antennas, it remains constant for the zeroth order antenna, where ω_{se} depends only on the lumped values L_R and C_L (or ω_{sh} on L_L and C_R). This property may be exploited to design electrically small or electrically large antennas. Table 11.2 shows that the

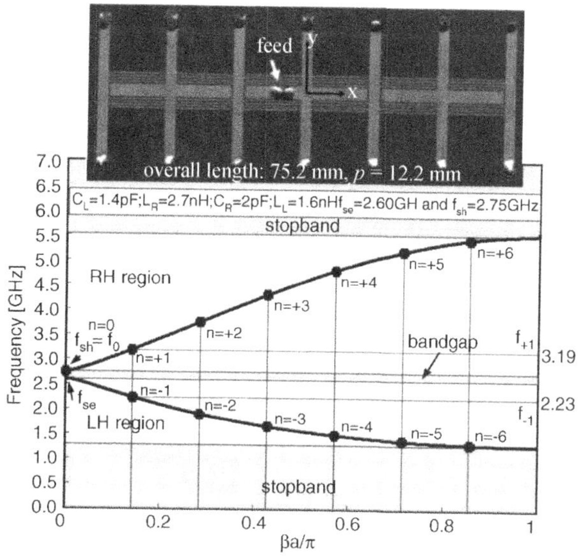

(a)

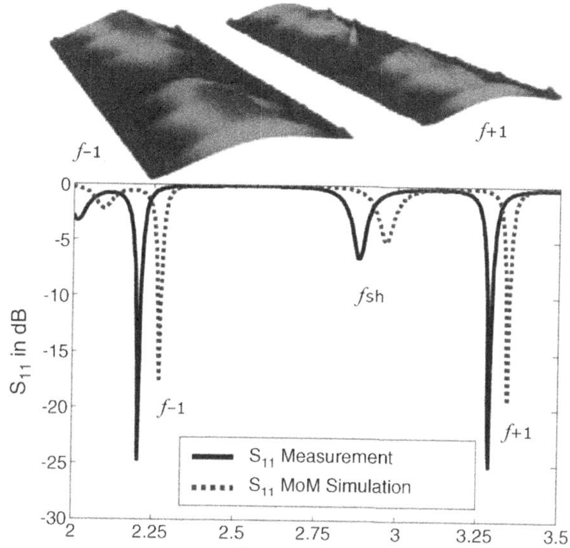

(b)

Figure 11.12 Half-wavelength ($\ell = \lambda_g/2$) dual-band ($m = \pm 1$) open-ended seven-cell CRLH microstrip resonant antenna (with symmetric L_L stubs for low cross-polarization), and corresponding dispersion relation including all possible resonance modes. Reproduced by permission of ©2005, 2008 IEEE

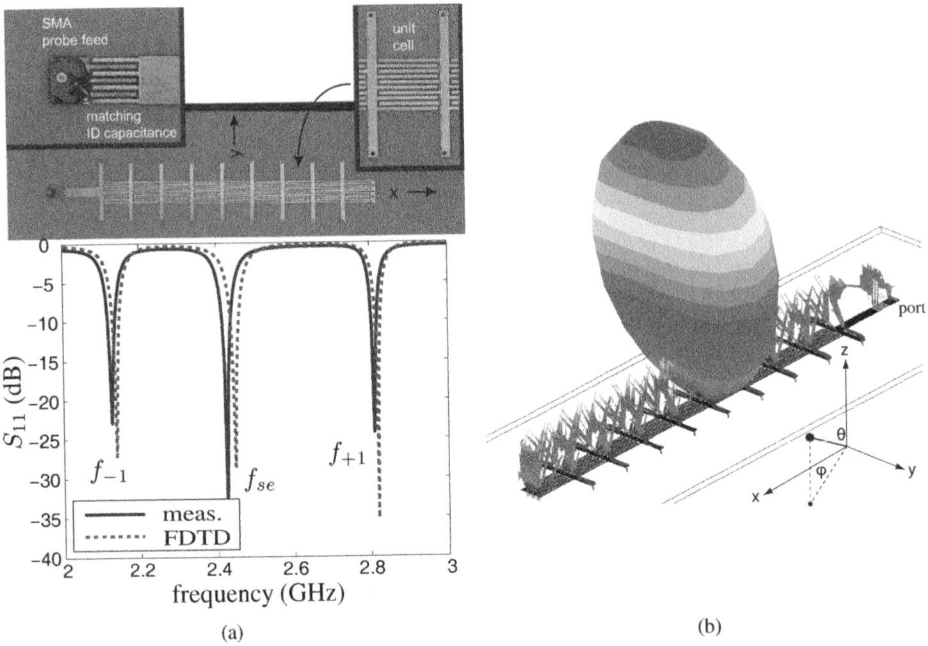

Figure 11.13 Zeroth order ($m = 0$) series-mode (ω_{se}) antenna. (a) Prototype with $N = 9$ unit cells (RO 4003 substrate with permittivity of 3.38, height of 1.52 mm and loss of $\tan\delta = 2.7 \times 10^{-3}$) and return loss. (b) Current distribution and broadside radiation pattern at $f_{se} = 2.4$ GHz (polarization: $\mathbf{E}\|\hat{x}$). Reproduced by permission of ©2006, 2008 IEEE

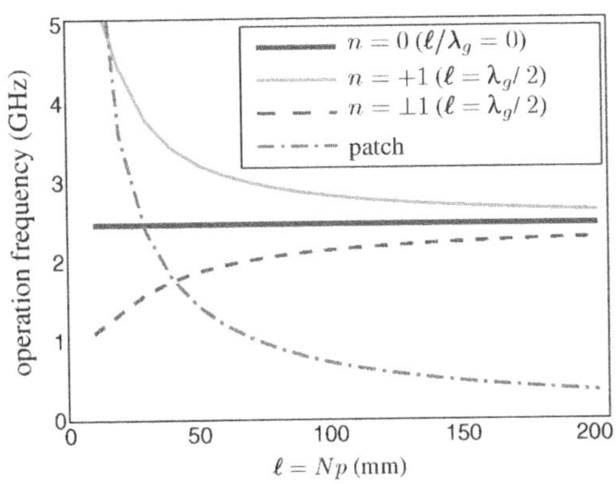

Figure 11.14 Resonance frequency comparison of a zeroth order CRLH antenna and a half-wavelength CRLH antenna for the structure of Figure 11.13 with a conventional patch antenna. Reproduced by permission of ©2006, 2008 IEEE

Table 11.2 Performance comparison for the three antennas of Figure 11.14. (Note: The larger directivity of the $n = 1$ CRLH antenna compared to the $n = 0$ CRLH antenna is due to the larger physical size of the former)

Type	CRLH $n = 0$	CRLH $n = +1$	Patch
f_{res} (GHz)	2.42	2.23	2.42
ℓ (mm)	102	118	33
ℓ/λ_0	0.83	0.88	0.27
D (dB)	8.8	9.6	6.3
η_{rad} (%)	72	39	76
G_{acc} (dB)	7.4	5.5	5.1

zeroth order antenna may achieve excellent efficiency, comparable to that of a conventional patch antenna, due to its uniform aperture field distribution.

The properties of CRLH zeroth order antennas might lead to optimal electrically small antennas [68]. For these, the fundamental reduction in efficiency may be mitigated by the perfectly uniform distribution of energy along the structure. Furthermore, the fundamental reduction in directivity may be mitigated by the corresponding maximized effective aperture.

11.4.5 High Directivity

A resonant CRLH antenna exhibits unique properties in terms of directivity. Compared to a conventional array, it has a single element, and does not require any corporate feeding network [37]. Compared to a leaky-wave antenna, it has a higher aperture efficiency, because it does not have an exponential decay of power along the aperture [67, 69].

Because its operating frequency is solely determined by its LC unit-cell elements (Figure 11.14), the size of a resonant CRLH antenna may be enlarged considerably at a fixed frequency. It may hence provide a super-high directivity, resulting from a very large effective aperture. This fact is illustrated in Table 11.2, where the CRLH antennas, operating at around 2.4 GHz, are more than three times larger than a patch antenna operating at the same frequency. They therefore exhibit a strongly enhanced directivity (here of over 2.5 dB), which could be even much further increased with further increased size. Figure 11.15 presents a CRLH zeroth-order antenna in three different sizes. Its increasing gain with increasing size (at fixed frequency) corroborates the previous statements.

A CRLH resonant antenna may thus virtually attain the directivity of any large-scale array, without requiring the array's large, cumbersome, and lossy feeding network. To attain such a high directivity, a leaky-wave antenna may require active elements for power regeneration (Section 11.3.7).

11.4.6 Electric/Magnetic Monopoles

CRLH resonant structures in loop configurations have several unique functionalities. For instance, they may be used as versatile multi-band and multi-polarization (linear/circular)

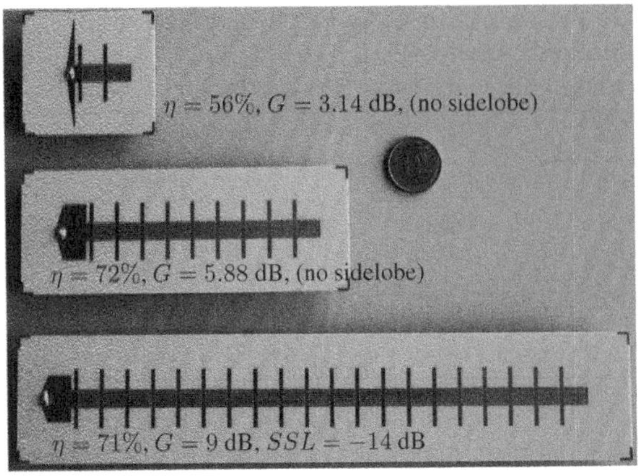

Figure 11.15 Three CRLH zeroth order (ω_{se}; shorting vias at the output and quarter-wave transformer to low impedance at input) microstrip resonant antennas of different sizes operating at the frequency of 2.4 GHz, with respective efficiencies (η), gains (G), and sidelobe levels (SSL). Reproduced by permission of ©2008, 2009 IEEE

antennas [70], or as electric/magnetic planar monopole (azimuthally symmetric radiation patterns) antennas [71]. Some of these are presented now.

A CRLH loop structure is obtained by folding a rectilinear CRLH structure so as to form a closed circular loop, as shown in the prototype of Figure 11.16(a), where the stubs are compactly placed in a radial manner within the loop area with a unique shorting via at the center of the structure. Possible backplane slot excitations are shown in Figure 11.16(b). Due to the additional (compared to the rectilinear case) azimuthal boundary condition, such a CRLH structure supports only modes of even (m) order, i.e. corresponding to loop circumferences multiple of the guided wavelength.

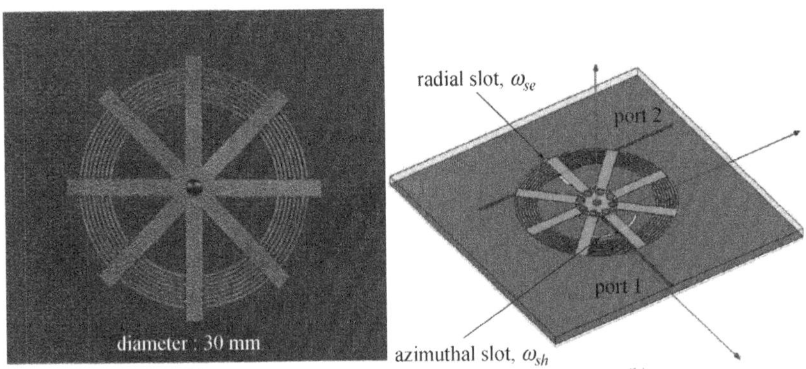

(a) (b)

Figure 11.16 CRLH loop resonant microstrip antenna. (a) Prototype. (b) Configuration, showing the two-port feeding structure for the simultaneous excitation of ω_{se} and ω_{sh} modes by backplane microstrip-fed slots, with one slot parallel to one stub and one slot perpendicular to another stub for ω_{se} and ω_{sh}, respectively. Reproduced by permission of ©2006, 2008 IEEE

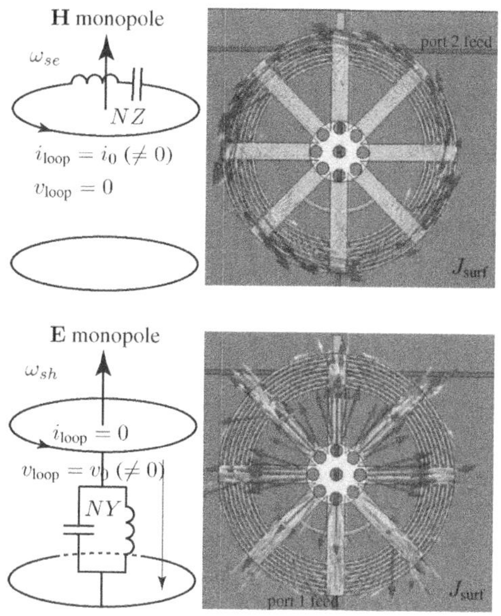

Figure 11.17 Zeroth order series and shunt resonances of a CRLH loop microstrip structure, leading to magnetic and electric dipole radiators, respectively. Reproduced by permission of ©2006, 2008 IEEE

Monopole-type radiation may by achieved in the zeroth order ($m = 0$) mode. The operation of this mode is described in Figure 11.17 with the help of full-wave computed surface current distributions. In the ω_{se} mode, $Z = 0$ [Equation (11.3a)], and therefore the series paths are seen as short circuits, leading to constant nonzero current associated with zero voltage along the loop; a magnetic monopole radiator is thus produced. In the ω_{sh} mode, $Y = 0$ [Equation (11.3b)], and therefore the shunt paths are seen as open circuits (the radial currents in the stubs represent only the L_L excited currents, but the overall shunt resonator current is zero), leading to zero loop current associated with constant nonzero voltage along the loop; an electric monopole radiator is thus produced. These two electric and magnetic monopole modes are independent (uncoupled) from each other. It should be noted that they may be excited simultaneously using two different slots (radial for ω_{se} and azimuthal for ω_{sh}), as shown in Figure 11.16(b).

Monopole radiators may also be obtained in the zeroth order CRLH resonance in "patch" configuration, providing a complement to the magnetic dipole of a conventional patch [37]. A magnetic monopole patch radiator operating in the CRLH zeroth order mode ω_{sh} is shown in Figure 11.18(a) along with its typical radiation patterns in Figure 1.18(b) [46, 72]. This monopole behavior results from the magnetic current loop $\mathbf{M}_s = -2\ \mathbf{n} \times \mathbf{E}$ created around the mushroom patch due to the uniformity of the vertical electric field. An electric monopole patch radiator operating in the CRLH zeroth order mode ω_{se} may also be conceived.

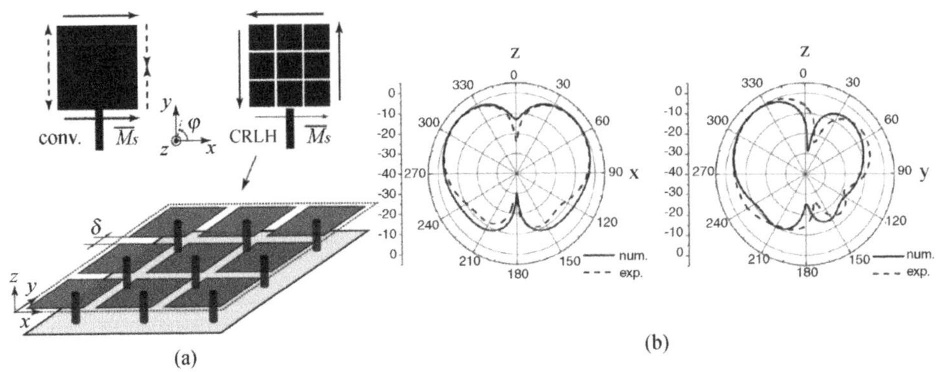

Figure 11.18 Magnetic monopole microstrip patch antenna using the CRLH zeroth order mode ω_{sh} in mushroom structure. (a) Configuration, showing also the case of the conventional patch for comparison. (b) Typical radiation patterns for the magnetic monopole antenna of Figure 11.18(a) (CRLH patch of 2×1 cells, size of $15 \times 7.5\,\text{cm}^2$, measured at 3.5 GHz). Reproduced by permission of ©2008 IEEE

11.5 Exotic Radiative Systems

11.5.1 Magnetic Resonance Imaging Coils

Magnetic resonance imaging (MRI) is an efficient non-ionizing medical imaging technique used in radiology to visualize the structure of the body by providing detailed images of it in any plane [73]. A strong DC magnetic field aligns the nuclear spins of the hydrogen protons of the water contained in the body, while a perpendicular RF magnetic field alters the alignment of these protons following prescribed pulse sequences. After each pulse, the protons drift back into alignment with the DC field, thereby emitting a detectable RF signal, which is picked up by coils and recorded. The MRI images with spatially varying contrast can then be constructed by various processing techniques based on the fact that protons in different tissues of the body (e.g. fat vs. muscle) realign at different speeds. A critical issue for high-resolution imaging in MRI is the capability to generate an extremely uniform (or homogeneous) RF field across the field of view.

Recently, a zeroth order antenna coil (ZORC) system for MRI was proposed in [74], as an alternative to conventional MRI coils, to achieve superior field uniformity over long distances, which may extend throughout the body of the patient, and which may be particular suitable to obtain one-shot sagittal and coronal section images. Figure 11.19(a) shows the proposed apparatus, which is comprised of several zeroth order resonant antennas (Section 11.4.4) distributed azimuthally around the body to image, while Figure 11.19(b) shows the high field uniformity achieved by the antenna, operating in its near-field as a transmitting and pick-up coil, in a phantom liquid emulating the body. So far, only preliminary results have been obtained. However, these results are encouraging. They exhibited an appropriate loaded/unloaded Q-factor ratio of around 1/2 and provided initial MRI images of good quality.

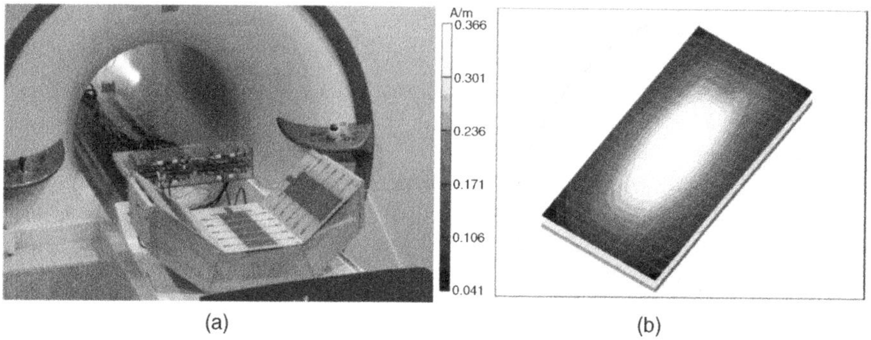

Figure 11.19 Magnetic resonance imaging (MRI) zeroth order antenna coil (ZORC) system. (a) General view of part of the octagonal system (3 zeroth order (series) CRLH resonant antenna out of the 8 possible are shown). (b) Measured magnetic field amplitude for one ZORC element inside a flat phantom (whose cross-section is represented by the elliptical shape around the structure) just above the ZORC.

11.5.2 Uniform Ferrite CRLH Leaky-Wave Antenna

The CRLH leaky-wave antenna described in Section 11.3.3 and in some subsequent sections is based on the lumped LC network implementation of the circuit shown in Figure 11.3(a). This section presents a recently discovered perfectly uniform ferrite waveguide CRLH leaky-wave structure,[8] acting as a full-space scanning antenna [75]. This antenna is much easier to design, due to its uniformity, and is automatically balanced, as its CRLH mode is inherently continuous at the spectral origin. In addition, it can be scanned at a fixed frequency by tuning the DC magnetic field, without requiring any chip components (such as varactors). With conventional ferrites, it requires a biasing magnet. However, it may be implemented in recently proposed integrated ferromagnetic nanowire substrates which exhibit the required remanence to suppress the need for a magnet [76].

Figure 11.20 (a) shows the uniform ferrite-loaded open waveguide leaky-wave antenna. The structure consists of a ferrite-filled rectangular waveguide where one side wall has been removed and where the bias magnetic field $\mathbf{H}_0$ is applied perpendicularly to the width of the waveguide (z-direction). It is thus an open waveguide structure. Due to the high permittivity ($\varepsilon_r \gg 1$) of the ferrite, most of the propagating energy is concentrated within the waveguide, but a small (perturbational) amount of energy leaks out of the structure. This leakage builds up the radiation.

The structure, being open and inhomogeneous, does not admit an analytical solution. However, due to the high permittivity of the ferrite (typically $\varepsilon_{r,\text{in}} \approx 15$), the electric field inside the waveguide is essentially tangential to the interface with air ($E_{n,\text{in}} = E_{n,\text{out}}/\varepsilon_{r,\text{in}} \approx 0$). Thus, the ferrite-air interface may be approximated as a perfect magnetic conductor (PMC). The resulting structure is closed and homogeneous, and therefore possesses a simple analytical solution. Due to the strong concentration of the field inside the waveguide, the actual leakage of the open structure represents only a perturbation of the closed waveguide, and the dispersion

[8] Being uniform, this structure is not a metamaterial structure. However, since it has the main attributes of CRLH metamaterial structures, we judged appropriate to include this topic in the present chapter.

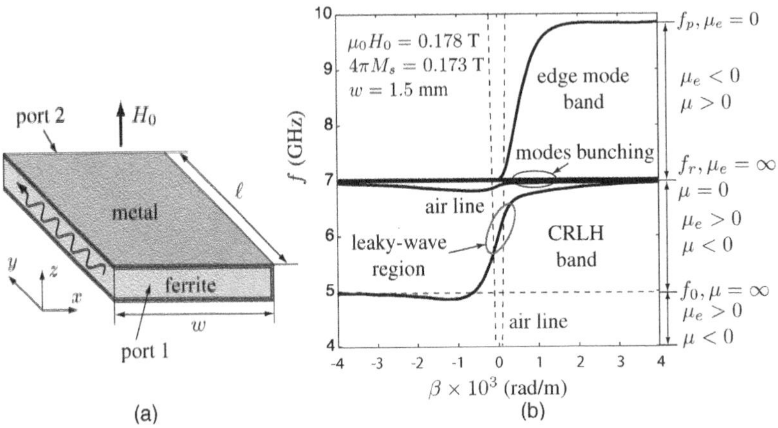

Figure 11.20 Uniform ferrite-loaded open waveguide structure with CRLH response. (a) Configuration. (b) Dispersion diagram computed by Equation (11.15) for typical YIG parameters. Reproduced by permission of ©2008 IEEE

relation of the closed waveguide therefore constitutes a good approximation to that of the open waveguide. The dispersion relation takes the simple form [75]

$$\tan\left(\frac{k_x w}{2}\right) = -\frac{\mu k_x}{\beta\kappa}, \tag{11.15}$$

where w is the width of the waveguide and, assuming the y-propagating waveform $E_z = f(x)e^{-j\beta y}$ corresponding to the dominant mode electric field

$$E_z = [A\sin(k_x x) + B\cos(k_x x)]\exp(-j\beta y), \tag{11.16}$$

we have

$$\beta = \sqrt{k^2 - k_x^2} = \sqrt{\omega^2\varepsilon\mu_e - k_x^2}, \tag{11.17}$$

where

$$\mu_e = \frac{\omega^2 - (\omega_0 + \omega_m)^2}{\omega^2 - \omega_0(\omega_0 + \omega_m)} = \frac{\omega^2 - \omega_p^2}{\omega^2 - \omega_r^2}, \tag{11.18a}$$

with the plasma frequency ($\mu_e = 0$)

$$\omega_p = \omega_0 + \omega_m, \tag{11.18b}$$

and the resonance frequency ($\mu_e = \infty$)

$$\omega_r = \sqrt{\omega_0(\omega_0 + \omega_m)}, \tag{11.19c}$$

with

$$\omega_0 = \gamma\mu_0 H_0, \tag{11.19a}$$

$$\omega_m = \gamma\mu_0 M_s, \tag{11.19b}$$

where H_0 is the bias magnetic field, M_s is the saturation magnetization, γ is the gyromagnetic ratio, ω_0 is the precession frequency, and μ an κ are elements of the Polder permeability tensor which reads for $+z$-directed bias [77]

$$\overline{\overline{\mu}} = \begin{bmatrix} \mu & j\kappa & 0 \\ -j\kappa & \mu & 0 \\ 0 & 0 & \mu_0 \end{bmatrix}, \tag{11.20}$$

with

$$\mu = \mu_0\left(1 + \frac{\omega_0\omega_m}{\omega_0^2 - \omega^2}\right), \tag{11.21a}$$

$$\kappa = \mu_0 \frac{\omega\omega_m}{\omega_0^2 - \omega^2}. \tag{11.21b}$$

The dispersion relation of Equation (11.15) is plotted in Figure 11.20(b) for typical parameters. Three zones may be distinguished in this (dominant mode) dispersion diagram. The band $[\omega_r, \omega_p]$ is the band of the well-known edge-mode isolator, where the energy is deflected to one edge of the waveguide, the open one in the passing direction and the closed shorting one in the stopping direction [77]. This mode is located outside of the radiation region and is therefore completely guided. In the region of the bulk[9] birefringent resonance frequency ω_r, an infinite number of strongly bunched and highly dispersive modes exist, which may not have a practical interest. Finally, the region of interest is the band $[\omega_0, \omega_r]$, where a CRLH, automatically balanced, response is achieved (Figure 11.4). This CRLH characteristic includes the leaky-wave region with full crossing of the dispersion curve, leading according to Equation (11.13) to full-space scanning, from backfire to endfire including broadside.

A detailed analytical, numerical and experimental study of this antenna is provided in [75]. Figure 11.21 shows a prototype of it, with its matching circuit and transition from microstrip to waveguide. Three frequency scanned beams of the antenna are presented in Figure 11.22. Bias field scanning at fixed frequency was also demonstrated in [75].

11.5.3 Uniform Ferrite CRLH Integrated Antenna-Duplexer

The uniform CRLH leaky-wave antenna of the previous section is a powerful device, which may find many applications. This section presents an integrated antenna-duplexer [78] as an illustration of such possibilities. The operation principle of this device is depicted in Figure 11.23. The two ends of the structure are terminated by ports with matching section, so that the structure effectively builds a three-port network if free space is considered as the third (RF) port. The transmit (Tx) port is set at one end and the receive (Rx) port is set at the

[9] Care must be paid here not to confuse μ_e with the effective permeability of the waveguide structure: μ_e is the effective permeability of the *unbounded* ferrite. The effective permeability of the waveguide is not related to this quantity.

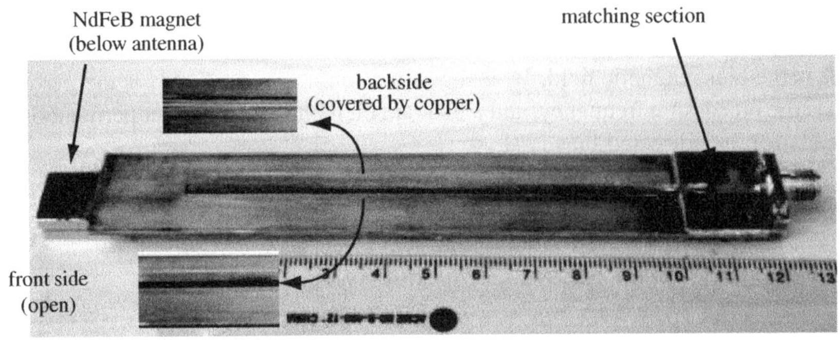

Figure 11.21 Uniform ferrite-loaded open waveguide structure (including a ground plane for unidirectionality) with CRLH corresponding to the parameters of Figure 11.20. Reproduced by permission of ©2009 IEEE

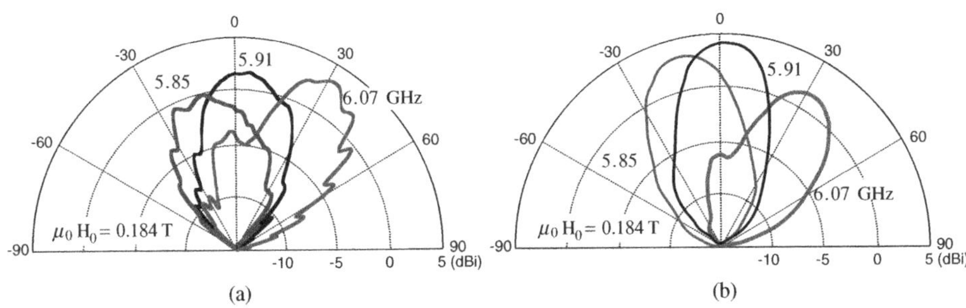

Figure 11.22 Frequency beam-scanning [*yz*-plane in Figure 11.20(a)]. (a) Experiment ($G = 2.4$ dBi). (b) FEM (HFSS) ($G = 4.6$ dBi, $\eta = 37\%$, $D = 8.9$ dB). Reproduced by permission of ©2009 IEEE

other end of the structure. The scanning property of the CRLH antenna is not used in this device, and the operation frequency is fixed at the broadside transition frequency.

The Tx signal is radiated by the leaky-wave antenna like in a conventional CRLH broadside leaky-wave antenna. By virtue of the leaky-wave radiation mechanism, the structure may be designed long enough so that all of the Tx signal power has radiated out of the structure before reaching the Rx port on the other side, thereby automatically preventing any Tx $\rightarrow$ Rx leakage and providing infinite Tx $\rightarrow$ Rx isolation. Moreover, the incoming RF signal picked up by the antenna can only propagate toward the Rx port due to the non-reciprocity of the structure [Figure 11.20(b)], and therefore infinite RF $\rightarrow$ Tx isolation is automatically achieved. The antenna can thus simultaneously transmit and receive without any interference or leakage between the Tx and Rx signals. Thus, it constitutes an integrated (or combined) antenna-duplexer device with excellent duplexing operation, since a single antenna simultaneously performs the Tx and Rx operations with perfect isolation. Another advantage of this system is the possibility to tune the operation frequency by the applied magnetic bias field, whereas such tuning is prevented both by the antenna and by the circulator in conventional designs.

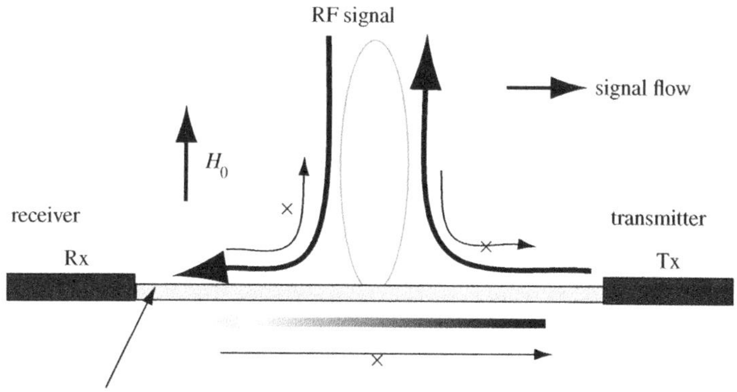

Figure 11.23 Operation principle of the proposed integrated leaky-wave antenna-duplexer.

The infinite Tx → Rx isolation property of the proposed integrated leaky-wave antenna-duplexer is very appreciable in practical communication or radar front-end systems to avoid demodulation or detection/ranging errors, or even receiver destruction, caused by the return loss of the antenna. This problem is illustrated in Figure 11.24. For instance, in an application using a typical antenna with a return loss of 15 dB and a Tx power of 50 dBm, a power of 35 dBm leaks into the Rx (neglecting the circulator's losses). Although a power limiter may be used to mitigate this problem, this techniques has its own limitations, including additional insertion loss, harmonic generation and power handling limitations. The proposed integrated leaky-wave antenna-duplexer solves these problems in a simple and elegant manner.

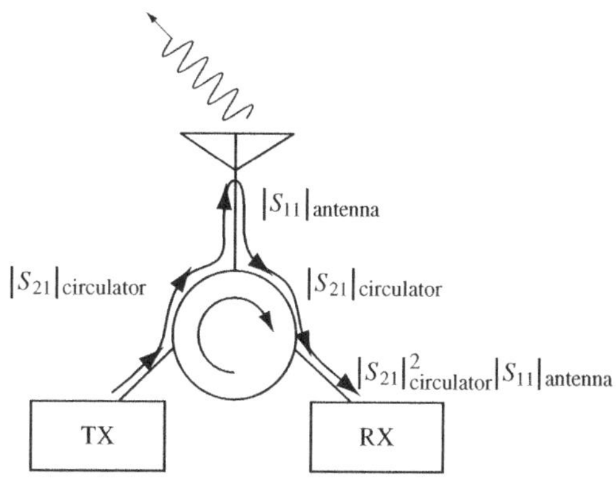

Figure 11.24 Limitation of isolation between Tx and Rx in a conventional antenna-circulator system due to the antenna return loss.

An integrated leaky-wave antenna-duplexer prototype is shown in Figure 11.25. This prototype was measured with a gain of 2.3 dBi with isolation of more than 15 dB at all ports. A detailed parametric analysis is available in [78].

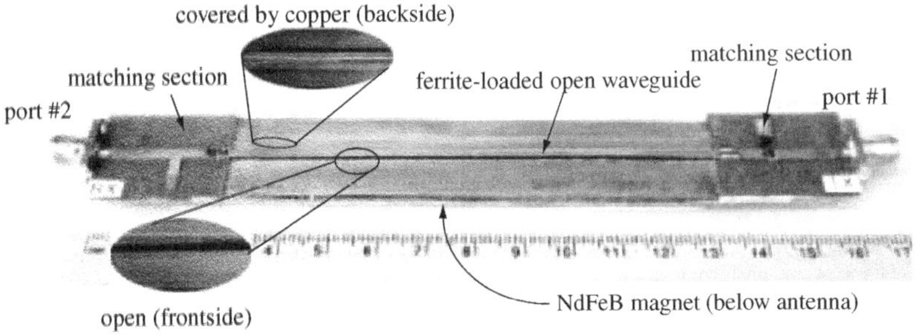

Figure 11.25 Integrated leaky-wave antenna-duplexer prototype.

11.5.4 Direction of Arrival (DOA) Estimator

The vast majority of the leaky-wave antennas reported literature have been considered only in the transmit mode. However, it pays off to exploit the assets (full-space scanning, high-directivity, absence of mutual coupling, compactness, low loss, cost effectiveness) of the CRLH leaky-wave antenna (Section 11.3) also in the receive mode. In particular, since this antenna exhibits the same full-space scanning capabilities as phased antenna arrays, they may be used at advantage in many array processing systems. This section presents the example of a simple direction of arrival (DOA) estimation system.

Figure 11.26(a) shows the schematic of the DOA estimation system, which operates in two modes: 1) analog mode (solid line), 2) digital mode (dashed line). The analog mode is based on comparing the received power at either power detector #1 or #2 as the beam is electronically steered from $-90°$ to $+90°$. Once the beam is aligned with the incident wave, the appropriate detector will exhibit maximum received power and the incident wave's angle can be subsequently determined. The digital mode is based on employing the beam-space MUSIC algorithm to the CRLH leaky-wave. From the received signal, the application of appropriate weight vectors to the CRLH leaky-wave forms several beams. Subsequently, the beam-space correlation matrix is obtained to which eigen-decomposition is applied to determine the incident wave's angle. Figure 11.26(b) shows typical resulting super-resolution angular spectrum for both modes of operation.

11.5.5 Real-Time Spectrum Analyzer (RTSA)

The analysis and characterization of complex non-stationary signals (such as radar, security and instrumentation, and EMI/EMC signals) require joint time-frequency representations, i.e. 2D plots where the 1D signal is represented as an image in a time-frequency plane, with the signal energy distribution coded a color-scale levels of the image [79]. Such representations are typically called spectrograms, and the systems which generate such spectrograms are called

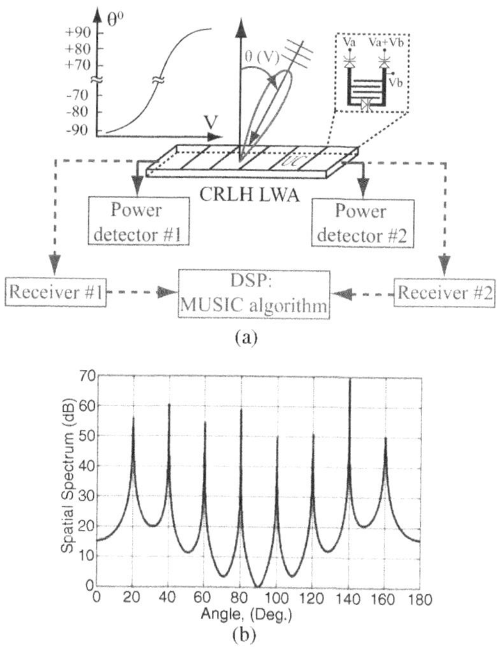

Figure 11.26 Proposed DOA estimation. (a) System schematic showing an electronically-scanned CRLH LWA, two power detectors for analog operation, and two receivers and DSP for digital operation. (b) Resultant angular spectrum for 8 incident waves separated by 20° angles.

real-time spectrum analyzers (RTSAs). Mathematically, the spectrogram of a signal $x(t)$ is calculated using

$$S(\tau,\omega) = \left| \int_{-\infty}^{\infty} x(t) g(t-\tau) e^{-j\omega t} dt \right|^2 \tag{11.22}$$

where $g(t)$ is a gate function.

The CRLH leaky-wave antenna RTSA system is depicted in Figure 11.27. This system was presented and fully characterized in [80]. According to Figure 11.4(a), the CRLH leaky-wave antenna may operate as an "antenna grating," performing a spatial-spectral decomposition of an incident pulse with a spectrum contained in its radiation bandwidth. The time variations of all the spectral components of the test pulse are then detected by a circular array of envelope demodulators. Finally, after some light digital signal processing treatment (dispersion rectification and calibration), the spectrogram of the input pulse is constructed and displayed. While commercial digital RTSA's are currently unable to accommodate ultra wide-band standards, this analog system can handle extremely broadband signals and may also be scaled to millimeter-wave frequencies.

In order to test the proposed RTSA, various time-domain functions are defined in a full-wave simulator (CST Microwave Studio) as the excitation signals of the CRLH leaky-wave antenna. First, a generalized modulated super-gaussian waveform with linear and quadratic chirps is defined by

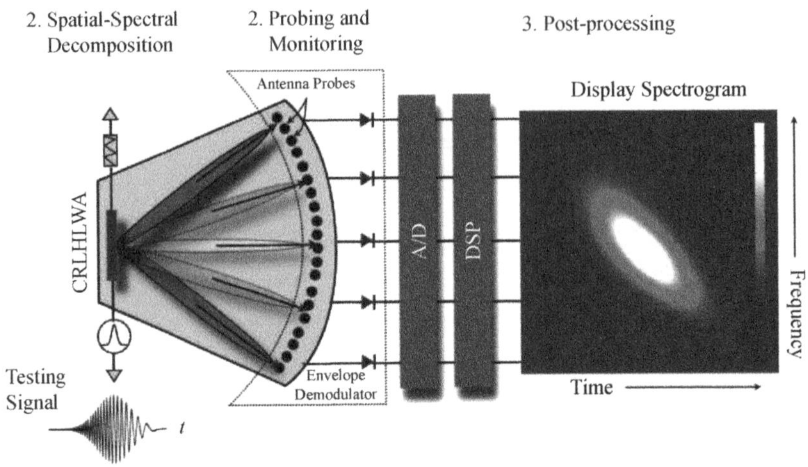

Figure 11.27 Real-time spectrum analysis (RTSA) system based on the spectral-spatial decomposition property of the CRLH leaky-wave antenna (LWA). Reproduced by permission of ©2009 IEEE

$$v(t) = \exp\left[-\frac{1}{2}\left(\frac{t-t_0}{\sigma_0}\right)^{2m}\right]\sin(2\pi f_0 t + c_1 t^2 + c_2 t^3), \quad m \in \mathbb{N}, \tag{11.23}$$

where c_1 and c_2 represent the linear and quadratic chirp parameters, respectively, and t_0 represents a time delay. Signals with multiple pulse components may be generated using different delays (t_0). Second, a convenient function to model the nonlinear phase effects, such as nonlinear self-phase modulation (SPM) [81], is defined by

$$v(t) = \mathrm{Re}[U(t)e^{j|U(t)|^2 z}], \tag{11.24}$$

Table 11.3 Time-domain waveform parameters for the spectrograms of Figure 11.28. Units: σ_0, T, t_0: [ns], f_0: [GHz]. (a-h) are parameters corresponding to Equation (11.23), (i) corresponds to Equation (11.24), and (j) corresponds to Equation (11.26)

a) Negative chirp	$\sigma_0 = 1, m = 1, f_0 = 4.11, C_1 = -10^{18}$
b) Modulated gaussian	$\sigma_0 = 1, m = 1, f_0 = 3.0$
c) Positive chirp	$\sigma_0 = 1, m = 1, f_0 = 1.9, C_1 = 10^{18}$
d) Modulated gaussian	$T = 1.41, m = 1, f_0 = 2, 2.45, 3, c_2 = 0.25 \cdot 10^{26}, t_0 = 4, 10, 16$
e) Modulated super-gaussian	$\sigma_0 = 3, m = 11, f_0 = 2.7$
f) Nonlinear cubic chirp	$\sigma_0 = 5, m = 1, f_0 = 2.27, c_2 = 0.25 \cdot 10^{26}$
g) Double chirp	$T = 1.8, m = 1, t_0 = 0, 3, f_0 = 2.5, c_1 = -3.08 \cdot 10^{17}, -3.08 \cdot 10^{17}$
h) Double chirp	$T = 1, m = 1, t_0 = 3.5, 6.0, f_0 = 4.11, 1.9, c_1 = -10^{18}, +10^{18}$
i) Self-phase modulated	$T = 5, m = 1, f_0 = 2.75, z = 10$
j) Dispersed pulse	$\sigma_0 = 1, m = 1, f_0 = 2.6$

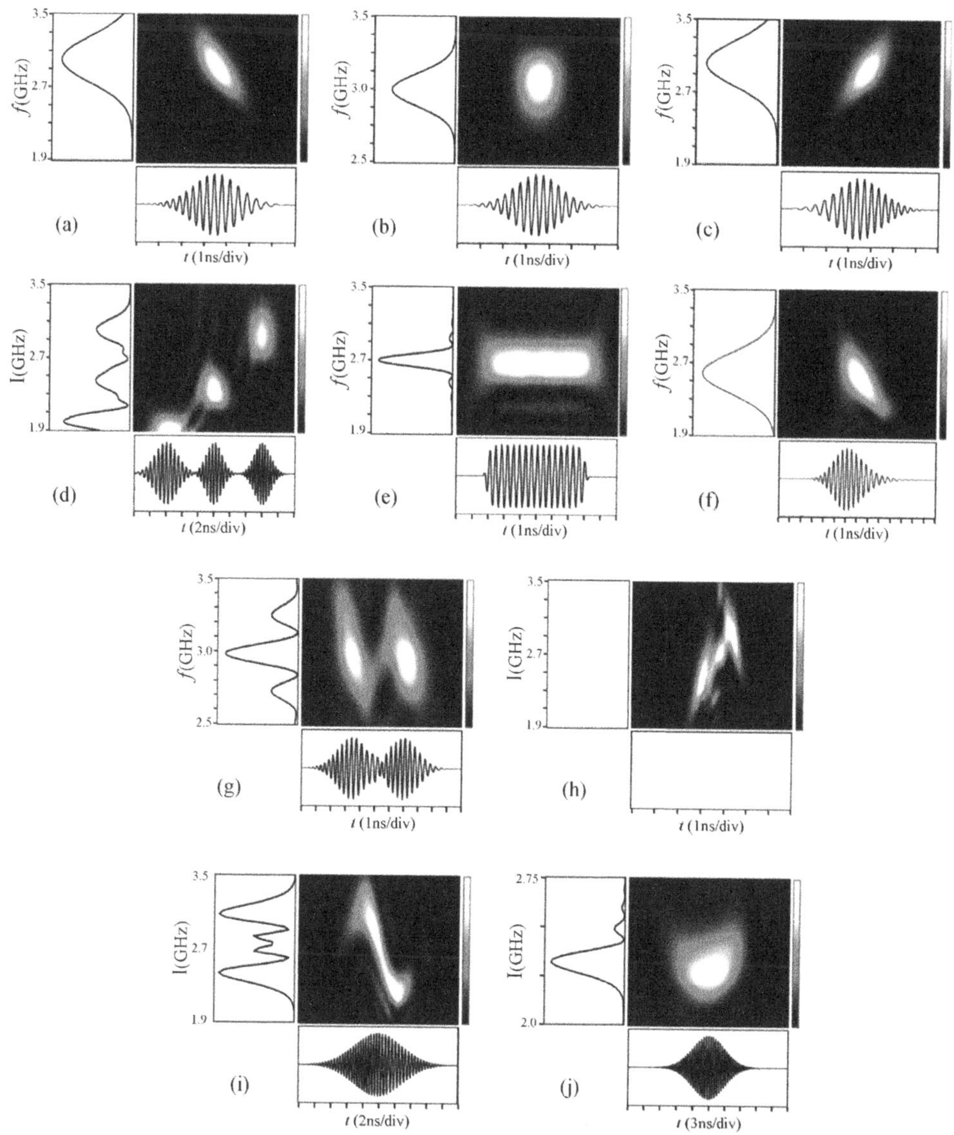

Figure 11.28 Full-wave computed spectrograms obtained using the setup in Figure 11.27. The input signals are generated by the time domain functions and parameters given in Table 11.3. a) Linear negatively chirped Gaussian pulse. b) Modulated (un-chirped) gaussian pulse. c) Linear positively chirped gaussian pulse. d) Multiple modulated gaussian pulses. e) Modulated (un-chirp) rectangular pulse. f) Nonlinear cubically chirped Gaussian pulse. g) Doubly negative chirped gaussian pulses. h) Oppositely chirped Gaussian pulses. i) Self-phase modulated pulses. j) Pulse dispersed through a CRLH transmission line. Reproduced by permission of ©2009 IEEE

where $U(t)$ is given by

$$U(t) = \exp\left[-\frac{1}{2}\left(\frac{t-t_0}{\sigma_0}\right)^2\right]\sin(2\pi f_0 t). \tag{11.25}$$

Finally, to model the pulse propagation in a dispersive medium such as a CRLH transmission line, the following Fourier transform approach is used [81]

$$v(t) = \Im^{-1}[\Im\{U(t)\}S_{21}(\omega)]. \tag{11.26}$$

The RTSA full-wave results, using the parameters in Table 11.3, are presented in Figure 11.28 for various non-stationary signals. In all cases, faithful representations of the instantaneous frequency distribution of the signals are obtained with the expected linear frequency variations across the temporal profile of the pulse.

The realized RTSA prototype is shown in Figure 11.29, where the system is designed for a 60° radiation sector centered at the broadside direction. The output of the antenna receivers can be subsequently envelope-detected and digitized to be sent to a computer for post-processing and spectrogram displaying. For simplicity, envelope detection and A/D conversion are achieved from the voltages induced at the antenna receivers using a digital oscilloscope.

The corresponding experimental results are displayed in Figure 11.30 for a few test signals. In all cases, faithful spectrograms are obtained, despite the rudimentariness of the apparatus used for the proof of concept. This system may be strongly improved in future by various microwave engineering optimizations.

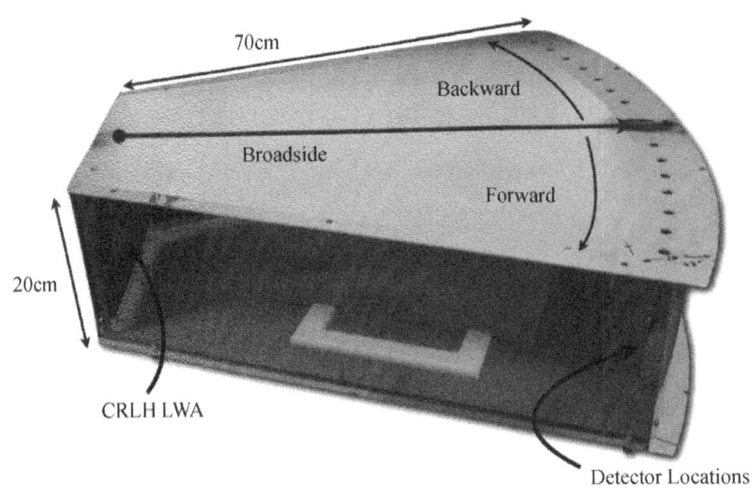

Figure 11.29 RTSA system prototype following the schematic in Figure 11.27 covering a 60° radiation space including broadside. Reproduced by permission of ©2009 IEEE

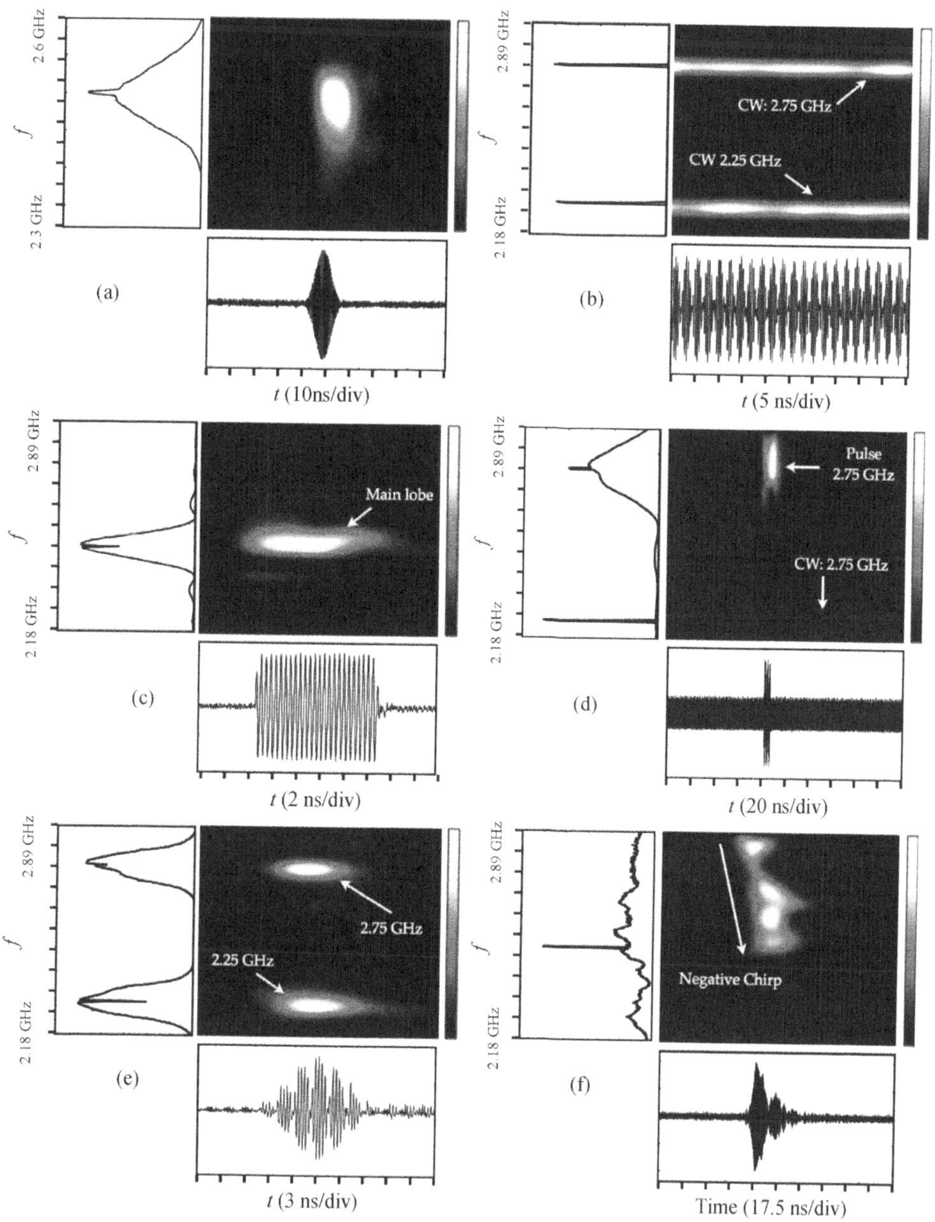

Figure 11.30 Measured spectrograms using the prototype shown in Figure 11.29. All the plots are normalized with a maximum value of 1. a) Modulated gaussian pulse. b) Sum of two tones at $f = 2.25$ GHz and $f = 2.60$ GHz. c) Modulated square pulse at $f = 2.47$ GHz. d) Combination of a sinusoidal signal at $f = 2.25$ GHz and a modulated square pulse with a pulse width of 10 ns and a modulation frequency of $f = 2.75$ GHz. e) gaussian Pulse mixed with a two-tone signal at $f = 2.3$ GHz and $f = 2.7$ GHz. f) Dispersed square pulse of 1 ns duration propagated along 6 CRLH transmission lines of 14 cells each. The time domain signals are measured with a digital sampling oscilloscope and the corresponding spectrum (linear normalized scale) is measured with a spectrum analyzer. Reproduced by permission of ©2009 IEEE

11.5.6 Talbot Spatial Power Combiner

We have have seen in the study of the analog RTSA (Section 11.5.5) that a CRLH leaky-wave antenna acted as a kind of diffraction grating. This fact was recently exploited in a novel spatial-temporal Talbot [82] combiner [83], which we present now.

The Talbot phenomenon is a self-imaging effect, occurring at the so-called Talbot distance, resulting from the constructive interference of spatial frequencies produced by a periodic spatial object (for instance, a diffraction grating) under monochromatic or polychromatic illumination. The temporal counterpart of this effect occurs when a signal which is periodic in time propagates along a first-order dispersive medium (such as an optical fiber), where the input pulse train is replicated at the Talbot distance [84]. In both cases, the Talbot effect is related to the dispersive features of the medium, which are spatial dispersion in the case of diffraction gratings and temporal dispersion in the case of optical fibers.

The novel CRLH Talbot combiner is shown in Figure 11.31. It is based on the combination of the conventional monochromatic spatial Talbot effect and the transient polychromatic effect of pulse radiation by a leaky-wave antenna. To produce this phenomenon, an array of CRLH leaky-wave antenas is fed simultaneously at all of its elements by a modulated pulse with center frequency located at the transition frequency of the antennas. The beams radiated by the different elements generate an interference pattern that self-image the spatial pulse distribution

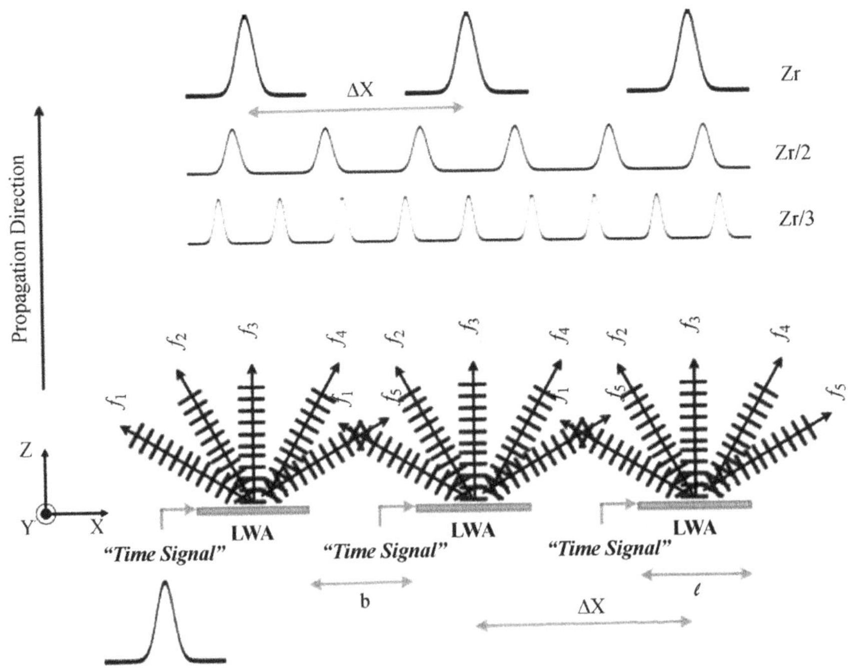

Figure 11.31 CRLH leaky-wave antenna array spatial-temporal Talbot system. Each antenna radiates the different frequency components of the input modulated pulse to different angles of space, which provides both temporal and spatial interference, leading to the spatial-temporal Talbot phenomenon. For the sake of representation simplicity, only the envelopes of the pulses are shown.

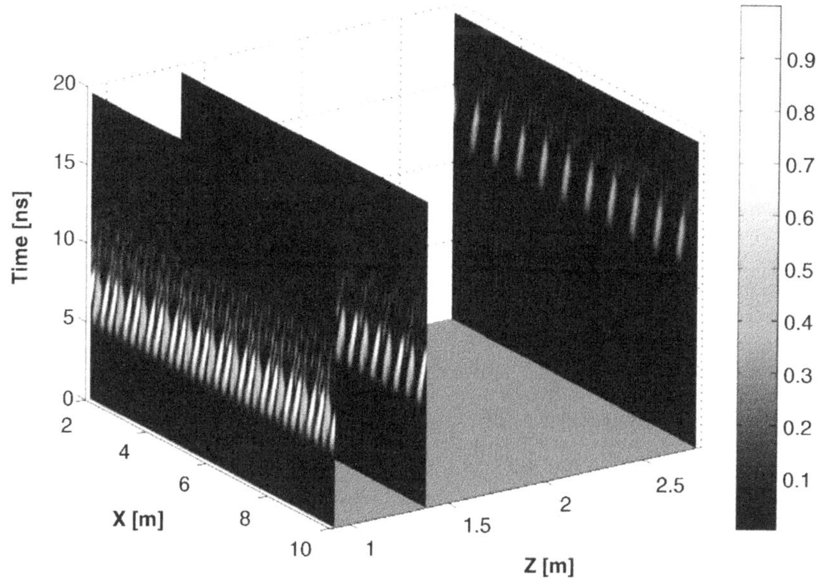

Figure 11.32 Field (magnitude) radiated by the Talbot system of Figure 1.31 (placed at $z = 0$ along the x-axis), for antenna element spacing of $b = 0.5$ m and modulated gaussian pulse excitation as a function of the position x and time at the propagation distances $z = z_T = 2.74$ m, $z = z_T/2 = 1.37$ m and $z = z_T/3 = 0.91$ m. The CRLH antenna parameters are: $N = 16$ unit cells, period $p = 1.56$ cm, $C_R = 4.5$ pF, $C_L = 2.5$ pf, $L_R = 4.5$ nH and $L_L = 2.5$ nH, corresponding to the transition frequency of $f_0 = 1.5$ GHz. The antenna is excited by an f_0-modulated gaussian pulse with full width at half maximum of 1.178 ns. Reprinted with permission from *J. App. Phys.*, vol. 104, pp. 104901:1–7, Nov. 200. © 2008, American Institute of Physics

along the antennas at the Talbot distance. Furthermore, an increase in the repetition rate of this spatial distribution occurs at the fractional Talbot distances. The CRLH antenna, which is sufficiently directive for a given pulse bandwidth, generates a paraxial diffraction i.e., radiation of the beams, leading to the spatial Talbot effect. This, combined with the transient nature of the pulsed antenna radiation, leads to the spatial-temporal Talbot phenomenon. In addition, the self-imaging effect replicates the spatial variation of the pulses as a function of time at each Talbot zone due to the pulses propagation along the CRLH leaky-wave antennas.

The complete theory of the spatio-temporal Talbot effect achieve in this combiner is available in [83]. Results, computed by a time-domain Green's function approach, are shown in Figure 11.32, where the first integer and two fractional Talbot images are clearly apparent. By using off-broadside radiation of the leaky-wave antennas, the Talbot distance may be tuned by varying the input pulse modulation frequency, as demonstrated both theoretically and experimentally in [85].

References

1. C. Caloz, A. Rennings, and T. Itoh, "CRLH metamaterial leaky-wave and resonant antennas," *Antenna Propagat. Mag*, vol. 50, pp. 25–39. 2008.

2. C. Caloz and T. Itoh, *Electromagnetic Metamaterials, Transmission Line Theory and Microwave Applications*, Wiley – IEEE Press, New York, 2005.
3. G. V. Eleftheriades and K. G. Balmain (eds.), *Negative Refraction Metamaterials: Fundamental Principles and Applications*, Wiley – IEEE Press, New York, 2005.
4. N. Engheta and R. W. Ziolkowski (eds.), *Electromagnetic Metamaterials: Physics and Engineering Explorations*, Wiley – IEEE Press, New York, 2006.
5. P. Markos and C. M. Soukoulis, *Wave Propagation: From Electrons to Photonic Crystals and Left-Handed Materials*, Princeton University Press, Princeon, NJ, 2008.
6. R. Marqués, F. Martin and M. Sorolla, *Metamaterials with Negative Parameters: Theory, Design and Microwave Applications*, Wiley Interscience, New York, 2008.
7. A. Sihvola, *Electromagnetic Mixing Formulae and Applications*, The Institution of Engineering and Technology, 2000.
8. R. E. Collin, *Field Theory of Guided Waves*, 2nd edn, IEEE Press, 1990.
9. S. B. Cohn, "Analysis of the metal-strip delay structure for microwave lenses," *J. App. Phys.* vol. 20, pp. 257–262, 1949.
10. W. E. Kock, "Analysis of the metal-strip delay structure for microwave lenses," *Bell Syst. Tech. J*, vol. 27, pp. 58–82, 1948.
11. V. Veselago, "The electrodynamics of substances with simultaneously negative values of " and μ," *Soviet Phys, Uspekhi*, vol. 10, pp. 509–514, (translation from Russian version published in 1967) 1968.
12. J. B. Pendry, A. J. Holden, W. J. Stewart, and I. Youngs, "Extremely low frequency plasmons in metallic mesostructures," *Phys. Rev. Lett*, vol. 76, pp. 4773–4776, 1996.
13. J. B. Pendry, A. J. Holden, D. J. Robbins, and W. J. Stewart, "Low frequency plasmons in thin-wire structures," *J. Phys. Condens. Matter*, vol. 10, pp. 4785–4809, 1998.
14. J. B. Pendry, A. J. Holden, D. J. Robbins, and W. J. Stewart, "Magnetism from conductors and enhanced nonlinear phenomena," *IEEE Trans. Microwave Theory Tech*, vol. 47, pp. 2075–2084, 1999.
15. R. A. Shelby, D. R. Smith, and S. Schultz, "Experimental verification of a negative index of refraction," *Science*, vol. 292, pp. 77–79, 2001.
16. D. R. Smith, W. J. Padilla, D. C. Vier, S. C. Nemat-Nasser, and S. Schultz, "Composite medium with simultaneously negative permeability and permittivity," *Phys. Rev. Lett*, vol. 84, pp. 4184–4187, 2000.
17. D. R. Smith, D. C. Vier, N. Kroll, and S. Schultz, "Direct calculation of permeability and permittivity for a left-handed metamaterial," *App. Phys.* Lett, vol. **77**, pp. 2246–2248, 2000.
18. R. W. Ziolkowski, and E. Heyman, "Wave propagation in media having negative permittivity and permeability," *Phys. Rev. E*, vol. 64, 056625: pp. 1–15, 2001.
19. C. Caloz, H. Okabe, T. Iwai, and T. Itoh, "Transmission line approach of left-handed (LH) materials," paper presented at *USNC/URSI*, San Antonio, 2002.
20. C. Caloz and T. Itoh, "Application of the transmission line theory of left-handed (LH) materials to the realization of a microstrip LH transmission line," in *Proc. IEEE Antennas Propagat. Symp*, San Antonio, 2002.
21. A. K. Iyer and G. V. Eleftheriades, "Negative refractive index metamaterials supporting 2D waves," in *Proc. IEEE Microwave Theory Tech. Symp*, Seattle, 2002.
22. A. A. Oliner, "A periodic structure negative refractive index medium without resonant elements," paper presented at *USNC/URSI*, San Antonio, 2002.
23. C. Caloz, "Perspectives on electromagnetic metamaterials," *Materials Today*, vol. 12, no. 3, pp. 12–20, 2009.
24. W. Rotman, "Plasma simulation by artificial dielectrics and parallel-plate media," *IRE Trans. Antennas Propagat*, vol. 10, pp. 82–95, 1962.
25. R. Marqués, F. Medina, and R. Rafii-El-Idrissi, "Role of bianisotropy in negative permeability and left-handed metamaterials," *Phys. Rev. B*, vol. 65, 144440: pp. 1–6. 2002.
26. J. A. Kong, *Electromagnetic Wave Theory*, EMW Publishing, 2008.
27. J. D. Jackson, *Classical Electrodynamics*, 2nd edn, Wiley, New York, 1999.
28. S. Ramo, J. R. Whinnery, and T. Van Duzer, *Fields and Waves in Communication Electronics*, 3rd edn, Wiley, New York, 1994.
29. C. Caloz and T. Itoh, "Novel microwave devices and structures based on the transmission line approach of metamaterials," in *Proc. IEEE Microwave Theory Tech. Symp*, Philadelphia, 2003.
30. G. Matthaei, E. M. T. Jones, and L. Young, *Microwave Filters, Impedance Matching Networks, and Coupling Structures*, Artech House Publishers, Boston, 1980.

31. J. R. Cameron, R. Mansour, and C. M. Kudsia, *Microwave Filters for Communication Systems: Fundamentals, Design and Applications*, Wiley Interscience, Chichester, 2007.
32. L. B. Felsen and N. Marcuvitz, *Radiation and Scattering of Waves*, Wiley-IEEE Press, New York, 1994.
33. L. Brillouin, *Wave Propagation in Periodic Structures*, Dover Publications, New York, 1946.
34. N. W. Ashcroft and N. D. Mermin, *Solid State Physics*, Brooks Cole, California, 1976.
35. J. D. Joannopoulos, S. G. Johnson, J. N. Winn, and R. D. Meade, *Classical Electrodynamics*, 2nd edn, Princeton University Press, Princeton, NJ, 2008.
36. S. Paulotto, P. Baccarelli, F. Frezza, and D. R. Jackson, "Full-wave modal dispersion analysis and broadside optimization for a class of microstrip CRLH leaky-wave antennas," *IEEE Trans. Microwave Theory Tech*, vol. 56, pp. 2826–2837, 2008.
37. W. L. Stutzman, and G. A. Thiele, *Antenna Theory and Design*, 2nd edn, John Wiley & Sons, Ltd, New York, 2000.
38. S. M. Rudolph, and A. R. Grbic, "Volumetric negative refractive index medium exhibiting broadband negative permeability," *J. App. Phys*, vol. 102, 013904: pp. 1–6, 2007.
39. R. E. Collin and F. J. Zucker (eds.), *Antenna Theory, Part I*, McGraw-Hill, New York, 1969.
40. A. A. Oliner and D. R. Jackson, Chap. 11 in J. L. Volakis (ed.) *Antenna Engineering Handbook*, 4th edn. McGraw-Hill, New York, 2007.
41. L. Liu, C. Caloz, and T. Itoh, "Dominant mode (DM) leaky-wave antenna with backfire-to-endfire scanning capability," *Electron. Lett*, vol. 38, pp. 1414–1416, 2002.
42. S. Lim, C. Caloz, and T. Itoh, "Metamaterial-based electronically-controlled transmission line structure as a novel leaky-wave antenna with tunable angle and beamwidth," *IEEE Trans. Microwave Theory Tech*, vol. 53, pp. 161–173, 2005.
43. G. Trentini (von), "Partially Reflecting Sheet Arrays," *IRE Trans. Antennas Propagat*, vol. 4, pp. 666–671, 1956.
44. N. G. Alexopoulos and D. R. Jackson, "Fundamental superstrate (cover) effects on printed circuit antennas," *Trans. Antennas Propagat*, vol. 32, pp. 807–816, 1984.
45. C. A. Allen, K. M. K. H. Leong, and T. Itoh, "2-D frequency-controlled beam-steering by a leaky/guided-wave transmission line array," in *Proc. IEEE Microwave Theory Tech. Symp*, San Francisco, 2006.
46. T. Kaneda, A. Sanada, and H. Kubo, "2D beam scanning planar antenna array using composite right/left-handed leaky wave antennas," *IEICE Trans. Electron*, vol. 89, pp. 1904–1911, 2006.
47. A. Lai, K. M. K. H. Leong, and T. Itoh, "Leaky-wave steering in a two-dimensional metamaterial structure using wave interaction excitation," in *Proc. IEEE Microwave Theory Tech. Symp*, San Francisco, 2006.
48. A. Lai, K. M. K. H. Leong, and T. Itoh, "Infinite wavelength resonant antennas with monopolar radiation pattern based on periodic structures," *IEEE Trans. Antennas Propagat*, vol. 55, pp. 868–876, 2006.
49. D. Lee, S. Lee, C. Cheon, and Y. Kwon, "A two-dimensional beam scanning antenna array using composite right/left-handed microstrip leaky-wave antennas," in *Proc. IEEEMicrowave Theory Tech. Symp*, Honolulu, 2007.
50. H. V. Nguyen, S. Abielmona, A. Rennings, and C. Caloz, "Pencil-beam, 2D scanning leaky-wave antenna array," *Int. Symp. Signals, Systems and Electronics (ISSSE)*, Montreal, 2007.
51. N. Yang, C. Caloz, H. V. Nguyen, S. Abielmona, K. Wu, "Non-radiative CRLH boxed stripline structure with high Q performances," In *Int. Symp. Electromagnetic Theory (EMTS)*, Digest, Ottawa, 2007.
52. M. A. Antoniades, and G. E. Eleftheriades, "A metamaterial series-fed linear dipole array with reduced beam squinting," in *Proc. IEEE Int. Symp. Antennas Propagat*, Albuquerque, 2006.
53. A. Lai, K. M. K. H. Leong, and T. Itoh, "A novel N-port series divider using infinite wavelength phenomena," in *Proc. IEEE Microwave Theory Tech. Symp*, San Francisco, 2005.
54. H. V. Nguyen and C. Caloz, "Arbitrary N-port CRLH infinite-wavelength series power divider," *Electron. Lett*, vol. 43, 2007.
55. H. V. Nguyen, S. Abielmona, and C. Caloz, "Highly efficient leaky-wave antenna array using a power-recycling series feeding network," *Antennas Wireless Propagat. Lett.*, vol. 8, pp. 441–444, 2009.
56. C. Caloz, F. P. Casares-Miranda, and Camacho-Peñalosa, "Active metamaterial structures and antennas," in *Proc. Mediterranean Electrotechnical Conf. (MELECON)*, Benalmadena, 2006.
57. F. P. Casares-Miranda, C. Camacho-Peñalosa, and C. Caloz, "High-gain active composite right/lefthanded leaky-wave antenna," *IEEE Trans. Antennas Propagat*, vol. 54, pp. 2292–2300, 2006.
58. M. Chen, H. Z. Chan, B. Houshmand, and T. Itoh, "Characterization of leaky wave antenna and active gain enhancement," in *Proc. European Microwave Conf*, Prague, 1996.
59. S. Abielmona, H. V. Nguyen, F. Casares-Miranda, Camacho-Peñalosa, and C. Caloz, "Real-time digital beam-forming active leaky-wave antenna," in *Proc. IEEE Antennas Propagat. Symp*, Honolulu, 2007.

60. J.-F. Frigon, C. Caloz, and Y. Zhao, "Dynamic radiation pattern diversity (DRPD) MIMO using CRLH leaky-wave antennas," in *Proc. IEEE Radio Wireless Symp, (RWS)*, Orlando, 2002.
61. C. Caloz, S. Abielmona, H. V. Nguyen, and A. Rennings, "Dual composite right/left-handed (DCRLH) leaky-wave antenna with low beam squinting and tunable group velocity," *Phys. Stat. Solidi (b)* vol. 244, pp. 1219–1226, 2007.
62. S. Otto, C. Caloz, A. Sanada, and T. Itoh, "A dual-frequency composite right/left-handed half-wavelength resonator antenna," in *Proc. IEEE Asia Pacific Microwave Conf*, New Delhi, 2004.
63. A. Rennings, T. Liebig, C. Caloz, and I. Wolff, "Double Lorentz transmission line metamaterials and their applications to triband devices," in *Proc. IEEE Microwave Theory Tech. Symp*, Honolulu, 2007.
64. A. Rennings, S. Otto, J. Mosig, C. Caloz, and I. Wolff, "Extended composite right/left-handed (ECRLH) metamaterial and its application as quadband quarter-wavelength transmission line," in *Proc. IEEE Asia Pacific Microwave Conf.*, Yokohama, 2006.
65. A. Rennings, T. Liebig, S. Abielmona, C. Caloz, and P. Waldow, "Tri-band and dual-polarized antenna based on (unpublished) CRLH transmission line," in *Proc. IEEE European Microwave Conf*, Munich, pp. 720–723, 2007.
66. A. Sanada, C. Caloz, and T. Itoh, "Zeroth order resonance in composite right/left-handed transmission line resonators," in *Asia Pacific Microwave Conf. (APMC) Dig*, Seoul, 2007.
67. A. Rennings, T. Liebig, S. Otto, C. Caloz, and I. Wolff, "Highly directive resonator antennas based on composite right/left-handed (CRLH) transmission lines," in *Proc. Int. ITG Conference on Antennas (INICA) Digest*, Munich, 2007.
68. R. C. Hansen, *Electrically Small, Superdirective, and Superconducting Antennas*, Wiley Interscience, New York, 2006.
69. A. Rennings, S. Otto, C. Caloz and P. Waldow, "Enlarged half-wavelength resonator antenna with enhanced gain," in *Proc. IEEE Antennas Propagat. Symp*, Washington, 2006.
70. A. Rennings, S. Otto, T. Liebig, C. Caloz, and I. Wolff, "Dual-band CRLH ring antenna with linear/circularpolarization capability," in *Proc. European Conf. Antennas Propagat, (EuCAP)*, Nice, 2006.
71. S. Otto, A. Rennings, C. Caloz, and P. Waldow, "Dual-mode zeroth order ring resonator with tuning capability and selective mode excitation," in *IEEE European Microwave Conf*, Paris, 2005.
72. A. Lai, K. M. K. H. Leong, and T. Itoh, "Infinite wavelength resonant antennas with monopolar radiation pattern based on periodic structures," IEEE Trans. Antennas Propagat., vol. 55, no. 3, pp. 868-876, March 2007.
73. R. H. Hashemi, G. W. Bradley, and J. C. Lisanti, *MRI: The Basics*, 2nd ed. Lippincott Williams, Wilkins, Philadelphia, 2004.
74. A. Rennings, J. Mosig, A. Bahr, C. Caloz, M. E. Ladd, and D. Erni, "A CRLH metamaterial based RF coil element for magnetic resonance imaging at 7 Tesla," in *Proc. European Conf. Antennas Propagat, (EuCAP)*, Berlin, 2009.
75. T. Kodera and C. Caloz, "Uniform ferrite-loaded open waveguide structure with CRLH response and its application to a novel backfire-to-endfire leaky-wave antenna," *Trans.Microwave Theory Tech*, vol. 57, no. 4, pp. 784–795, April 2009.
76. L.-P. Carignan, T. Kodera, A. Yelon, C. Caloz, and D. Ménard, "Integrated and self-biased planar magnetic microwave circuits based on ferromagnetic nanowire substrates," in *Proc. European Microwave Conf*, Rome, pp. 743–746, 2009.
77. B. Lax, and K. J. Button, *Microwave Ferrites and Ferrimagnetics*, McGraw-Hill, Basingstoke, 1962.
78. T. Kodera and C. Caloz, "Integrated leaky-wave antenna duplexer using CRLH a uniform ferriteloaded open waveguide," *Trans. Microwave Theory Tech*, vol. 58, 2010, *to be published.*
79. L. Cohen, "Time-frequency distributions-a review," *Proc. IEEE* **77**, pp. 941–981, 1989.
80. S. Gupta, S. Abielmona, and C. Caloz, "Microwave analog real-time spectrum analyzer (RTSA) based on the spatial-spectral decomposition property of leaky-wave structures," *IEEE Trans. Microwave Theory Tech*, vol. 57, no. 12, 2009.
81. G. P. Agarwal, *Nonlinear Fiber Optics*, Academic Press, New York, 2005.
82. H. F. Talbot, "Facts relating to optical science," *Philos. Mag*, vol. 9, pp. 1–4, 1836.
83. J. S. Gómez-Díaz, A. A. Alvarez-Melcon, S. Gupta, and C. Caloz, "Spatio-temporal Talbot phenomenon using metamaterial composite right/left-handed leaky-wave antennas," *J. App. Phys*, 104, 104901: pp. 1–7, 2008.
84. J. Azaña, and M. A. Muriel, "Temporal self-imaging effects: theory and application for multiplying pulse repetition rates," *IEEE J. Sel. Top. Quantum Electron*, vol. **7**, pp. 728–744, 2001.
85. J. S. Gómez-Díaz A. A. Alvarez-Melcon, S. Gupta, and C. Caloz, "Tunable Talbot imaging distance using an array of beam-steered metamaterial leaky-wave antennas," *J. App. Phys.*, vol. 106, no. 8, pp. 084908:1–8, 2009.

12

Defected Ground Structure for Microstrip Antennas

Debatosh Guha[1], Sujoy Biswas[2] and Yahia M. M. Antar[3]
[1]*Institute of Radio Physics and Electronics, University of Calcutta, India*
[2]*Institute of Technology and Marine Engineering, India*
[3]*Royal Military College, Canada*

12.1 Introduction

Defected Ground Structure (DGS) is relatively a new area of research and applications associated with printed circuits and antennas. DGS use for antennas is probably considered, for the first time, as a separate book chapter here. We, therefore, introduce some basic insight into the subject to give a better understanding of its applications. The evolution of DGS from the photonic band gap (PBG) structure, the basic geometries, and their analytical models will be discussed. Most of the possible applications of DGS to microstrip antennas are covered, based on the results available in the open literature. A chronological development of DGS indicating some major advances, especially related to microstrip antennas is presented.

12.2 Fundamentals of DGS

This section addresses the fundamentals of Defected Ground Structures from its evolution and modeling to the state-of-the-art applications.

12.2.1 Evolution

The concept of Defected Ground Structures (DGS) evolved in recent years primarily from the studies of Photonic Band Gap (PBG) structures in electromagnetics. The PBGs, employed in electromagnetic (EM) applications, are now referred to as Electromagnetic Band Gap (EBG) structures [1]. They are actually artificial periodic structures exhibiting an unusual property of

Microstrip and Printed Antennas: New Trends, Techniques and Applications. Edited by Debatosh Guha and Yahia M.M. Antar

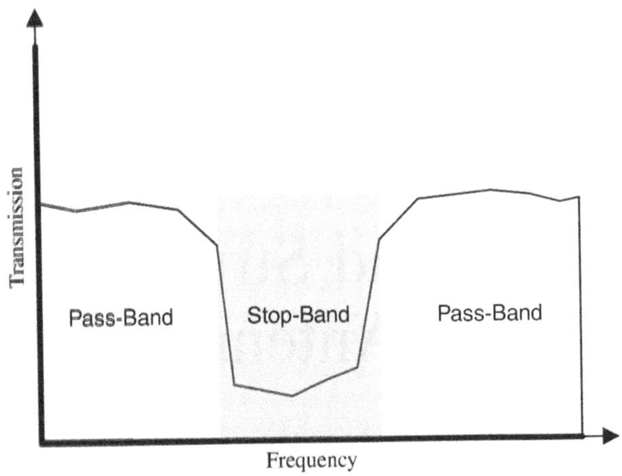

Figure 12.1 Representative diagram showing a stopband in transmission characteristics of a typical EBG structure

preventing EM waves from propagating through them over a range of frequencies, termed as "stopband" and allowing EM waves to propagate through them over a range of frequencies, termed as "passband." Figure 12.1 indicates the "band-gap" in transmission caused by an EBG structure. The pioneering studies with PBG date back to 1987 [2, 3] and were in optical frequencies. This gradually became popular in microwave and millimeter wave applications. Various geometries evolved through a series of investigations and a brief review of basic EBG structures leading to realizing DGS is provided here.

Structures having periodic arrangements of metallic [4], dielectric [5] or metallo-dielectric bodies [6, 7] with a lattice period $p = n\lambda_g/2$, λ_g being the guide wavelength, are found to exhibit EBG behavior. Their periodicity may be broadly categorized into three groups: (1) three-dimensional (3D) [5]; (2) two-dimensional (2D) [8–10]; and (3) one-dimensional (1D) [11].

A 2D planar EBG structure [8], shown in Figure 12.2(a), is actually a repetitive pattern of circular *unit cells* etched out on the ground plane of a printed transmission line. A simplified 1D variation of Figure 12.2(a) is shown in Figure 12.2(b).

For a printed microwave circuit and transmission line, its ground plane appears to be the most suitable choice to implement an EBG as a printed pattern. In 1999, a group of researchers further simplified the geometry and discarded the periodic nature of the pattern. They simply used a unit cell of dumbbell shape to achieve considerable stopband in C and X-bands for a microstrip line and in their introductory paper [12], they called it a "PBG unit structure." In their subsequent article [13], the same structure was termed as "Defected Ground Structure" (DGS). Therefore, a DGS may be regarded as a simplified variant of a printed EBG on a ground plane.

12.2.2 *Definition and Basic Geometries*

Defected Ground Structure (DGS), as the name implies, refers to some compact geometries, commonly known as a "unit cell" etched out as a single defect or in periodic configuration

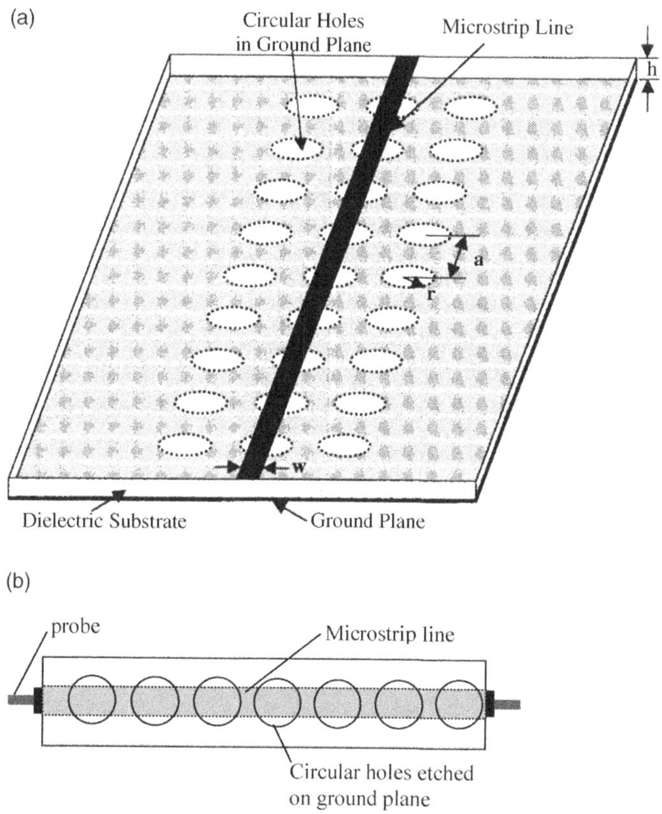

Figure 12.2 (a) Periodic two-dimensional (2D) etched circular pattern on the ground plane of a microstrip transmission line. Reproduced by permission of ©1998 The IEEE [8] (b) periodic one-dimensional (1D) etched pattern on the ground plane of a microstrip transmission line. Reproduced by permission of ©2003 The IET [11]

with small period number on the ground plane of a microwave printed circuit board (M-PCB) to attribute a feature of stopping wave propagation through the substrate over a frequency range. Thus a DGS can be described as a unit cell EBG or an EBG with limited number of cells and a period. The DGS slots are resonant in nature. They have different shapes and sizes with different frequency responses and equivalent circuit parameters. The presence of a DGS under a printed transmission line actually perturbs the current distribution in the ground plane and thus modifies the equivalent line parameters over the defected region. Thus it influences the guided wave characteristics and is found to exhibit (1) bandgap properties as revealed due to EBG structures; and (2) a slow wave effect, which helps in compacting the printed circuits. As mentioned earlier, the study of DGS can be classified into two categories depending on their configurations: (1) single unit cell DGS; and (2) uniform or non-uniform periodic arrangement of unit cells. Figure 12.3 shows some basic DGS classifications and geometries.

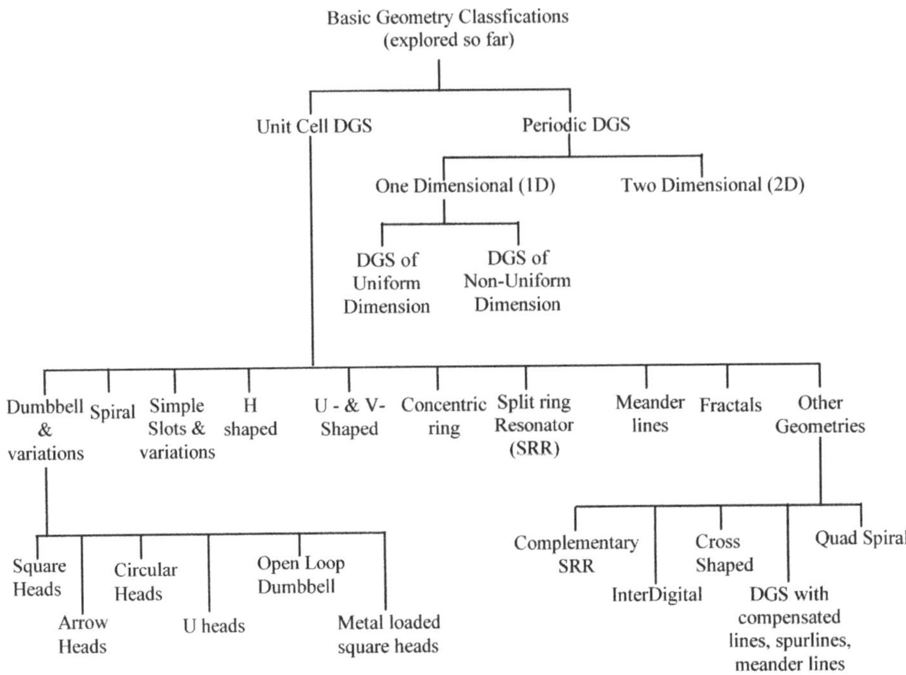

Figure 12.3 Some basic DGS classifications and geometries

12.2.2.1 Unit Cell DGS

The DGS geometries reported so far include simple shapes such as rectangular dumbbell [13], circular dumbbell [14], spiral [15], "U" [16], "V" [16], "H" [17], cross [18], concentric rings [19], etc. Also different complex structures like split ring resonators [20, 21], or fractals [22] have been examined. Some simple and complex type unit cell DGSs are shown in Figure 12.4. They are utilized to implement filters, suppress unwanted surface waves, control harmonics in microstrip antennas, compact microwave circuits and other microwave applications. Different geometries have been explored with the aim of achieving improved performance in terms of stopband and passbands, compactness and ease of design.

Dumbbell-Shaped DGS

The first unit cell, that was reported by Park et al. [12] and subsequently by Kim et al. [13], was a simple dumbbell-shaped DGS. It consists of two rectangular slots connected by a narrow slot, thus resembling a dumbbell-shaped lattice, etched out on the metallic ground plane of a 50 ohm microstrip line, as shown in Figure 12.5(a). The S-parameters as functions of the dimensions of the DGS are shown in Figure 12.5(b). It is evident that like EBG, a simple unit cell DGS shows a distinct stop-band property for the wave propagating through the microstrip line.

Figure 12.4 Different DGS geometries: (a) dumbbell-shaped (b) spiral-shaped (c) H-shaped (d) U-shaped (e) arrow head dumbbell (f) concentric ring shaped (g) split-ring resonators (h) interdigital (i) cross-shaped (j) circular head dumbbell (k) square heads connected with U slots (l) open loop dumbbell (m) fractal (n) half-circle (o) V-shaped (p) L-shaped (q) meander lines (r) U-head dumbbell (s) double equilateral U (t) square slots connected with narrow slot at edge

Spiral-Shaped DGS

After the dumbbell-shaped DGS was successfully implemented and used to design low pass filters, the researchers explored many other patterns to achieve improved performance. Spiral DGS was reported in [15] and is shown in Figure 12.6. If a spiral and a dumbbell DGS, both occupying same surface area, are etched on identical substrates, it is observed that the attenuation pole for spiral DGS occurs at a much lower frequency than that due to the dumbbell DGS. This means that a spiral DGS needs a much smaller space for a given frequency response. It also gives comparatively steeper rejection characteristics.

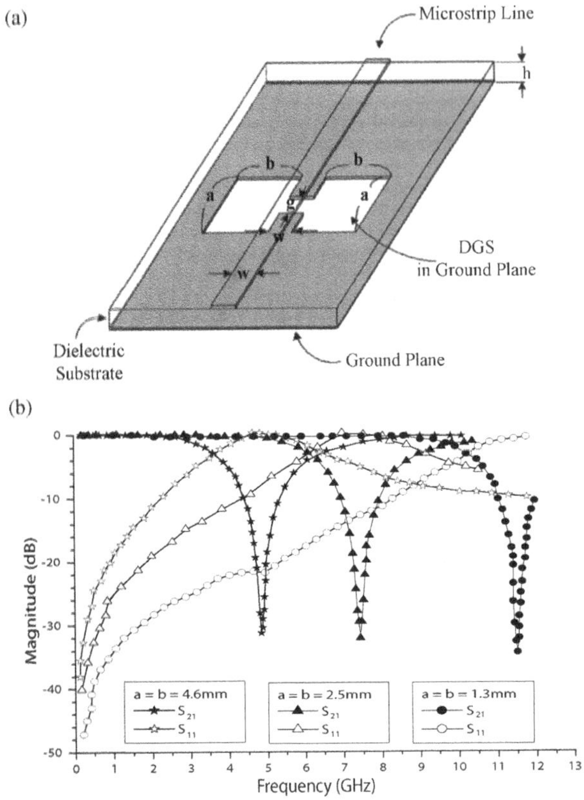

Figure 12.5 (a) Dumbbell-shaped DGS with rectangular slot dimensions "*a*" and "*b*" and gap "*g*" integrated with 50 Ω microstrip transmission line fabricated on a 62mil TACONIC substrate with dielectric constant $\varepsilon_r = 10$, and a microstrip line having width $w = 1.46$ mm; (b) simulated S-parameters for the DGS in (a) with gap "*g*" = 0.2 mm for all values of "*a*" and "b". Reproduced by permission of ©2000 IEEE [13]

H-Shaped DGS

A decrease in the surface area required by a DGS without compromising its frequency response and bandstop characteristics is a challenge to researchers. An H-shaped DGS [17], as shown in Figure 12.7(a), is more compact in shape and provides a solution to this problem. This configuration has a large external Q and a sharp transition from pass to stopband. A comparative study of square head dumbbell, circular head dumbbell and H-shaped DGSs was presented in [17]. Figure 12.7(b) also reveals H-DGS has a sharper passband to stopband transition compared to other kinds.

U- and V-Shaped DGSs

DGSs with sharp and deep band rejection properties are often required to suppress spurious signals. Other than H-shaped, U- and V- DGSs exhibit steep rejection and high Q values.

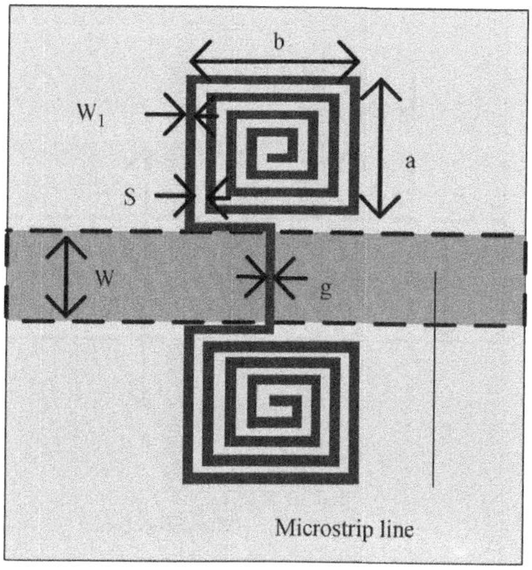

Figure 12.6 Top view of spiral-shaped DGS integrated with 50 Ω microstrip transmission line fabricated on a 31mil thick RT/Duroid 5880 substrate with $\varepsilon_r = 2.2$ and $a = b = 5$ mm, $w_1 = g = 0.4$ mm, $w = 2.4$ mm, $s = 0.2$ mm. Reproduced by permission of ©2002 The IET [15]

Figures 12.4(d) and 12.4(o) show schematic views of U- and V- DGSs, respectively. Conventional dumbbell and spiral-shaped DGSs do not provide such steep rejection characteristics owing to their low Q. Although a spiral DGS shows steeper rejection compared to a dumbbell DGS, its Q is no more than 10, and as such it is not capable of producing very sharp rejection as required. The Q-factor of U- or V- DGSs can be increased by decreasing the distance between the two arms of "U"-shaped or reducing the angle between the two arms of "V"-shaped slots. A comparison between U-shaped and spiral DGSs [16] shows that for the same resonance frequency, the former provides a Q value of 36.05 whereas that due to the latter is only 7.478.

Circular Ring-Shaped DGS

A concentric ring-shaped DGS [19] shown in Figure 12.4(f) was found to show a wide stopband around 10 GHz. Half-ring geometries shown in Figures 12.8(a) and 12.8(b) were also examined in [19]. The bandstop properties can also be controlled by varying their radii and thicknesses. Metallic backing behind the DGS has also been explored in [19] with a view to suppressing leakage or back radiation through the defects. The diagram and the results are shown in Figures 12.8(c) and 12.8(d) respectively. The measurements indicate that the introduction of metallic backing at a suitable distance from the DGS provides better bandstop characteristics.

Other DGS Geometries

Apart from the DGSs discussed above, many other forms and shapes have been explored for different attractive features and applications. A cross-shaped DGS shown in Figure 12.4(i) is a good example. Interdigital slots discussed in [23] and shown in Figure 12.4(h) allow us to vary

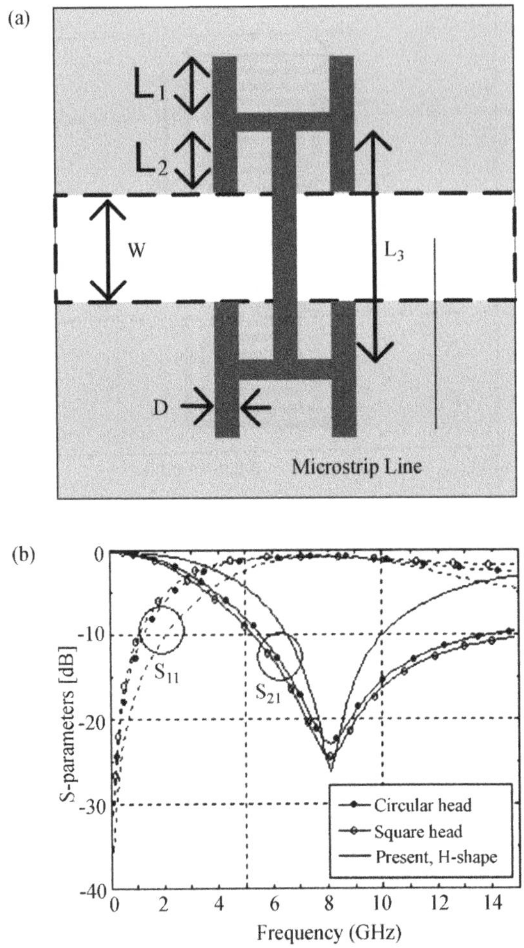

Figure 12.7 (a) Top view of H-shaped DGS integrated with 50 Ω microstrip transmission line 0.381 mm thick RT/Duroid 5880 substrate with $\varepsilon_r = 2.2$ with $D = 0.5$ mm, $L_1 = L_2 = 2.5$ mm, $L_3 = 6.2$ mm with resonance frequency at 8.1 GHz; (b) simulated S parameters of square and circular head DGS shapes on identical substrates having same resonant frequency and comparison with S parameters of H-shaped DGS. Reproduced by permission of ©2006 IEEE [17]

the resonant frequency by changing the finger lengths without changing the occupied area. An improved and extended stopband is also obtained using mutually coupled dumbbell-shaped DGS [24] and half-circle DGS [25] [Figure 12.4(n)]. A steeper transition from passband to stopband may also be obtained using a split ring resonator (SRR) [20]. This is a good candidate to achieve filters with elliptic response. Many other DGSs are also found to show a multi stopband feature, which is of interest for many circuit applications. The L-shaped defect shown in Figure 12.4(p) and the metal-loaded dumbbell DGS studied in [26] are appropriate examples.

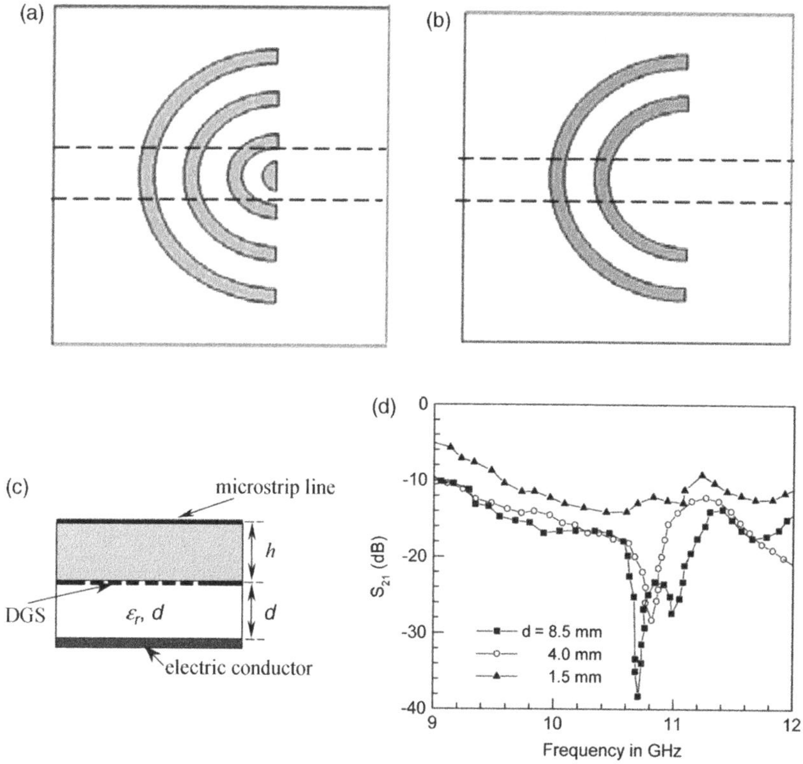

Figure 12.8 (a) Top view of three-ring HR-DGS integrated with 50 Ω microstrip transmission line on a 1.575 mm thick Taconic substrate with $\varepsilon_r = 2.2$; (b) top view of two ring HR-DGS integrated with 50 Ω microstrip transmission line on identical substrate; (c) cross-sectional view of a microstrip transmission line integrated with concentric ring DGS backed by metallic sheet with the intermediate space being filled up with dielectric medium; (d) measured transmission characteristics of the microstrip line on concentric ring DGS backed by metallic sheet with variable distance from DGS and intermediate space filled with air. Reproduced by permission of ©2006 IEEE [19]

12.2.2.2 Periodic DGS

Repetition of unit cells in a periodic fashion sometimes helps achieve deeper, steeper and wider stopband characteristics. A DGS thus formed by repetition of unit cells is referred to as a Periodic DGS. Figure 12.9(a) [27] shows a five-cell periodic DGS fabricated on Taconic substrate with a 50 Ω microstrip line printed on its other side. Its transmission characteristics, presented in Figure 12.9(b), show a much wider stopband compared to that produced by a single identical cell. This also indicates that the stopband cutoff depends on the cell dimensions. The period marginally affects the center of the band. Rather, the number of cells determines the width and rejection depth of S_{21} versus frequency plot.

So far our discussions have been restricted to Uniform periodic DGS. Another category is Non-Uniform periodic DGS as discussed here. Better performance in terms of enhancing the

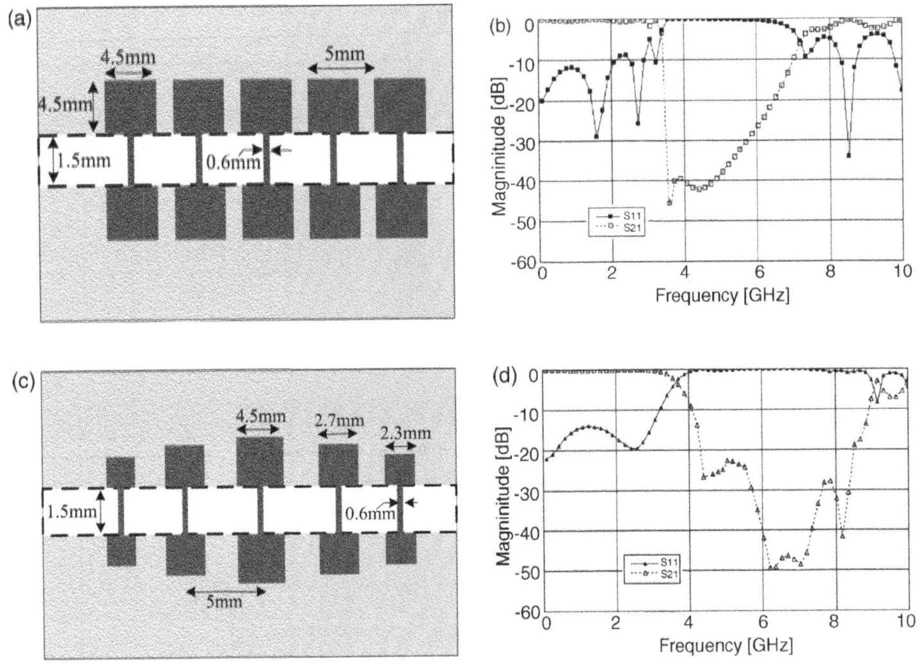

Figure 12.9 (a) Top view of one-dimensional 5-cell uniform dumbbell-shaped DGS on a 50 Ω microstrip line fabricated on a 1.575 mm thick TACONIC-CER 10 substrate with $\varepsilon_r = 10$; (b) measured S parameters for the uniform periodic DGS shown in (a); (c) top view of one-dimensional 5-cell non-uniform dumbbell-shaped DGS on a 50 Ω microstrip line fabricated on a 1.575 mm thick TACONIC-CER 10 substrate with $\varepsilon_r = 10$; (d) measured S parameters for the non-uniform periodic DGS shown in (b). Reproduced by permission of ©2004 IEEE [27]

stop-bandwidth and suppression of ripples is obtained by using a non-uniform periodic structure. In Figure 12.9(c), the cell dimensions are varied as $e^{1\backslash n}$ relative to the central unit cell, where n is a positive integer indicating the number of the elements on either side including the central one. For example, in Figure 12.9(c), n = 3 and the amplitude distribution of the units are [27]:

$$e^{1/3}\, e^{1/2}\, e^{1}\, e^{1/2}\, e^{1/3} \approx 1.396,\ 1.649,\ 2.718,\ 1.649,\ 1.396.$$

Here, the central dumbbell, chosen to have its amplitude "e^1" has 4.5 mm side length. Then the other dumbbell cells have side lengths of 2.7 mm and 2.5 mm, respectively following the exponential distribution mentioned above. Compared to the uniform periodic DGS, shown in Figure 12.9(a), the non-uniform variant provides a much wider stopband feature along with suppressed ripples in the passband, as shown in Figure 12.9(d).

Most of the periodic DGS use the 1D arrangement of unit cells. Some applications may need 2D periods. One interesting example is shown in Figure 12.10 [28], where a number

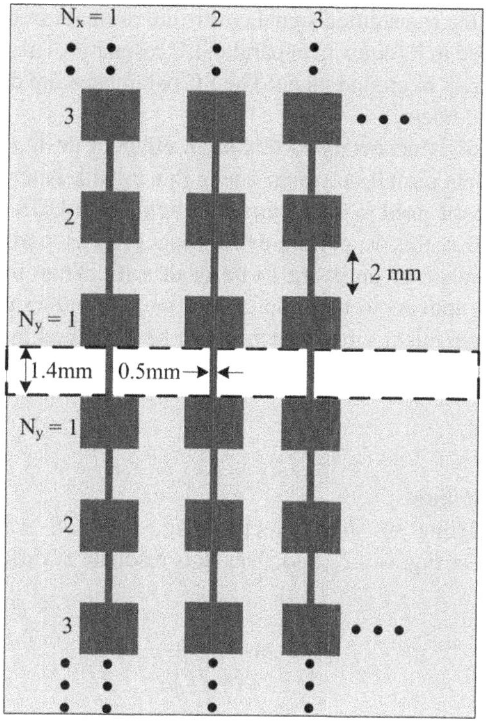

Figure 12.10 Planar two-dimensional arrangement of dumbbell-shaped DGS integrated with microstrip transmission line on a substrate of thickness 0.75 mm and dielectric constant $\varepsilon_r = 4.6$. Reproduced by permission of ©2002 IEEE [28]

of cells are periodically arranged in a vertical column and then each column is periodically arranged along the horizontal direction. This study was mainly focused to achieve the increased slow-wave factor for microstrip line and coplanar waveguide transmissions, compared to that achievable using EBG structures or 1D periodic DGS. The increased slow-wave factor is effectively used to reduce the size and compacting microwave circuits [28].

12.2.3 Modeling of DGS

A microstrip line placed on a DGS shows stopband(s) in its transmission characteristics. It is of interest to find out why. Both qualitative and quantitative analyses are essential to understand the working principle of a DGS. A defect indeed changes the current distribution in the ground plane of a microstrip line, giving rise to an equivalent inductance and capacitance. Thus a DGS behaves like an L-C resonator circuit coupled to the microstrip line. When an RF signal is transmitted through a DGS-integrated microstrip line, strong coupling occurs between the line and the DGS around the frequency where the DGS resonates.

Obviously, it happens if the transmitted signal covers the resonant frequency of the DGS, and most of the signal is stored in its equivalent parallel LC resonator. This indirectly indicates the bandstop feature of a defect in ground plane. The LC parameters are determined by the shape and size of the defect geometry.

A quantitative analysis is necessary to obtain an efficient design specified for different frequencies. An equivalent circuit may help one in this regard. Alternatively, a commercial full wave simulator may be used to characterize a circuit with DGS and to optimize it on a trial and error basis. But this is a time-consuming process particularly when a large structure or a large number of units are to be dealt with. Therefore, efficient modeling using equivalent circuit appears to be a useful solution to handle this issue in a simplified way. Modeling using equivalent circuit method can be classified into three categories: (1) transmission line modeling; (2) LC and RLC circuit modeling; and (3) quasi-static modeling.

Transmission Line Modeling

Transmission line modeling of simple rectangular slot DGS was introduced in [29]. The geometry is shown in Figure 12.11(a). The slots resonate at different frequencies given by

$$f_m \cong m \frac{c_0}{2d\sqrt{\varepsilon_{eff}^{slot}}} \tag{12.1}$$

where, m = 1,2,3,4... and as usual do not allow energy to propagate around those frequencies. The slots are modeled as a transmission line having characteristic impedance Z_0^{slot} and electrical length $\theta = \theta_1 + \theta_2 = \beta_m^{slot} d$ as shown in Figure 12.11(b). Here $\theta_1 = \theta_2$ if the position of the microstrip line is symmetric with respect to d.

The coupling between the slots and microstrip line is represented by an ideal transformer with a turn ratio [29]

$$n \cong \sqrt{\frac{Z_0^{\mu strip}}{Z_0^{slot}}} \tag{12.2}$$

LC and RLC Equivalent Circuit Modeling

The transmission line model, discussed above for a simple slot DGS, needs to evaluate the impedance Z_0^{slot}. Since this parameter critically depends on frequency, it may not be an easy task to model any DGS shape. Therefore, a more general approach of representing a DGS in terms of equivalent parallel LC or RLC circuit has been explored as discussed below.

A simple example is given using a dumbbell-shaped DGS shown in Figure 12.5(a). The larger rectangular defect on either side of the line causes effective series inductance L and the narrow slot beneath the line produces a gap capacitance C in parallel with L, as shown in Figure 12.12(a). Once the equivalent L and C values are known, the determination of the DGS characteristics is straightforward.

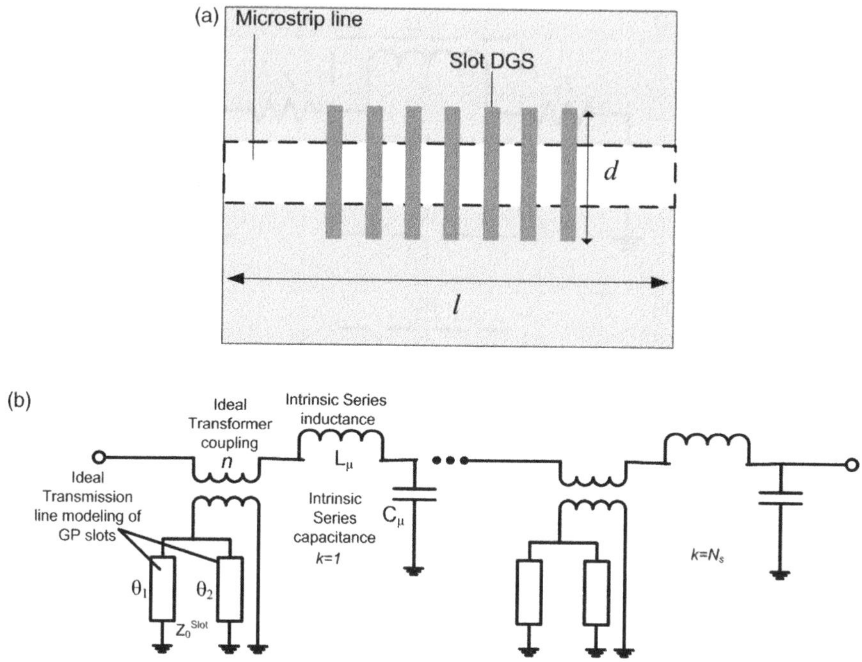

Figure 12.11 (a) Top view schematic of simple 7-cell uniform periodic slot DGS with slot length $d = 10.7$ mm, slot width $w = 0.51$ mm and center-to-center slot gap $p = 1.52$ mm with a 50 Ω microstrip line fabricated on a 25mil thick RT/Duroid 6010 substrate with $\varepsilon_r = 10.2$; (b) equivalent circuit of slot DGS shown in (a) based on a transmission line model. Reproduced by permission of ©2004 IEEE [29]

The extraction of equivalent L and C values [30] is described below.

An EM simulator may be employed to determine the S-parameters. For the dumbbell-shaped DGS in Figure 12.5(a), the attenuation pole is located at 8 GHz with 3 dB cut-off at 3.5 GHz. The result resembles the response of a single pole LPF and as such can be fitted with the response of one-pole Butterworth LPF as shown in Figure 12.12(b). The comparison directly correlates the Butterworth elements with the unknown L and C values.

The reactance of the equivalent circuit of Figure 12.12(a) is

$$X_{LC} = \frac{1}{\omega_0 C}\left(\frac{\omega_0}{\omega} - \frac{\omega}{\omega_0}\right) \tag{12.3}$$

where ω_0 is the angular resonant frequency.

The reactance of LPF in Figure 12.12(b) is given by

$$X_L = \omega_1 Z_0 g_1 \tag{12.4}$$

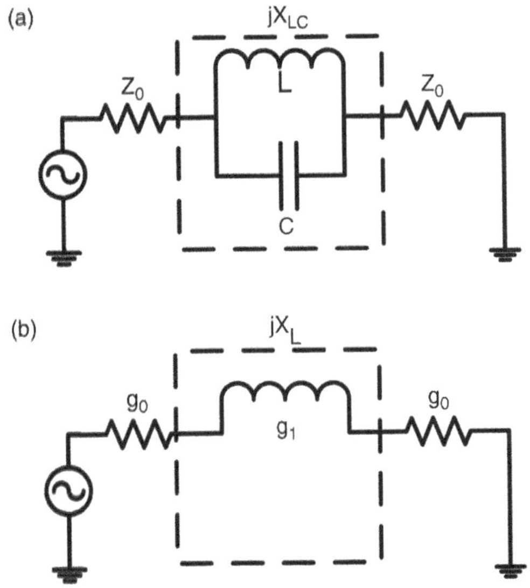

Figure 12.12 (a) LC equivalent circuit of single cell dumbbell-shaped DGS; (b) one-pole Butterworth prototype Low Pass filter. Reproduced by permission of ©2001 IEEE [30]

where ω_1 is the normalized angular frequency, Z_0 is input and output port impedances, and g_1 is the 'prototype element' as described in [31].

Equating (12.3) and (12.4) at the cut-off frequency, we have,

$$X_{LC}\big|_{\omega=\omega_c} = X_L\big|_{\omega_1=1} \tag{12.5}$$

$$C = \frac{\omega_c}{Z_0 g_1}\left(\frac{1}{\omega_0^2 - \omega_c^2}\right) \tag{12.6}$$

$$L = \frac{1}{4\pi^2 f_0^2 C} \tag{12.7}$$

where f_0 is the resonant frequency for DGS as well as the attenuation pole of the Butterworth prototype. Various possible design parameters can be extracted based on this study and a comprehensive view is presented in Table 12.1.

The LC modeling does not account for any losses caused by radiation or conductor/dielectric losses. A more realistic model takes an equivalent loss resistance R into account as shown in Figure 12.13(a). This loss resistance R can be extracted from simulated S_{11} employing the following relation [32]:

$$R = \frac{2Z_0}{\sqrt{\frac{1}{|S_{11}(\omega)|^2} - \left(2Z_0\left(\omega C - \frac{1}{\omega L}\right)\right)^2 - 1}} \tag{12.8}$$

where,

$$S_{11}(\omega) = \frac{Z_{in} - Z_0}{Z_{in} + Z_0} \tag{12.9}$$

and the equivalent L and C are expressed in Equations (12.7) and (12.6), respectively.

The order of accuracy of this RLC model has been further improved adding more circuit segments [33]. One such example is shown in Figure 12.13(b), where the added networks take care of the fringing fields occurring at the step discontinuities in the defects.

Table 12.1 Circuit elements and characteristics of the equivalent circuits in Figure 12.12 obtained for different dimensions of a dumbbell shaped DGS [30]

Equivalent circuit elements and characteristics	Dumbbell-shaped DGS: Dimensions in mm					
	g = 0.2			*a* = *b* = 2.5		
	a = *b* = 1.3	*a* = *b* = 2.5	*a* = *b* = 4.6	*g* = 0.2	*g* = 0.4	*g* = 0.6
Inductance (nH)	0.3675	0.86594	1.97725	0.81051	0.90712	0.96825
Capacitance (pF)	0.51222	0.52845	0.53794	0.60286	0.43306	0.34247
Cut-off frequency (GHz)	10.15	6.085	3.62	6	6.4	6.72
Attenuation pole (GHz)	11.6	7.44	4.88	7.2	8.03	8.74

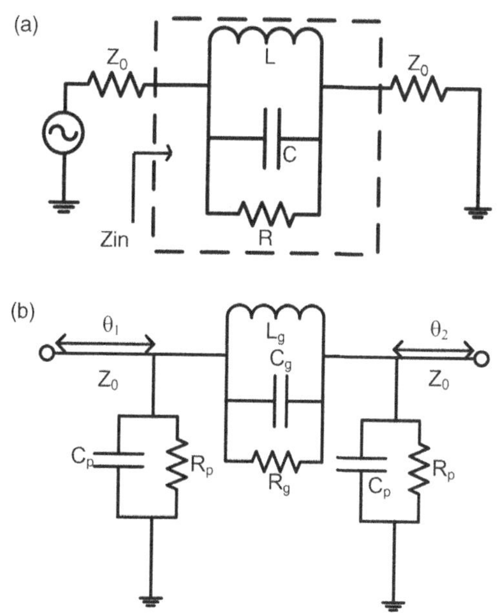

Figure 12.13 (a) LCR equivalent circuit of single cell dumbbell-shaped DGS; (b) Pi-type equivalent circuit of single cell dumbbell-shaped DGS, reproduced by permission of ©2002 IEEE [33]

Quasi-Static Modeling

This is another efficient technique to model a DGS [34]. The quasi-static approach uses an equivalent circuit considering the physical dimensions of the defects in the very beginning. The basic steps to be followed are shown in Figure 12.14. This technique is applicable to comparatively simpler geometries where each DGS segment could be represented by its equivalent circuit component. To understand the process, we again consider a dumbbell-shaped DGS shown in Figure 12.5(a). The nature of the current distribution on the ground plane surrounding the defect is to be visualized first. Commercial EM simulator can be efficiently used for that, as is shown in Figure 12.15 [34]. This indicates an equivalent current ribbon with different segments having different widths. The width of the ribbon is estimated from simulated bunching of current lines.

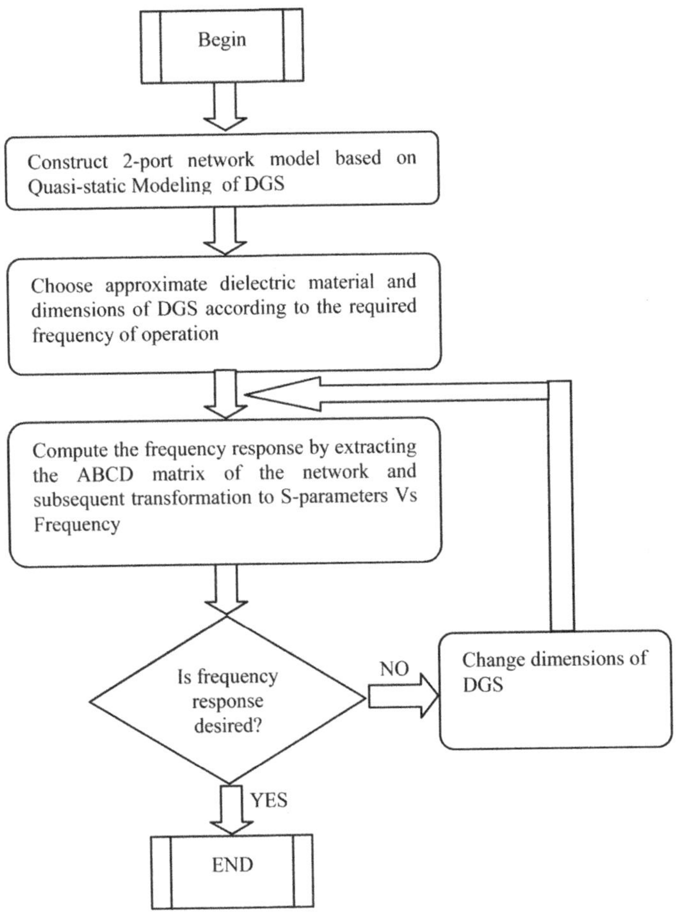

Figure 12.14 Steps followed for design of DGS using quasi-static modeling

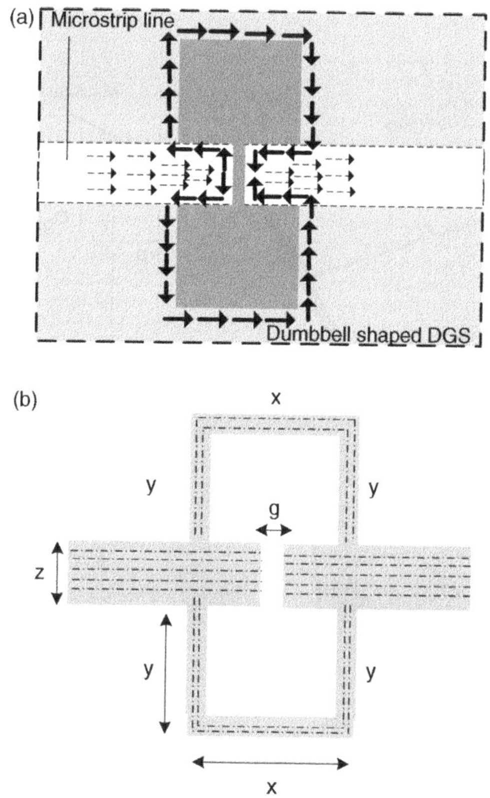

Figure 12.15 (a) Schematic of perturbed current distribution around the periphery of the defect obtained using commercial EM simulators; (b) schematic showing the equivalent current ribbon. Reproduced by permission of ©2006 IEEE [34]

Careful observation of the ribbon model reveals that it is a combination of two microstrip cross-junctions. Equivalent circuits representing a microstrip gap and a cross-junction are shown in Figures 12.16 [34]. This concept translates a single dumbbell DGS to an equivalent circuit shown in Figure 12.17. Equivalent L and C values may be directly expressed in terms of DGS dimensions [34] and successive design steps are described in Figure 12.14.

Comments on the Design Approaches

Of the three techniques discussed above, the last one appears to be more practical. The major disadvantage of LC/RLC method is that there is no direct relationship between the physical dimensions of the defect and the equivalent L or C. Thus, for a given frequency response, the DGS dimensions have to be guessed and varied iteratively using an EM simulator to determine the best suitable value. This is very time-consuming and not efficient at all. On the contrary, the quasi-static approach is straightforward. The quasi-static equivalent inductance and capacitance have a direct relationship with DGS dimensions and thus allow one to design a DGS with the help of theoretical calculations.

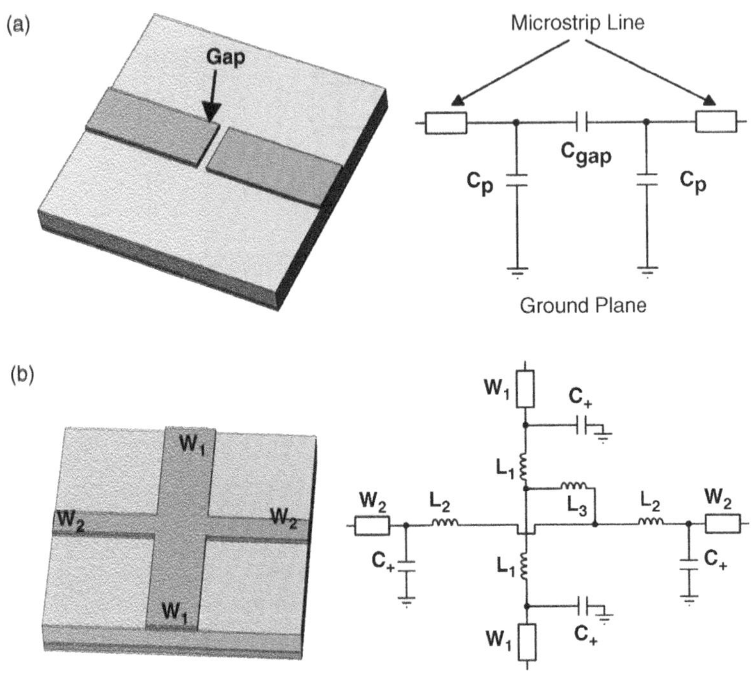

Figure 12.16 (a) Microstrip gap and its equivalent circuit; (b) microstrip cross and its equivalent circuit. Reproduced by permission of ©2006 IEEE [34]

12.2.4 Popular Applications to Printed Circuits

DGS have been widely used for various microwave applications, which may be broadly categorized as: (1) DGS for microwave printed circuits; and (2) DGS for printed antennas. Since this chapter is focused on antenna applications, we will discuss this in more detail later. Now some basic circuit applications are mentioned here.

DGS for Printed Filters

This is probably the most popular application of DGS and different shapes have been explored to improve filter properties such as low insertion loss, deeper and wider stopband, less ripple in the passband and sharp rejection characteristics. A few examples are discussed below:

The first three pole LPF designs using dumbbell-shaped DGSs [30] are shown in Figure 12.18 and their S-parameters in Figure 12.19. From the reported results in open literature, it is found that most of the proposed DGSs have been tested for filter applications.

DGS also helps to improve the performance of a BPF. Conventional parallel coupled line BPF exhibits asymmetric transitions from passband to stopband with unwanted passband at higher frequencies. DGS was successfully applied to eliminate such spurious passband and also

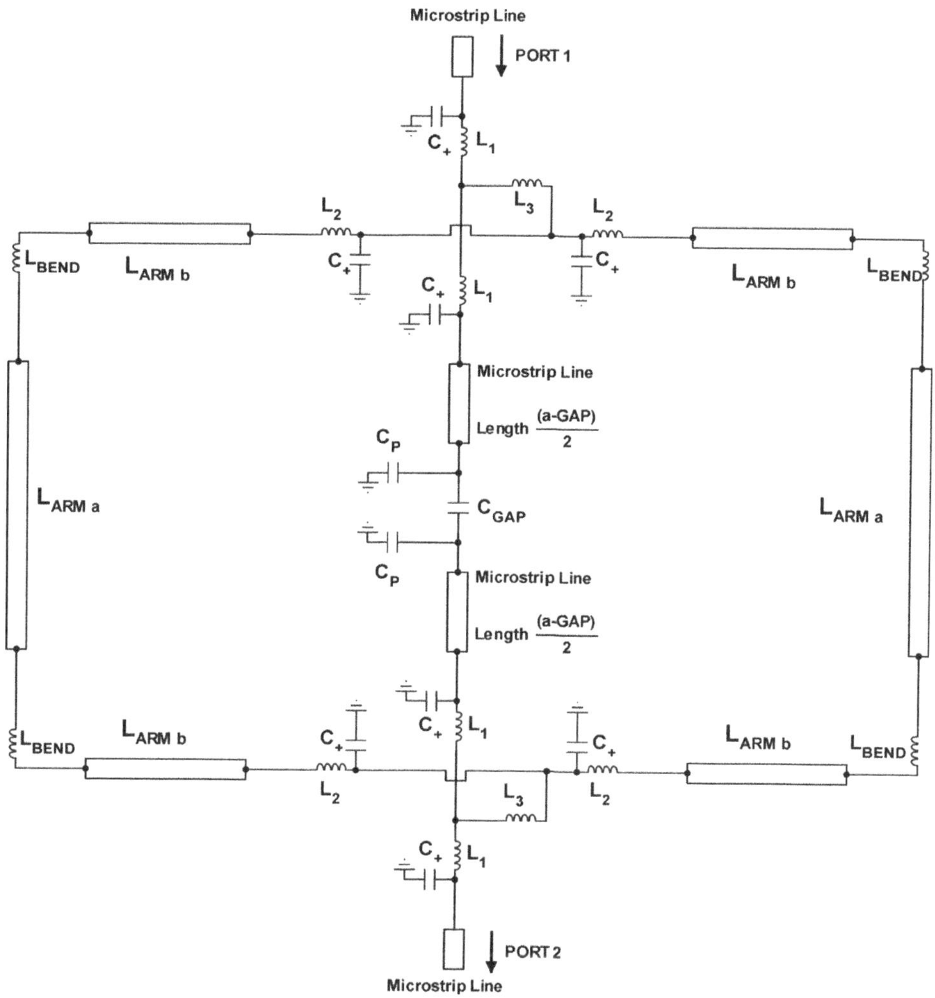

Figure 12.17 Equivalent circuit of unit cell dumbbell-shaped DGS based on the quasi-static model. Reproduced by permission of ©2006 IEEE [34]

to achieve a sharp transition from passband to stopband as shown in Figure 12.20 [35]. The function of each DGS unit used in Figure 12.20(a) is indicated in the figure caption. Improvements in selectivity and suppression of spurious passband signals have also been achieved using U-shaped defects [36] and Split Ring Resonators (SRR) [37].

Another interesting application is to achieve a tunable bandstop filter. A varactor diode inserted into a DGS was examined in [38, 39] to control the equivalent capacitance leading to a tunable characteristics of the filter.

DGS for Printed Circuit Components

The bandstop property of a DGS has also been used to make microstrip diplexers [40], rat race couplers [41], and Wilkinson power dividers [42]. Figure 12.21 shows a microstrip diplexer

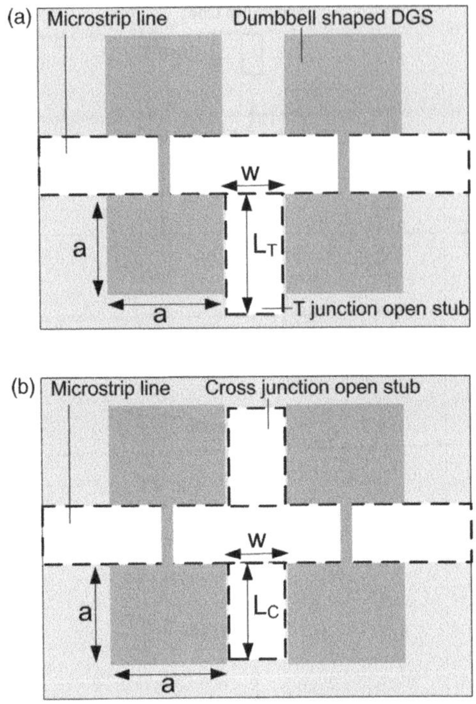

Figure 12.18 Schematics of three pole LPFs constructed using two DGS units each with sides $a = 5$ mm, connecting slot $g = 0.5$ mm on a 31-mil thick substrate with dielectric constant 2.2 and also using (a) T-junction open stub for parallel capacitance having width $W = 5$ mm and $L_T = 10$ mm; (b) cross-junction open stub for parallel capacitance having width $W = 5$ mm and $L_C = 6$ mm. Reproduced by permission of ©2001 IEEE [30]

designed using DGS [40]. A reduction of 2nd and 3rd harmonic in case of a power amplifier was also reported using a DGS integrated $\lambda/4$ bias line [43].

DGS for High Impedance Microstrip Line

The characteristic impedance of a microstrip line depends on the w/h ratio, where w is the width of the line etched on substrate having thickness h and dielectric constant ε_r. Using a commercially available microwave substrate with a limited range of ε_r and h values, it is practically impossible to realize a high impedance line particularly beyond 150 Ω. Very high impedance means very small w for a given h, which indeed becomes unrealizable above 150 Ω. A DGS, integrated with a microstrip line, results in an additional inductance leading to increased line impedance. Figure 12.22(a) shows a 4 : 1 Wilkinson power divider [44] with a requirement of 158 Ω microstrip line. This uses 0.4 mm-wide DGS-integrated microstrip line etched on RT Duriod 5880 substrate with $h = 31$ mil. A conventional microstrip would need 0.17 mm-wide line to achieve the same impedance of 158 Ω. A similar design of high impedance DGS-integrated lines, used for branch-line couplers [45] is shown Figure 12.22(b).

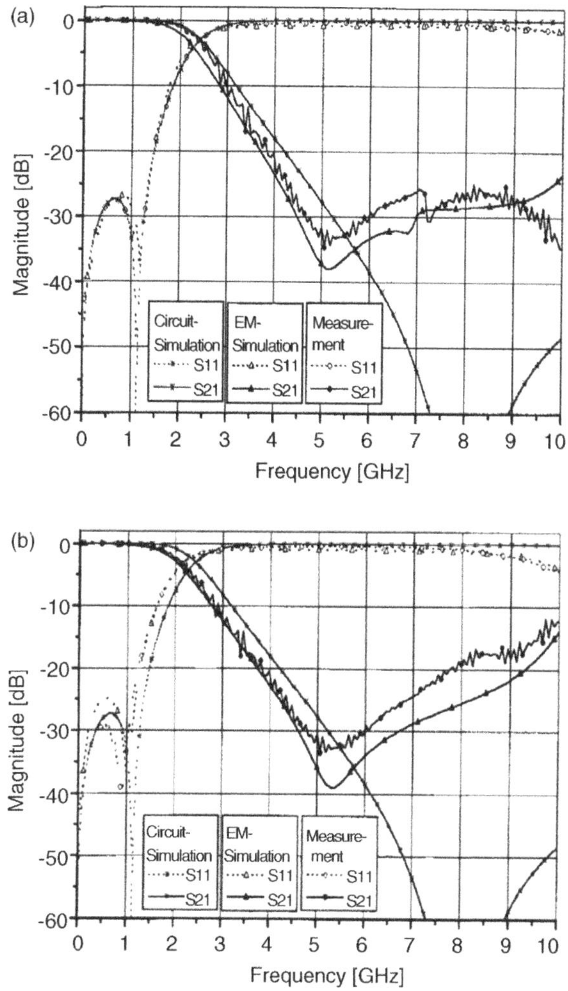

Figure 12.19 (a) Measured S-parameter results for the fabricated DGS LPF shown in Figure 12.18(a); (b) measured S-parameter results for the fabricated DGS LPF shown in Figure 12.18(b). Reproduced by permission of ©2001 IEEE [30]

DGS for Compacting Microwave Circuits

A DGS, placed beneath a microstrip line, causes a slow-wave phenomenon due to the increased inductance in the line. Considerably larger electrical length is thus achieved within a smaller space. This in turn allows one to reduce the size of microwave circuits and compact them. The Wilkinson power divider discussed above uses $\lambda/4$ transmission lines, which by using DGS is reduced by 17% of its usual length. This feature has also been used in rat race coupler [41] and amplifiers [28, 46] to reduce their size.

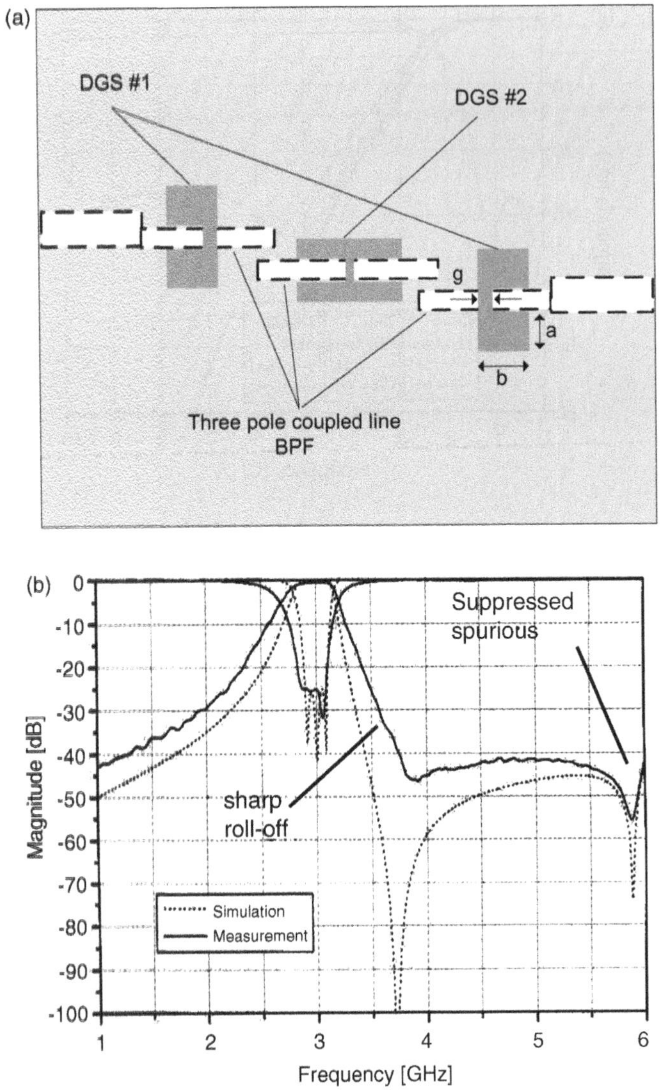

Figure 12.20 (a) Schematic diagram of coupled line bandpass filter fabricated on a 1.27 mm thick Duroid substrate with $\varepsilon_r = 10.2$ and DGS#1 for sharp roll off and DGS#2 to suppress unwanted passband; (b) simulated and measured results of the coupled-line bandpass filter shown in (a). Reproduced by permission of ©2002 IEEE [35]

12.3 DGS for Controlling Microstrip Antenna Feeds and Front-End Characteristics

12.3.1 Basic Idea

Defects in the ground plane were first used for microstrip antennas by Horii and Tsutsumi in 1999 [47], but in those days, the defects were termed as PBG or EBG material. They used the

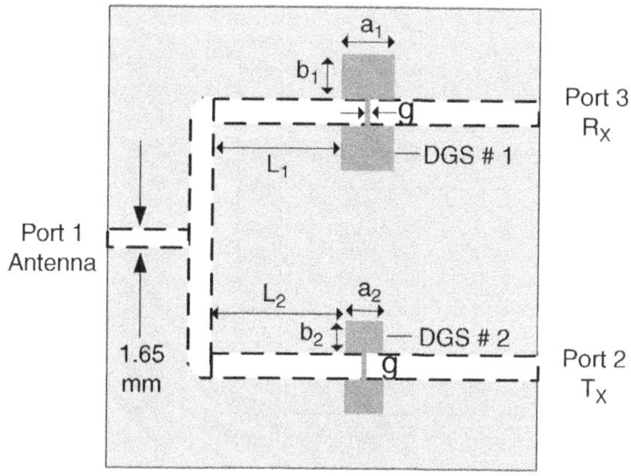

Figure 12.21 Schematic diagram of a Diplexer fabricated on a 1.575 mm thick substrate with $\varepsilon_r = 10$ with dimensions $a_1 = b_1 = 5$ mm, $g = 0.2$ mm, $L_1 = 5$ mm, $a_2 = b_2 = 2$ mm, $L_2 = 5$ mm. Reproduced by permission of ©2005 IEEE [40]

"stopband" characteristics of DGS-integrated microstrip transmission line. The first DGS for microstrip antenna in [47] was based on two earlier works [8, 48] published in 1998. In [47], a microstrip line was placed on circular hole-shaped defects and a sharp wide stopband around 11 GHz was achieved. The geometry is shown in Figure 12.2(a). Subsequent investigations applying DGS to control or improve antenna feed and front-end characteristics are discussed in this section.

12.3.2 Harmonic Control in Active Microstrip Antennas

In most of the active microstrip antennas, the oscillator and amplifier circuits are directly integrated with the radiating elements. Here, the microstrip patch acts not only as a radiator, but also as a frequency selective circuit. It is, therefore, prone to generating higher harmonics resulting in spurious radiations.

To stop the higher harmonics at the input of a microstrip element, proper bandstop characteristics in the feeding network have been explored using different DGS configurations since 1999 [47]. The first experiment [47] used a ground plane with 3 × 4 circular defects for a microstrip-fed rectangular patch as shown in Figure 12.23(a). The defects were described as a PBG lattice, as it was commonly called those days. But such a limited number of defects, later on, were simply called "DGS" [13]. Measured return loss characteristics of the patch with and without DGS are shown in Figure 12.23(b). The antenna resonates at its dominant mode near 900 MHz, which translates to $\lambda = 33.33$ cm, i.e. more than 18 times larger than the defect diameter. This smoothly allows 900 MHz signal to propagate and shows a stopband over 1760–2720 MHz with S_{21}<-20dB. Two higher harmonics at 1800 MHz and 2700 MHz are, therefore, effectively suppressed by the DGS-integrated feed line.

Then what is the role of the defects located under the patch? They actually take care of suppressing signals beyond 2700 GHz, but the reason or physical insight into this has not been explicitly discussed in [47].

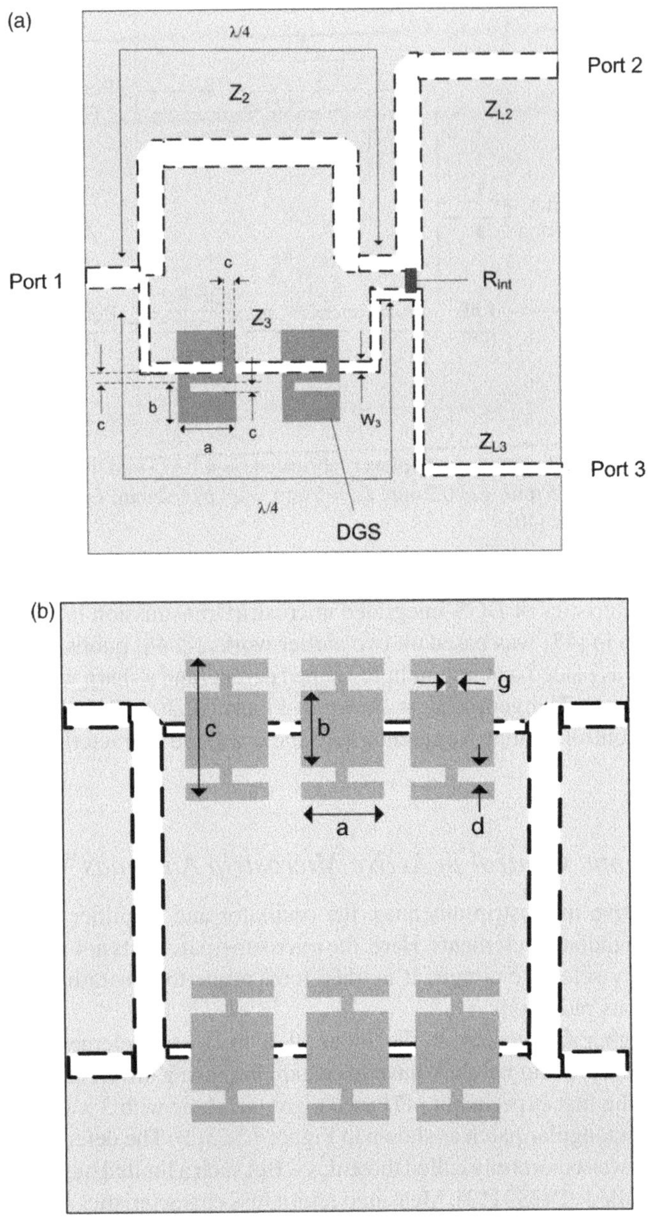

Figure 12.22 (a) Schematic diagram of a 4 : 1 unequal Wilkinson power divider with a 158 Ω microstrip line with DGS of width $W_3 = 0.4$ mm fabricated on a *31mil* thick substrate with $\varepsilon_r = 2.2$. The dimensions of the DGS are $a = b = g = 6$ mm, $c = 0.4$ mm, reproduced by permission of ©2001 IEEE [44]; (b) schematic diagram of a 10 dB 90° branch line coupler with DGS integrated microstrip line fabricated on a 31mil thick RT/Duroid 5880 substrate with $\varepsilon_r = 2.2$. The dimensions of DGS are $a = b = 6$ mm, $c = 13$ mm, $d = 2$ mm, $g = 0.5$ mm and center-to-center separation of the DGS units is 8 mm, reproduced by permission of ©2000 The IET [45]

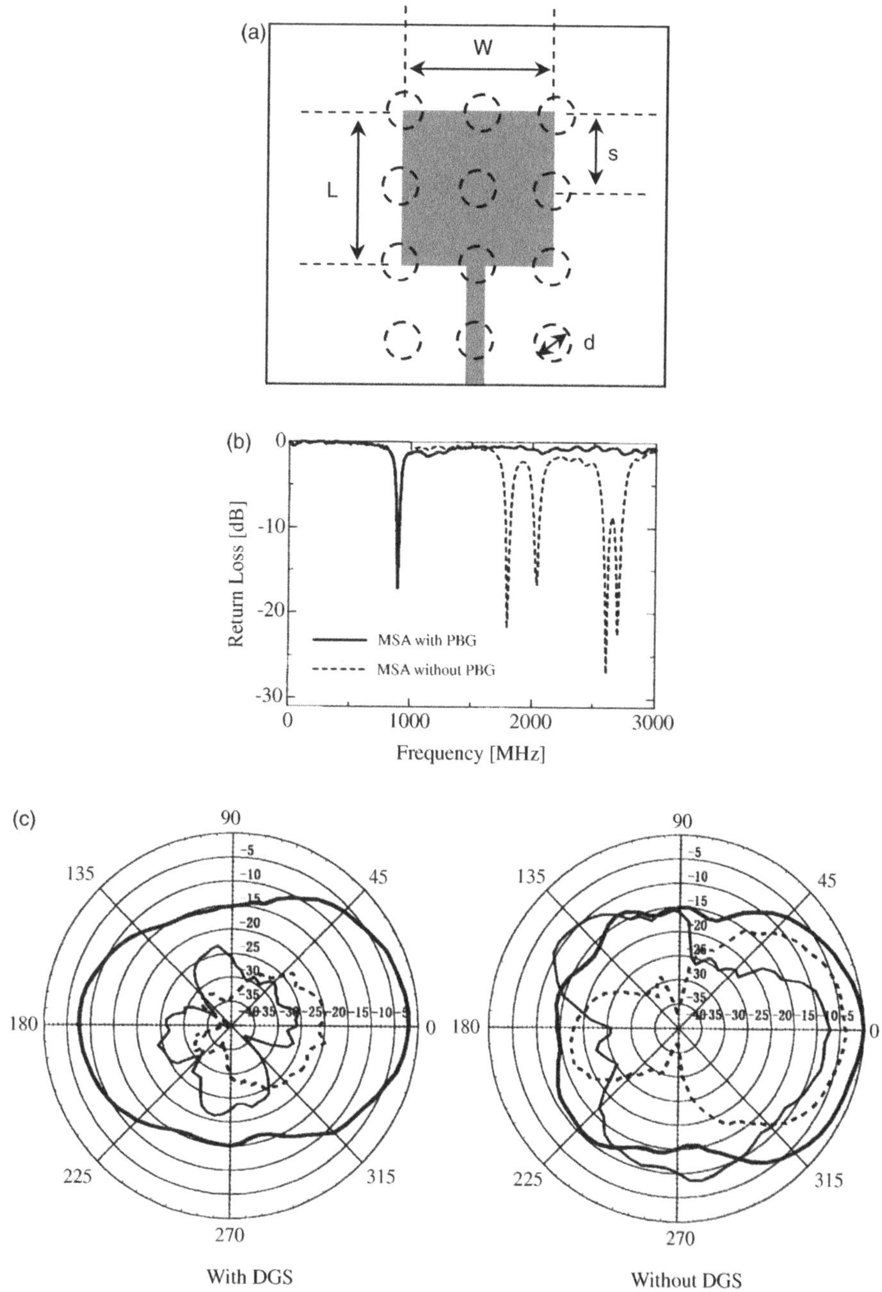

Figure 12.23 (a) Microstrip line-fed square patch with defected ground plane (PBG structure in the form of circular defects). $L = W = 76\,\text{mm}$, $d = 18\,\text{mm}$, $s = 38\,\text{mm}$, $w = 2.8\,\text{mm}$, substrate thickness $= 1.6\,\text{mm}$, $\varepsilon_r = 4.8$; (b) Measured return loss of the square patch compared with that of an identical one using normal ground plane (c) Measured co-polarized radiation patterns of the antenna in H-plane with and without DGS. Bold line: 900 MHz, thin line: 1800 MHz, broken line: 2700 MHz. Reproduced by permission of ©1999 IEEE [47]

Now let us look at the measured radiation characteristics presented in Figure 12.23(c) for the operating frequency of the antenna and at its first two higher harmonics. The feed and patch without DGS allow all the frequencies to radiate with small relative variation in the peak gain. But in the presence of DGS, only a 900 MHz signal radiates effectively. It is interesting to note that the DGS causes an increase in back radiation [47]. One can explain this due to possibly leaking EM waves through the defects present beneath the patch. This is not fully true. We shall discuss the reason later with proper evidence.

The same DGS integrated square patch, shown in Figure 12.23(a) was investigated in [48] with a small change in the DGS shape located below the feed-line. This is shown in Figure 12.24(a) [49]. Two circular defects below the feed are replaced by two square head dumbbell-shaped DGSs. This enhances the width of the stopband in the transmission characteristics of the feed shown in Figure 12.24(b). Thus it can take care of suppressing higher frequencies over a wider range and the defects under the patch can easily be removed. Such a configuration looks like that in Figure 12.25 and was examined in [32], and earlier in [48]. Chang and Lee [32] theoretically showed the possibility of using their geometry (Figure 12.25) for harmonic controls over a wide frequency band.

So far we have experienced the harmonic control up to the second harmonic, and of all the configurations (Figure 12.23(a) [47], Figure 12.24(a) [49], and Figure 12.25 [32]), Figure 12.25 appears to be the simplest one engaging the minimum number of DGS cells.

Further reduction in the number of DGS cells was explored by Sung and Kim [50]. They employed single dumbbell-shaped DGS beneath a microstrip feed as shown in Figure 12.26(a). Its radiation characteristics in Figure 12.26(b) indicate considerable suppression of the first harmonic. The radiated power of the first harmonic is about 15 to 20 dB down compared to that due to the dominant mode. Considerable back radiations at 1.79 GHz are evident, although no DGS is present beneath the patch. Therefore, the back radiation is not necessarily caused by defects below the patch only. The DGS below the feed is equally responsible and therefore proper measures should be taken when such configurations are used in integrated circuit design.

Another variant of unit cell DGS is shown Figure 12.27(a) [51]. This is relatively more compact in size and is placed beneath the neck of an inset-fed patch. As a result, the DGS cell could be located under the patch without using any extra space. This also takes care of suppressing the first harmonic of the operating frequency. Simulated electric fields on the top side of the substrate are depicted in Figure 12.27(b). They pictorially show the nature of excitation of the first harmonic with and without DGS.

A different concept to improve the stopband of a feed line was examined by Mandal et al. [52]. Their design includes a circular head open stub attached to the microstrip feed line, represented by black shading in Figure 12.28(a) and a pair of circular head dumbbell DGS, shown using gray shading, below the feed. Figures 12.28(b) shows a prototype operating at 3.12 GHz. Excellent rejection covering up to the fourth harmonic of fundamental resonance has been experimentally demonstrated.

12.3.3 Isolation at Microstrip Antenna Front-Ends

In the previous section, we saw how the bandstop characteristics of a DGS-integrated microstrip feed could be used to suppress higher harmonics for a microstrip antenna. A similar DGS-integrated feed was used to isolate the ports of a dual frequency dual polarized

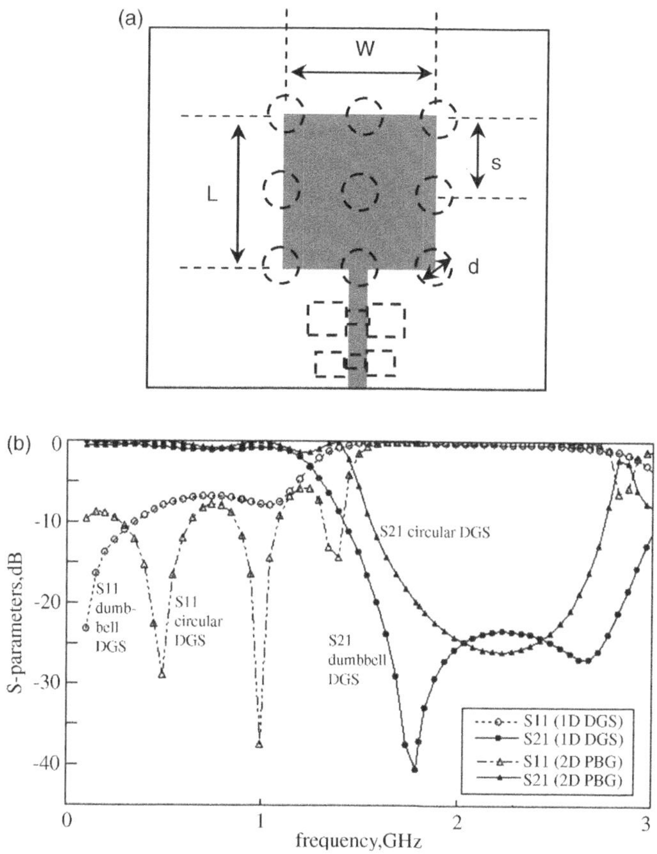

Figure 12.24 (a) Microstrip line-fed square patch antenna derived in [49] as a variant of that shown in Figure 12.23(a); DGS:15 mm × 15 mm and 10 mm × 10 mm dumbbell heads connected by 0.1 mm wide slots; other parameters as in Figure 12.23(a); (b) Transmission and return loss characteristics of a microstrip lines integrated with dumbbell-shaped DGS [1D as in Figure 12.24(a)] and circular DGS [2D as in Figure 12.23(a)]. Reproduced by permission of ©2005 IEEE [49]

rectangular patch in [53, 54]. The basic scheme is shown in Figure 12.29(a). The patches operate at 2.5 GHz in transmitting (Tx) mode and at 2.0 GHz in receiving (Rx) mode. The Rx port is connected to a low noise amplifier (LNA) and therefore needs proper isolation from the Tx port for signal purity and also to save the LNA from high power. This is accomplished using a spiral-shaped DGS integrated into the receiving port, which behaves like a conventional microstrip line at 2.0 GHz and as a bandpass filter around 2.5 GHz.

The Tx port is connected to a power amplifier through a microstrip line and therefore is prone to excitation with harmonics. To stop these unwanted frequencies above the dominant mode, a single dumbbell-shaped DGS is integrated with the feed line as was shown in the previous section. It allows up to 2.5 GHz and stops higher frequencies. Figures 12.29(b) and (c) show the photograph of a prototype and the measured transmission characteristics between the Tx and Rx ports, respectively. More than 20 dB isolation between the ports is evident at 2.5 GHz.

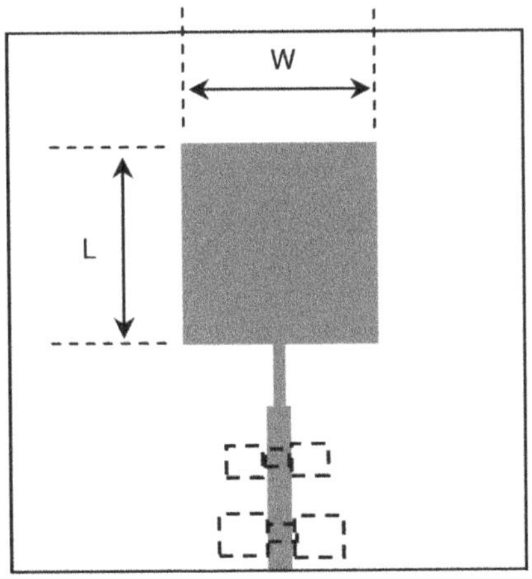

Figure 12.25 Microstrip patch fed by a DGS integrated microstrip line. Reproduced by permission of ©2002 IEEE [32]

12.3.4 Impedance Matching for Microstrip Feed Design

A microstrip line feeding a microstrip patch may be regarded as a two-port device whose image impedance [55] can easily be controlled by placing a DGS below it. We have learnt that the DGS dimensions modify the effective inductance and capacitance of the line and thus may help in achieving proper impedance matching.

An application is shown in Figure 12.30(a) [56], which actually represents a microstrip line-fed circularly polarized patch antenna. The white part indicates its metallized area on a copper-clad FR4 substrate and the gray part the etched out portions. A 33 mm × 30 mm patch is fed by two feed lines connected to the outputs of a 90° hybrid coupler. The patch dimensions support two close resonances at 2.5 GHz and 2.2 GHz respectively. One input of the coupler is connected to a 50 Ω SMA connector and the other one is terminated by a matched load.

The most challenging part is the dual feeding of the circularly polarized antenna at its edges where the input impedance is very high. So the concept of image impedance is used and two H-shaped DGSs are integrated to achieve impedance matching without any change in the feed dimensions. The basic physical concept behind this has already been discussed in Section 12.2.4 in connection with high impedance microstrip line. The measured input VSWR in Figure 12.30(b) shows excellent impedance matching. The antenna also provides optimum right-hand circularly polarized (RHCP) radiation near 2.3 GHz.

12.4 DGS to Control/Improve Radiation Properties of Microstrip Patch Antennas

12.4.1 Basic Idea

A microstrip patch antenna is actually an open resonator bounded by two electric walls at the top and the bottom. The ground plane serves as the bottom boundary and as such a defect on the

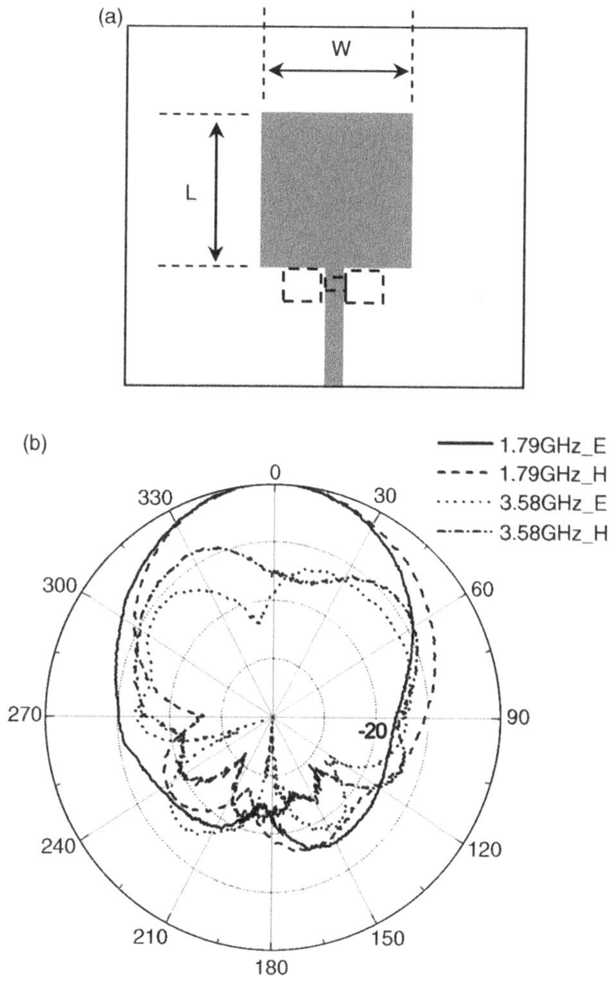

Figure 12.26 (a) Microstrip patch fed by a DGS integrated microstrip line; (b) Co-polarized radiation properties of the antenna at operating and first harmonic frequencies. Reproduced by permission of ©2005 IEEE [50]

ground plane under the patch should perturb the excited EM fields. This in turn should cause several changes in field profile or their orientation under the patch. This basic idea has been implemented to improve or modify the radiation properties of microstrip patches. Several applications are discussed in this section.

12.4.2 Suppression of Cross-Polarized Radiations from Microstrip Patches

Simple defects on the ground plane of a microstrip patch can be used to suppress the cross-polarized (XP) radiations normally occurring in any microstrip radiator. Guha et al. [57] first

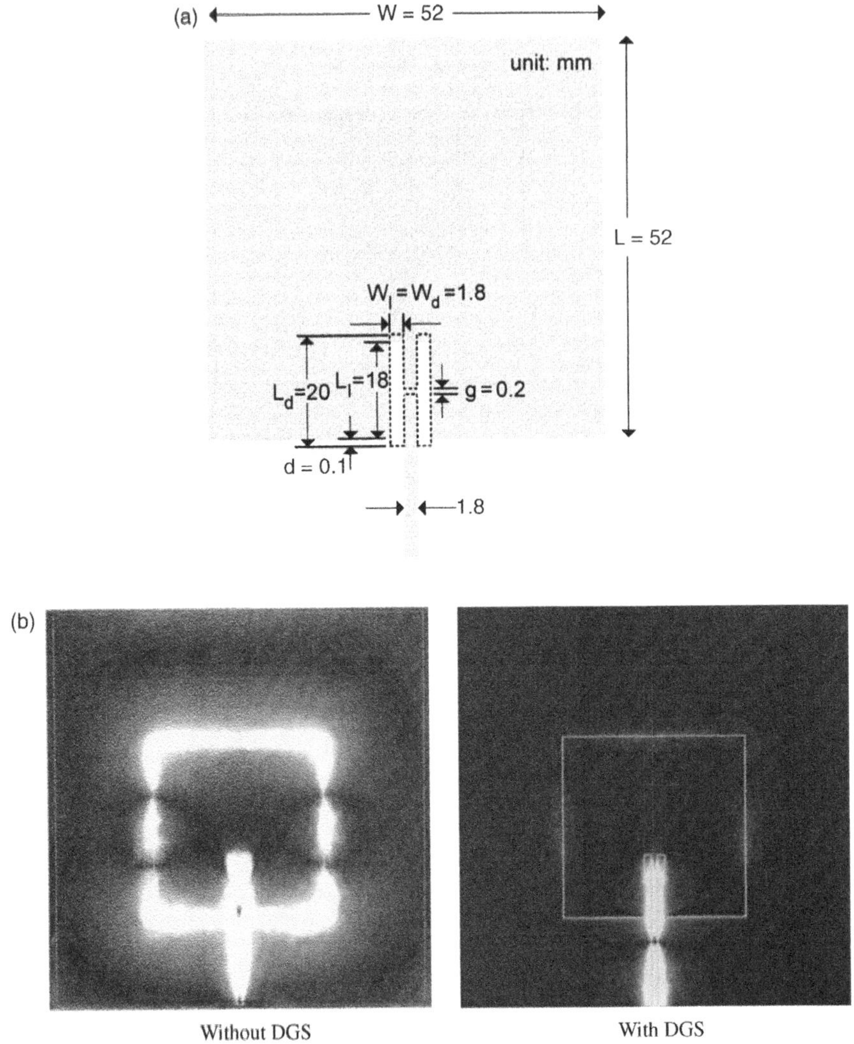

Figure 12.27 (a) A square patch fed by a microstrip line with H-shaped DGS. Substrate thickness is 0.508 mm and $\varepsilon_r = 2.2$ (b) Electric field distributions in the substrate with and without DGS obtained at the first harmonic. Reproduced by permission of ©2003 IEEE [51]

conceived of this idea and experimentally demonstrated it employing a pair of circular dot-shaped DGSs under a circular patch. The schematic diagram is shown in Figure 12.31(a), where the DGSs are strategically located on the H-plane of the antenna. The center of each circular defect is aligned with the patch boundary so that nearly half of the defect remains outside and the rest inside the patch, if looked vertically from top. This was initially conceived as a technique to perturb favorable condition of a resonance for a field in H-plane under the

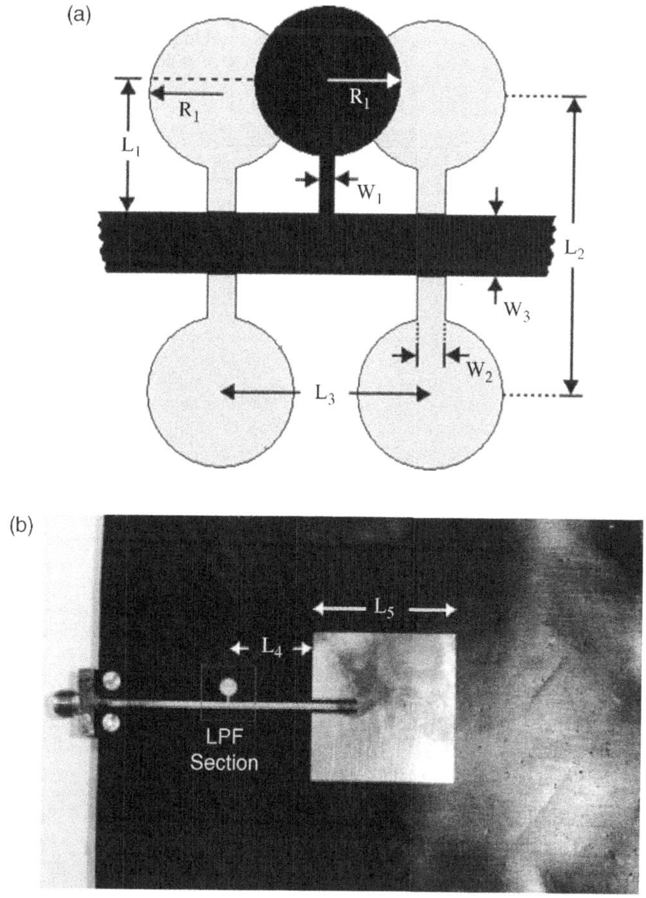

Figure 12.28 (a) Schematic diagram of a section of DGS integrated microstrip line, designed to feed a square patch operating at 3.12 GHZ. $W_1 = 0.5$, $W_2 = 0.5$, $W_3 = 1.2$, $L_1 = 3.4$, $L_2 = 6.0$, $L_3 \approx 5.0$, $R_1 = R_2 = 2.0$ (all dimensions in mm); (b) Photograph of a prototype etched on a 0.381 mm thick RT-duroid 5880 substrate. $L4 = 18.5$ mm and $L5 = 32.0$ mm. Reproduced by permission of ©2007 John Wiley & Sons Inc. [52]

patch. Usually a weak second order resonance in H-plane is responsible for producing the cross-polarized fields.

The size and location of the DGSs were so chosen that they could hardly affect the dominant resonant mode. Therefore the return loss behavior of the patch with and without DGS, shown in Figure 12.31(b), was examined and the defect diameter causing minimum deviation was chosen as the primary design parameter. The radiation characteristics of the DGS-integrated patch are compared with those due to an identical conventional antenna in Figure 12.31(c). Since the DGS is considerably distant from the major field variations of the dominant mode, almost no change in co-polarized patterns is observed. But the XP level is suppressed by about

(a)

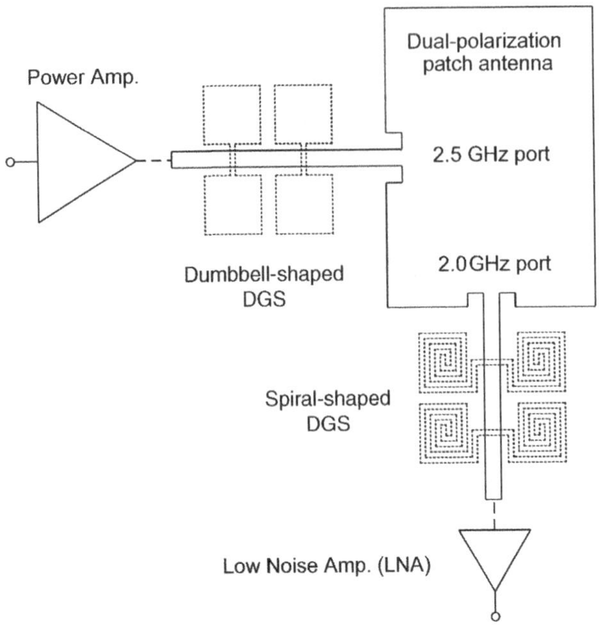

(b)

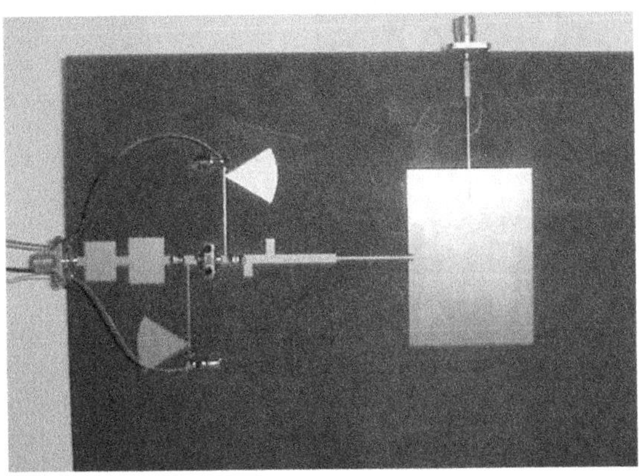

Figure 12.29 (a) DGS integrated RF front-end proposed for a dual polarized dual frequency microstrip antenna; (b) Photograph of a physically realized dual polarized dual frequency rectangular patch; (c) Measured S_{21} between the Tx and Rx ports showing isolation between them. Reproduced by permission of ©2004 IEEE [54]

5 dB. It is also interesting to note that, even for normal ground plane, the H-plane XP variations do not follow the usual pattern: a dip near 0°, then gradually increasing on either side showing peaks near 30°–40° and finally decreasing as azimuth increases. The E-plane XP values using normal ground plane are also considerably high. The reason is examined in Figure 12.31(d) with the help of simulated electric fields present in the substrate. A resonant field, set in orthogonally along the *y*-axis, is evident and its profile resembles the nature of the dominant

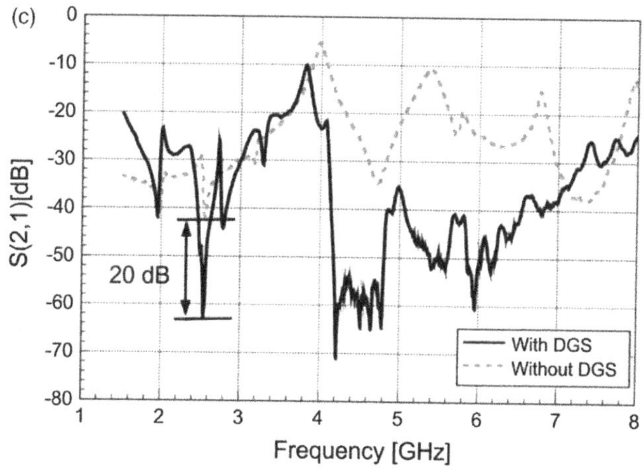

Figure 12.29 (*Continued*)

mode. This attributes to the high XP values occurring in the E-plane and also the change in the H-plane XP pattern. It is quite relevant to note again that usually the XP radiation is caused by the first higher mode, which in the present study is so weak that it is not pictorially visible. We will introduce a few more results later in this section. The varying nature of orthogonal fields causing XP radiations and the possibilities of their appearance have recently been studied in [58]. The DGS in Figure 12.31(d) effectively removes the trace of the orthogonal fields under the patch.

Further improvements in the XP radiations have been investigated by the same group using two new DGS geometries. They are in the form of an annular ring and circular arcs respectively, as depicted in Figure 12.32(a) and (b) [59, 60]. Here also the defect parameters are optimized carefully so that they minimally affect the dominant resonant mode. This is verified in Figure 12.33(a). The presence of DGS shows no change in the resonant frequency, other than the input impedance of the patch as indicated by the values of S_{11} minima.

Measured radiations [59, 60] show usual variations of XP in H-plane and indicate that the first higher mode is the producing factor. The ring-DGS [Figure 12.32(a)] improves the H-plane XP values by 5–7 dB, but there is no change in E-plane. The arc-shaped DGS is more effective as revealed from the measured and simulated data presented in Figure 12.33(b). The XP values are suppressed by about 12 dB, resulting in 30 dB isolation between the peak co- to cross-polarized levels. The E-plane XP level is normally 30 dB down and no further change is caused by the DGS.

The arc-DGS spreads over a wide area and thus can be favorably located to interact with the orthogonal fields, but without affecting the dominant mode. This is not possible with a ring-DGS. Thus an arc-DGS works more efficiently compared to a ring [Figure 12.32(a)] or circular dot [Figure 12.31(a)].

The performance of the DGS is valid over the entire operating bandwidth ($S_{11} < -10$ dB) and this is verified experimentally in Figure 12.33(c). The reduction in XP level is consistent at three different frequencies covering the entire impedance bandwidth.

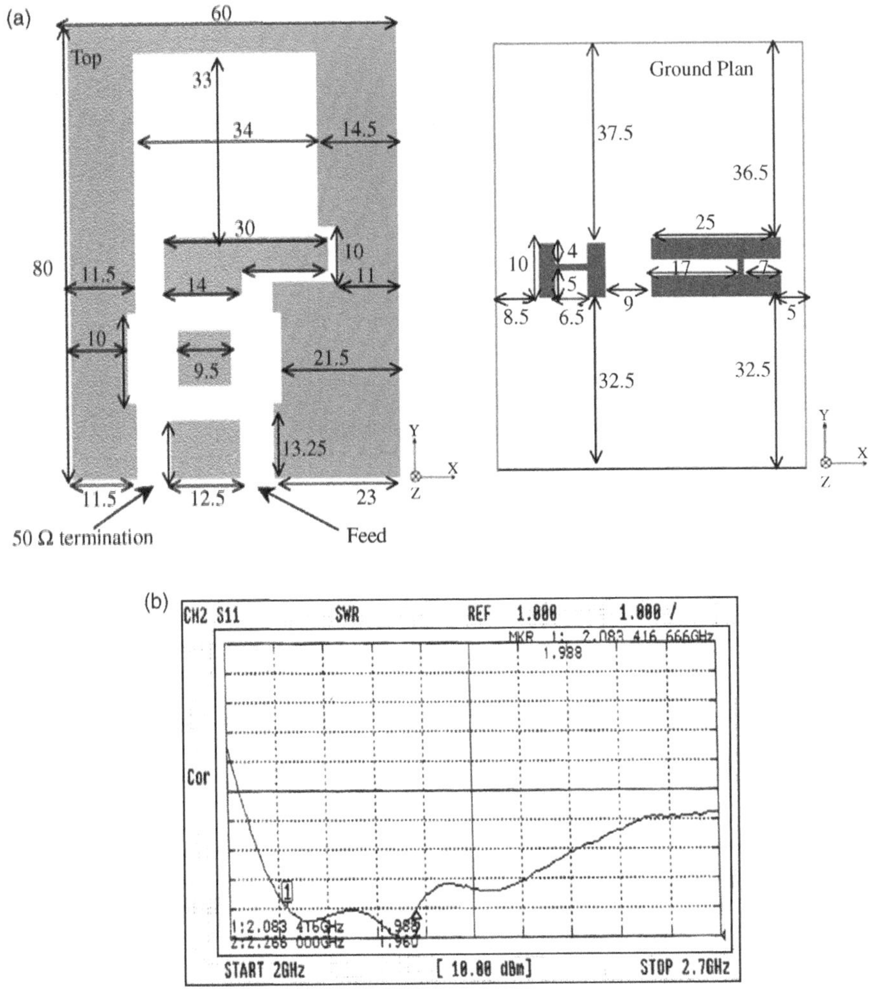

Figure 12.30 (a) Microstrip line-fed circularly polarized (CP) rectangular patch antenna. White shading: copper; Gray shading: etched out area. All dimensions are in mm. Substrate: 3.2 mm thick FR4 with $\varepsilon_r = 4.4$; (b) Measured VSWR of the CP antenna. Reproduced by permission of ©2006 IEEE [56]

12.5 DGS for Reduced Mutual Coupling Between Microstrip Array Elements and Associated Improvements

12.5.1 Basic Idea

So far, we have learned about DGS-integrated microstrip feeds and/or patches, where they directly control the individual element. But placing a DGS in between two microstrip patches also has some useful applications in reducing mutual coupling between them. Indeed, the surface waves propagating through the grounded substrate result in coupling between the

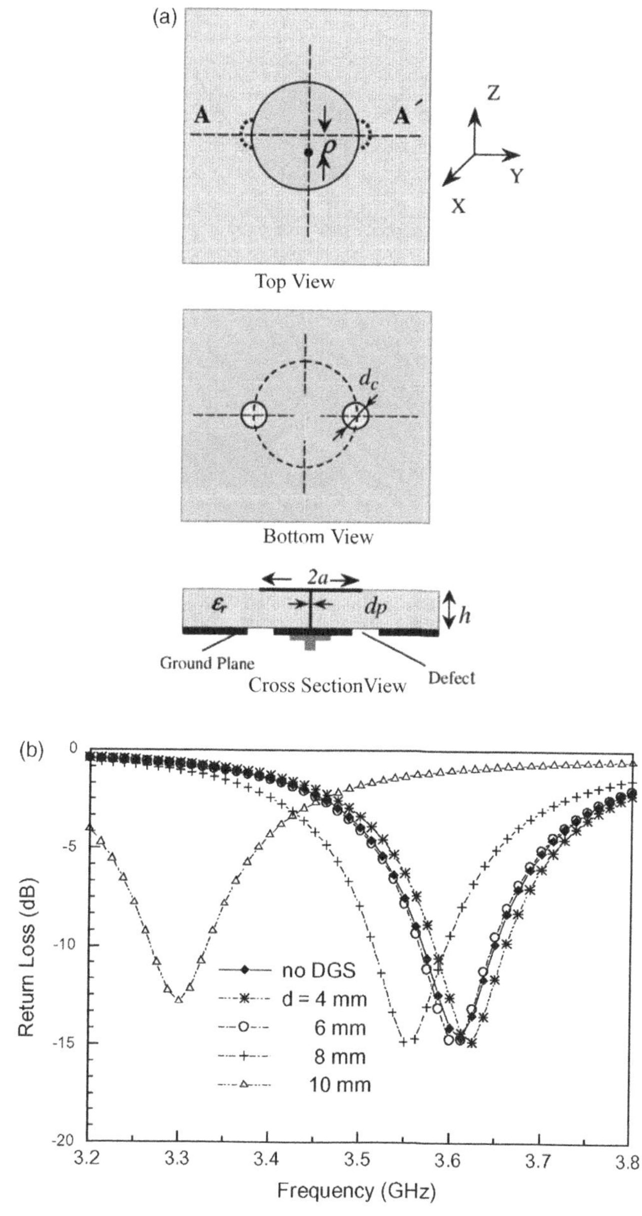

Figure 12.31 (a) Probe-fed circular microstrip patch with dot-shaped DGS; (b) Return loss characteristics of the DGS integrated circular patch for different DGS diameters. $a = 15$ mm, $h = 1.575$ mm, $d_p = 0.76$ mm, $\rho = 5$ mm, $\varepsilon_r = 2.32$, ground plane: $0.7\lambda \times 0.7\lambda$; (c) Measured radiation patterns with and without DGS obtained at $f = 3.6$ GHz; (d) Simulated electric fields under the patch at $f = 3.6$ GHz. Reproduced by permission of ©2005 IEEE [57]

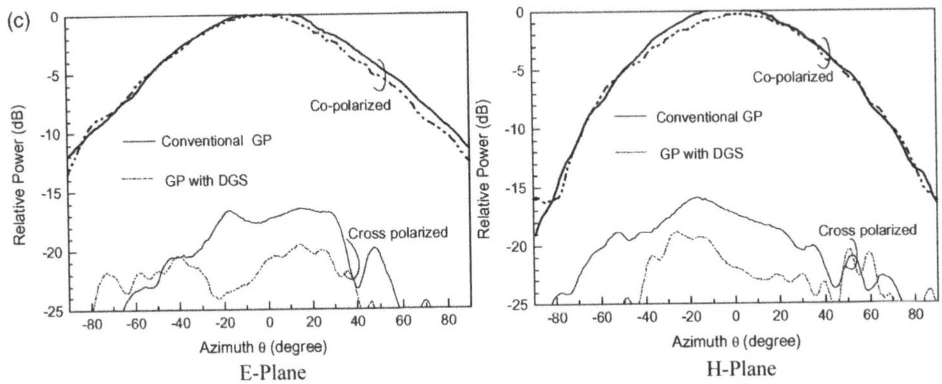

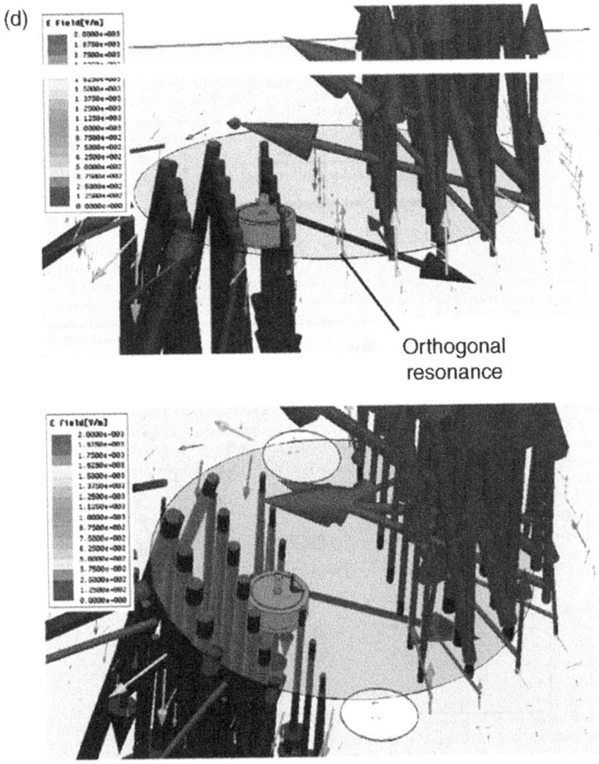

Figure 12.31 (*Continued*)

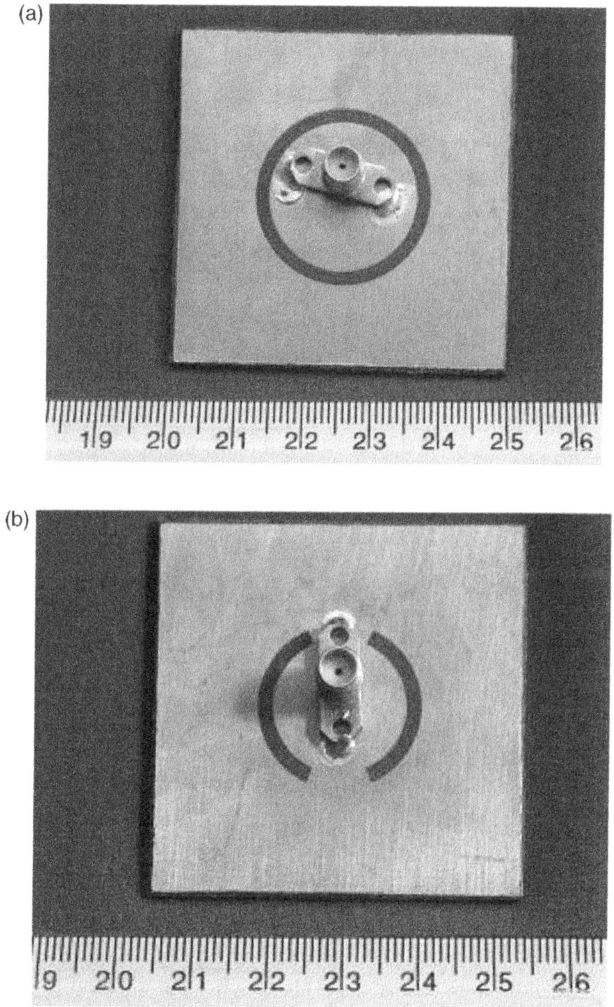

Figure 12.32 DGS shapes examined to reduce cross-polarized radiation of a circular patch; (a) ground plane with annular ring-shaped defect; (b) ground plane with arc-shaped defect. Reproduced by permission of ©2009 IEEE [59] and [60]

adjacent microstrips. This is highly detrimental to array performance such as high side lobe levels or pattern anomalies associated with scan blindness.

The bandstop feature of a DGS should also interact with surface waves propagating across it and help in reducing its intensity over certain frequency bands. This idea was first implemented by two research groups simultaneously in 2006 [19, 61] using two different geometries. Subsequently, this has been explored by different researchers with a view to improving the array performance, which are discussed in this section.

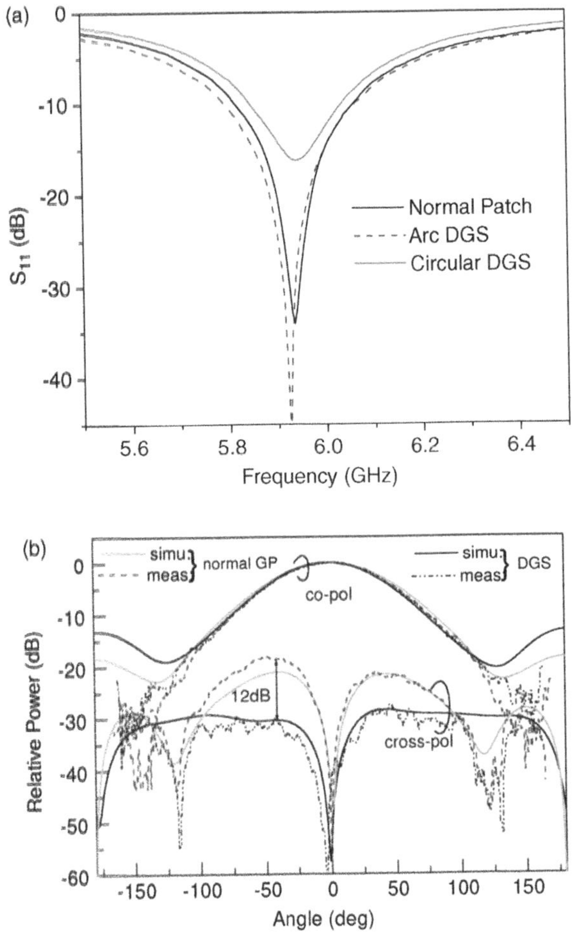

Figure 12.33 (a) Measured return loss of a circular patch using different ground plane configurations [59]. Patch radius = 9 mm; substrate height = 1.575 mm, $\varepsilon_r = 2.33$; probe diameter = 1.3 mm, feed location from center = 2.8 mm; DGS: inner radius = 9 mm, outer radius = 2 mm, optimum angle subtended by arc = 66°; ground plane: $1\lambda \times 1\lambda$; (b) H-plane radiation patterns of a circular patch with arc-DGS [Figure 12.32(b)] compared with those due to conventional ground plane at $f = 5.93$ GHz [60]. Parameters as in Figure 12.33(a); (c) H-plane cross-polarized radiations of a circular patch with arc-DGS occurring at three different frequencies [60]. Parameters as in Figure 12.33(a). Reproduced by permission of ©2009 IEEE [60]

12.5.2 Ring-Shaped DGS for Circular Patch Array

Guha et al. [19] proposed the idea of using a DGS to reduce mutual coupling between two microstrip patches in connection with developing concentric ring-shaped DGSs for microstrip applications. The ring DGS configurations were already shown in Figure 12.4. The

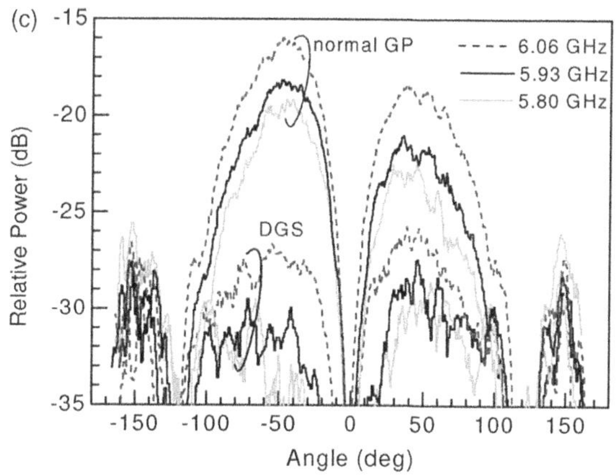

Figure 12.33 *(Continued)*

proposed geometry for two-element E-plane circular patch array is shown in Figure 12.34(a) [19], and it uses a single circular ring surrounding any of two circular patches. Figure 12.34(b) shows a prototype etched on a 62 mil Taconic substrate with $\varepsilon_r = 2.3$ and connected to a network analyzer. The measured S_{21} against frequency is shown in Figure 12.34(c) [65]. About 4–5 dB reduction in mutual coupling near the operating frequency $f = 10$ GHz is evident.

For an H-plane array, DGS also helps in reducing mutual coupling, but the effect is not that significant. However, a ring DGS can serve altogether maximum five elements: one located centrally surrounded by the ring, two on both of its sides along E-plane and two on both of its sides along H-plane. In such a configuration, the DGS does not affect the radiation properties of individual element. Only that one, surrounded by DGS, may have benefited in terms of H-plane XP radiations.

A similar ring DGS was also tested experimentally to reduce mutual coupling between two cylindrical Dielectric Resonator Antennas [63]. An identical response in reducing the mutual coupling was documented.

12.5.3 Dumbbell-Shaped DGS for Rectangular Patch Array

The DGS configuration shown in Figure 12.34(a) is not symmetrically located with respect to either patch. A more symmetric and simple configuration was examined simultaneously with [19] by Salehi et al. [61, 64] employing dumbbell-shaped DGS for a two-element rectangular microstrip array. Figure 12.35(a) shows the configuration examined in [61, 64] using a high dielectric constant substrate. Higher dielectric constant usually supports a larger amount of surface waves and as such a DGS shows significant effect on the transmission or coupling properties. The simulated result is very promising, as indicated in Figure 12.35(b), particularly when the separation between the patches is nearly half of the resonant wavelength. A similar configuration has been employed later when designing sensors for landmine detection [65].

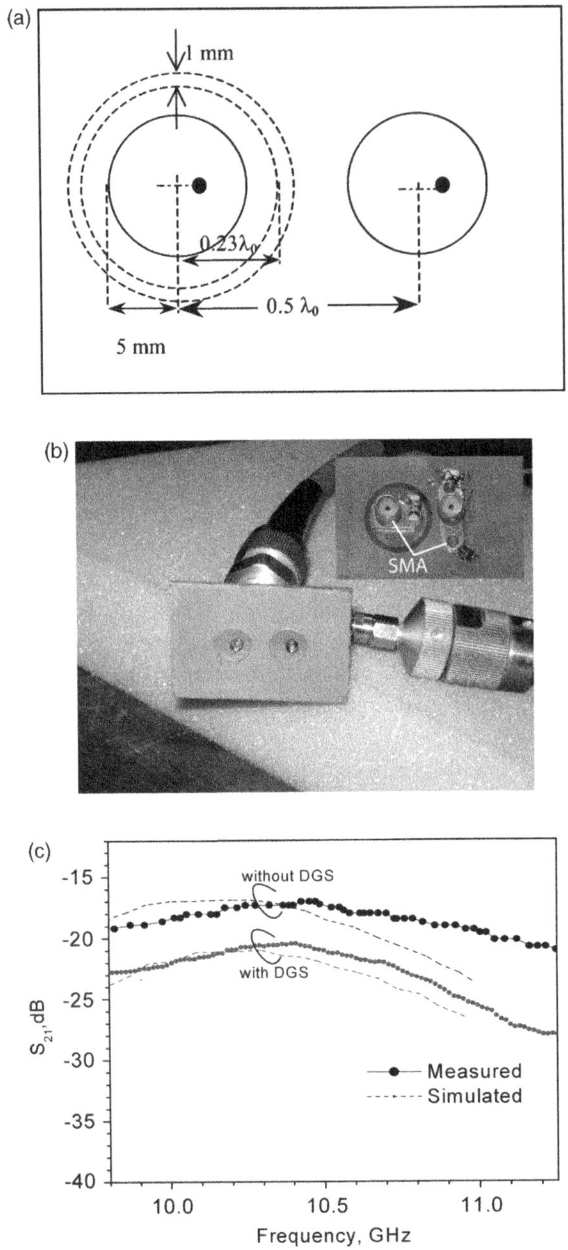

Figure 12.34 (a) Single ring-shaped DGS for a two-element E-pane microstrip patch array. Reproduced by permission of ©2006 IEEE [19]; (b) A prototype, operating at 10.4 GHz, under test. Inset figure: photograph taken from ground plane side. Reproduced by permission of ©2009 IEEE [62]; (c) Measured and simulated S_{21} versus frequency. Reproduced by permission of ©2009 IEEE [62]

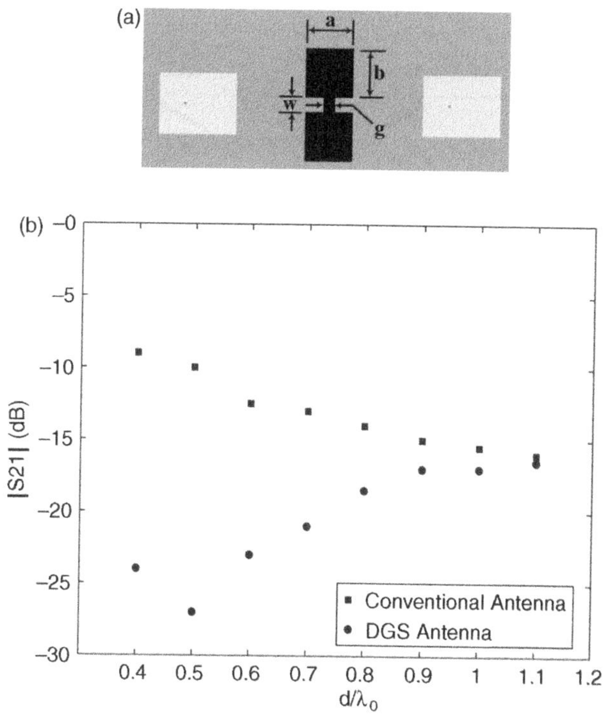

Figure 12.35 (a) Single dumbbell-shaped DGS for a two-element rectangular patch array; (b) Mutual coupling between the rectangular patches versus separation between them normalized with respect to resonant wavelength. Reproduced by permission of ©2006 Elsevier [61]

Multiple dumbbell-shaped DGS cells have been investigated recently [66] for a two-element multiband array. The overall configuration is a bit different. Figure 12.36(a) [66] shows two microstrip elements electromagnetically coupled to microstrip feed etched on the bottom layer. The defects are located on the ground plane of the printed feed line. Each radiating element is of composite nature operating at 2.3 GHz, 3.3 GHz, and 5.8 GHz. Four DGS cells are symmetrically located on the ground plane and here, they are aligned along the H-plane of the antenna elements. The cell numbers are optimized based on simulation studies. It should be restricted to a limited number from the practical aspect. One set of representative results obtained at 3.3 GHz is depicted in Figure 12.36(b) showing the change in mutual coupling with the number of DGS cells. An increase in the cell number from 1 to 4 gradually improves the suppression level, but with 5 cells, it drastically changes the situation. Measured results showed an improvement up to 5 dB using 4 DGS cells.

12.5.4 *Elimination of Scan Blindness of Microstrip Phased Array*

Scan blindness is an undesirable feature in the radiation characteristics of a phased array antenna. This actually means a decrease in gain at specific frequencies and over specific angles;

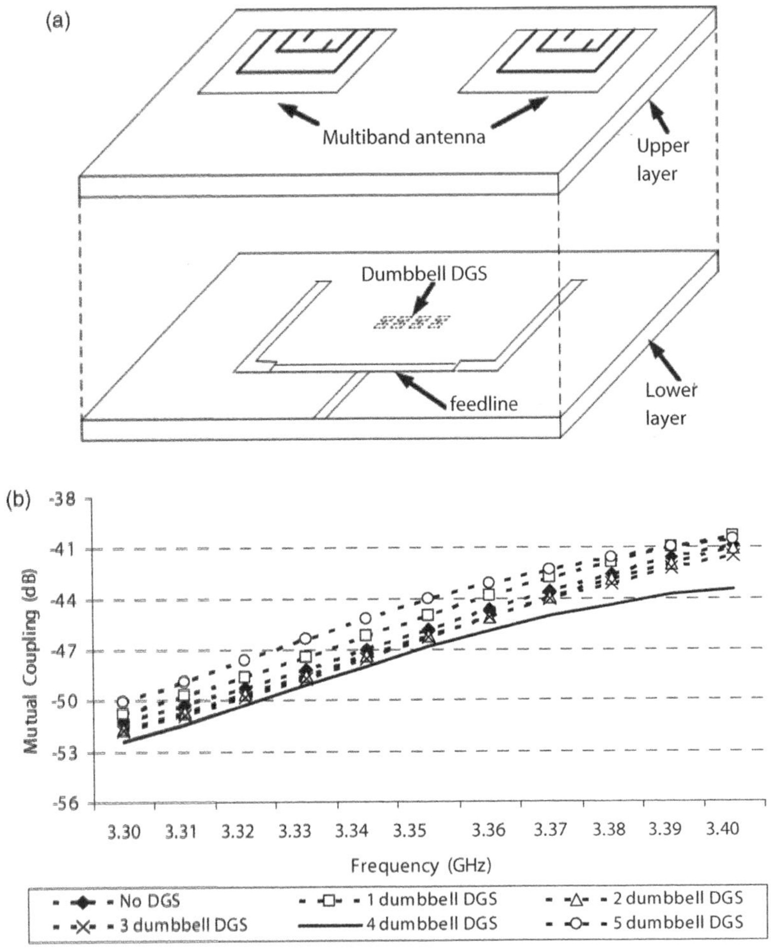

Figure 12.36 (a) Schematic diagram of multiple DGSs applied to a two-element electromagnetically coupled multiband rectangular patch array; (b) Mutual coupling between the patch elements for different numbers of DGS cells around resonance frequency. Reproduced by permission of ©2010 EMW Publishing [66]

and therefore, the antenna array becomes incapable of functioning in those directions. The mutual coupling between the array elements is the primary reason behind this phenomenon. Minimizing this effect, therefore, is of great practical interest and some researchers have employed EBG structures earlier [67, 68] to handle this shortcoming.

In 2007, Moghadas et al. [69] first used DGS in a six-element microstrip array as depicted in Figure 12.37(a). The geometry looks like an extension of Figure 12.35(a), examined in [61]. The array uses a high dielectric constant substrate with $\varepsilon_r = 10.2$ and operates near 12 GHz. The scan characteristics of the array are shown in Figure 12.37(b). Normalized power with respect to that radiated along the bore sight is plotted against angular variation. The scan

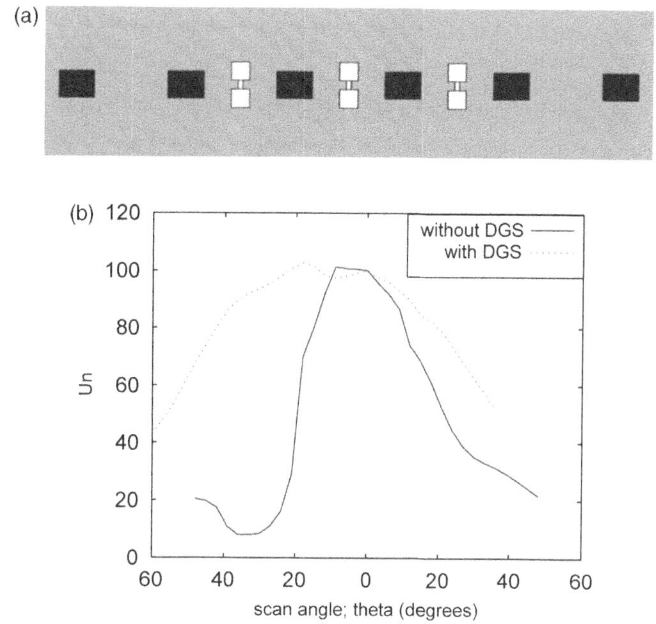

Figure 12.37 (a) Six element array of rectangular patches using three dumbbell-shaped DGSs in the interleaving space; (b) Scan characteristics of the array etched on a high dielectric constant ($\varepsilon_r = 10.2$) substrate. Reproduced by permission of ©2008 Elsevier [69]

blindness is apparent near 33° when the array uses a conventional ground plane. The design using DGS removes the dip or the no-gain angle from the radiation pattern. The element to element mutual coupling with and without DGS also showed an interesting result: the minimum reduction was 5 dB, occurring between two extreme elements #1 and #6, the maximum being 12 dB between elements #2 and #5.

Another investigation explored dumbbell DGS with an H-shaped head [70] as shown in Figure 12.38(a). This also uses $\varepsilon_r = 10.2$ substrate for a two-element array operating near 5.3 GHz. The E-plane radiation patterns shown in Figure 12.38(b) are self-explanatory. It indicates considerable improvements in the null and side lobe performances. The H-plane pattern is free from scan blindness and as such will not be discussed here. The same configuration extended to a 7×3 array was theoretically examined in [70]. The active element pattern of the central radiator showed prominent blind spots without DGS, but achieved a ripple-free no null pattern when the DGS was considered in the design.

12.6 Conclusion

Since its inception, the DGS has been popular mainly for printed circuit applications. Initial DGS applications for microstrip antennas exploited the filter properties associated with the microstrip feed line. DGS-integrated patch antenna was first examined in 2005 to improve the

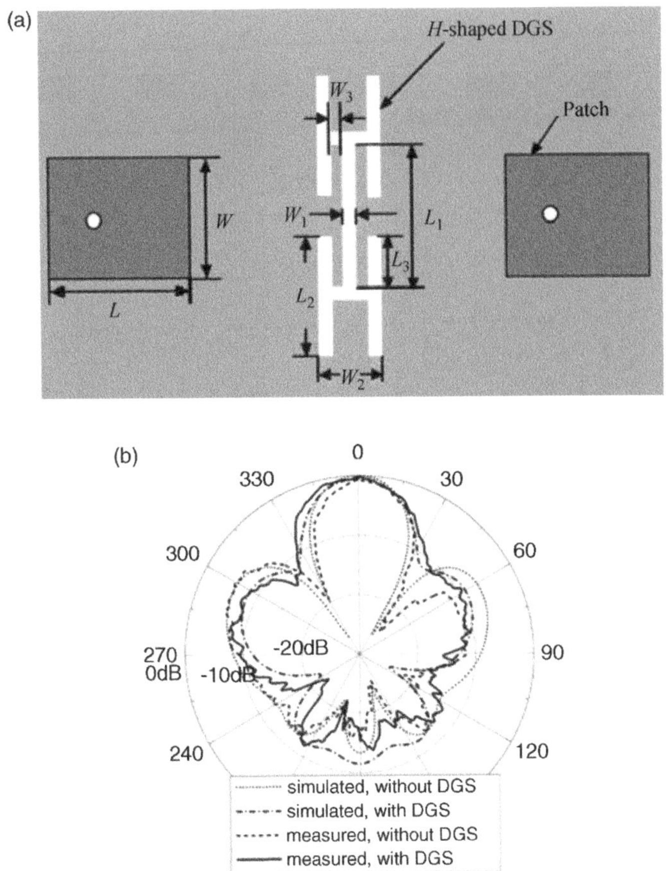

Figure 12.38 (a) Two-element microstrip array with single H-shaped dumbbell DGS; (b) E-plane co-polarized radiation patterns of fully excited array. Reproduced by permission of ©2009 The IET [70]

radiation properties of a single element. After that, during the next five years, the DGS has been explored in different shapes and forms. A comprehensive review indicating the major developments and applications of DGS has been presented here. The number of investigations focusing on DGS for antennas is very limited, although it is quite a fertile area of research in view of the need for compact and efficient wireless antennas. The scope for further investigations, therefore, appears to be very bright in a true sense.

Appendix: A Brief DGS Chronology

1998–99 PBG as defected ground structure for harmonic control of printed antennas [47, 48]

2000 The term "DGS" in place of PBG/EBG used to describe dumbbell-shaped defects [13]

2000	DGS for designing printed branch-line coupler [45]
2001	Dumbbell-shaped DGS analyzed and employed to design Low Pass Filter [30].
2002	DGS to achieve reduced size circuits [46]
2005	DGS to suppress cross-polarized radiation of a microstrip antenna [57]
2006	DGS to reduce mutual coupling between two microstrip array elements [19, 61]
2008	DGS to eliminate scan blindness of microstrip array antenna [69]

References

1. Y. Rahmat-Samii and H. Mosallaei, "Electromagnetic band-gap structures: classification, characterization and applications," *Proceedings of IEE-ICAP Symposium*, vol. 2, pp. 5601–5644, April 2001.
2. E. Yablonovitch, "Inhibited spontaneous emission in solid-state physics and electronics," *Physical Review Letters*, vol. 58, no. 20, pp. 2059–2062, 1987.
3. S. John, "Strong localization of photons in certain disordered dielectric superlattices," *Physical Review Letters*, vol. 58, no. 23, pp. 2486–2489, 1987.
4. A. S. Barlevy and Y. Rahmat-Samii, "Characterization of electromagnetic band-gaps composed of multiple periodic tripods with interconnecting vias: concept analysis, and design," *IEEE Trans. Antennas Propagat.*, vol. 49, no. 3, pp. 343–353, March 2001.
5. E. Ozbay, A. Abeyta, G. Tuttle, M. Tringides, R. Biswas, T. Chan, C. M. Soukoulis, and K. M. Ho, "Measurement of a three-dimensional photonic band gap in a crystal structure made of dielectric rods," *Phys. Rev. B, Condens. Matter*, vol. 50, no. 3, pp. 1945–1948, July. 1994.
6. H. Contopanagos, L. Zhang, and N. G. Alexopoulos, "Thin frequency-selective lattices integrated in novel compact MIC, MMIC, and PCA architectures," *IEEE Trans. on Microwave Theory and Techniques*, vol. 46, no. 11, pp. 1936–1948 1998.
7. F. Yang and Y. Rahmat-Samii, "Microstrip antennas integrated with electromagnetic band-gap (EBG) structures: a low mutual coupling design for array applications," *IEEE Trans. Antennas Propagat.*, vol. 51, no. 10, 2936–2946, Oct. 2003.
8. V. Radisic et al., "Novel 2-D photonic bandgap structure for microstrip lines," *IEEE Microwave and Guided Wave Letters*, vol. 8, no. 2, pp. 69–71, Feb. 1998.
9. D. Sievenpiper, L. Zhang, R. F. J. Broas, N. G. Alexopolus, and E. Yablonovitch, "High-impedance electromagnetic surfaces with a forbidden frequency band," *IEEE Trans. Microwave Theory Tech.*, vol. 47, no. 11, pp. 2059–2074, Nov. 1999.
10. I. Rumsey, et al. "Photonic bandgap structure used as filters in microstrip circuits," *IEEE Microwave and Guided Wave Letters*, vol. 8, no. 10, pp. 336–338. Oct. 1998.
11. C. C. Chiau, X. Chen, and C. Parini, "Multiperiod EBG structure for wide stopband circuits," *IEE Proc. Microw. Antennas Propag.*, vol. 150, no. 6, pp. 489–492, Dec. 2003.
12. J. I. Park, C. S. Kim, J. Kim, J. S. Park, Y. Qian, D. Ahn, and T. Itoh, "Modeling of a photonic bandgap and its application for the low pass filter design," *Proceedings Asia Pacific Microwave Conference*, pp. 331–334, Dec. 1999.
13. C. S. Kim, J. S. Park, D. Ahn, and J. B. Lim, "A novel 1-D periodic defected ground structure for planar circuits," *IEEE Microwave Wireless Components. Lett.*, vol. 10, no. 4, pp. 131–133, Apr. 2000.
14. A. B. Abdel-Rahman, A. K. Verma, A. Boutejdar, and A. S. Omar, "Control of bandstop response of hi-lo microstrip low-pass filter using slot in ground plane," *IEEE Trans. Microwave Theory Tech.*, vol. 52, no. 3, pp. 1008–1013, Mar. 2004.
15. C. S. Kim, J. S. Lim, S. Nam, K. Y. Kang, and D. Ahn, "Equivalent circuit modelling of spiral defected ground structure for microstrip line," *Electronic Lett.*, vol. 38, no. 19, pp. 1109–1110, Sept. 2002.
16. D. J. Woo, T. K. Lee, J. W. Lee, C. S. Pyo, and W. K. Choi, "Novel U-slot and V-slot DGSs for bandstop filter with improved Q factor," *IEEE Trans. Microwave Theory Tech.*, vol. 54, no. 6, pp. 2840–2847, June 2006.
17. M. K. Mandal and S. Sanyal, "A novel defected ground structure for planar circuits," *IEEE Microwave Wireless Components. Lett.*, vol. 16, no. 2, pp. 93–95, Feb. 2006.
18. H. J. Chen, T. H. Huang, C. S. Chang, L. S. Chen, N. F. Wang, Y. H. Wang, and M. P. Houng, "A novel cross-shape DGS applied to design ultra-wide stopband low-pass filters," *IEEE Microwave Wireless Components. Lett.*, vol. 16, no. 5, pp. 252–254, May 2006.

19. D. Guha, S. Biswas, M. Biswas, J. Y. Siddiqui, and Y. M. M. Antar, "Concentric ring-shaped defected ground structures for microstrip applications," *IEEE Antennas Wireless Propagation Lett.*, vol. 5, pp. 402–405, 2006.
20. Z. Z. Hou, "Novel wideband filter with a transmission zero based on split-ring resonator DGS," *Microwave Opt. Technol. Lett.*, vol. 50, no. 6, pp. 1691–1693, Mar 2008.
21. S. N. Burokur, M. Latrach, and S. Toutain, "A novel type of microstrip coupler utilizing a slot split ring resonators defected ground plane," *Microwave Opt. Technol. Lett.*, vol. 48, no. 1, pp. 138–141, Nov. 2005.
22. H. W. Liu, Z. F. Li, and X. W. Sun, "A novel fractal defected ground structure and its application to the low-pass filter," *Microwave Opt. Technol. Lett.*, vol. 39, no. 6, pp. 453–456, Dec. 2003.
23. A. Balalem, A. R. Ali, J. Machac, and A. Omar, "Quasi-elliptic microstrip low-pass filters using an interdigital DGS slot," *IEEE Microwave Wireless Components. Lett.*, vol. 17, no. 8, pp. 586–588, Aug. 2007.
24. A. Boutejdar, M. Makkey, A. Elsherbini, O. Luxor, and A. Omar, "Design of compact extended-stopband microstrip low-pass filters by employing new mutual-coupling technique for defected ground structures (DGSs)," *Proceedings of the 37th European Microwave Conference*, pp. 71–74, Oct. 2007.
25. A. Boutejdar, A. Sherbini, W. Ali, S. Fouad, L. Ahmed, and A. Omar, "Design of compact microstrip lowpass filters using coupled half-circle defected ground structures (DGSs)," *IEEE Antennas and Propagation Society International Symposium*, pp. 1–4, July 2008.
26. J. S. Hong and B. M. Karyamapudi, "A general circuit model for defected ground structures in planar transmission lines," *IEEE Microwave Wireless Components. Lett.*, vol. 15, no. 10, pp. 706–708, Oct. 2005.
27. H. W. Liu, Z. F. Li, X. W. Sun, and J. F. Mao, "An improved 1-D periodic defected ground structure for microstrip line," *IEEE Microwave Wireless Components. Lett.*, vol. 14, no. 4, pp. 180–182, Apr. 2004.
28. J. S. Lim, Y. T. Lee, C. S. Kim, D. Ahn, and S. Nam, "A vertically periodic defected ground structure and its application in reducing the size of microwave circuits," *IEEE Microwave Wireless Components. Lett.*, vol. 12, no. 12, pp. 479–481, Dec. 2002.
29. C. Caloz, H. Okabe, T. Iwai, and T. Itoh, "A simple and accurate model for microstrip structures with slotted ground plane," *IEEE Microwave Wireless Components. Lett.*, vol. 14, no. 4, pp. 133–135, Apr. 2004.
30. D. Ahn, J. S. Park, C. S. Kim, J. Kim, Y. Qian, and T. Itoh, "A design of the low-pass filter using the novel microstrip defected ground structure," *IEEE Trans. Microwave Theory Tech.*, vol. 49, no. 1, pp. 86–93, Jan. 2001.
31. J. S. Hong and M. J. Lancaster, *Microstrip Filters for RF/Microwave Applications*, John Wiley & Sons, Inc., New York, 2007.
32. I. Chang and B. Lee, "Design of defected ground structures for harmonic control of active microstrip antenna," *IEEE Antennas and Propagation Society International Symposium*, vol. 2, pp. 852–855, 2002.
33. J. S. Park, J. H. Kim, J. H. Lee, S. H. Kim, and S. H. Myung, "A novel equivalent circuit and modeling method for defected ground structure and its application to optimization of a DGS lowpass filter," *IEEE MTT-S Int. Microwave Symp. Digest*, vol. 1, 417–420, 2002.
34. N. C. Karmakar, S. M. Roy, and I. Balbin, "Quasi-static modeling of defected ground structure," *IEEE Trans. Microwave Theory Tech.*, vol. 54, no. 5, pp. 2160–2168, May 2006.
35. J. S. Park, J. S. Yun, and D. Ahn, "A design of the novel coupled-line bandpass filter using defected ground structure with wide stopband performance," *IEEE Trans. Microwave Theory Tech.*, vol. 50, no. 9, pp. 2037–2043, Sept. 2002.
36. D. Piscarreta and S. W. Ting, "Microstrip parallel coupled line bandpass filter with selectivity improvement using U-shaped defected ground structure," *Microwave Opt. Technol. Lett.*, vol. 50, no. 4, pp. 911–915, Apr. 2008.
37. J. G. García, F. Martín, F. Falcone, J. Bonache, I. Gil, T. Lopetegi, M. A. G. Laso, M. Sorolla, and R. Marqués, "A spurious passband suppression in microstrip coupled line band pass filters by means of split ring resonators," *IEEE Microwave Wireless Components. Lett.*, vol. 14, no. 9, pp. 416–418, Sept. 2004.
38. X. H. Wang, B. Z. Wang, H. Zhang, and K. J. Chen, "A tunable bandstop resonator based on a compact slotted ground structure," *IEEE Trans. Microwave Theory Tech.*, vol. 55, no. 9, pp. 1912–1918, Sept. 2007.
39. A. M. E. Safwat, F. Podevin, P. Ferrari, and A. Vilcot, "Tunable bandstop defected ground structure resonator using reconfigurable dumbbell-shaped coplanar waveguide," *IEEE Trans. Microwave Theory Tech.*, vol. 54, no. 9, pp. 3559–3564, Sept. 2006.
40. H. Liu, T. Yoshimasu, S. Kurachi, J. Chen, Z. Li, and X. Sun, "A novel microstrip diplexer design using defected ground structure," *Proceedings International Conference on Communication, Circuits & Systems*, vol. 2, pp. 1099–1100, May 2005.

41. Y. J. Sung, C. S. Ahn, and Y.-S. Kim, "Size reduction and harmonic suppression of rat-race hybrid coupler using defected ground structure," *IEEE Microwave Wireless Components. Lett.*, vol. 14, no. 1, pp. 7–9, Jan. 2004.
42. D. K. Woo and T. K. Lee, "Suppression of harmonics in Wilkinson power divider using dual-band rejection by asymmetric DGS," *IEEE Trans. Microwave Theory Tech.*, vol. 53, no. 6, pp. 2139–2144, June 2006.
43. Y. C. Jeong, S. G. Jeong, J. S. Lim, and S. Nam, "A new method to suppress harmonics using $\lambda/4$ bias line combined by defected ground structure in power amplifiers," *IEEE Microwave Wireless Components. Lett.*, vol. 13, no. 12, pp. 538–540, Dec. 2003.
44. J. S. Lim, S. W. Lee, C. S. Kim, J. S. Park, D. Ahn, and S. Nam, "A4: 1 unequal Wilkinson power divider," *IEEE Microwave Wireless Components. Lett.*, vol. 11, no. 3, pp. 124–126, Mar. 2001.
45. J. S. Lim, C. S. Kim, J. S. Park, D. Ahn, and S. Nam, "Design of 10dB 90^0 branch line coupler using microstrip line with defected ground structure," *Electronic Lett.*, vol. 36, no. 21, pp. 1784–1785, Oct. 2000.
46. J. S. Lim, J. S. Park, Y. T. Lee, D. Ahn, and S. Nam, "An application of defected ground structure in reducing the size of amplifiers," *IEEE Microwave Wireless Components. Lett.*, vol. 12, no. 7, pp. 261–263, Mar. 2002.
47. Y. Horri and M. Tsutsumi, "Harmonic control by photonic bandgap on microstrip patch antenna," *IEEE Microwave Guided Lett.*, vol. 9, pp. 13–15, Jan. 1999.
48. V. Radisic, Y. Qian, and T. Itoh, "Broadband power amplifier integrated with slot antenna and novel harmonic tuning structure," in *1998 IEEE MTT-S Dig.*, pp. 1895–1898, 1998.
49. H. Liu, Z. Li, X. Sun, and J. Mao, "Harmonic suppression with photonic bandgap and defected ground structure for a microstrip patch antenna," *IEEE Microw. Compon. Lett.*, vol. 15, no. 2, pp. 55–56, Feb. 2005.
50. Y. J. Sung and Y.-S. Kim, "An improved design of microstrip patch antennas using photonic bandgap structure," *IEEE Trans. Antennas Propagat.*, vol. 53, no. 5, pp. 1799–1803, May, 2005.
51. Y. J. Sung, M. Kim, and Y.-S. Kim, "Harmonic reduction with defected ground structure of a microstrip patch antenna," *IEEE AntennasWireless Propag. Lett.*, vol. 2, pp. 111–113, 2003.
52. M. K. Mandal, P. Mondal, S. Sanyal, and A. Chakrabarty, "An improved design of harmonic suppression for microstrip patch antennas," *Microwave and Opt. Technol. Lett.*, vol. 49, no. 1, pp. 103–105, Jan. 2005.
53. Y. Chung, S. S. Jeon, D. Ahn, J. I. Choi, and T. Itoh, "High isolation dual polarized patch antenna using integrated defected ground structure," *IEEE Microw. Compon. Lett.*, vol. 14, no. 1, pp. 4–6, Jan. 2004.
54. Y. Chung, S. S. Jeon, S. Kim, D. Ahn, J. I. Choi, and T. Itoh, "Multifunctional microstrip transmission lines integrated with defected ground structure for RF front-end application," *IEEE Trans. Microwave Theory Tech.*, vol. 52, no. 5, pp. 1425–1432, May 2004.
55. Image Impedance: Derivation: http://en.wikipedia.org/wiki/Image_impedance.
56. J. P. Thakur and J. Park, "An advance design approach for circular polarization of the microstrip antenna with unbalance DGS feedlines," *IEEE Antennas Wireless Propag. Lett.*, vol. 5, pp. 101–103, 2006.
57. D. Guha, M. Biswas, and Y. M. M. Antar, "Microstrip patch antenna with defected ground structure for cross polarization suppression," *IEEE Antennas Wireless Propag. Lett.*, vol. 4, pp. 455–458, 2005.
58. C. Kumar and D. Guha, "A new look into the cross-polarized radiation form of a circular microstrip antenna and suppression using dot-shaped DGS," accepted in" *IEEE AP-S Symp.*, Toronto, 2010.
59. C. Kumar and D. Guha, "New defected ground structures (DGSs) to reduce cross-polarized radiation of circular microstrip antennas," *Applied Electromagnetics Conf. AEMC 2009*, pp. 1–4, Kolkata, India, 2009.
60. D. Guha, C. Kumar, and S. Pal, "Improved cross-polarization characteristics of circular microstrip antenna employing arc-shaped defected ground structure (DGS)," *IEEE Antennas Wireless Propag. Lett.*, vol. 8, pp. 1367–1369, 2009.
61. M. Salehi and A. Tavakoli, "A novel low mutual coupling microstrip antenna array design using defected ground structure," *Int. J. Electron Commun.*, vol. 60, pp. 718–723, 2006.
62. D. Guha, S. Biswas, and C. Kumar, "Annular ring shaped DGS to reduce mutual coupling between two microstrip patches," *Applied Electromagnetics Conf. AEMC 2009*, pp. 1–4, Kolkata, India, 2009.
63. D. Guha, S. Biswas, T. Joseph, and M. T. Sebastian, "Defected ground structure to reduce mutual coupling between cylindrical dielectric resonator antennas," *Electronic Lett.*, vol. 44, no. 14, pp. 836–837, July 2008.
64. M. Salehi, A. Motevasselian, A. Tavakoli, and T. Heidari, "Mutual coupling reduction of microstrip antennas using defected ground structure," *10th IEEE International Conference on Communication Systems (ICCS 2006)*, pp. 1–5, Oct. 2006.
65. S. H. Zainud-Deen, M. E. S. Badr, E. El-Deen, K. H. Awadalla, and H. A. Sharshar, "Microstrip antenna with defected ground plane structure as a sensor for landmines detection," *Progress in Electromagnetics Research B*, vol. 4, pp. 27–39, 2008.

66. F. Y. Zulkili, E. T. Rahardjo, and D. Hartanto, "Mutual coupling reduction using dumbbell defected ground structure for multiband microstrip antenna array," *Progress in Electromagnetics Research Letters*, vol. 13, pp. 29–40, 2010.
67. Y. Fu and Y. Naichang, "Elimination of scan blindness in phased array of microstrip patches using electromagnetic bandgap materials," *IEEE Antennas Wireless Propagation Letters*, vol. 3, no. 4, pp. 63–65, 2004.
68. L. Zhang, J. A. Castaneda, and N. G. Alexopoulos, "Scan blindness free phased array design using PBG materials," *IEEE Trans. Antennas Propagat.*, vol. 52, no. 8, pp. 2000–2007, Aug. 2004.
69. H. Moghadas, A. Tavakoli, and M. Salehi, "Elimination of scan blindness in microstrip scanning array antennas using defected ground structure," *Int. J. Electron. Commun. (AEÜ)*, vol. 62, pp. 155–158, 2008.
70. D.-B. Hou, et al., "Elimination of scan blindness with compact defected ground structures in microstrip phased array," *IET Microwave Antennas Propag.*, vol. 3, no. 2, pp. 269–275, 2009.

13

Printed Leaky Wave Antennas

Samir F. Mahmoud[1] and Yahia M.M. Antar[2]
[1]*University of Kuwait, Kuwait*
[2]*Royal Military College, Canada*

13.1 Introduction

Leaky wave antennas form one type of traveling wave antennas in which an aperture is illuminated by the fields of a traveling wave. Usually a leaky wave stems from a closed guiding structure that supports traveling waves but has some means of continuous power leakage into the exterior region. The illuminated aperture extends over several wavelengths and is limited by wave attenuation caused by power leakage. A leaky wave antenna structure is depicted in Figure 13.1 where an incident mode travels inside a closed waveguide except that one of the walls allows power leakage, causing some perturbation to the propagating mode. Therefore, the mode longitudinal wave number β_{zc} in a completely closed waveguide will be slightly changed to, say, β_z, and there appears, in addition, an attenuation factor α_z. Therefore, a leaky mode is one having necessarily a complex wave number $k_z = \beta_z - j\alpha_z$, where β_z is less than the free space wave number k_0 rendering the leaky mode to be a fast wave. A leaky mode will radiate in a direction θ given by $\theta = \sin^{-1}(\beta_z/k_0)$. Since β_z is a function of frequency, it follows that the radiation beam can be steered by frequency scanning between broadside and end fire. The beamwidth depends mainly on the attenuation rate α_z. As α_z is reduced, the illuminated aperture extends over larger area resulting in higher directivity and lower beamwidth.

The basic properties of leaky wave antennas were founded in the pioneering work of Tamir and Oliner back in the early1960s [1, 2] and reviewed later by Tamir [3]. Recently, the need for high gain microstrip antennas has revived interest in leaky waves, resulting in a great number of papers on printed leaky wave antennas. In this chapter, we present a self-contained study of leaky waves and their supporting structures. In the following sections, we shall describe a leaky wave mathematically as a complex plane wave. The leaky mode supporting structure is treated as a perturbation of a closed waveguide. A planar antenna configuration with a partially reflecting screen will be studied in detail as an example of a leaky wave antenna structure.

Microstrip and Printed Antennas: New Trends, Techniques and Applications. Edited by Debatosh Guha and Yahia M.M. Antar

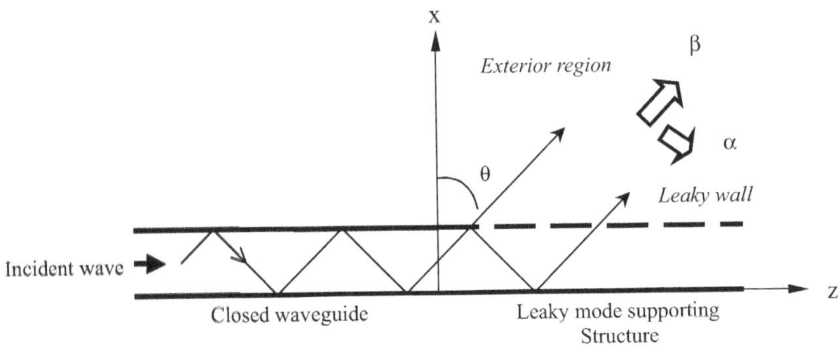

Figure 13.1 Transition from a closed waveguide to a leaky mode supporting structure

Another example of such structure is a multilayered planar antenna with the capability of gain enhancement. Analysis of these two structures reveals the main properties of leaky wave antennas. In addition, we also consider the design of one-dimensional and two-dimensional leaky wave antennas that radiate fan-shaped beams, and conical or pencil beams respectively.

13.2 The Leaky Wave as a Complex Plane Wave

The leaky wave mode is characterized by a complex wave number along the direction of wave propagation. For time harmonic fields behaving as *exp* $(j\omega t)$, the leaky mode field dependence on time and z is expressed as *exp* $(j\omega t\text{-}j\beta_z z\text{-}\alpha_z z)$. With reference to Figure 13.1, the fields in the outer region have to match this behavior in order to satisfy the boundary condition along the (leaky) interface; $(x=d)$. In the transverse direction, the fields will also have a complex wave number, say, $\beta_x - j\alpha_x$. Hence, the complete field expression in the exterior region (assumed free space) is of the form

$$F = \exp[j\omega t - j\beta_z z - \alpha_z z]\,\exp[-j\beta_x x - \alpha_x x] \tag{13.1}$$

where

$$(\beta_z - j\alpha_z)^2 + (\beta_x - j\alpha_x)^2 = k_0^2 \tag{13.2}$$

and k_0 is the free space wave number. Equating the imaginary parts on both sides yields,

$$\alpha_z\beta_z + \alpha_x\beta_x = 0 \tag{13.3}$$

Since α_z is positive, α_x must be negative (both β_z and β_x are positive for outward propagation), which means that the fields increase rather than decrease with x. The fields of a leaky mode do violate the radiation condition. The leaky mode should therefore be categorized as an improper mode. However, the fields of a leaky wave remain finite in a certain wedge region. Let $x = r\cos\theta$ and $z = r\sin\theta$, then in the region defined by

$$\theta > \theta_c = \tan^{-1}(-\alpha_x/\alpha_z) = \tan^{-1}(\beta_z/\beta_x), \tag{13.4}$$

the leaky wave fields are finite. We may define vector wave number and vector attenuation rate as $\vec{\beta} = \beta_x\hat{x} + \beta_z\hat{z}$, $\vec{\alpha} = \alpha_x\hat{x} + \alpha_z\hat{z}$. The $\vec{\beta}$ and $\vec{\alpha}$ vectors are shown in Figure 13.1 by double arrows; they are mutually orthogonal. The direction of $\vec{\beta}$ entails power leakage into the exterior region. The direction of $\vec{\alpha}$ reflects the improper behavior of the leaky wave as the wave increases in the outward transverse direction.

It is useful to define the direction of propagation of the leaky wave through a complex angle w, as

$$k_z = \beta_z - j\alpha_z = k_0 \sin w, \; k_x = \beta_x - j\alpha_x = k_0 \cos w \tag{13.5}$$

In terms of the real and imaginary parts (w_r, w_i) of w

$$k_z/k_0 = \sin w_r \cosh w_i + j \cos w_r \sinh w_i \tag{13.6}$$

$$k_x/k_0 = \cos w_r \cosh w_i - j \sin w_r \sinh w_i \tag{13.7}$$

It follows that the angle θ_c defining the finite field wedge in (13.4) is simply $\theta_c = w_r$.

13.3 Radiation Pattern of a Leaky Wave

We can derive the radiation pattern of a leaky wave from the aperture traveling wave field. Depending whether the leaky wave is traveling in the $+$ z direction only or in both $+$ z and $-$z, the radiation is unidirectional or bidirectional. In the following section, we derive the radiation pattern for both cases.

13.3.1 Unidirectional Leaky Wave

With reference to Figure 13.1, an incident traveling wave from the closed guide excites a z-directed leaky wave in the leaky waveguide. Let us consider a TE wave having E_y field on the $x = d$ plane. The aperture E-field is given by

$$E_y(z) = E_0 \exp(-k_{zL}z); \quad 0 \le z < \infty \tag{13.8}$$

where $k_{zL} = \beta_{zL} - j\alpha_{zL}$ is the complex propagation constant of the leaky wave. The field is independent of the y-coordinate. The extent of the aperture is limited by the attenuation rate α_{zL}, so that we may consider the aperture to lie between $z = 0$ and $z = L$, such that $\alpha_{zL}L \gg 1$. The aperture field may be considered as a continuous array of line sources. For a line source of infinitesimally narrow width Δ along z, the field can take the form

$$e_y(z) = E_0\Delta\,\delta(z) = (E_0\Delta/2\pi)\int_{-\infty}^{\infty} e^{-j\beta z}\,d\beta \tag{13.9}$$

The second equality arises from an identity for the Dirac delta function δ(z). Using the wave equation, the above field can be extended to the $x > d$ region; so with $k_x = (k_0^2 - \beta^2)^{1/2}$, and $x' = x - d$,

$$e_y(x, z) = (E_0\Delta/2\pi)\int_{-\infty}^{\infty} e^{-j(k_x x' + \beta z)}\,d\beta \tag{13.10}$$

At this point, we invoke the following identity for the Hankel function of zero order [4]

$$H_0^{(2)}(k_0 r) = (1/\pi) \int_{-\infty}^{\infty} e^{-j(k_x x' + \beta z)} d\beta / k_x, \tag{13.11}$$

where $r = \sqrt{(x-d)^2 + z^2}$. Taking the derivative of both sides with respect to x, we are able to express the integral term in Equation (13.10) as $j\pi \, \partial H_0^{(2)}(k_0 r)/\partial x = -j\pi k_0 H_1^{(2)}(k_0 r) \cos\theta$, hence

$$e_y(x, z) = (-jk_0 E_0 \Delta/2)\, H_1^{(2)}(k_0 r) \cos\theta \tag{13.12}$$

In the far zone, $k_0 r \gg 1$, the large argument approximation of the Hankel function is used [5] to get

$$e_y(r, \theta) \cong -\frac{jk_0 E_0 \Delta}{2} \sqrt{\frac{2}{\pi k_0 r}} \exp(-jk_0 r - j3\pi/4) \cos\theta \tag{13.13}$$

The final step in deriving the radiation pattern of the leaky wave mode is to multiply the last equation by the array factor (AF) due to the aperture distribution $\exp(-jk_{zL}z)$ of the leaky mode

$$\begin{aligned} AF(\theta) &= \int_0^L \exp(-jk_{zL}z) \exp(jk_0 z \sin\theta)\, dz \\ &= \frac{1/jk_0}{\sin\theta - \sin w_L} \end{aligned} \tag{13.14}$$

where w_L is the complex angle defined by $k_{zL} = k_0 \sin w_L$ and the condition $\exp(-\alpha_{zL} L) \ll 1$ is used. Thus, combining Equations (13.13) and (13.14), the power radiation pattern (PRP) of the leaky wave takes the form (apart from a constant)

$$\begin{aligned} PRP(\theta) &= \frac{\cos^2\theta}{|\sin\theta - \sin w_L|^2} = \\ &\frac{\cos^2\theta}{(\sin\theta - \sin w_{rL} \cosh w_{iL})^2 + \cos^2 w_{rL} \sinh^2 w_{iL}} \end{aligned} \tag{13.15}$$

Another useful approach to find the radiation pattern of the leaky wave is to use the spectral representation of the field in the outer region, which takes the general Sommerfeld integral

$$E_y(x, z) = \int_{-\infty}^{\infty} f(\beta) \exp[-jk_x x' - j\beta z]\, d\beta \tag{13.16}$$

The integration is taken over all real numbers of the longitudinal wave number β and as before, $k_x^2 + \beta^2 = k_0^2$. The function $f(\beta)$ should be derived for the given structure under study. In general, it contains a number of proper poles in addition to a leaky mode. The proper modes account for the discrete spectrum of surface waves. The leaky mode, if it has a small attenuation, can well approximate the continuous spectrum or the radiation fields. In this case $f(\beta)$ is given by

$$f(\beta)=\frac{A}{\beta-k_{zL}}+\sum_{p=1,2..}\frac{B_p}{\beta-k_{zp}} \tag{13.17}$$

The first term is the leaky wave pole contribution. The summation term accounts for the discrete spectrum whose residues give the bounded surface wave fields. We concentrate here on the leaky mode pole that accounts for the radiation fields. The stationary phase method of integration can be used to evaluate the integral in the far zone where $k_0 r\gg 1$. With $\theta=\tan^{-1}(z/x')$, $\beta=k_0\sin w$, and $k_x=k_0\cos w$, the exponential term in Equation (13.16) can be written as $\exp[-jk_0 r\cos(\theta-w)]$, which is stationary around the point $w=\theta$ and is highly oscillatory away from this point. Meanwhile $f(\beta)$ is a slowly varying function. Thus Equation (13.16) is well approximated by [4]

$$E_y(r,\theta)=f(k_0\sin\theta)\,k_0\cos\theta\int_C\exp[-jk_0 r\cos(w-\theta)]dw \tag{13.18}$$

The contour C lies in the w-plane that maps the real axis of k_z. The reader can verify that C runs from $-j\infty-\pi/2$ to $-\pi/2$ parallel to the imaginary axis, to $+\pi/2$ on the real axis, and then to $+\pi/2+j\infty$ parallel to the imaginary axis. The integral term is identically equal to $\pi H_0^{(2)}(k_0 r)$ [4]. Using the first term of Equation (13.17) in Equation (13.18)

$$E_y(r,\theta)=\frac{A\pi\cos\theta}{\sin\theta-\sin w_L}H_0^{(2)}(k_0 r) \tag{13.19}$$

where $k_{zL}=k_0\sin w_L$ has been used. It follows that the power radiation pattern is identical to that derived earlier using Equation (13.15).

Exercise 1: Fill in the missing steps in deriving Equation (13.19).

The leaky mode can generate a high directivity beam as long as it has low attenuation. From (13.15) the reader can readily show that the peak radiation occurs at an angle θ_p given by

$$\sin\theta_p=\sin w_{rL}/\cosh w_{iL} \tag{13.20}$$

Further, the peak value of PRP at θ_p is given by $(1/\sinh^2 w_i)$. When $|w_i|\ll 1$, the 3 dB beamwidth is found to be given by $BW\approx 2w_{iL}\approx 2\alpha_{zL}/\beta_{xL}$ (Radians) [6]. It is worth noting that θ_p is different, but close to θ_c. The latter relates to the direction of the beam in the closed waveguide.

Exercise 2: Prove (13.20) and find the peak value of the $PRP(\theta_p)$.

Exercise 3: Show that the 3 dB beamwidth of the PRP in (13.15) is given as stated above when $w_i\ll 1$. Ignore the variation of the $\cos^2\theta$ term around θ_p.

The directivity is reported in [3] as

$$\text{Directivity} \cong e^{|w_i|}/\sin\text{h}|w_{iL}| \approx 1+1/|w_{iL}| \approx 1+\beta_{xL}/\alpha_{zL} \qquad (13.21)$$

The last equality is an approximation, valid for $w_{iL} \ll 1$. As an example, for a leaky mode with $w_L = \pi/6 - j0.05$, the radiation pattern has a directivity = 21 and a peak at the angle $\theta_p = 29.96°$. The beamwidth (BW) is given by $BW = 0.1$ radians.

It is interesting to examine the possibility of having broadside and endfire radiation from a unidirectional leaky mode. For an endfire beam ($\theta_p = \pi/2$), Equation (13.20) dictates that $w_{rL} = \pi/2$ and $w_{iL} = 0$. This means that the leaky mode attenuation $\alpha_{zL} = \alpha_{xL} = 0$, and the leaky mode would have infinite power. Such a mode cannot be excited by any physical source and therefore an endfire beam cannot be generated. On the other hand, a broadside beam ($\theta_p = 0$) requires that $w_{rL} = 0$ and $|w_{iL}| > 0$, or $\beta_{zL} = 0$ and $\alpha_{zL} > 0$. This corresponds to a mode under cutoff, which carries no power. Therefore, a unidirectional leaky mode cannot radiate a broadside beam. However, it is possible to have a beam very close to broadside when both w_{rL} and w_{iL} are $\ll 1$. We will show below that, in contrast, a bidirectional leaky mode can readily have broadside radiation.

13.3.2 Bidirectional Radiation Pattern

In most cases of leaky wave excitation, there is a localized source that excites leaky modes traveling in both $+z$ and $-z$ directions. In this case, we will have two leaky wave poles at $\pm k_{zL}$. The radiation pattern will be bidirectional and symmetrical about $\theta = 0$. With $k_{zL} = k_0 \sin w_L$, Equation (13.14) will then be replaced by

$$AF(\theta)|_{bi} = \frac{1/jk_0}{\sin\theta - \sin w_L} - \frac{1/jk_0}{\sin\theta + \sin w_L} = \frac{2\sin w_L/jk_0}{\sin^2\theta - \sin^2 w_L} \qquad (13.22)$$

Combining this with Equation (13.13), the corresponding form of the power radiation pattern becomes

$$PRP(\theta)|_{bi} = \frac{\cos^2\theta}{|\sin^2\theta - \sin^2 w_L|^2} \qquad (13.23)$$

It can be shown that the peak radiation occurs at the angles $\pm\theta_p$ given by

$$\sin\theta_p = \pm\sqrt{\sin^2 w_r - \sinh^2 w_i} \qquad (13.24)$$

Exercise 4: Prove Equation (13.24) and show that the *PRP* at $\theta_p = 1/(4\sin^2 w_{rL}\sinh^2 w_{iL})$.

The radiation pattern is obviously a bidirectional one provided that $\sin w_{rL} > \sinh w_{iL}$. For a low attenuation leaky mode; $w_{iL} \ll w_{rL}$, the peak radiation angle $\theta_p \cong w_{rL}$ and it can be shown that the 3 dB beamwidth, $BW \cong 2|w_{iL}| \cong 2\alpha_{zL}/\beta_{xL}$.

As $\sin\theta_p$ is reduced, the two peaks at $\pm\theta_p$ approach each other until they join in one peak in the broadside direction ($\theta_p = 0$). When $\sin w_{rL} = \sinh w_{iL}$, it follows that for small w_L (both w_{rL} and $w_{iL} \ll 1$), $\beta_{zL} = \alpha_{zL}$. Under this condition, the radiation is in the broadside direction and it has an optimum directivity [6, 7], and the beamwidth equal to $2\sqrt{2}|w_i|$. Finally, if $\sin w_{rL} < \sinh w_{iL}$, the radiation is still broadside ($\theta_p = 0$), but it will have reduced directivity because of increased attenuation.

As an example, if a bidirectional leaky mode has $k_{zL} = (0.1 - j0.099)\ k_O$, one gets $w_L = 0.0997 - j0.0993$. Equation (13.24) gives $\theta_p \approx 0$ (Broadside radiation). The beamwidth is equal to 0.28 radians = 16^0. Next, if the attenuation rate increases to 0.15 k_O, i.e., $k_z = (0.1 - j0.15)\ k_O$, then $w = 0.99 - j0.15$. The angle θ_p is purely imaginary, which means that the peak radiation is still broadside. However, the beamwidth increases to 24^0.

13.4 Examples of Leaky Mode Supporting Structures

Two examples of leaky mode supporting structures will be studied in this section. As indicated earlier, a leaky guide stems from a closed waveguide, which is perturbed to leak power in the outer region. The first leaky mode supporting guide to be considered is shown in Figure 13.2. It is a two parallel plate guide having a ground plate on one side and a partially reflecting screen (PRS) on the other side. The other leaky guide that will be considered consists of two layers; a substrate and a superstrate layer between a ground plane and an air half-space region as depicted in Figure 13.5. Both structures have been used as high gain printed antennas as published in recent literature.

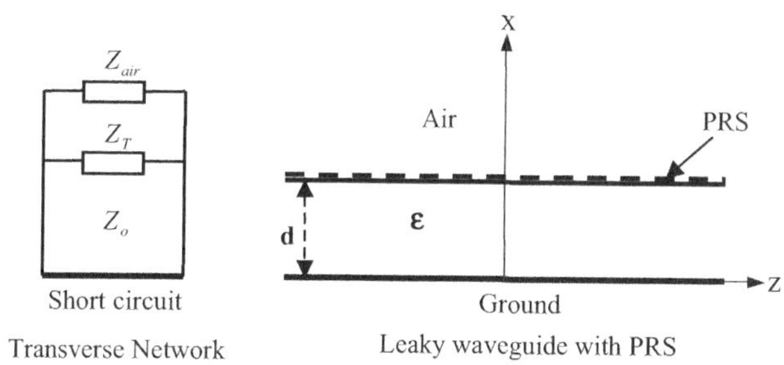

Z_T= Transfer Impedance

$Z_o = \dfrac{\omega\mu_o}{k_x}$ $\quad Z_{air} = \dfrac{\omega\mu_o}{k_{xo}}$ (for TE_x modes)

Figure 13.2 A two parallel plate waveguide with a partially reflecting screen (PRS) and its transverse equivalent network

13.4.1 A Two Parallel Plate Leaky Waveguide

The structure of Figure 13.2 consists of a grounded substrate and a partially reflecting screen at $x = d$. The partially reflecting screen (PRS) is characterized by an effective transfer impedance $Z_T = jX$ defined as the ratio of the tangential E field to the discontinuity of the tangential magnetic field across the PRS [8]. A PRS may be realized by screen of periodic patches or a thin layer of high permittivity superstrate, which acts as a capacitive screen [9]. An inductive braided metal screen can also be used as a PRS. When $X = 0$, the structure is a closed waveguide with a set of discrete modes. For a finite X, there is a leaky mode with complex propagation constant. With reference to Figure 13.2, to design a leaky wave antenna with a beam directed at a given angle (approximately θ_c) from broadside, we consider first the closed waveguide ($X = 0$) and choose a frequency such that $\beta_{zc} = k_O \sin\theta_c$ for a chosen mode. For the TE_m mode, the corresponding transverse wave number k_{xc} in the closed guide $= m\pi/d$; m being an integer > 0, hence the required frequency of operation is given by

$$k_0 = \omega/c = \frac{m\pi/d}{\sqrt{\varepsilon_r - \sin^2\theta_c}} \tag{13.25}$$

The transverse resonance method [3] can now be used to determine the complex wave number k_x corresponding to the desired leaky mode. This method is based on the uniformity of the structure longitudinally along z so that resonance of the normal modes occurs in the transverse direction. The transverse network representing the leaky structure is given in Figure 13.2, where an equivalent transmission line of length "d" is shorted at one end and connected at the other end to the transfer reactance jX, in parallel with the impedance reflected by the outer medium. Considering TE to x modes, the transverse wave impedances for the transmission line and the outer medium are $\omega\mu/k_x$ and $\omega\mu/k_{xO}$ respectively. The resonance condition requires that the impedances looking up and down be equal in magnitude and opposite in sign. This is actually a statement of the continuity of the tangential field components. Applying the resonance condition at $x = d$, we get, after simple manipulations

$$\tan k_x d = \chi, \text{ with}$$

$$\chi = -\frac{\bar{X}\, k_x/k_0}{1 + j\bar{X}\, k_{x0}/k_0} \cong -\frac{\bar{X}\,\sqrt{\varepsilon_r - \sin^2\theta_c}}{1 + j\bar{X}\cos\theta_c} \tag{13.26}$$

In the above expression $\bar{X} = X/\eta_0$ where η_0 is the free space wave impedance ($=120\pi\ \Omega$) and $k_{x0} = \sqrt{k_x^2 - (\varepsilon_r - 1)k_0^2}$. The last equality in (13.26) comes about by approximating k_x and k_{xO} by their closed guide values; namely $k_x/k_O = \sqrt{\varepsilon_r - \sin^2\theta_c}$ as obtained from (13.25) and $k_{xO} = k_O \cos\theta_c$. Equation (13.26) is the modal equation for the leaky mode.

For $\bar{X} \ll 1$, an approximate solution to (13.26) yields $k_x d = m\pi + \delta$, where $\delta \ll 1$ and is given by

$$\tan\delta \cong \delta \cong -\frac{\bar{X}\sqrt{\varepsilon_r - \sin^2\theta_c}}{1 + j\bar{X}\cos\theta_c} \tag{13.27}$$

The corresponding complex leaky mode wave number $k_{zL} = \beta_{zL} - j\alpha_{zL}$ is given by $k_{zL} = \pm k_0\sqrt{\varepsilon_r - (m\pi + \delta)^2/(k_0 d)^2}$. The $\pm$ signs refer to possible forward and backward leaky modes. With $|\delta| \ll 1$, we get $k_{zL}/k_0 \cong \pm\sqrt{\sin^2\theta_c - 2\delta(\varepsilon_r - \sin^2\theta_c)/m\pi}$, which can be used as an initial guess in any root finding algorithm to get α_{zL} and β_{zL}. When the source excites two leaky modes in the $\pm$z, the peak radiation occurs at angles $\pm\theta_p$ given by Equation (13.24).

We solve Equation (13.26) for fixed $\bar{X} = -0.1$ and $\varepsilon_r = 2.2$ and varying θ_c or $k_O d$. The frequency is scanned between $k_0 d = \pi/\sqrt{\varepsilon_r}$ and $k_0 d = \pi/\sqrt{\varepsilon_r - 1}$ which corresponds to beam steering between $\theta_c = 0$ to 90^0 (see Equation (13.25) with $m = 1$). The TE leaky mode β_{zL}/k_0 and α_{zL}/k_0 are plotted versus $k_O d$ in Figure 13.3 (a). The corresponding angle of

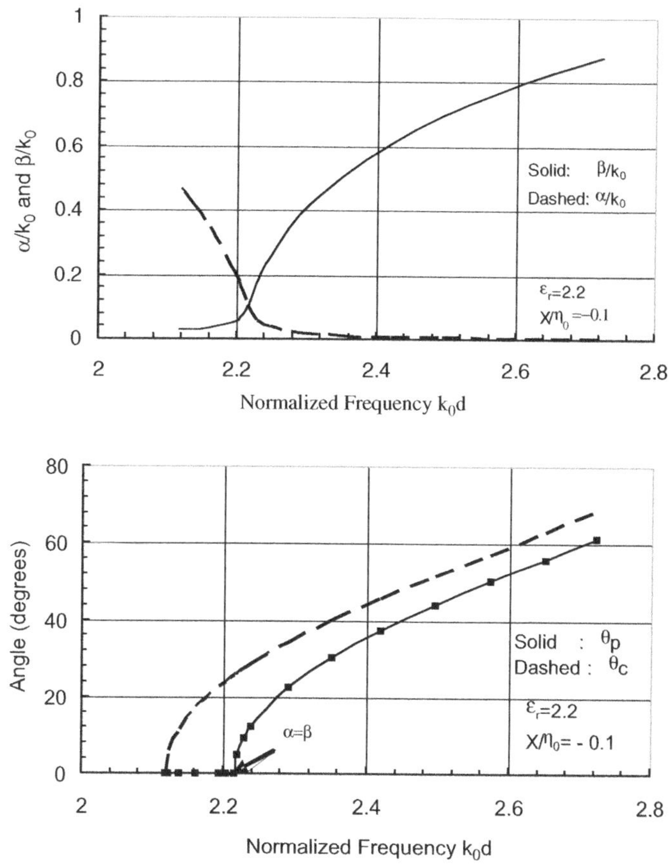

Figure 13.3 (a) Leaky mode phase and attenuation rates on a two parallel plate guide with a PRS versus normalized frequency. TE_x mode is considered. The frequency is scanned between $\pi/\sqrt{\varepsilon_r}$ and $\pi/\sqrt{\varepsilon_r - 1}$ (b) Angle of peak radiation θ_p and θ_c versus $k_0 d$ for the TE_x leaky mode on the structure of Figure 13.2

peak radiation θ_p and the angle θ_c are plotted in Figure 13.3 (b). The frequency scanning property of the leaky mode is obvious. It is noticed that θ_p is lower than θ_c. This is explained by the capacitive nature of the PRS which causes the waveguide width to appear smaller than "*d*." This causes the angle of the ray leaving the guide to be smaller than θ_c. If the PRS turns to be inductive (positive $\bar{X}$), θ_p would be $>\theta_c$. The frequency at which $\beta_{zL} = \alpha_{zL}$ corresponds to the transition from two fan beams at $\pm\theta_p$ to one broadside beam at $\theta_p = 0$. At lower frequencies the radiation remains in broadside, but with lower directivity (due to increased attenuation).

The TM_x mode can be treated in a similar way as the TE_x. In this case the nonzero field components are H_y, E_x and E_z. The transverse equivalent network in Figure 13.2 is still valid except that the transverse wave impedances change to $k_x/\omega\varepsilon_0\varepsilon_r$ and $k_{x0}/\omega\varepsilon_0$.

Exercise 5: Show that for TM_x mode, the mode equation derived from the transverse equivalent network is given by

$$\tan k_x d = -\frac{\bar{X}\,\varepsilon_r k_0/k_x}{1+j\bar{X}\,k_0/k_{x0}} \cong -\frac{\bar{X}\varepsilon_r\,/\sqrt{\varepsilon_r-\sin^2\theta_c}}{1+j\bar{X}/\cos\theta_c} \tag{13.28}$$

The phase and attenuation rates for the TMx leaky wave are obtained by solving Equation (13.28) numerically and they are plotted in Figure 13.4. In addition to the leaky mode, there is one TM surface wave mode in the frequency range considered. The phase constant β_{sw} of the surface wave mode is also shown in the same figure. Although the screen transfer impedance Z_T is taken constant in Figure 13.4, it is important to stress the fact that Z_T is a function of frequency. This can have a significant effect on the frequency dependence of the surface wave. This point will be discussed in more detail in Section 13.6.

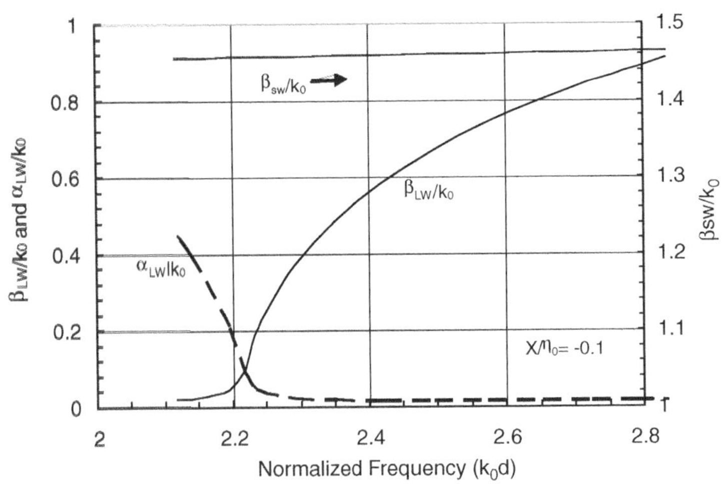

Figure 13.4 TM Leaky mode attenuation and phase rates in a parallel plate waveguide with a partially reflecting top screen. The phase rate of the excited TM surface wave mode is also shown

13.4.2 A Two-Layer Leaky Wave Structure

A conventional microstrip antenna is composed of a patch on a low permittivity substrate. Although it has several advantages including low profile and conformality to curved bodies, such a microstrip antenna suffers from low gains being around 6 dB. A method for enhancing the gain of printed antennas has been introduced and analyzed in [6, 10–15]. The method involves the addition of a superstrate layer of high ε_r on top of the patch. An extension to multiple superstrates has been studied [12]. The effect of the superstrate, when properly designed, is to support a low attenuation leaky mode with a high directivity radiation beam.

In the following, we consider the two layer structure depicted in Figure 13.5 as a leaky mode supporting structure. The substrate of thickness d_l has a low or moderate dielectric constant ε_{1r}

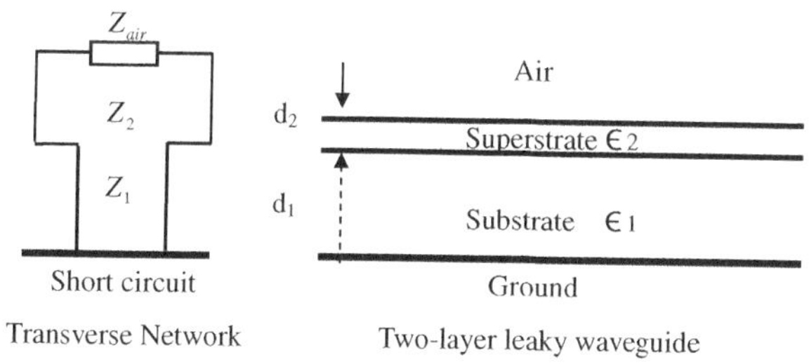

$$Z_1 = \frac{\omega\mu_o}{k_{x1}} \qquad Z_2 = \frac{\omega\mu_o}{k_{x2}} \qquad Z_{air} = \frac{\omega\mu_o}{k_{xo}}$$

Figure 13.5 A two dielectric layer leaky structure and its transverse equivalent network. TE_x mode is assumed

and the superstrate has a high $\varepsilon_{2r} \gg 1$ and thickness d_2. The outside air region reflects a low admittance loading the superstrate.

The closed waveguide model of this structure is considered to be the two layers between the perfectly conducting wall at $x=0$ and a virtual magnetic wall $(Y=0)$ at $x=d_1+d_2$. The transverse resonance of the closed waveguide occurs by the choice $k_{x1c}d_1=\pi$ and $k_{x2c}d_2=\pi/2$ [11], where the subscript c relates to the closed guide. To have a radiation beam at (approximately) angle θ_c from broadside, $\beta_{zc}=k_0\sin\theta_c$, hence the applied frequency is governed by

$$k_0 = \frac{\pi/d_1}{\sqrt{\varepsilon_{1r}-\sin^2\theta_c}} = \frac{\pi/2d_2}{\sqrt{\varepsilon_{2r}-\sin^2\theta_c}} \tag{13.29}$$

Next, the leaky guide may be considered as a perturbation to the closed guide and is represented by the transverse equivalent network given in Figure 13.5. The network is composed of two transmission lines of lengths d_1 and d_2, characteristic impedances $k_{x1}/\omega\varepsilon_1$ *and* $k_{x2}/\omega\varepsilon_2$, short-circuited at one end and loaded by impedance $k_{x0}/\omega\varepsilon_0$ representing the outside air medium. We have assumed TM_x wave excitation. Obviously all k_{xi}, $i=0,1,2$ are complex and given by $k_{xi}=\sqrt{\varepsilon_{ir}k_0^2-k_{zL}^2}$ and k_{zL} is the complex z-wave number $=\beta_{zL}-j\alpha_{zL}$ of the leaky mode. Application of the transverse resonance condition at $z=d_1$, leads to

$$\tan(k_{x1}d_1) = j\frac{\varepsilon_{1r}k_{x2}}{\varepsilon_{2r}k_{x1}}\frac{-j\cot(k_{x2}d_2)/k_{x2}+1/\varepsilon_{2r}k_{x0}}{1/k_{x2}-j\cot(k_{x2}d_2)/\varepsilon_{2r}k_{x0}} \tag{13.30}$$

Since the leaky guide is a perturbed version of the closed guide when $\varepsilon_{2r} \gg 1$, we approximate k_{xi} by their closed guide values plus a small quantity, namely $k_{x1}d_1 \simeq \pi+\delta_1$ and $k_{x2}d_2 \simeq \pi/2+\delta_2$, where $\delta_1,\delta_2 \ll 1$. Approximate values for δ_1 and δ_2 are derived as

$$\delta_1 \cong j\frac{\varepsilon_{1r}[\varepsilon_{2r}-\sin^2(\theta_c)]}{\varepsilon_{2r}^2\cos\theta_c\sqrt{\varepsilon_{1r}-\sin^2(\theta_c)}} \tag{13.31}$$

$$\delta_2 \cong \frac{\delta_1}{2}\frac{\varepsilon_{1r}-\sin^2\theta_c}{\varepsilon_{2r}-\sin^2\theta_c} \cong \frac{j\varepsilon_{1r}}{2\varepsilon_{2r}^2}\frac{\sqrt{\varepsilon_{1r}-\sin^2\theta_c}}{\cos\theta_c} \tag{13.32}$$

Both δ_1 and δ_2 are $\ll 1$ as $\varepsilon_{2r} \gg 1$. The corresponding longitudinal phase and attenuation rates are approximately

$$\beta_{zL} - j\alpha_{zL} \cong k_0\sqrt{\sin^2\theta_c - 2\delta_1(\varepsilon_{1r} - \sin^2\theta_c)/\pi} \tag{13.33}$$

Of particular interest is the observation that for broadside radiation ($\sin\theta_c = 0$), $\beta_{zL} = \alpha_{zL}$. The exact phase and attenuation rates are obtained by solving Equation (13.30) numerically, and are plotted versus the normalized scanned frequency, for two values of ε_{2r}, in Figure 13.6 (a). As expected, the attenuation rate is lower for the higher ε_{2r}. The corresponding θ_c and θ_p are plotted in Figure 13.6 (b), and it is observed that, unlike the structure of Figure 13.2, there is a small

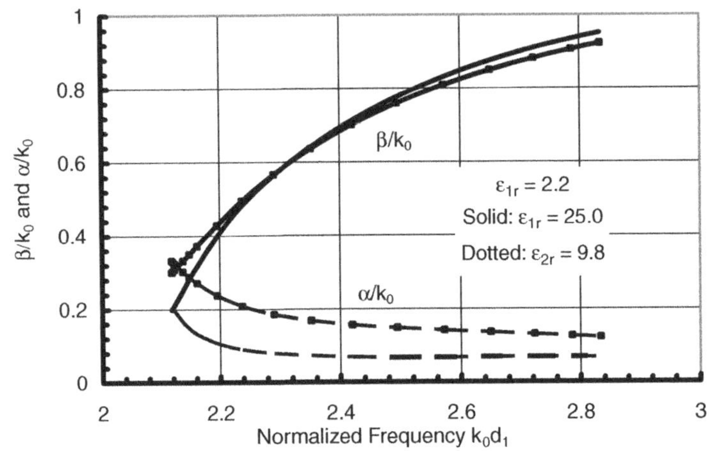

(a)

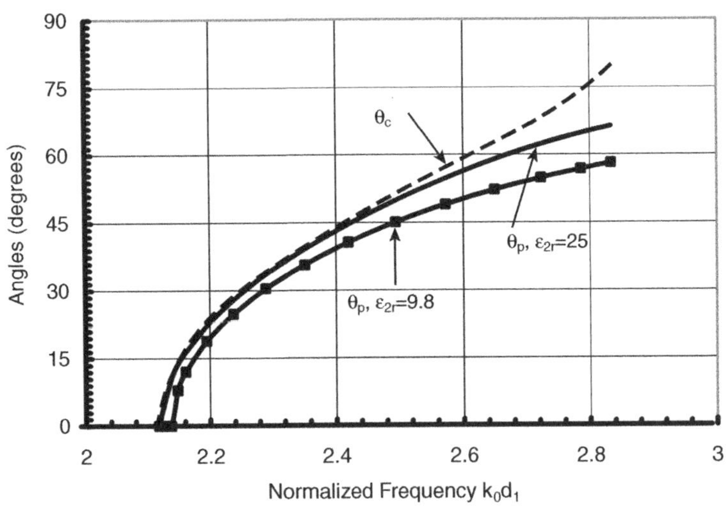

(b)

Figure 13.6 (a) Leaky mode phase and attenuation rates versus normalized frequency on a two-layer leaky structure for two values of the substrate relative permittivity (25 and 9.8). The substrate $\varepsilon_{1r} = 2.2$. The frequency is scanned between $k_0 d_1 = \pi/\sqrt{\varepsilon_{1r}}$ and $\pi/\sqrt{\varepsilon_{1r} - 1}$. (b) Angle of peak radiation θ_p and θ_c versus $k_0 d$ for two values of the superstrate $\varepsilon 2_r$. $\varepsilon_{1r} = 2.2$

difference between θ_c and θ_p. This is explained by the absence of a reactive load in the transverse network of Figure 13.5. Finally, we note that a similar treatment to the above can be applied to the TE_x leaky modes.

13.5 The Excitation Problem

So far, we have been studying the properties of leaky waves and their derivation through the application of the transverse resonance method. No mention, however, has been made of the excitation of leaky modes. In order to design a leaky wave antenna, two conditions need to be fulfilled. The first is to have a guiding structure that can support a leaky wave mode with the desired radiation character. As shown in the last section, such a structure is basically a closed waveguide, which is perturbed to allow power leakage. The second condition is to efficiently excite the leaky wave mode. The problem of excitation of a leaky mode supporting structure by a given source is considered in this section. To this end, we reconsider the leaky guide of Figure 13.2 where a simple electric line source is used for excitation. The line source is assumed to run parallel to the y-axis at $x = h$ and $z = 0$ (refer to Figure 13.7). With a total current $I\, exp\, (j\omega t)$, the source current density is expressed as

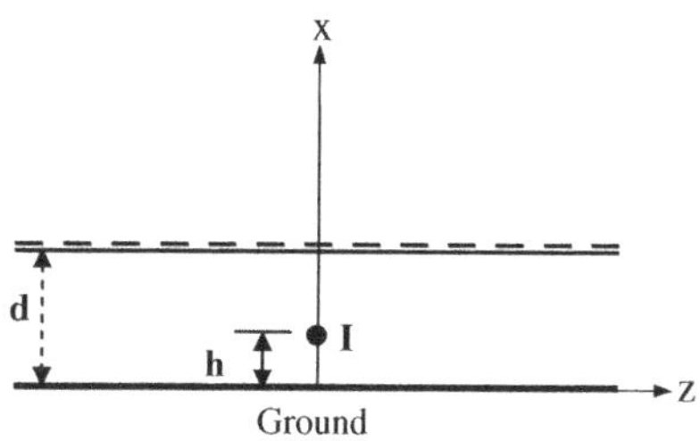

Figure 13.7 A two parallel plate waveguide with a PRS fed by a current line source

$$J(x,z) = I\,\delta(x-h)\,\delta(z) \tag{13.34}$$

This source excites TE_{xz} waves so that E is totally y-directed. Let us apply the following Fourier transform to the source and all fields, for example

$$E_y(x,z) = \int_{-\infty}^{\infty} \tilde{E}_y(x)\, e^{-j\beta z} d\beta \tag{13.35}$$

where $\tilde{E}_y$ is the z-transformed field. Applying this transform to Equation (13.34), we get $\tilde{J}(x) = (I/2\pi)\,\delta(x-h)$. Now we can write the following expressions for $\tilde{E}_y$ in the different regions

$$\begin{aligned} \tilde{E}_y(x) &= a(\beta)\sin k_x x; && 0 \le x < h \\ &= b(\beta)\sin k_x(x-d) + c(\beta)\cos k_x(x-d); && h < x \le d \\ &= c(\beta)\exp[-jk_{x0}(x-d)]; && x \ge d \end{aligned} \tag{13.36}$$

In the above $k_{x0} = \sqrt{k_0^2 - \beta^2}$ and $k_x = \sqrt{\varepsilon_r k_0^2 - \beta^2}$. Note that the continuity of $\tilde{E}_y$ at $x = d$ is readily satisfied. The continuity at $x = h$ requires that

$$a\,(\beta)\sin k_x h = -b\,(\beta)\sin k_x s + c\,(\beta)\cos k_x s \qquad (13.37)$$

where $s = d - h$. In Equation (13.37), $a(\beta)$, $b(\beta)$ and $c(\beta)$ are, so far, unknown coefficients to be determined from the boundary conditions. The source discontinuity at $x = h$ reads

$$\tilde{H}_z(x = h^-) - \tilde{H}_z(x = h^+) = I/2\pi \qquad (13.38)$$

At $x = d$, the transfer impedance Z_T imposes the following

$$\tilde{E}_y(x = d) = Z_T\,[\tilde{H}_z(x = d^-) - \tilde{H}_z(x = d^+)] \qquad (13.39)$$

Using Maxwell's equations, $\tilde{H}_z = (j/\omega\mu_0)\,\partial\tilde{E}_y/\partial x$. Using this in Equations (13.38) and (13.39), we have three linear equations;(13.37)–(13.39) , for the three coefficients $a(\beta)$,$b(\beta)$ and $c(\beta)$. Of particular interest is $c(\beta)$. After some simple manipulations, we get

$$c\,(\beta) = \frac{j\omega\mu_0 I}{2\pi k_x}\frac{\chi \sin k_x h}{\sin k_x d - \chi\cos k_x d} \qquad (13.40)$$

where

$$\chi = \frac{j\bar{Z}\,k_x/k_0}{1 + \bar{Z}\,k_{x0}/k_0}, \text{ and } \bar{Z} = Z_T/120\pi$$

Exercise 6: Fill in the missing steps in deriving (13.40).

The electric field in the air region is obtained from Equations (13.36) and (13.40) after taking the inverse Fourier transform

$$E_y(x,z) = \int_{-\infty}^{\infty} c\,(\beta)\exp[-j(k_{x0}(x-d) + \beta z)]d\beta \qquad (13.41)$$

The integration can be performed in the far zone $k_0 r \gg 1$ using the stationary phase method [4] to get

$$E_y(r,\theta) \cong \frac{j\omega\mu_0 I}{2\pi}\left[\frac{k_0\chi\,\sin(k_x h)\,\cos\theta}{k_x(\sin k_x d - \chi\cos k_x d)}\right]_{\beta = k_0\sin\theta}\sqrt{\frac{2}{\pi k_0 r}}\,e^{-jk_0 r - j3\pi/4} \qquad (13.42)$$

This gives the radiation pattern. A few remarks can be made on the pattern. First of all, as expected, there is no radiation if $\chi = 0$; this corresponds to a perfectly reflecting screen ($Z_T = 0$). The factor sin ($k_x h$) depends on the source position and may be considered as the source excitation factor. In view of Equation (13.25) for $m = 1$, the maximum excitation occurs at θ_c when $h = d/2$ (and $k_x d \approx \pi$). Finally, we observe that $c(\beta)$ or the square bracketed term has the same leaky mode pole defined by Equation (13.26).

13.5.1 Radiation Field in Terms of the Leaky Mode Pole

Alternatively, the far field can be obtained in terms of the leaky mode pole. We recall that $c(\beta)$ has a leaky mode pole at $\beta = \pm k_{zL} = \pm\sqrt{\varepsilon_r k_0^2 - k_x^2}$, with k_x given by (13.26). In the case where radiation fields are dominated by the leaky mode, $c(\beta)$ can be expressed by

$$c\,(\beta) \cong \frac{A^+}{\beta - k_{zL}} + \frac{A^-}{\beta + k_{zL}} \tag{13.43}$$

In view of Equation (13.40), $c(\beta)$ has the form $N(k_x)/D(k_x)$. We can then get $A^{\pm}$ as

$$A^{\pm} = \frac{N(k_x)}{[\partial D(k_x)/\partial k_x]\,\partial k_x/\partial \beta}|_{\beta=\pm k_{zL}} \tag{13.44}$$

Noting that $\partial k_x/\partial\beta = -\beta/k_x = \mp k_{zL}/k_x$, we infer that $A^+ = -A^-$. Therefore, Equation (13.43) becomes

$$c\,(\beta) \cong \frac{2k_{zL}A^+}{\beta^2 - k_{zL}^2} \tag{13.43a}$$

Plugging this into Equation (13.41) and evaluating the integral by the stationary phase method [4], we get the radiation pattern based on the leaky mode dominance as

$$E_y(r,\theta) \cong \left[\frac{2k_{zL}A^+\cos\theta}{k_0^2\sin^2\theta - k_{zL}^2}\right]\sqrt{\frac{2}{\pi k_0 r}}\,e^{-jk_0 r - j3\pi/4} \tag{13.45}$$

This should be numerically compared to Equation (13.42) as will be done below. The total power radiated can be obtained by numerical integration of $|E_y|^2$ in Equation (13.42) over θ. We can thus obtain the radiation resistance and antenna directivity.

Exercise 7: Using Equation (13.42), show that the radiation resistance per unit length (along y) is given by

$$R_{rad} = \frac{\eta_0 k_0}{2\pi^3}\int_{-\pi/2}^{\pi/2} \left|\frac{k_0\chi\sin(k_x h)\cos\theta}{k_x(\sin k_x d - \chi\cos k_x d)}\right|^2 \beta = k_0\sin\theta d\theta \tag{13.46}$$

13.5.2 A Numerical Example

As a numerical example, we plot the radiation pattern in dB as computed from Equations (13.42) and (13.45) in Figure 13.8. The screen transfer impedance Z_T is chosen equal to $-j0.1\,\eta_0$. The vertical axis is adjusted to read the directivity as a function of θ. Two cases that correspond to peak radiation angle of $\pm 15^0$ and 0^0 (broadside) are given. The latter correspond to the condition $\alpha_{zL} = \beta_{zL}$. It is seen that the two methods of computing the radiation pattern give almost identical results. This means that the leaky mode in these cases dominates the radiation fields.

The radiation resistance per one wavelength along y is plotted in Figure 13.9 versus $j\bar{Z}(= -\bar{X})$ for three values of θ_c. In order to make sense of the numbers, it is useful to note

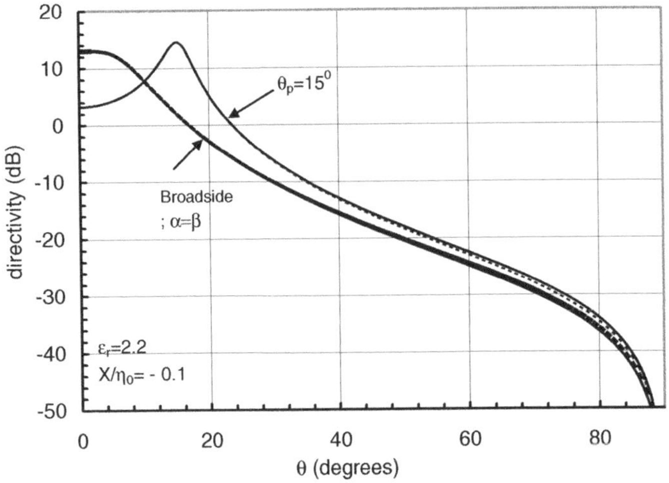

Figure 13.8 Radiation pattern for $\theta_p = 15^0$ and 0^0 (broadside) for the leaky structure of Figure 13.7. The vertical scale gives the directivity. The pattern is symmetrical about the $\theta = 0$ direction. Solid lines correspond to Equation (13.42) and dashed lines correspond to leaky mode pattern given by Equation (13.45)

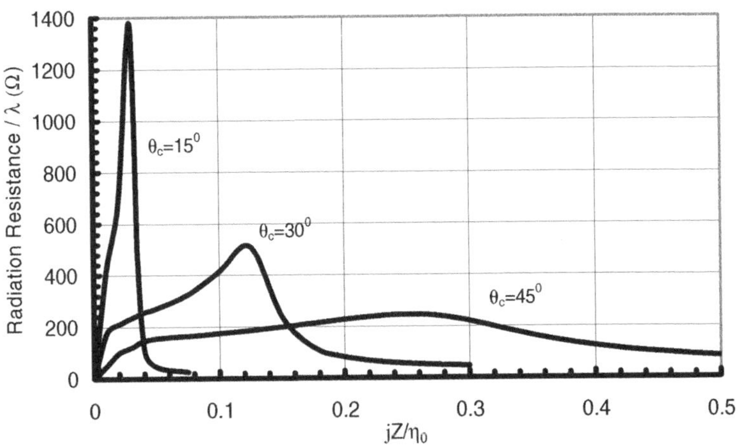

Figure 13.9 Radiation resistance of the antenna in Figure 13.7 versus the (capacitive) reactance of the PRS, with θ_c as a parameter

that the radiation resistance of a line source in free space is equal to $(\pi/2)\eta_0\ \Omega/\lambda_0$; that is 592 Ω/λ_0. We infer that there is an optimum value of $\bar{X}$ for maximum radiated power that depends on θ_c or the direction of peak radiation. Note however that θ_p is not constant for a given θ_c since θ_p is a function of both θ_c and $\bar{X}$.

13.6 Two-Dimensional Leaky Waves

So far, we have been considering leaky waves, which propagate in one direction (the z-direction). The aperture fields are invariant along y. These leaky waves radiate fan-shaped beams. In this section, we consider the more practical case in which the leaky waves propagate in two dimensions or as cylindrical waves. In this case, the radiation fields will form conical beams or a pencil beam (in the broadside direction). Study of cylindrical leaky waves in general are given in [13] and implemented in several papers [6, 11–15]. It is shown that vertically resonant structures perform as high gain antennas. We have already shown in Section 13.4 that a half wavelength grounded substrate topped by a quarter-wavelength superstrate of high ε_r acts as a high gain antenna. It has been shown in [12] that multiple superstrates can be used to achieve higher antenna gains. In this case, the superstrate layers should have alternately high and low impedance [15]. Radial leaky waves can be excited by a small horizontal dipole inside the substrate in Figure 13.2 or Figure 13.5. In this case the radiation beam is either conical (for finite θ_p) or pencil beam at broadside.

The radiation fields have been obtained by using the reciprocity principle between the given dipole source and a plane wave incident from the exterior air region [6, 11]. This approach is considered a short cut to the far zone fields since it avoids the derivation of the fields in the near zone. A different approach in which a full wave analysis is pursued is given in [14] for a three-layer structure. This allows one to get the surface wave fields besides the radiation fields as well as the near fields to the source. Therefore, the radiation power, the surface wave power, the radiation impedance and the input impedance can thus be obtained. In what follows, we adopt this approach to study radiation from a leaky wave structure comprising a periodic screen.

13.6.1 Leaky Wave Antennas with Periodic Screen

In this section, we reconsider the leaky wave antenna of Figure 13.7, where a single grounded dielectric layer is covered by a periodic leaky screen. Instead of the line source, we use a small horizontal electric dipole to excite a two-dimensional leaky mode as shown in Figure 13.10. The periodic screen is now assumed to contain a two-dimensional array of crossed printed dipoles (oriented along x and y directions). The dipole of moment $Il\, e^{j\omega t}\hat{x}$ (Ampere-m) is assumed to be concentrated at $\rho = 0$ and $z = h$. Such a dipole will excite radial waves having *cos*

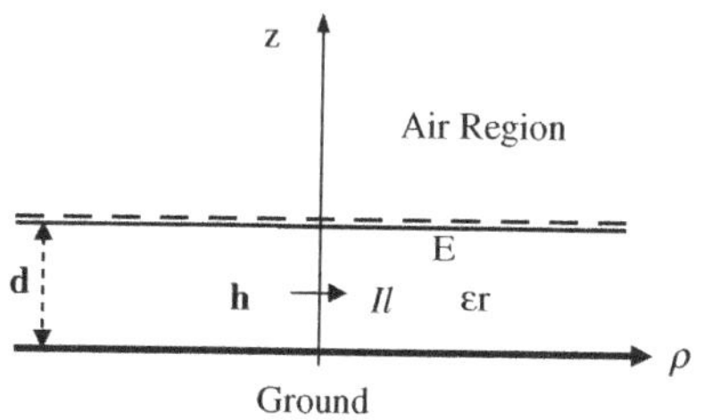

Figure 13.10 A two parallel plate waveguide with a top PRS excited by a small x-oriented electric dipole. The dipole is located at height 'h' above the ground

ϕ or $sin\ \phi$ behavior. The radiation will be in the form of a conical beam when θ_p is finite or a pencil beam at broadside.

Before deriving the excited fields, we mention some design considerations of the present antenna. As stated earlier in Section 13.4.1, the leaky wave guide is considered as a perturbation of a closed parallel plate waveguide with TE_m (or TM_m) mode having $k_z = m\pi/d$. In order to have a radiation beam in the direction θ_c off broadside, the frequency is chosen as in Equation (13.25), which is repeated here for convenience

$$k_0 = \omega/c = \frac{m\pi/d}{\sqrt{\varepsilon_r - \sin^2\theta_c}} \tag{13.25}$$

For a given m, as the frequency is increased, the beam is scanned from broadside ($\theta_c = 0$) to endfire ($\theta_c = 90^0$). If it is required to have only one beam (with no grating lobes), then the frequency of broadside radiation for $m = 2$, should be greater than the frequency of endfire radiation for $m = 1$. This means that

$$2/\sqrt{\varepsilon_r} > 1/\sqrt{\varepsilon_r - 1}, \tag{13.47}$$

which dictates that $\varepsilon_r > 4/3$.

If the screen consists of a two-dimensional periodic metal strips of period 'r', there are, in general, grating lobes corresponding to the existence of Floquet modes of orders $p = 0, \pm 1, ..$ To avoid such grating lobes, then as the $p = 0$ beam is scanned in the forward endfire ($\theta_c = 90^0$), the $p = -1$ mode should not have emerged in the backward endfire ($\theta_c = -90^0$). In symbols, when $\beta_{p=0} = k_0$, we should have $\beta_{p=-1} = k_0 - 2\pi/r < -k_0$. This, immediately, implies that the period $r < \lambda_0/2$. This result and that in Equation (13.47) have been derived in [9].

Now, we derive the fields within the dielectric layer and the exterior air region. As stated above, the source is an electric dipole of moment $Il\,e^{j\omega t}\hat{x}$, that is concentrated at $\rho = 0$ and $z = h$. We first obtain the primary fields of the dipole and break them into TE_z and TM_z wave components. The TE_z and TM_z waves are not coupled by the planar interfaces, hence they can be treated separately to get their corresponding scattered fields. The primary fields produced by the dipole are derived in terms of an x-oriented magnetic vector potential $A_x = Il\exp(-jkR)/4\pi R$, where $R = \sqrt{\rho^2 + (z-h)^2}$. This can be recast in the form of a cylindrical wave spectrum using a well known identity [16]

$$A_x = \frac{Il}{4\pi}\int_0^\infty J_0(\beta\rho)\exp(-jp|z-h|)\frac{\beta}{jp}\,d\beta, \tag{13.48}$$

where $p = \sqrt{k^2 - \beta^2}$, *and* $k = k_0\sqrt{\varepsilon_r}$. The primary z-field components are [17]

$$H_{zp} = -\partial A_x/\partial y = -\frac{Il}{4\pi}\frac{\partial}{\partial y}\int_0^\infty J_0(\beta\rho)\exp(-jp|z-h|)\frac{\beta}{jp}\,d\beta \tag{13.49}$$

$$E_{zp} = \frac{\partial^2 A_x}{j\omega\varepsilon\,\partial x\partial z} = -\frac{Il}{4\pi j\omega\varepsilon}\frac{\partial}{\partial x}\int_0^\infty \pm J_0(\beta\rho)\ \exp(-jp|z-h|)\,\beta\,d\beta \tag{13.50}$$

The ± sign apply for $z > h$ and $z < h$ respectively. These fields correspond to the TE_z and TM_z waves respectively. The TE_z and TM_z fields are uncoupled at the layer interfaces, hence can be treated separately. Inside the dielectric layer the total field is composed of the primary field plus downward and upward traveling waves that represent the waves reflected at the boundaries $z = 0$ and $z = d$. The boundary condition at the ground plane entails that $H_z = 0$ for TE_x waves and $\partial E_z/\partial z = 0$ for TM_x waves. Skipping details we get the fields in the dielectric layer; $0 \leq z < d$, as

$$H_z = -\frac{Il}{4\pi}\frac{\partial}{\partial y}\int_0^\infty J_0(\beta\rho)\, g(\beta; z)\frac{\beta}{jp}\, d\beta \tag{13.51}$$

$$E_z = -\frac{Il}{4\pi j\omega\varepsilon}\frac{\partial}{\partial x}\int_0^\infty J_0(\beta\rho) f(\beta; z)\beta\, d\beta \tag{13.52}$$

where

$$\begin{aligned} f(\beta; z) &= 2a(\beta)\cos pz + 2\cos p(z-h) \\ g(\beta; z) &= -2jb(\beta)\sin pz - 2j\sin p(z-h) \end{aligned} \tag{13.53}$$

valid for $h < z < d$. In the region below the source, $0 < z < h$, only the first term on the right-hand side remains. In the above, $a(\beta)$ and $b(\beta)$ are unknown function coefficients to be determined from the boundary conditions at the screen ($z = d$).

The fields in the exterior (air) region; $z > d$ take similar forms as in Equations (13.51) and (13.52); namely for $z > d$,

$$H_z = -\frac{Il}{4\pi}\frac{\partial}{\partial y}\int_0^\infty J_0(\beta\rho)\, g_{air}(\beta)\exp[-jp_0(z-d)]\frac{\beta}{jp}\, d\beta \tag{13.54}$$

$$E_z = -\frac{Il}{4\pi j\omega\varepsilon}\frac{\partial}{\partial x}\int_0^\infty J_0(\beta\rho) f_{air}(\beta)\exp[-jp_0(z-d)]\beta\, d\beta \tag{13.55}$$

where $f_{air}(\beta)$ and $g_{air}(\beta)$ are to be determined from boundary conditions and $p_0 = \sqrt{k_0^2 - \beta^2}$.

13.6.1.1 Boundary Conditions

The partially reflecting screen at $z = d$ is modeled as a transfer impedance $Z_T(\beta)$ that accounts for the discontinuity of the magnetic field across the screen. The boundary conditions at the screen dictate the continuity of the tangential electric field (E_x or E_y) and the discontinuity of the tangential magnetic field (H_y or H_x).

Explicitly, the latter condition reads $H_y(d^-) - H_y(d^+) = E_x(d)/Z_T(\beta)$. At this point, it is important to note that Z_T is generally different for TE_z and TM_z modes. So we denote the transfer impedance by $Z_{T,TE}$ and $Z_{T,TM}$ for the two wave types. Now the above two boundary conditions apply to each spectral component, and lead to the determination of the coefficients $g_{air}(\beta)$ and $b(\beta)$ for the TE_z fields and $f_{air}(\beta)$ and $a(\beta)$ for the TM_z fields.

$$g_{air}(\beta) = \frac{2\sin ph}{\sin pd - \chi_{TE}\cos pd}\frac{\bar{Z}_{T,TE}(\beta)\,p/k_0}{1+\bar{Z}_{T,TE}(\beta)\,p_0/k_0} \tag{13.56}$$

$$b(\beta) = \frac{\chi_{TE}\cos p(d-h) - \sin p(d-h)}{\sin pd - \chi_{TE}\cos pd} \tag{13.57}$$

$$f_{air}(\beta) = \frac{2\sin ph}{\sin pd - \chi_{TM}\cos pd}\frac{\varepsilon_r \bar{Z}_{T,TM}(\beta)}{\bar{Z}_{T,TM}(\beta) + p_0/k_0} \tag{13.58}$$

$$a(\beta) = \frac{\chi_{TM}\cos p(d-h) - \sin p(d-h)}{\sin pd - \chi_{TM}\cos pd} \tag{13.59}$$

where

$$\chi_{TE} = \frac{j\bar{Z}_{T,TE}(\beta)p/k_0}{1+\bar{Z}_{T,TE}(\beta)p_0/k_0} \tag{13.60}$$

$$\chi_{TM} = \frac{j\varepsilon_r \bar{Z}_{T,TM}(\beta)k_0/p}{1+\bar{Z}_{T,TM}(\beta)k_0/p_0} \tag{13.61}$$

and $\bar{Z}_T(\beta) = Z_T(\beta)/\eta_0$ is the normalized transfer impedance relative to the free space wave impedance. Note that the poles of $g_{air}(\beta)$ and $f_{air}(\beta)$ are the same as those obtained earlier by the transverse resonance method (Equations (13.26) and (13.28).

13.6.1.2 Radiated Power and Surface Wave Power

The radiation fields are obtained by evaluating Equations (13.54) and (13.55) in the far zone using the stationary phase procedure

$$E_\phi(\theta,\phi) = \eta_0 H_z/\sin\theta = \frac{jIl\eta_0}{4\pi}\frac{g_{air}(k_0\sin\theta)\cos\theta\sin\phi}{\sqrt{\varepsilon_r - \sin^2\theta}}\frac{k_0 e^{-jk_0R}}{R} \tag{13.62}$$

$$E_\theta(\theta,\phi) = -E_z/\sin\theta = \frac{-jIl\eta_0}{4\pi\varepsilon_r} f_{air}(k_0\sin\theta)\cos\theta\cos\phi\frac{k_0 e^{-jk_0R}}{R} \tag{13.63}$$

with $R = \sqrt{\rho^2 + (z-d)^2}$. We note that the radiation pattern is shaped mainly by the terms g_{air} and f_{air}, which in turn are controlled by the leaky mode poles. From Equations (13.56) and (13.58), the leaky mode complex roots $\pm\beta_{LW}$ are given by $\tan pd = \chi_{TE}$ or χ_{TM} for the TE and TM mode respectively. As a fast wave, the leaky mode has $\mathrm{Re}[\beta_{LW}] < k_0$. When the attenuation of the leaky mode is small, the factors g_{air} and f_{air} are dominated by the leaky mode contribution, hence take the form $2\beta_{LW}/(k_0^2\sin^2\theta - \beta_{LW}^2)$ as derived in Section 13.5.1.

The radiated power is obtainable from Equations (13.62) and (13.63) by integrating the Poynting's vector over the surface of a large hemisphere leading to

$$\begin{aligned} P_{rad} &= \int_{\phi=0}^{2\pi}\int_{\theta=0}^{\pi/2}\left[(|E_\theta|^2 + |E_\phi|^2)/\eta_0\right]R^2\sin\theta\, d\theta d\phi \\ &= \frac{(Ilk_0)^2\eta_0}{16\pi}\int_{\theta=0}^{\pi/2}\left[\frac{g_{air}^2}{\varepsilon_r - \sin^2\theta} + \frac{f_{air}^2}{\varepsilon_r^2}\right]\cos^2\theta\sin\theta d\theta \end{aligned} \tag{13.64}$$

The surface wave poles are the roots of the same equation as the leaky modes; $\tan pd = \chi_{TE}$ or χ_{TM}. As a slow wave, a surface wave mode has a root β_{sw} that is real and lies between k_0 and $k = k_0\sqrt{\varepsilon_r}$. In this range of β, the wave number p is real while p_0 is purely imaginary. This implies that both χ_{TE} and χ_{TM} are purely real, hence β_{sw} are real (as expected for a lossless structure).

Now, we derive the input power, P_{in}, which is the total power delivered by the source. This is given by

$$P_{in} = -Il\,\mathrm{Re}[E_x(\rho = 0, z = h)] \tag{13.65}$$

In terms of H_z and E_z in (13.51) and (13.52), E_x is given by $\beta^2 E_x = \partial^2 E_z/\partial x\partial z - j\omega\mu_0\,\partial H_z/\partial y$. Substituting in the above equation and noting that $\partial^2 J_0(\beta\rho)/\partial x^2 = \partial^2 J_0(\beta\rho)/\partial y^2 = -\beta^2/2$ as ρ tends to zero, we get

$$P_{in}/(Il)^2 = (4\pi)^{-1}\mathrm{Im}\left\{\int_0^{\infty}\left[\frac{a(\beta)}{\omega\varepsilon} + \frac{\omega\mu_0\,b(\beta)}{p^2}\right]p\sin(ph)\,\beta d\beta\right\} \tag{13.66}$$

In the absence of dielectric and metallic losses, this power is converted to radiation plus surface waves. The radiation power is that part of the integration with the limits $0 \le \beta \le k_0$ and is identically equal to P_{rad} of Equation (13.64) as can be verified numerically. The surface wave power is the sum the residues at the surface wave poles, which lie in the range $k_0 < \beta_{sw} < k$. Finally, the part of the integration where $\beta > k$ has no contribution since then $a(\beta)$ and $b(\beta)$ are purely real.

As a numerical example, we plot the radiation fields of a broadside pencil beam in the H and E planes for $\bar{Z}_T = -j0.1$ and $-j0.15$ in Figure 13.11. The vertical axis is adjusted to read the directivity in dB versus the elevation angle θ. It is seen that the peak directivity achieved exceeds 21 dB for $\bar{Z}_T = -j0.1$ and 18 dB for $\bar{Z}_T = -j0.15$. It should be noted that one or more

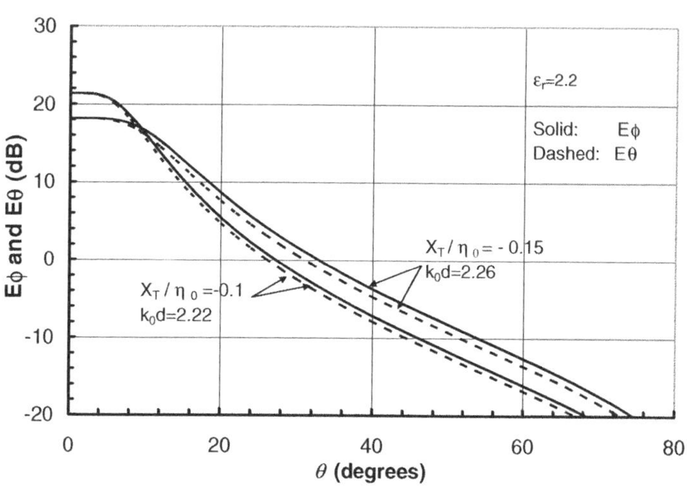

Figure 13.11 Radiation pattern in H and E planes for the Leaky wave antenna of Figure 13.10. Two values of the transfer reactance X_T are considered. The vertical scale is adjusted to read the directivity. The radiation is a pencil beam in the broadside direction

surface wave mode may be excited. The power taken by these surface wave modes will reduce the radiation efficiency. However, surface waves can be avoided if the operating frequency lies in the electromagnetic band gap (EBG) of surface waves, as will be discussed in the next section.

13.6.2 Characterization of the Periodic Screen as an EBG Structure

Now we try to characterize the periodic screen in Figures 13.2 or 13.10 as a surface with frequency dependent surface impedance or admittance. In fact, any periodically shaped screen can be considered as a frequency selective surface (FSS). The surface admittance will also depend on the polarization of the incident wave, so we expect different behavior depending on the incident wave, being a TE_x or TM_x wave and the angle of incidence [9, 18]. Numerical data can be obtained for the reflection coefficient of the screen under different frequency and polarization of the incident plane wave and may be used to model the screen. Maci et al. [19] have expressed the screen admittance Y_{FSS} in terms of its zeros and poles in the frequency domains. Thus, for a given polarization Y_{FSS} is expressed as a function of frequency and the angle of incidence

$$Y_{Fss}(\omega, k_x) = j\omega C_0 \frac{[\omega^2-\omega_{z1}^2(k_x)][\omega^2-\omega_{z2}^2(k_x)]\ldots.}{[\omega^2-\omega_{p1}^2(k_x)][\omega^2-\omega_{p2}^2(k_x)]\ldots.} \tag{13.67}$$

where k_x is the x-wave number (normal to the screen) in the air region and C_0 is a constant. $\omega_{zi}(k_x)$ and $\omega_{pi}(k_x)$ are the ith zero and pole frequencies of Y_{FSS}. As stated above, it is important to note that this set of zero and pole frequencies depends on the exact polarization of the incident plane wave and whether it is TE or TM wave. So there are different expressions of Y_{FSS} for TE_x and TM_x waves [19]. Obviously the zeros of Y_{FSS} are identified as those frequencies (ω_{zi}) at which the screen acts as an open circuit, hence it appears invisible to an incident wave. In other words, ω_{zi} are the frequencies at which the reflection coefficient from a dielectric layer with a screen is the same as the reflection from the dielectric layer without the screen. On the other hand, the poles correspond to the screen acting as an electric conductor ($Y_{FSS} = \infty$). The zero and pole frequencies of Y_{FSS} for TE_x or TM_x wave type can be obtained by numerically computing the reflection from the screen at different frequencies and angles of incidence.

The transverse equivalent network corresponding to the leaky wave antenna of Figure 13.10 is the same as that in Figure 13.2. The input admittance looking down into the screen is equal to $Y_{FSS}^{TE}(\omega, k_x) - j(k_{x1}/\omega\mu_0)\cot(k_{x1}d)$ for TE_x waves, and $Y_{FSS}^{TM}(\omega, k_x) - j(\omega\varepsilon_1/k_{x1})\cot(k_{x1}d)$ for TM_x waves (with x is normal to the screen). It is thus possible to see an effective magnetic conductor (zero admittance) looking down into the screen at a certain frequency. The important point to make here is that there is a frequency range in which surface waves are forbidden on the antenna structure. This electromagnetic band gap (EBG) lies between the frequency of an effective magnetic conductor and a pole of Y_{FSS} [19]. Designing the leaky wave antenna to operate in this band gap ensures high antenna efficiency due to the disappearance of surface waves. Further discussions and results on leaky wave antennas in the band gap frequency range are found in [20, 21]. A leaky wave antenna based on a superstrate on an FSS screen has been investigated by Foroozesh and Shafai [22] and has provided a high gain, reaching 20 dB over a broad input impedance bandwidth.

13.7 Further Advances on a Class of Periodic Leaky Wave Antennas

In this section, we consider a class of periodic leaky wave antennas in which radiation arises from a fast spatial harmonic, corresponding to a suitable Floquet mode. Examples of this class of antennas are depicted in Figures 13.12 and 13.13. In both figures, the waveguide on the right-hand side (RHS) has a leaky periodic screen and supports a leaky mode that is excited by an incident bounded mode from the left-hand side (LHS) guide. In Figure 13.12, the LHS guide is a closed two parallel plate guide that supports a TEM mode with longitudinal phase constant β^{un} (unperturbed β) equal to the dielectric wave number $k = k_0\sqrt{\varepsilon_r}$. The leaky screen on the RHS guide consists of periodically placed narrow slots, which may be considered as a perturbation to the closed two parallel plate guide. The Floquet modes on the RHS guide have longitudinal wave numbers $\beta_n = \beta_0 + 2\pi n/d$, where d is the period of the screen and $n = 0, \pm 1, \pm 2\ldots$ Associated with β_n, there must be an attenuation factor "α" that accounts for power leakage. The zero order β_O is expected to differ slightly from β^{un} of the LHS guide, since the perturbation caused by the screen is assumed small. Since the incident mode is a slow mode; $\beta^{un} > k_0$, it follows that a leaky (fast) mode in the RHS guide will necessarily correspond to a Floquet mode with $n < 0$ and $|\beta_n| < k_O$. Similarly, the LHS guide in Figure 13.13 is an open guide, with no upper conductor, hence it supports a surface wave mode whose longitudinal wave number $\beta^{un} > k_O$. The leaky screen on the RHS guide onsists of periodic narrow strips separated by wider slots to form a perturbation to the open guide on the LHS. Again a fast leaky mode will correspond to a Floquet mode with $n < 0$ and $|\beta_n| < k_O$. In general, there can exist more than one fast spatial harmonic with power leakage.

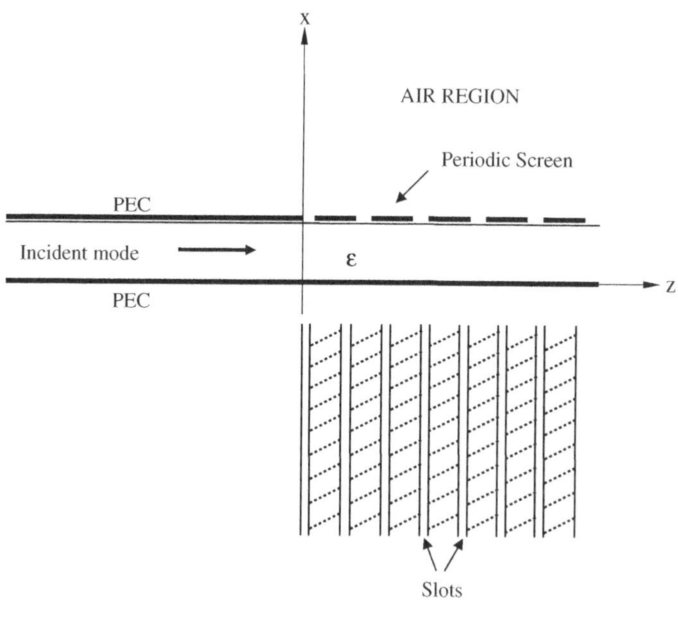

Figure 13.12 A periodic leaky wave structure. A bounded incident TEM mode is incident from the LHS guide to the leaky guide on the RHS. The periodic screen contains periodically placed narrow slots

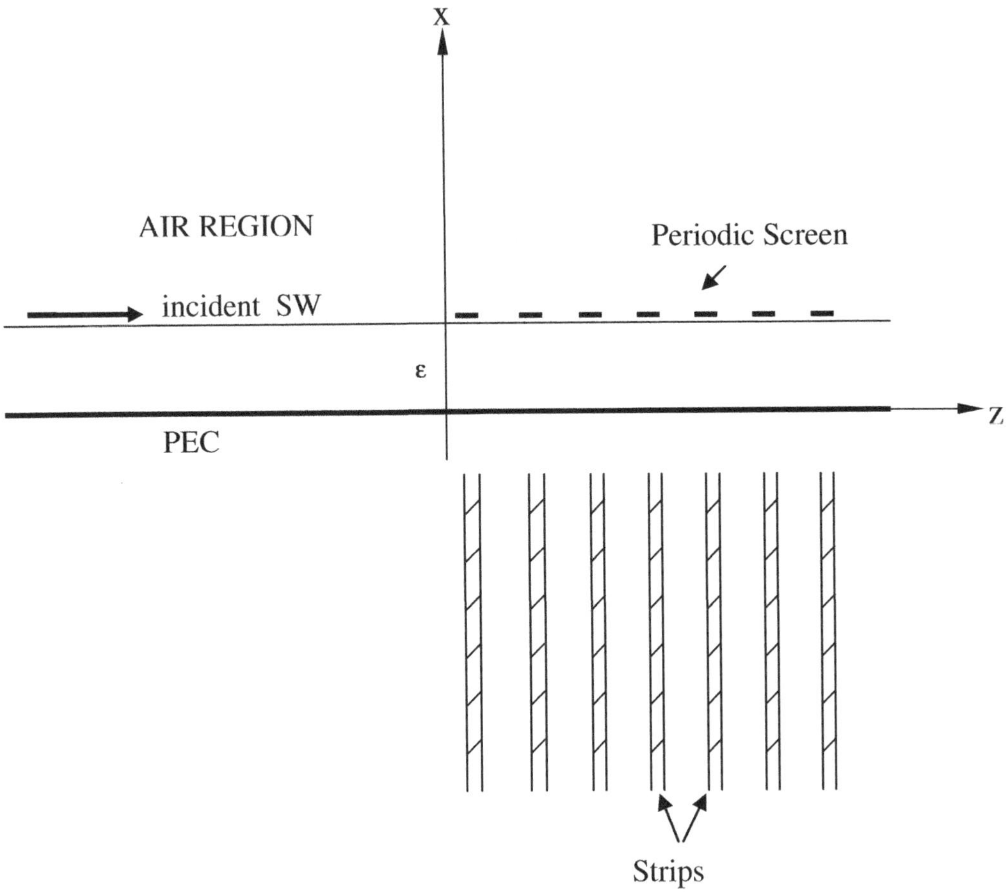

Figure 13.13 A periodic leaky wave structure. An incident surface wave mode is incident from the LHS guide to the leaky guide on the RHS. The periodic screen contains periodically placed narrow strips

In order to characterize the periodic screen, a transverse equivalent network (TEN) may be used, where the screen is represented by a transfer impedance as used in Section 13.6. The transfer impedance is generally determined by numerical methods such as the Moment method or its derivatives [9, 18]. The transverse equivalent network can thus be used to find the leaky wave parameters and the radiation pattern of periodic antennas such as those of Figures 13.12 and 13.13. However, in this section, we shall follow another approach that has been suggested in [23], where a longitudinal equivalent network is used to model the periodic leaky wave structure. Thus, with reference to Figure 13.14, one period of the structure is modeled by two transmission lines separated by a series impedance. Each transmission line is of length $d/2$ (half period), characteristic impedance Z_c and phase constant β^{un} of the unperturbed (LHS) guide. The series impedance accounts for the discontinuity of the screen; that is the narrow slot in the screen of Figure 13.12, or the narrow strip in the screen of Figure 13.13. Since the discontinuity is associated with power leakage, the series impedance Z_s must be complex; $Z_s = R_s + jX_s$

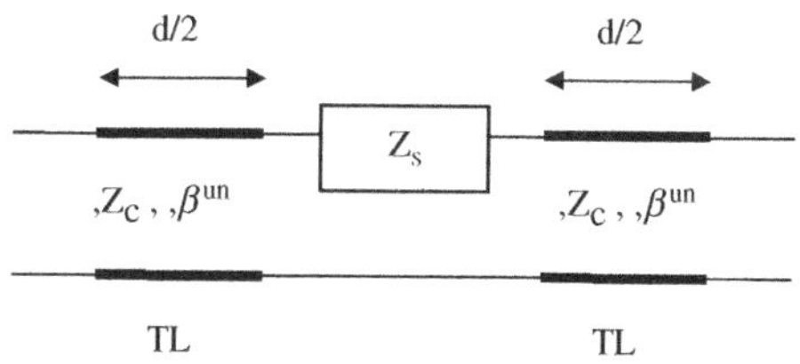

Figure 13.14 Equivalent circuit of one period of the leaky wave guide of Figures 13.12 or 13.13

where R_s accounts for the power leakage. It is worth noting that the equivalent circuit of Figure 13.14 is valid for TM_z incident modes. In case of TE_z modes, the series impedance should be replaced by a shunt admittance.

Denoting the equivalent input voltage and current for the circuit of Figure 13.14 by (V_i, I_i), the output must equal to $(V_i, I_i)\exp(-j\beta_0 d)$ to satisfy the structure periodicity. Hence using the transmission (ABCD) matrices of the two transmission lines and the series Z_s, we arrive at the modal equation for β_0; namely

$$\cos(\beta_0 d) = \cos(\beta^{un} d) + (j/2)\,\bar{Z}_s \sin(\beta^{un} d), \tag{13.68}$$

where $\bar{Z}_s$ is now normalized to Z_c. Since $\beta_n d = \beta_0 d + 2n\pi$, one can replace $\beta_0 d$ in the above equation by $\beta_n d$.

13.7.1 Broadside Radiation

To have a broadside radiation, we require that $\beta_{-1} \cong 0$, which is equivalent to $\beta^{un} d \cong 2\pi$. To be exact we set $\beta^{un} d = 2\pi + \delta^{un}$; $\delta^{un} \ll 1$, and $\beta_{-1} d = \delta \ll 1$. Using these approximations in (13.68), we get

$$\delta^2/2 \cong (\delta^{un})^2/2 - (j/2)(\bar{R}_s + j\bar{X}_s)\,\delta^{un} \tag{13.69}$$

At the frequency where $\delta^{un} = -\bar{X}_s$, we get $\delta^2 = j\bar{R}_s \bar{X}_s$, which leads to

$$\beta_{-1} d = \pm\sqrt{\bar{R}_s|\bar{X}_s|/2} - j\sqrt{\bar{R}_s|\bar{X}_s|/2} \tag{13.70}$$

Note that the phase rate and the attenuation rate are equal, as is usually the case for optimum broadside radiation of a leaky mode. This occurs at the frequency corresponding to $\beta^{un} d = 2\pi - \bar{X}_s$. As an example, if $\bar{R}_s = 0.05$ and $\bar{X}_s = -0.1$, the attenuation and phase constants of the $n = -1$ spatial mode are equal to $(0.05/d)$. The radiation pattern is then given by Equation (13.15) with $k_{zL} = \beta_{-1}$.

13.7.2 Frequency Scanning

It is seen from the above argument that a broadside beam is attained at a frequency corresponding to $\beta^{un} d = 2\pi - \bar{X}_s$. If the incident mode in the structure of Figure 13.12 is the TEM mode, then $\beta^{un} = k_0\sqrt{\varepsilon_r}$. For the incident surface wave mode in Figure 13.13,

$\beta^{un} = k_0\sqrt{\varepsilon_{eff}}$, where ε_{eff} is an effective dielectric constant that lies between unity and ε_r. As the frequency increases above the value required for broadside radiation, the beam is steered to forward directions with peak radiation at $\theta_p > 0$. To demonstrate this frequency scanning possibility, we have solved Equation (13.68) for increasing values of $\beta^{un}d$ to obtain the complex β_{-l}. In doing so we have assumed that $\bar{Z}_s = 0.05 - j0.1$. The peak radiation angle is then given by $\sin\theta_p = \mathrm{Re}\,(\beta_{-1}/k_0)$. We plot the angle θ_p in degrees versus the normalized frequency $k_0 d$ in Figure 13.15. Here we have $k_0 d = \beta^{un}d/\sqrt{\varepsilon_r}$ pertaining to the antenna in Figure 13.12 with incident TEM mode. It is seen that the radiation beam is scanned from broadside to $\theta_p = 85^0$ as $k_0 d$ varies from 2.03 (corresponding to $\beta^{un}d = 2\pi - \bar{X}_s$) to 2.96. The directivity, as computed from Equation (13.21) with $w_{iL} = \mathrm{Im}\,[\sin^{-1}(\beta_{-1}/k_0)]$, is also plotted in the same figure. Directivities reaching 19 dB are achieved. Backward beam scanning (not shown) is also

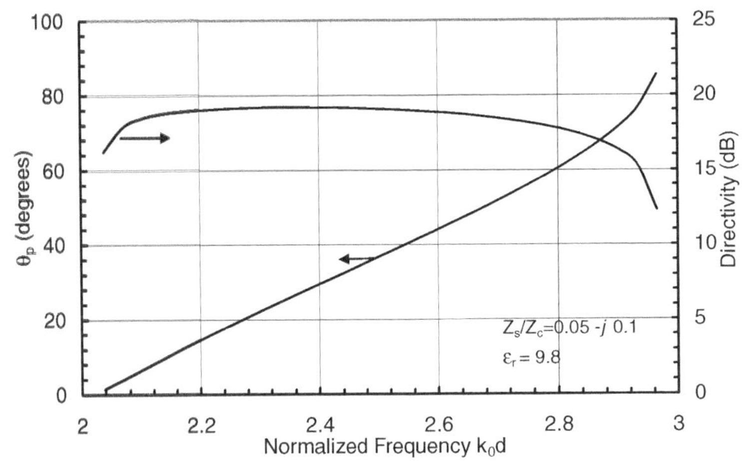

Figure 13.15 Frequency scanning of the leaky mode radiation beam for the structure of Figure 13.12. θ_p in degrees and beam directivity in dB are plotted versus $k_0 d$

possible as the $\beta^{un}d$ is less than 2π.

In order to avoid any grating lobes, we require that the $n = -2$ spatial mode does not emerge as the $n = -1$ mode scans the space from back to front endfire. So, as this mode reaches the front endfire ($\theta_p = 90^0$), the $n = -2$ mode should not emerge in the back endfire ($\theta_p = -90^0$). This dictates that $\beta^{un}d\text{-}4\pi < -k_0 d$ while $\beta^{un}d - 2\pi = k_0 d$ which is equivalent to $\beta^{un}/k_0 > 3$. For the case of a TEM incident mode, this imposes that ε_r be >9 and for the surface wave case (structure of Figure 13.15), $\varepsilon_{eff} > 9$. However, it should be noted that this simple argument does not take into account the role of the series impedance at the two spatial modes and the excitation efficiency of each mode. Actually, one expects that this condition will be relaxed to lower values of ε_r.

13.7.3 Cylindrical Leaky Wave Antenna Design

The leaky wave antennas depicted in Figures 13.12 and 13.13 produce a one-dimensional radiation beam (fan beam) since fields are independent of one coordinate. Two-dimensional

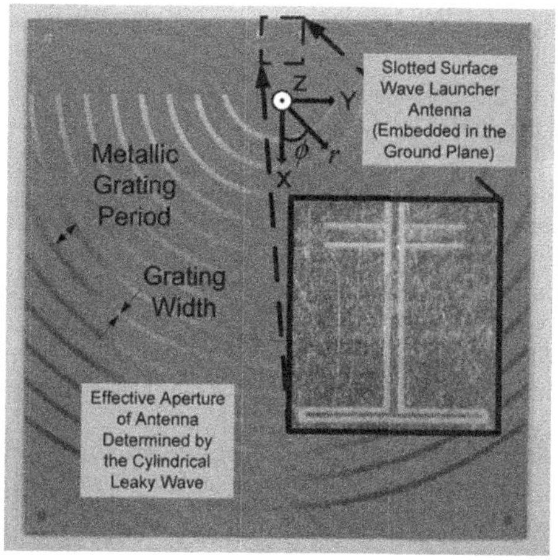

Figure 13.16 Realized antenna structure for directive radiation at broadside. The periodic screen is realized by a segmented circular strip grating, defining the antenna aperture. Shown also is the surface wave launcher made up of a main slot antenna with secondary and tuning slots. Reproduced by permission of ©2008 IEEE [24]

periodic leaky structures have been proposed [21, 24, 25]. Podilchak et al. [24, 25] have introduced a two-dimensional periodic leaky wave antenna as shown in Figure 13.16, which is obtained from reference [24]. The antenna is similar to the structure of Figure 13.13 except that the periodic screen is made of circular metallic narrow strip gratings. A cylindrical TM_0 surface wave is efficiently launched on the grounded dielectric slab using a coplanar line as described in [26] and [27]. With a grounded dielectric layer of height 1.27 mm, $\varepsilon_r = 10.2$ and $\tan\delta = 0.0023$, a TM_0 surface wave launched at $f = 21.2$ GHz produces a broadside pencil beam. The gratings have metallic strips of width 1.25 mm with radial periodicity of $d = 7$mm and ϕ–period of 14 mm. The measured antenna gain at broadside was 23.5 dB. For a detailed design and construction of this antenna, the reader is referred to [25].

References

1. T. Tamir and A. A. Oliner, "Guided complex waves, Part 1, relation to radiation patterns," *Proc. Inst. Elec. Eng.*, vol. 110, pp. 310–324. Feb. 1963.
2. T. Tamir and A. A. Oliner, "Guided complex waves, Part 2, fields at an interface," *Proc. Inst. Elec. Eng.*, vol. 110, pp. 325–334. Feb. 1963.
3. T. Tamir, "Leaky–wave antennas," in *Antenna Theory*, eds. R. E. Collin and F. J. Zucker, Ch. 20. McGraw-Hill, Basingstoke, 1969.
4. P. C. Clemmow, *The Plane Wave Spectrum Representation of Electromagnetic Fields*, Pergamon Press, Oxford, sec. 2.2.3, 1966.
5. N. W. McLachlan, *Bessel Functions for Engineers*, 2nd edn., Clarendon Press, Oxford, p. 198, 1955.
6. D. R. Jackson and A. A. Oliner, "A leaky wave analysis of the high gain printed antenna configurations," *IEEE Trans. Antennas Propagat.*, vol. 36, pp. 905–910, July, 1988.
7. G. Lovat, P. Burhhignoli, and D. R. Jackson, "Fundamental properties and optimization of broadside radiation from uniform leaky wave antenna," *IEEE Trans. Antennas Propagat*, vol. 54, pp. 1442–1452, May, 2006.
8. J. R. Wait, *Electromagnetic Wave Theory*, Harper & Row Publishers, New York, sec. 5.3, 1985.

9. T. Zhao, D. R. Jackson, J.T. Williams, H. Y. D. Yang, and A. A. Oliner, "2D periodic leaky wave antennas-Part I: Metal patch design," *IEEE Trans. Antennas Propagat*, vol. 53, pp. 3505–3514, Nov. 2005.
10. N. G. Alexopoulos and D. R. Jackson, "Fundamental superstrate effects on printed circuit antennas", *IEEE Trans. Antennas Propagat*, vol. 32, pp. 807–816, Aug. 1984.
11. D. R. Jackson and N. G. Alexopoulos, "Gain enhancement methods for printed circuit antennas," *IEEE Trans. Antennas Propagat*, vol. 33, pp. 976–987, Sept. 1985.
12. H. Y. Yang and N. G. Alexopoulos, "Gain enhancement methods for printed antennas through multiple superstrates," *IEEE Trans. Antennas Propagat*, vol. 35, pp. 860–863, July 1987.
13. I. P. Antonio and D. R. Jackson, "Radiation from cylindrical leaky waves," *IEEE Trans. Antennas Propagat*, vol. AP-38 pp. 482–488, April 1990.
14. S. F. Mahmoud, F. Samir, Y. M. M. Antar, A. Etrugh, and A. P. Freundorfer, "Gain enhancement of a slot dipole in a three dielectric layers structure," *IEEE AP-S International Symposium on Antennas and Propagation and URSI/ URSI, Radio Science*, Albuquerque, 9–14 July 2006.
15. A. Foroozesh and Lotfollah Shafai "Size reduction of a microstrip antenna with dielectric superstrate using metamaterials: artificial magnetic conductors versus magneto-dielectrics," *IEEE Antennas and Propagation Society International Symposium 2006*, pp. 11–14, July 2006.
16. J. R. Wait, *Geo-Electromagnetism*, Academic Press, New York, p. 128, 1982.
17. R. F. Harrington, *Time Harmonic Electromagnetic Fields*, McGraw-Hill Inc, New York, p. 130, 1961.
18. T. Zhao, D. R. Jackson, and J. T. Williams, "2D periodic leaky wave antennas Part II: slot design," *IEEE Trans. Antennas Propagat*, vol. 53, pp. 3515–3524, Nov. 2005.
19. S. Maci, M. Caiazzo, A. Cucini, and M. Casaletti, "A pole-zero matching method for EBG surfaces composed of a dipole FSS printed on a grounded dielectric slab," *IEEE Trans. Antennas Propagat*, vol. 53, pp. 70–81, Jan. 2005.
20. N. Llombart, A. Neto, G. Gerini, and P. De-Maagt, "Planar circularly symmetric EBG structures for reducing surface waves in printed antennas," *IEEE Trans. Antennas Propagat*, vol. 53, pp. 3210–3218, Oct. 2005.
21. M. Ettorre, S. Bruni, G. Gerini, A. Neto, N. Llombart, and S. Maci. "Sector PCS-EBG antenna for low-cost high-directivity applications," *IEEE Antennas and Wireless Propagation Letters*, vol. 6, pp. 537–539, 2007.
22. A. Foroozesh and L. Shafai, "Investigation into the effects of the patch-type FSS superstrate on the high-gain cavity resonance antenna design," *IEEE Trans. Antennas Propagat*, vol. 58, pp. 258–270, Feb. 2010.
23. P. Burghignoli, G. Lovat, and D. R. Jackson, "Analysis and optimization of leaky-wave radiation at broadside from a class of 1-D periodic structures," *IEEE Trans. Antennas Propagat*, vol. 54, pp. 2593–2604, Sept. 2006.
24. S. K. Podilchak, A. Freundorfer, and Y. M. M. Antar, "Planar leaky wave antenna designs offering conical-sector beam scanning and broadside radiation using surface wave launchers," *IEEE Antennas and Wireless Propagation Letters*, vol. 7, pp. 155–158, Jan. 2008.
25. S. K. Podilchak, "Planar leaky-wave antenna designs for directive beam scanning," M.Sc. thesis, Royal Military College and Queen's University, Kingston, Ontario, Canada, 2007.
26. H. Hammad, Y. M. M. Antar, A. Freundorfer, and M. Sayer, "A new dielectric grating antenna at millimeter wave frequency," *IEEE Trans. Antennas Propagat*, vol. 52, pp. 36–44, Jan. 2004.
27. S. F. Mahmoud, Y. M. M. Antar, H. Hammad, and A. Freundorfer, "Theoretical considerations in the optimization of surface waves on a planar structure," *IEEE Trans. Antennas Propagat*, vol. 52, pp. 2057–2063, April 2004.

Appendix I

Preliminary Ideas: PTFE-Based Microwave Laminates and Making Prototypes

A1.1 PTFE Laminates

PTFE is a popular abbreviation representing a very useful high frequency material, whose chemical nomenclature is *polytetrafluoroethylene*. It shows some unique properties such as very low loss-tangent (<0.002), high chemical resistance, good tensile strength (~6,240 psi), low weight, excellent thermal stability (−55° to 260°C), thermal expansion coefficient identical to that of copper, and a high melting point (330° C). All these are suitable for using the material at microwave frequencies.

PTFE sheets having a thickness varying over a wide range from 0.005″ to 0.375″, are used as microwave substrate. These are available with $^1/_4$oz to 2oz electrodeposited copper cladding on the both sides.

Some popular commercial products commonly used in laboratories and industries are provided in Table A1.1.

For detailed technical and commercial information, readers can consult the following websites:

1. http://www.mbelectronique.fr/PAGE_Article.awp?P1=US&P2=131
2. http://www.rogerscorp.com/documents/607/acm/The-Advantage-of-Nearly-Isotropic-Dielectric-Constant-of-RT-duroid-5870-5880-Glass-Microfiber-PTFE.aspx
3. http://www.rogerscorp.com/acm/products/12/RT-duroid-6002-6202-6006-6010-PTFE-Ceramic-Laminates.aspx
4. http://www.rogerscorp.com/documents/641/acm/PTFE-Safety-Notes-on-Using-RT-duroid-Composite-Material.aspx
5. http://www.taconic-add.com/en--products--material-view.php
6. http://www.standardpc.com/Microwave Laminates Comparison Chart.pdf

Microstrip and Printed Antennas: New Trends, Techniques and Applications. Edited by Debatosh Guha and Yahia M.M. Antar

Table A1.1 Some commercial PTFE composite laminates commonly used for making prototypes (1–18 GHz) and laboratory measurements

Product	ε_r at 10 GHz	Tanδ	Comment
RT/duroid 5870*	2.33	0.0012	Glass microfiber reinforced PTFE composites, developed in the 1960s. Ideal for high moisture environments as its water absorption is extremely low.
RT/duroid 5880*	2.2	0.0009	
RT/duroid 6006*	6.15	0.0027	Ceramic-PTFE composites, developed in late 1970s. High ε_r to provide option to reduce the size of the circuit elements.
RT/duroid 6010*	10.2	0.0023	
RT/duroid 6002*	2.94	0.0012	Introduced in the late 1980s
RT/duroid 6202*	2.94	0.0015	Ideal for phased array/GPS antennas.
RT/duroid 6202PR*	0.005″ laminate 2.90	0.0020	Suitable for planar and non-planar antennas and multilayer circuits.
'PR' stands for 'Palanar Resistor'	0.010″ laminate 2.98		
	0.015″ laminate 3.0		
	0.020″/0.030″ laminate 2.9		
TLY**	2.17–2.40	0.0009	Fiberglass/PTFE resin laminate
TLX**	2.45–2.65	0.0019	

Notes: *Manufactured by Rogers Corporation (www.rogerscorp.com).
**Manufactured by Taconic (www.taconic-add.com).

A1.2 Making Prototype Boards

Any prototype of a printed antenna or circuit is created by etching. This means removal of selected unwanted conductor from the copper-clad microwave substrates. Two standard techniques are usually adopted: *chemical etching*, and *mechanical etching*. They are briefly described below:

A1.2.1 Chemical Etching

The basic principle is to cover the "wanted metallic part" by a chemical coating and then the unwanted part is exposed to a corrosive liquid. The liquid dissolves the unwanted copper, leaving the "wanted pattern" on the substrate as the prototype printed antenna or circuit element.

The basic steps usually followed in an in-house process in a laboratory are described below:

- Copper surface should be cleaned and ungreased.
- The surface is coated with a photoresist. Photoresist is a chemical paste (synthetic polymer of methyl methacrylate or phenol formaldehyde resin) which is sensitive to ultraviolet (UV) light and reacts to it in two ways. One type of photoresist, on being exposed to UV light, gets hard and does not dissolve in the photoresist developer. This is known as *negative*

photoresist. The other type, when exposed to light, becomes soluble to the photoresist developer while the unexposed part remains insoluble. This is called *positive* photoresist.

- A proper mask of the prototype artwork is prepared depending on the type of photoresist to be used and is carefully aligned on the photoresist-coated surface of the PTFE board.
- It is exposed to UV light for a considerable time as per specification.
- The mask sheet is peeled off the exposed surface.
- The photoresist image is rinsed gently. This leaves a pattern of developed photoresist which may be the exact replica of the wanted microstrip element (if a *negative* photoresist is used) or just its "negative"(in terms of photographic language) (if a *positive* photoresist is used).
- The board is dipped in the etching solution (ferric chloride with hydrochloric acid). It is carefully observed and flipped frequently. As soon as the chemical etching is completed, the board is removed from the tray.
- The board is rinsed using distilled water to remove the photoresist layer and is dried for experiments.

A1.2.2 Mechanical Etching

This technique involves a highly precise computer-controlled small cutting machine. It uses a special milling cutter, which moves horizontally along a linear axis (say, x or y) over a circuit board and extracts copper strips. Multi-axis horizontal movements of the board help to remove unwanted copper-clad areas leaving a fine and precise prototype geometry.

The printed circuits with accurate dimensions are primarily drawn using a commercial CAD tool, like autoCAD. This is fed to the etching machine from a computer through proper interface. All mechanical movements of the board and cutter are controlled accordingly.

The instrument is quite expensive, but does not have to undertake the many intermediate time-consuming processes involved in chemical etching. Moreover, it also does not use any toxic chemicals. These factors have made mechanical etching very popular and attractive in different settings from microwave laboratory to PCB prototyping shops.

Mechanical milling needs high quality cutters or bits, which are available in different configurations. The bit diameter typically varies from 0.01″ to 0.05″ and it looks like a sharp lead pencil. Figure A1.1 shows a commercially available mechanical etching machine, which can be used for prototyping printed antennas and circuits operating up to about 26 GHz. In Figure A1.1, the horizontal platform is holding a PTFE board, on which a semi-finished printed antenna is shown. The rectangular deep shaded box is attached to a cylindrical body and it moves in a left–right direction. The vertical cylinder holds the cutter and moves linearly along the edge of the black shaded box. Thus a two-dimensional motion over the PCB is achieved mechanically. The typical weight of such a machine is about 30–40 kilograms.

The overall process is simple and vendor-specific. Once learned, it is not a hard task to implement. Figure A1.2 shows two 2.45 GHz microstrip arrays etched in the laboratory. These are as good as chemically etched prototypes. Some basic steps are noted below:

- Circuit layout is done using a PCB layout package.
- A post-processor generates the boundary paths to be followed by the cutter.
- A PCB board, cut to the proper size, is mounted on the machine.

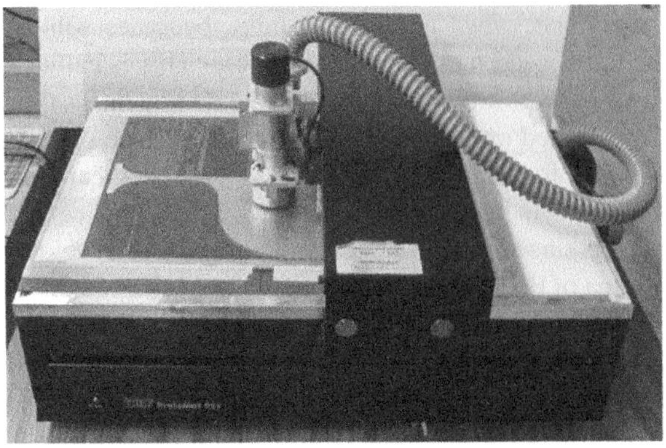

Figure A1.1 Commercial PCB-making machine using the mechanical etching technique

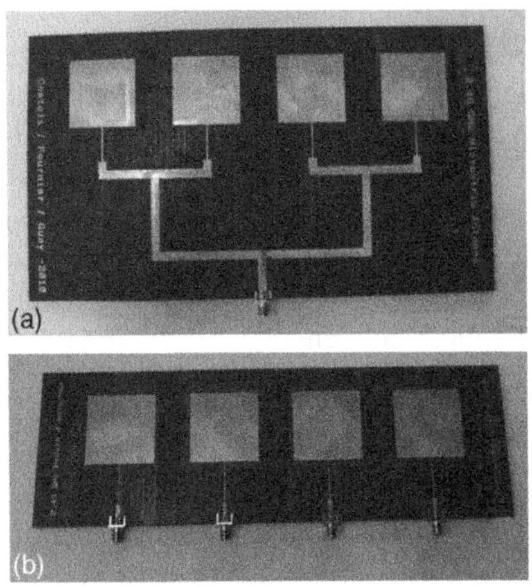

Figure A1.2 Printed prototypes etched on PTFE substrates using the machine shown in Figure A1.1 (a) sample 1; (b) sample 2

- Vertical drilling, needed to insert a coaxial probe or interconnects, is done first.
- The etching bit is chosen and fitted to the holder.
- The cutting depth is adjusted so that just the copper foil is pilled, not the substrate. The tip angle of the cutter plays an important role and proper knowledge is required prior to using it.
- The machine is allowed to operate and the process is a bit slow. It may take considerable time to finish.
- The finished board is immediately ready for experiments.

A1.3 Making Simple Printed Antennas or Circuits Using a Laser Printer

A simple in-house technique may be tried by the students for their laboratory experiments. Its basic principle and process are described below:

- A laser printer does not use any ink, we know it is simply a plastic toner, that is, a type of black powder. This plastic toner is resistant to etching solutions. Therefore, if we can print a circuit on a copper-clad substrate, the rest follows the same process as chemical etching to make a prototype board.
- The question is: How to take a printout on copper board? The intelligent technique of *toner-transfer* is to be employed. On applying heat, laser toner melts to a sticky paste, like most plastics. So a normal print-out taken on paper may be placed with the front face-down on the copper board. Application of mild heat and uniform pressure on the paper using a household iron will make the toner melt to be glued to the copper surface.
- Now if the paper is removed safely, the printed layout is transferred onto the PCB. This process is easy if a clever technique is applied. A glossy photo paper or magazine paper is an ideal choice. The toner hardly penetrates the glossy layer and thus the toner transfer is efficient. Also the glossy coatings dissolve in water easily and thus water helps to remove the paper from the PCB surface.
- Once the toner-transfer is done properly, we get the circuit layout on the board and the exposed metals are removed to get the final circuit.
- Like chemical etching, ferric chloride solution may safely be used to dissolve the exposed copper.
- The board is rinsed in water and finally the toner is removed using a thinner. Nail polish remover serves the purpose very well.
- The circuit is now ready for measurements.

Readers may consult the following websites for detailed information:

1. http://sfprime.net/pcb-etching/index.htm
2. http://www.thinktink.com/stack/volumes/volvi/etching.htm
3. http://www.wikihow.com/Create-Printed-Circuit-Boards
4. http://blog.makezine.com/archive/2007/01/how_to_use_a_laser_to_etc.html
5. http://www.riccibitti.com/pcb/pcb.htm
6. http://www.qsl.net/ve2emm/pcb/pcb2e.html

Appendix II

Preliminary Ideas: Microwave Connectors for Printed Circuits and Antennas

A microwave connector is basically a "coaxial agent" which supports a low VSWR connection to a coaxial port installed into a microstrip circuit board or microwave housing. Two standard connectors fitted with coaxial cables are shown in Figure A2.1.

A2.1 Specification

A user specifies a connector in terms of *frequency* (implying maximum frequency of operation), *impedance* (implying characteristics impedance of the connecting port or transmission cable), *type* (implying *male* or *female* type of threading) and also, *size* (implying physical dimension, which is technically important for different applications).

A2.1.1 Frequency Limit

Unlike waveguides, the coaxial connectors do not have any lower cut-off frequency. But there is definitely a frequency limit for a coaxial transmission line up to which the desired TEM mode can propagate through it.

With the changing requirements, microwave industries have developed a series of "connectors" with varying features and nomenclature. A list of commonly used connectors, sources of their acronyms, frequency limits, years of development, etc. are provided in Table A2.1. Here it is interesting to note that as we move from lower to higher frequency, the dielectric used as filling material is replaced with air with gradual increase in cost. This shows that air is more expensive than PTFE. In reality, air dielectric needs critical engineering to hold the central conductor within the air medium which eventually increases the cost of precise engineering.

Microstrip and Printed Antennas: New Trends, Techniques and Applications. Edited by Debatosh Guha and Yahia M.M. Antar

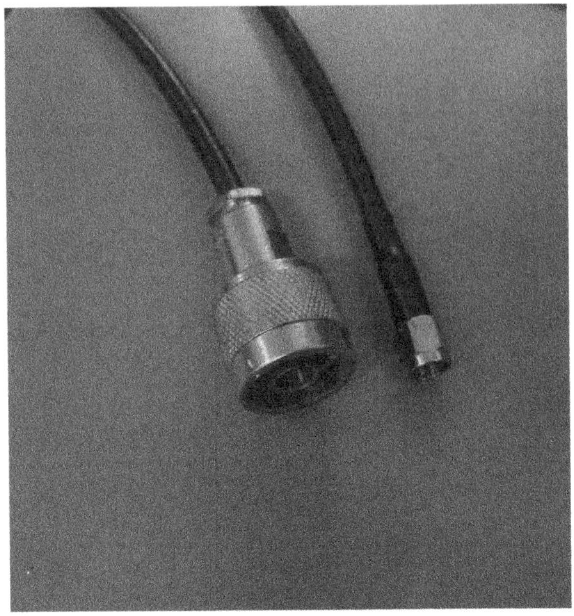

Figure A2.1 Coaxial cables fitted with two microwave connectors. Left: N-type, Right: SMA-type

A2.1.2 Impedance

Most RF and microwave systems use transmission lines and ports with a characteristic impedance of 50 ohm while some use 75 ohms. However, for connectors like the N-type, some are available with both 50 ohm and 75 ohm. If not specifically stated, for general microwave purposes, a connector is of 50 ohm specification.

The characteristic impedance of a coaxial transmission line with a circular cross-section is determined by three parameters: (1) radius of outer metallic boundary (outer radius, r_o); (2) radius of the central conductor (inner radius, r_i); (3) and the dielectric constant (ε_r). This can be calculated using a simple relation

$$Z_0 = \frac{138}{\sqrt{\varepsilon_r}} \log_{10} \frac{r_0}{r_i} \Omega$$

A2.1.3 Type

The terminal of any coaxial cable or port is normally threaded in two standard configurations of "Plug" or "Jack." They are more popularly known by the terms of "male-type" and "female-type," respectively. For example, in Figure A2.1, both are male-type.

A2.2 Adapters

Adapters are special kind of connectors providing the transition from one *series* to another *series* (example: N-to-SMA or SMA-to-N: Figures A2.2 and A2.3) or from one *type* to another *type* in the same series (example: SMA male-to-SMA male, SMA male-to-SMA female, and SMA female-to-SMA female: Figure A2.4).

Table A2.1 Some common connectors* used in microwave frequencies

Connector series	Used up to	Dielectric used	Remarks
BNC *Bayonet type-N connector* or *Bayonet Neill-Concelman* (Named after Paul Neill and Carl Concelman)	4 GHz	PTFE	Developed at Bell Laboratory in 1950s. Paul Neill was with Bell Laboratory and Carl Concelman was with Amphenol
SMB *Sub-miniature type B*	4 GHz	PTFE	
SMC *Sub-miniature type C*	10 GHz	PTFE	
TNC *Threaded Neill-Concelman*	15 GHz	PTFE	This is actually a BNC with threads to reduce vibrational hazards
N (Named after Paul Neill)	11/18 GHz	PTFE	Developed in the 1940s
APC-7, 7 mm *Amphenol Precision Connector*	18 GHz	PTFE	Developed in the 1960s Sexless connector
OSP *Omni-Spectra Push-on*	22 GHz	PTFE	
SMA *Sub-miniature type A*	25 GHz	PTFE	Developed in the 1960s Mate with 3.5 mm and 2.92 mm connectors
3.5 mm	26.5 GHz	Air	Mate with SMA and 2.92 mm connectors
OSSP *Omni-Spectra subminiature push-on*	28 GHz	PTFE	Smaller version of OSP connector
SSMA	38 GHz	PTFE	Smaller version of SMA
2.92 mm	40 GHz	Air	Developed in 1974 Mate with SMA and 3.5 mm connectors
K	40 GHz	Air	Covers from DC to all of the K frequency bands
GPO, OSMP *Gilbert push-on, Omni-spectra microminiature push-on*	40 GHz	PTFE	
2.4 mm	50 GHz	Air	Developed in 1986 Mechanically compatible with 1.85 mm connectors
1.85 mm	60 GHz	Air	Mechanically compatible with 2.4 mm connectors
1 mm	110 GHz	Air	Developed at Hewlett Packard in 1989. It's the costliest in the series

Note: *Of all these listed above, N-, SMA, and 3.5 mm type connectors are most popular for laboratory and industry uses up to Ku-band.

SMA adapters, shown in Figure A2.4, are also called by different names:

- Male-to-male: barrel adapter (Figure A2.4 (a))
- Male-to-female or vice versa: connector saver (Figure A2.4 (b))
- Female-to-female: bullet adapter (Figure A2.4 (c)

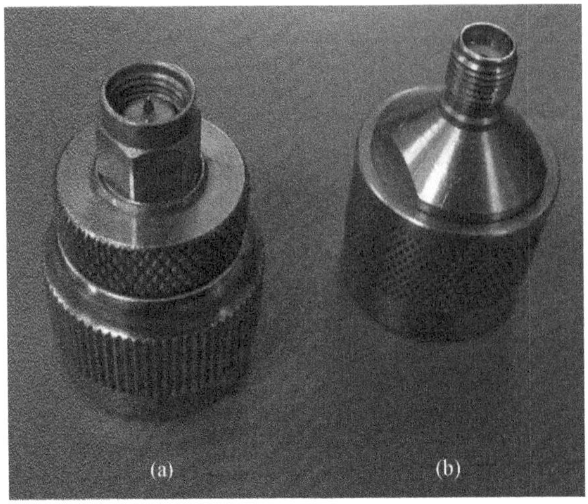

Figure A2.2 N-to-SMA adapters: (a) N male to SMA male; (b) N male to SMA female

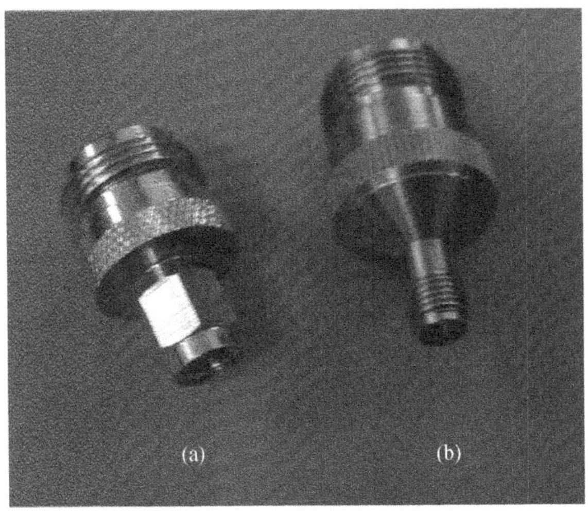

Figure A2.3 N-to-SMA adapters: (a) N female to SMA male; (b) N female to SMA female

A connector saver is actually more useful in the situation when the connect-disconnect process has to be repeated frequently with a costly instrument. Indeed, there is every possibility of a sophisticated port being damaged. Such an adapter thus "saves"' the connecting port from damage.

An elbow connector is also a type of adapter providing a 90 degree bend. This is very useful for ease of connection without damaging a port, where bending is unavoidable. Figure A2.5 shows three elbow connectors giving different combinations of transitions. A commercially available cable fitted with a straight and elbow-shaped SMA connectors is shown in Figure A2.6.

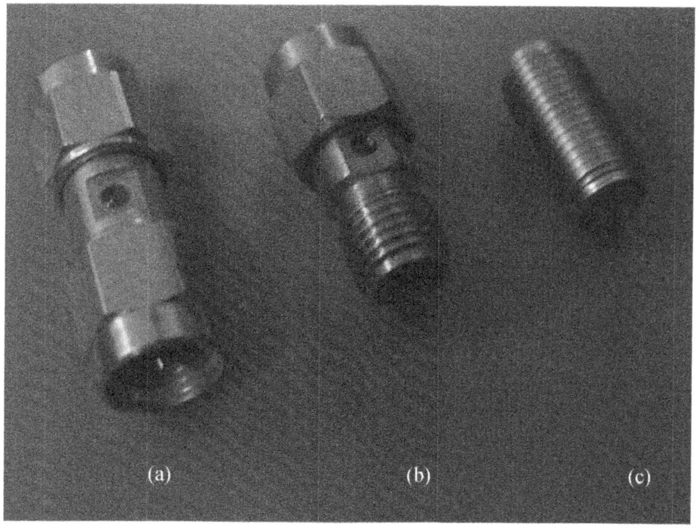

Figure A2.4 SMA to SMA adapters. (a) SMA male to SMA male: barrel adapter; (b) SMA male to SMA female: connector saver; (c) SMA female to SMA female: bullet adapter

Figure A2.5 Elbow connectors: (a) SMA male to SMA male; (b) SMA male to SMA female; (c) SMA female to SMA female

A2.3 SMA Probe/Launcher for Microstrip Circuits and Antennas

SMA is a very common, popular and easily available connector, which is used to excite a microstrip line or a microstrip antenna. They are a mainly 50 ohm probe with a central conductor extended to connect the microstrip element. Depending on the size of the antenna or substrate, the connector dimensions may be different. Figure A2.7 shows different SMA probes for microstrip antennas and circuits. They have different dimensions of central conductor and surrounding dielectric sleeve. The probes, shown in Figures A2.7 (a)–(c) are suitable to excite a microstrip patch or dielectric resonator antennas. A smaller probe like Figure A2.7 (a) is suitable for using within and above X-band. The probe shown in Figure A2.7 (d) is designed to excite a microstrip transmission line or a printed circuit component horizontally. A clip-type

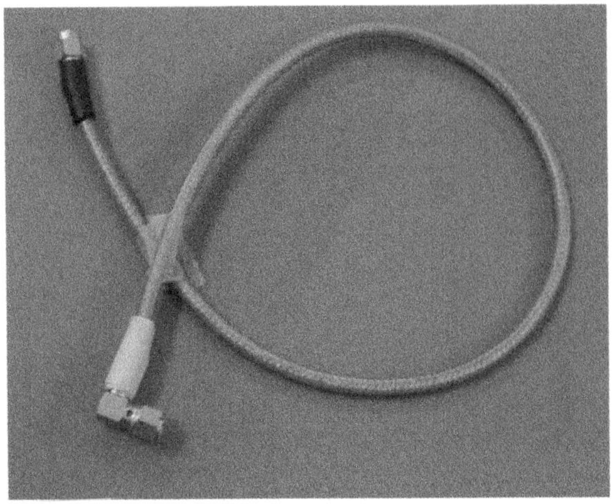

Figure A2.6 Coaxial cable fitted with an elbow type at one end and a conventional straight SMA connector at the other end

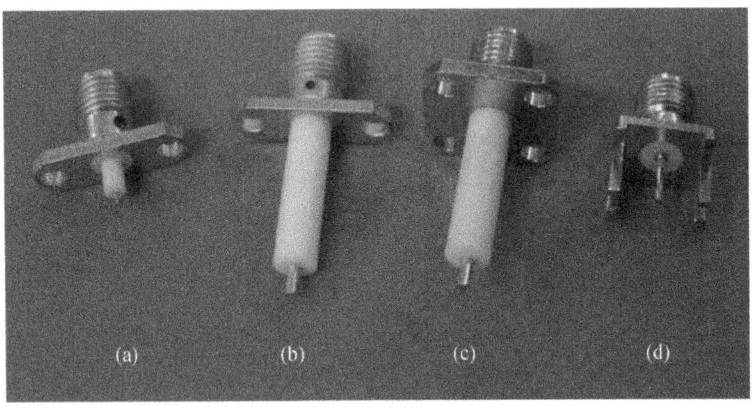

Figure A2.7 SMA probes: (a) narrow flange, small size, suitable for X-Ku bands; (b) narrow flange and long probe, suitable for C band and below; (c) square flange and long probe, suitable for C band and below; (d) square flange, specially designed for microstrip circuits

holder is attached to the flange and this is inserted through the substrate to give mechanical rigidity to the connector. A practical example is shown in Figure A2.8. An SMA probe exciting a circular microstrip is shown in Figure A2.9.

A2.4 Maintenance

All microwave connectors should be protected from dust and heat and should always be protected with dust covers. Frequent cleaning is suggested. Cotton swabs wet in the proper liquid such as isopropyl alcohol are normally used to clean the connector threads. Detailed

Figure A2.8 SMA probe used to launch electromagnetic fields in a microstrip transmission line etched on a 0.508 mm thick PTFE substrate

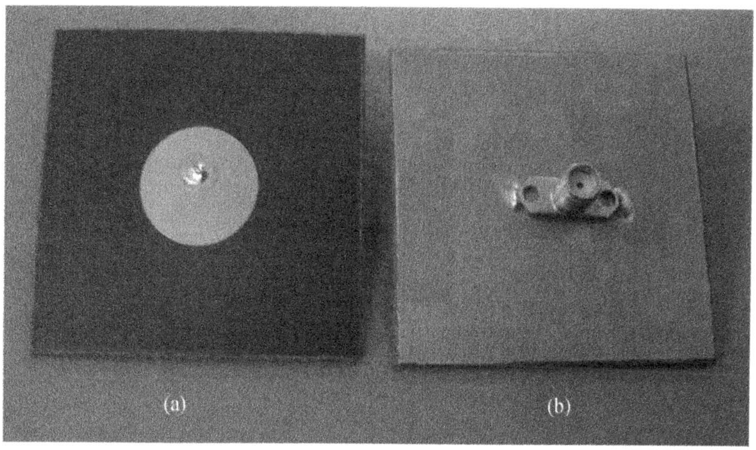

Figure A2.9 SMA probe used to excite a circular microstrip patch operating at C-band: (a) top view showing the central conductor soldered to the patch; (b) viewed from the back showing the flange flushed with the ground plane

instructions may be found in several commercial website documents. For interested readers, a list of such websites is provided as a ready reference. For detailed commercial information, readers may visit:

1. http://www.microwaves101.com/encyclopedia/connectors.cfm
2. http://nocat.net/connectors.html
3. http://www.hdcom.com/rfadapt.html
4. http://www.maurymw.com/support/faqs/faqs/faq5.html

5. http://www.fairviewmicrowave.com/ntosmaadapters.htm
6. http://www.wiredco.com/so239-female-jack-male-plug-adapter-radio-gold-p-166.html
7. http://parts.digikey.com/1/parts/360338-adapter-sma-male-male-4283.html
8. http://www.pasternack.com
9. http://www.radiall.com

Index

Microstrip and Printed Antennas: New Trends, Techniques and Applications. Edited by Debatosh Guha and Yahia M.M. Antar

Printed and bound by CPI Group (UK) Ltd, Croydon, CR0 4YY
18/09/2026
14965015-0006